EXERGY

Energy, Environment and Sustainable Development

EXERGY

Energy, Environment and Sustainable Development

Ibrahim Dincer and Marc A. Rosen

ELSEVIER

Amsterdam • Boston • Heidelberg • London • New York • Oxford
Paris • San Diego • San Francisco • Sydney • Tokyo

Elsevier
Linacre House, Jordan Hill, Oxford OX2 8DP, UK
30 Corporate Drive, Suite 400, Burlington, MA 01803, USA

First edition 2007

British Library Cataloguing in Publication Data
A catalogue record for this book is available from the British Library

Library of Congress Cataloging-in-Publication Data
A catalog record for this book is available from the Library of Congress

ISBN: 978-0-08-044529-8

For information on all Elsevier Science publications
visit our web site at books.elsevier.com

Typeset by Charon Tec Ltd (A Macmillan Company), Chennai, India
www.charontec.com

Printed and bound in China
07 08 09 10 10 9 8 7 6 5 4 3 2 1

Working together to grow
libraries in developing countries

www.elsevier.com | www.bookaid.org | www.sabre.org

ELSEVIER   BOOK AID
International   Sabre Foundation

PREFACE

Exergy analysis is a method that uses the conservation of mass and conservation of energy principles together with the second law of thermodynamics for the analysis, design and improvement of energy and other systems. The exergy method is a useful tool for furthering the goal of more efficient energy-resource use, for it enables the locations, types and magnitudes of wastes and losses to be identified and meaningful efficiencies to be determined.

During the past two decades we have witnessed revolutionary changes in the way thermodynamics is taught, researched and practiced. The methods of exergy analysis, entropy generation minimization and thermoeconomics are the most visible and established forms of this change. Today there is a much stronger emphasis on exergy aspects of systems and processes. The emphasis is now on system analysis and thermodynamic optimization, not only in the mainstream of engineering but also in physics, biology, economics and management. As a result of these recent changes and advances, exergy has gone beyond thermodynamics and become a new distinct discipline because of its interdisciplinary character as the confluence of energy, environment and sustainable development.

This book is a research-oriented textbook and therefore includes practical features in a usable format often not included in other, solely academic textbooks. This book is essentially intended for use by advanced undergraduate or graduate students in several engineering and non-engineering disciplines and as an essential tool for practitioners. Theory and analysis are emphasized throughout this comprehensive book, reflecting new techniques, models and applications, together with complementary materials and recent information. Coverage of the material is extensive, and the amount of information and data presented is sufficient for exergy-related courses or as a supplement for energy, environment and sustainable development courses, if studied in detail. We believe that this book will be of interest to students and practitioners, as well as individuals and institutions, who are interested in exergy and its applications to various systems in diverse areas. This volume is also a valuable and readable reference for anyone who wishes to learn about exergy.

The introductory chapter addresses general concepts, fundamental principles and basic aspects of thermodynamics, energy, entropy and exergy. These topics are covered in a broad manner, so as to furnish the reader with the background information necessary for subsequent chapters. Chapter 2 provides detailed information on energy and exergy and contrasts analysis approaches based on each. In Chapter 3, extensive coverage is provided of environmental concerns, the impact of energy use on the environment and linkages between exergy and the environment. Throughout this chapter, emphasis is placed on the role of exergy in moving to sustainable development.

Chapter 4 delves into the use of exergy techniques by industry for various systems and processes and in activities such as design and optimization. This chapter lays the foundation for the many applications presented in Chapters 5 to 16, which represent the heart of the book. The applications covered range from policy development (Chapter 5), psychrometric processes (Chapter 6), heat pumps (Chapter 7), drying (Chapter 8), thermal storage (Chapter 9), renewable energy systems (Chapter 10), power plants (Chapter 11), cogeneration and district energy (Chapter 12), cryogenic systems (Chapter 13), crude oil distillation (Chapter 14), fuel cells (Chapter 15) and aircraft systems (Chapter 16).

Chapter 17 covers the relation between exergy and economics, and the exploitation of that link through analysis tools such as exergoeconomics. Chapter 18 extends exergy applications to large-scale systems such as countries, regions and sectors of an economy, focusing on how efficiently energy resources are utilized in societies. Chapter 19 focuses the utilization of exergy within life cycle assessment and presents various applications. Chapter 20 discusses how exergy complements and can be used with industrial ecology. The book closes by speculating on the potential of exergy in the future in Chapter 21.

Incorporated throughout are many illustrative examples and case studies, which provide the reader with a substantial learning experience, especially in areas of practical application.

The appendices contain unit conversion factors and tables and charts of thermophysical properties of various materials in the International System of units (SI).

Complete references are included to point the truly curious reader in the right direction. Information on topics not covered fully in the text can, therefore, be easily found.

We hope this volume allows exergy methods to be more widely applied and the benefits of such efforts more broadly derived, so that energy use can be made more efficient, clean and sustainable.

Ibrahim Dincer and Marc A. Rosen

June 2007

ACKNOWLEDGMENTS

We gratefully acknowledge the assistance provided by Dr. Mehmet Kanoglu, a visiting professor at UOIT, in reviewing and revising several chapters, checking for consistency and preparing questions/problems.

We are thankful to Mehmet Fatif Orhan, a graduate student at UOIT, for preparing some figures and tables, and arranging references.

Some of the material presented in the book derives from research that we have carried out with distinguished individuals who have been part of our research group or collaborated with us over the years. We are extremely appreciative of the efforts of these colleagues:

- Dr. Mikhail Granovskiy, Senior Researcher, UOIT, Canada
- Dr. Feridun Hamdullahpur, Professor, Carleton University, Canada
- Dr. Arif Hepbasli, Professor, Ege University, Turkey
- Dr. Frank C. Hooper, Professor Emeritus, University of Toronto, Canada
- Mr. Mohammed M. Hussain, Ph.D. Student, University of Waterloo, Canada
- Dr. Mehmet Kanoglu, Associate Professor, Gaziantep University, Turkey
- Dr. Mehmet Karakilcik, Assistant Professor, Cukurova University, Turkey
- Dr. Xianguo Li, Professor, University of Waterloo, Canada
- Dr. Adnan Midilli, Associate Professor, Nigde University, Turkey
- Mr. Husain Al-Muslim, Engineer, ARAMCO and Ph.D. Student, KFUPM, Saudi Arabia
- Dr. Zuhal Oktzy, Associate Professor, Balikesir University, Turkey
- Dr. Ahmet D. Sahin, Associate Professor, Istanbul Technical University, Turkey
- Dr. Ahmet Z. Sahin, Professor, KFUPM, Saudi Arabia
- Dr. David S. Scott, Professor Emeritus, University of Victoria, Canada
- Dr. Iyad Al-Zaharnah, Instructor, KFUPM, Saudi Arabia
- Dr. Syed M. Zubair, Professor, KFUPM, Saudi Arabia.

In addition, the contributions of several past undergraduate and graduate students are acknowledged, including Can Coskun, Jason Etele, Lowy Gunnewiek, Gerald Kresta, Minh Le, Norman Pedinelli and Raymond Tang.

Last but not least, we thank our wives, Gülşen Dinçer and Margot Rosen, and our children Meliha, Miray, İbrahim Eren and Zeynep Dinçer, and Allison and Cassandra Rosen. They have been a great source of support and motivation, and their patience and understanding throughout this project have been most appreciated.

İbrahim Dinçer and Marc A. Rosen
June 2007

ABOUT THE AUTHORS

Ibrahim Dincer is a full professor of Mechanical Engineering in the Faculty of Engineering and Applied Science at UOIT. Renowned for his pioneering works he has authored and co-authored several books and book chapters, over 350 refereed journal and conference papers, and numerous technical reports. He has chaired many national and international conferences, symposia, workshops and technical meetings and is the founding chair or co-chair of various prestigious international conferences, including the International Exergy, Energy and Environment Symposium. He has delivered over 70 keynote and invited lectures. He is an active member of various international scientific organizations and societies, and serves as editor-in-chief (for International Journal of Energy Research by Wiley and International Journal of Exergy by Inderscience), associate editor, regional editor and editorial board member on various prestigious international journals. He is a recipient of several research, teaching and service awards, including the Premier's research excellence award in Ontario, Canada in 2004. He has made innovative contributions to the understanding and development of exergy analysis of advanced energy systems for his so-called: five main pillars as better efficiency, better cost effectiveness, better environment, better sustainability and better energy security. He is the chair of a new technical group in ASHRAE, named Exergy Analysis for Sustainable Buildings.

IBRAHIM DINCER

Marc A. Rosen is founding Dean of Engineering and Applied Science at the University of Ontario Institute of Technology in Oshawa, Canada. A Past-President of the Canadian Society for Mechanical Engineering, Dr. Rosen received an Award of Excellence in Research and Technology Development from the Ontario Ministry of Environment and Energy, and is a Fellow of the Engineering Institute of Canada, the American Society of Mechanical Engineers, the Canadian Society for Mechanical Engineering, and the International Energy Foundation. He has worked for Imatra Power Company in Finland, Argonne National Laboratory and the Institute for Hydrogen Systems, near Toronto. Dr. Rosen is a registered Professional Engineer in Ontario. He became President-elect of the Engineering Institute of Canada in 2006 and has received that Institute's Julian C. Smith medal for achievement in the development of Canada. Prior to 2002, he was a professor in the Department of Mechanical, Aerospace and Industrial Engineering at Ryerson University in Toronto, Canada for 16 years. During his tenure at Ryerson, Dr. Rosen served as department Chair and Director of the School of Aerospace Engineering. With over 50 research grants and contracts and 400 technical publications, Dr. Rosen is an active teacher and researcher in thermodynamics and energy conversion (e.g., cogeneration, district energy, thermal storage, renewable energy), and the environmental impact of energy and industrial systems. He has been a key contributor to and proponent of advanced exergy methods and applications for over two decades.

MARC A. ROSEN

TABLE OF CONTENTS

Chapter 1

THERMODYNAMIC FUNDAMENTALS

1.1. Introduction

Energy, entropy and exergy concepts stem from thermodynamics and are applicable to all fields of science and engineering. This chapter provides the necessary background for understanding these concepts, as well as basic principles, general definitions and practical applications and implications. Illustrative examples are provided to highlight the important aspects of energy, entropy and exergy.

The scope of this chapter is partly illustrated in Fig. 1.1, where the domains of energy, entropy and exergy are shown. This chapter focuses on the portion of the field of thermodynamics at the intersection of the energy, entropy and exergy fields. Note that entropy and exergy are also used in other fields (such as statistics and information theory), and therefore they are not subsets of energy. Also, some forms of energy (such as shaft work) are entropy-free, and thus entropy subtends only part of the energy field. Likewise, exergy subtends only part of the energy field since some systems (such as air at atmospheric conditions) possess energy but no exergy. Most thermodynamic systems (such as steam in a power plant) possess energy, entropy and exergy, and thus appear at the intersection of these three fields.

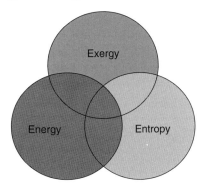

Fig. 1.1. Interactions between the domains of energy, entropy and exergy.

1.2. Energy

Energy comes in many forms. Thermodynamics plays a key role in the analysis of processes, systems and devices in which energy transfers and energy transformations occur. The implications of thermodynamics are far-reaching and applications span the range of the human enterprise. Throughout our technological history, our ability to harness energy and use it for society's needs has improved. The industrial revolution was fueled by the discovery of how to exploit energy in a large scale and how to convert heat into work. Nature allows the conversion of work completely into heat, but heat cannot be entirely converted into work, and doing so requires a device (e.g., a cyclic engine). Engines attempt to optimize the conversion of heat to work.

1.2.1. Applications of energy

Most of our daily activities involve energy transfer and energy change. The human body is a familiar example of a biological system in which the chemical energy of food or body fat is transformed into other forms of energy such as heat and work. Engineering applications of energy processes are wide ranging and include power plants to generate electricity, engines to run automobiles and aircraft, refrigeration and air-conditioning systems, etc.

Many examples of such systems are discussed here. In a hydroelectric power system, the potential energy of water is converted into mechanical energy through the use of a hydraulic turbine. The mechanical energy is then converted into electric energy by an electric generator coupled to the shaft of the turbine. In a steam power generating plant, chemical or nuclear energy is converted into thermal energy in a boiler or a reactor. The energy is imparted to water, which vaporizes into steam. The energy of the steam is used to drive a steam turbine, and the resulting mechanical energy is used to drive a generator to produce electric power. The steam leaving the turbine is then condensed, and the condensate is pumped back to the boiler to complete the cycle. Breeder reactors use uranium-235 as a fuel source and can produce some more fuel in the process. A solar power plant uses solar concentrators (parabolic or flat mirrors) to heat a working fluid in a receiver located on a tower, where a heated fluid expands in a turbogenerator as in a conventional power plant. In a spark-ignition internal combustion engine, chemical energy of fuel is converted into mechanical work. An air–fuel mixture is compressed and combustion is initiated by a spark device. The expansion of the combustion gases pushes against a piston, which results in the rotation of a crankshaft. Gas turbine engines, commonly used for aircraft propulsion, convert the chemical energy of fuel into thermal energy that is used to run a gas turbine. The turbine is directly coupled to a compressor that supplies the air required for combustion. The exhaust gases, on expanding in a nozzle, create thrust. For power generation, the turbine is coupled to an electric generator and drives both the compressor and the generator. In a liquid-fuel rocket, a fuel and an oxidizer are combined, and combustion gases expand in a nozzle creating a propulsive force (thrust) to propel the rocket. A typical nuclear rocket propulsion engine offers a higher specific impulse when compared to chemical rockets. A fuel cell converts chemical energy into electric energy directly making use of an ion-exchange membrane. When a fuel such as hydrogen is ionized, it flows from the anode through the membrane toward the cathode. The released electrons at the anode flow through an external load. In a magnetohydrodynamic generator, electricity is produced by moving a high-temperature plasma through a magnetic field. A refrigeration system utilizes work supplied by an electric motor to transfer heat from a refrigerated space. Low-temperature boiling fluids such as ammonia and refrigerant-134a absorb thermal energy as they vaporize in the evaporator causing a cooling effect in the region being cooled.

These are only some of the numerous engineering applications. Thermodynamics is relevant to a much wider range of processes and applications not only in engineering, but also in science. A good understanding of this topic is required to improve the design and performance of energy transfer systems.

1.2.2. Concept of energy

The concept of energy was first introduced in mechanics by Newton when he hypothesized about kinetic and potential energies. However, the emergence of energy as a unifying concept in physics was not adopted until the middle of the 19th century and is considered one of the major scientific achievements in that century. The concept of energy is so familiar to us today that it seems intuitively obvious to understand, yet we often have difficulty defining it precisely.

Energy is a scalar quantity that cannot be observed directly but can be recorded and evaluated by indirect measurements. The absolute value of the energy of a system is difficult to measure, whereas the energy change is relatively easy to evaluate.

Examples of energy use in life experiences are endless. The sun is the major source of the earth's energy. It emits a spectrum of energy that travels across space as electromagnetic radiation. Energy is also associated with the structure of matter and can be released by chemical and atomic reactions. Throughout history, the emergence of civilizations has been characterized by the discovery and effective application of energy to help meet society's needs.

1.2.3. Forms of energy

Energy manifests itself in many forms, which are either internal or transient. Energy can be converted from one form to another. In thermodynamic analysis, the forms of energy can be classified into two groups: macroscopic and microscopic.

Macroscopic forms of energy: are those which an overall system possesses with respect to a reference frame, e.g., kinetic and potential energies. For example, the macroscopic energy of a rising object changes with velocity and elevation. The macroscopic energy of a system is related to motion and the influence of external effects such as gravity, magnetism, electricity and surface tension.

The energy that a system possesses as a result of its motion relative to some reference frame is *kinetic energy*. Kinetic energy refers to the energy of the system because of its 'overall' motion, either translational or rotational. Overall is used here to specify that we refer to the kinetic energy of the entire system, not the kinetic energy of the molecules in the system. If the system is a gas, for example, the kinetic energy is the energy due to the macroscopic flow of the gas, not the motion of individual molecules.

The *potential energy* of a system is the sum of the gravitational, centrifugal, electrical and magnetic potential energies. The energy that a system possesses as a result of its elevation in a gravitational field is called gravitational potential energy (or commonly just potential energy). For example, a 1 kg mass, 100 m above the ground, has a greater potential energy than the same mass on the ground. Potential energy can be converted into other forms of energy, such as kinetic energy, if the mass is allowed to fall.

Kinetic and potential energy depend on the environment in which the system exists. In particular, the potential energy of a system depends on the choice of a zero level. For example, if ground level is considered to be at zero potential energy, then the potential energy of the mass 100 m above the ground has a positive potential energy equal to the mass (1 kg) multiplied by the gravitational constant ($g = 9.807$ m/s^2) and the height above the ground (100 m). Its potential energy will be 980.7 (kgm^2)/s^2 (or 980.7 Newton-meters (Nm), or 980.7 J). The datum plane for potential energy can be chosen arbitrarily. If it had been chosen at 100 m above the ground level, the potential energy of the mass would have been zero. Of course, the difference in potential energy between the mass at 100 m and the mass at ground level is independent of the datum plane.

Microscopic forms of energy: are those related to the molecular structure of a system and the degree of molecular activity, and are independent of outside reference frames. The sum of all the microscopic forms of energy of a system is its *internal energy*. The internal energy of a system depends on the inherent qualities, or properties, of the materials in the system, such as composition and physical form, as well as the environmental variables (temperature, pressure, electric field, magnetic field, etc.). Internal energy can have many forms, including mechanical, chemical, electrical, magnetic, surface and thermal. Some examples are considered for illustration:

- A spring that is compressed has a higher internal energy (mechanical energy) than a spring that is not compressed, because the compressed spring can do work on changing (expanding) to the uncompressed state.
- Two identical vessels, each containing hydrogen and oxygen, are considered that have different chemical energies. In the first, the gases are contained in the elemental form, pure hydrogen and pure oxygen, in a ratio of 2:1. The second contains an identical number of atoms, but in the form of water. The internal energies of these systems differ. A spark may set off a violent release of energy in the first container, but not in the second.

The structure of thermodynamics involves the concept of equilibrium states and postulates that the change in the value of thermodynamic quantities, such as internal energy, between two equilibrium states of a system does not depend on the thermodynamic path the system takes to get from one state to the other. The change is defined by the final and initial equilibrium states of the system. Consequently, the internal energy change of a system is determined by the parameters that specify the system in its final and initial states. The parameters include pressure, temperature, magnetic field, surface area, mass, etc. If a system changes from state 1 to state 2, the change in internal energy ΔU is $(U_2 - U_1)$, the internal energy in the final state is less than in the initial state. The difference does not depend on how the system gets from state 1 to state 2. The internal energy thus is referred to as a state function, or a point function, i.e., a function of the state of the system only, and not its history.

The thermal energy of a system is the internal energy of a system which increases as temperature is increased. For instance, we have to add energy to an iron bar to raise its temperature. The thermal energy of a system is not referred to as heat, as heat is energy in transit between systems.

1.2.4. The first law of thermodynamics

The first law of thermodynamics (FLT) is the *law of the conservation of energy*, which states that, although energy can change form, it can be neither created nor destroyed. The FLT defines internal energy as a state function and provides a formal statement of the conservation of energy.

However, it provides no information about the direction in which processes can spontaneously occur, i.e., the reversibility aspects of thermodynamic processes. For example, the FLT cannot indicate how cells can perform work while existing in an isothermal environment. The FLT provides no information about the inability of any thermodynamic process to convert heat fully into mechanical work, or any insight into why mixtures cannot spontaneously separate or un-mix themselves. A principle to explain these phenomena and to characterize the availability of energy is required to do this. That principle is embodied in the second law of thermodynamics (SLT) which we explain later.

1.2.5. Energy and the FLT

For a control mass, the energy interactions for a system may be divided into two parts: dQ, the amount of heat, and dW, the amount of work. Unlike the change in total internal energy dE, the quantities dQ and dW are not independent of the manner of transformation, so we cannot specify dQ and dW simply by knowing the initial and final states. Hence it is not possible to define a function Q which depends on the initial and final states, i.e., heat is not a state function. The FLT for a control mass can be written as follows:

$$dQ = dE + dW \qquad (1.1)$$

When Eq. (1.1) is integrated from an initial state 1 to a final state 2, it results in

$$Q_{1-2} = E_2 - E_1 + W_{1-2} \quad \text{or} \quad E_2 - E_1 = Q_{1-2} - W_{1-2} \qquad (1.2)$$

where E_1 and E_2 denote the initial and final values of the energy E of the control mass, Q_{1-2} is the heat transferred to the control mass during the process from state 1 to state 2, and W_{1-2} is the work done by the control mass during the process from state 1 to state 2.

The energy E may include internal energy U, kinetic energy KE and potential energy PE terms as follows:

$$E = U + KE + PE \qquad (1.3)$$

For a change of state from state 1 to state 2 with a constant gravitational acceleration g, Eq. (1.3) becomes

$$E_2 - E_1 = U_2 - U_1 + m(V_2^2 - V_1^2)/2 + mg(Z_2 - Z_1) \qquad (1.4)$$

where m denotes the fixed amount of mass contained in the system, V the velocity and Z the elevation.

The quantities dQ and dW can be specified in terms of the rate laws for heat transfer and work. For a control volume an additional term appears from the fluid flowing across the control surface (entering at state i and exiting at state e). The FLT for a control volume can be written as

$$\dot{Q}_{cv} = \dot{E}_{cv} + \dot{W}_{cv} + \sum \dot{m}_e \hat{h}_e - \sum \dot{m}_i \hat{h}_i \quad \text{or} \quad \dot{E}_{cv} = \dot{Q}_{cv} - \dot{W}_{cv} + \sum \dot{m}_i h_i - \sum \dot{m}_e \hat{h}_e \qquad (1.5)$$

where $\dot{m}$ is mass flow rate per unit time, $\hat{h}$ is total specific energy, equal to the sum of specific enthalpy, kinetic energy and potential energy, i.e., $\hat{h} = h + V^2/2 + gZ$.

1.2.6. Economic aspects of energy

Although all forms of energy are expressed in the same units (joules, megajoules, gigajoules, etc.), the financial value of energy varies enormously with its grade or quality. Typically, electrical and mechanical energy are the most costly, followed by high-grade thermal energy. At the other extreme, thermal energy which is only a few degrees from ambient has virtually no commercial value. These examples highlight the weakness of trying to equate the energy contained in steam or the heat content of geothermal fluids with the high-grade energy obtainable from fossil fuels or nuclear reactions. Economics usually suggests that one should avoid using energy at a significantly higher grade than needed for a task. For example, electrical energy, which has a high energy grade, should be used for such purposes as mechanical energy generation, production of light, sound and very high temperatures in electrical furnaces. Electric space heating, on the other hand, in which electricity is used for raising the temperature of ambient air only to about 20°C, is an extremely wasteful use of electricity. This observation applies in both domestic and industrial contexts. In many jurisdictions, there is excess electricity generation capacity at night and therefore some of the nighttime electricity is sold at reduced prices for space heating purposes, even though this is inherently wasteful. It is often more advantageous and efficient in such situations to utilize energy storage such as flywheels, compressed air or pumped water, which leads to reduced thermodynamic irreversibility.

In industry settings, tasks often require energy, but at different grades. The opportunity often exists to use the waste heat from one process to serve the needs of another in an effective and efficient manner. Sometimes a cascade of tasks can be satisfied in this manner; for example, a typical glass works releases waste heat at between 400°C and 500°C, which is sufficient for raising intermediate-pressure steam for running back-pressure turbines to produce electricity or raising low-pressure steam at about 120°C for other purposes, or for heating operations at temperatures as high as almost 400–500°C. The heat exhausted from a steam turbine can, in turn, be used to evaporate moisture from agricultural products. The water vapor obtainable from such processes can be condensed to provide warm water at about 60°C, which can be employed for space heating or for the supply of heat to fish farms or greenhouses. In this example, the original supply of high-grade energy obtained by burning coal, oil or natural gas performs four separate tasks:

1. The various glass constituents are melted after being heated to above their solidification temperature (about 1500°C).
2. Medium-pressure steam is used to produce electricity (500°C).
3. The exhaust-steam from the back-pressure turbine is used for crop drying (120°C).
4. The condensed water vapor heats water for use in space heating, fish farms or greenhouses (60°C).

1.2.7. Energy audit methods

Energy management opportunities often exist to improve the effectiveness and efficiency with which energy is used. For instance, energy processes in industrial, commercial and institutional facilities, including heating, cooling and air conditioning, can often be improved. Many of these opportunities are recognizable during a walk-through audit or more detailed examination of a facility. Such an audit is usually more meaningful if someone from outside the facility but generally familiar with energy management is involved. Typical energy saving items noted during a walk-through audit include steam and water leaks at connections and other locations, damaged insulation, excessive lighting, etc. Alert management and operating staff and good maintenance procedures can, with little effort, reduce energy usage and save money.

Not all items noted in a walk-through audit are easy to analyze. For example, a stream of cooling water may be directed to a drain after being used for a cooling application, even though some thermal energy remains in the water. The economics of recovering this heat needs to be investigated to determine if it is worth recovering. Some relevant questions to consider in such an assessment include the following:

- How much thermal energy is available in the waste stream?
- Is there a use for this energy?
- What are the capital and operating costs involved in recovering the energy?
- Will the energy and associated cost savings pay for the equipment required to recover the energy?

A diagnostic audit is required to determine the thermal energy available in a waste stream, how much energy can be recovered, and if there is a use for this recovered energy within or outside the facility. The cost savings associated with recovering the energy are determined and, along with the cost to supply and install the heat recovery equipment, the simple payback period can be evaluated for the measure to establish its financial viability.

1.2.8. Energy management

Energy management refers to the process of using energy carefully so as to save money or achieve other objectives. Energy management measures can be divided into the following categories: maintenance (or housekeeping), low cost (or simple) and retrofit. Many energy management measures are outlined here along with their potential energy savings. This list is not intended to be comprehensive (e.g., it does not cover all opportunities available for heating, cooling and air-conditioning equipment), but rather to help those involved in management, operations and maintenance to identify energy savings opportunities specific to a particular facility. Other energy management opportunities exist.

Energy management is best approached in an open manner that allows previously accepted inefficient practices to be explored. Improved awareness on the part of the staff managing, operating or maintaining a facility, combined with imagination and/or expert assistance, can yield large dividends in terms of energy use and cost reductions. Several practical energy management measures are covered below.

Maintenance opportunities: Maintenance measures for energy management are those carried out on a regular basis, normally no more than annually, and include the following:

- Sealing leaks at valves, fittings and gaskets.
- Repairing damaged insulation.
- Maintaining temperature and pressure controls.
- Maintaining steam traps.
- Cleaning heat transfer surfaces.
- Ensuring steam quality is adequate for the application.
- Ensuring steam pressure and temperature ranges are within the tolerances specified for equipment.
- Ensuring steam traps are correctly sized to remove all condensate.
- Ensuring heating coils slope from steam inlet to steam trap to prevent coils from flooding with condensate.

Low-cost opportunities: Low-cost energy management measures are normally once-off actions for which the cost is not considered great:

- Shutting equipment when not required.
- Providing lockable covers for control equipment such as thermostats to prevent unauthorized tampering.
- Operating equipment at or near capacity whenever possible, and avoiding running multiple units at reduced capacity.
- Adding thermostatic air vents.
- Adding measuring and monitoring equipment to provide the operating data needed to improve system operation.
- Assessing the location of control devices to ensure best operation.

Retrofit opportunities: Retrofit energy management measures are normally once-off actions with significant costs that involve modifications to existing equipment. Many of these measures require detailed analysis and are beyond the scope of this chapter. Worked examples are provided for some of the listed energy management opportunities, while in other cases there is only commentary. Typical energy management measures in this category follow:

- Converting from direct to indirect steam heated equipment and recovery of condensate.
- Installing/upgrading insulation on equipment.
- Relocating steam heated equipment from central building areas to areas with exterior exposures so that heat loss from the equipment can assist in heating the area.
- Reviewing general building heating concepts as opposed to task heating concepts.
- Modifying processes to stabilize or reduce steam or water demand.
- Investigating scheduling of process operations in an attempt to reduce peak steam or water demands.
- Evaluating waste water streams exiting a facility for heat recovery opportunities.

1.3. Entropy

In this section, basic phenomena like order and disorder as well as reversibility and irreversibility are discussed. Entropy and the SLT are also covered, along with their significance.

1.3.1. Order and disorder and reversibility and irreversibility

Within the past 50 years our view of Nature has changed drastically. Classical science emphasized equilibrium and stability. Now we observe fluctuations, instability and evolutionary processes on all levels from chemistry and biology to cosmology. Everywhere we observe irreversible processes in which time symmetry is broken. The distinction between reversible and irreversible processes was first introduced in thermodynamics through the concept 'entropy.'

The formulation of entropy is in the modern context fundamental for understanding thermodynamic aspects of self-organization and the evolution of order and life that we observe in nature. When a system is isolated, the entropy of a system continually increases due to irreversible processes, and reaches the maximum possible value when the system attains a state of thermodynamic equilibrium. In the state of equilibrium, all irreversible processes cease. When a system begins to exchange entropy with its surroundings then, in general, it is driven away from the equilibrium state it reached when isolated, and entropy-producing irreversible processes begin. An exchange of entropy is associated with the exchange of heat and matter. When no accumulation of entropy within a system occurs, the entropy flowing out of the system is always larger than the entropy flowing in, the difference arising due to the entropy produced by irreversible processes within the system. As we shall see in the following chapters, systems that exchange entropy with their surroundings do not simply increase the entropy of the surroundings, but may undergo dramatic spontaneous transformations to 'self-organization.' Irreversible processes that produce entropy create these organized states. Such self-organized states range from convection patterns in fluids to organized life structures. Irreversible processes are the driving force that creates this order.

Much of the internal energy of a substance is randomly distributed as kinetic energy at the molecular and sub-molecular levels and as energy associated with attractive or repulsive forces between molecular and sub-molecular entities, which can move closer together or further apart. This energy is sometimes described as being 'disordered' as it is not accessible as work at the macroscopic level in the same way as is the kinetic energy or gravitational potential energy that an overall system possesses due to its velocity or position in a gravitational field. Although some energy forms represent the capacity to do work, it is not possible directly to access the minute quantities of disordered energy possessed at a given instant by the entities within a substance so as to yield mechanical shaft work on a macroscopic scale. The term disorder refers to the lack of information about exactly how much and what type of energy is associated at any moment with each molecular or sub-molecular entity within a system.

At the molecular and sub-molecular level there also exists 'ordered energy' associated with the attractive and repulsive forces between entities that have fixed mean relative positions. Part of this energy is, in principle, accessible as work at the macroscopic level under special conditions, which are beyond the scope of this book.

Temperature is the property that reflects whether a system that is in equilibrium will experience a decrease or increase in its disordered energy if it is brought into contact with another system that is in equilibrium. If the systems have different temperature, disordered energy will be redistributed from the system at the higher temperature to the one at the lower temperature. The process reduces the information about precisely where that energy resides, as it is now dispersed over the two systems.

Heat transfer to a system increases its disordered energy, while heat transfer from a system reduces its disordered energy. Reversible heat transfer is characterized by both the amount of energy transferred to or from the system and the temperature at which this occurs. The property entropy, whose change between states is defined as the integral of the ratio of the reversible heat transfer to the absolute temperature, is a measure of the state of disorder of the system. This 'state of disorder' is characterized by the amount of disordered energy and its temperature. Reversible heat transfer from one system to another requires that both systems have the same temperature and that the increase in the disorder of one be exactly matched by a decrease in disorder of the other. When reversible adiabatic work is done on or by a system, its ordered energy increases or decreases by exactly the amount of the work and the temperature changes correspondingly, depending on the substances involved. Reversible work is characterized by the amount of energy transferred to or from the system, irrespective of the temperature of the system. Irreversible work, such as stirring work or friction work between subsystems, involves a change in the disorder of the system and, like heat transfer to a system, has the effect of increasing the entropy.

1.3.2. Characteristics of entropy

We now introduce the thermodynamic property entropy, which is a measure of the amount of molecular disorder within a system. A system possessing a high degree of molecular disorder (such as a high-temperature gas) has a high entropy and vice versa. Values for specific entropy are commonly listed in thermodynamic tables along with other property data (e.g., specific volume, specific internal energy, specific enthalpy). A fundamental property related to the SLT, entropy has the following characteristics:

- The entropy of a system is a measure of its internal molecular disorder.
- A system can only generate, not destroy, entropy.
- The entropy of a system can be increased or decreased by energy transports across the system boundary.

Heat and work are mechanisms of energy transfer. They can cause changes in the internal energy in a body as energy is transferred to or from it. Work is accomplished by a force acting through a distance. Heat requires a difference in temperature for its transfer. The definition of heat can be broadened to include the energy stored in a hot gas as the average kinetic energy of randomly moving molecules. This description helps explain the natural flow of heat from a hot to a cooler substance. The concept of random motion can be translated into the notion of order and disorder, and leads to a relation between order and disorder and probability. Energy transfers associated with a system can cause changes in its state. The natural direction of the change in state of a system is from a state of low probability to one of higher probability. Since disordered states are more probable than ordered ones, the natural direction of change of state of a system is from order to disorder. Entropy is a measure of order that helps explain the natural direction for energy transfers and conversions. The entropy of a system at a specific state depends on its probability. Thus the SLT can be expressed more broadly in terms of entropy in the following way:

In any transfer or conversion of energy within a closed system, the entropy of the system increases. The consequences of the second law can thus be stated as (1) the spontaneous or natural direction of energy transfer or conversion is toward increasing entropy or (2) all energy transfers or conversions are irreversible. More loosely, the FLT implies 'You can't win' because energy is conserved so you cannot get more energy out of a system than you put in, while the SLT states 'You can't break even' because irreversibilities during real processes do not allow you to recover the original quality of energy you put into a system.

Low-entropy energy sources are normally desired and used to drive energy processes, since low-entropy energy is 'useful.' Energy sources can be rated on an entropy or usefulness scale, with zero-entropy energy forms like work and kinetic and gravitational potential energy being the most useful, and high-entropy forms like heat being less useful.

This broader interpretation of the SLT suggests that real 'energy conservation' should consider the conservation of both energy quantity and quality. For high thermodynamic efficiency, energy transfers or conversions should be arranged, all else being equal, so that the change in entropy is a minimum. This requires that energy sources be matched in entropy to energy end use.

1.3.3. Significance of entropy

The entropy of a system at some state is a measure of the probability of its occurrence, with states of low probability having low entropy and states of high probability having high entropy. From the previous section, it is seen that the entropy of a system must increase in any transfer or conversion of energy, because the spontaneous direction of the change of state of a closed system is from a less to a more probable state. Consequently, a simple statement of the second law is 'In any energy transfer or conversion within a closed system, the entropy of the system increases.'

In open systems, energy conversions can occur which cause the entropy of part or all of a system to decrease. Charging a storage battery, freezing ice cubes, and the growth of living entities are examples. In each of these examples, the order of the system increases and the entropy decreases. If the combination of the system and its surroundings is considered, however, the overall net effect is always to increase disorder. To charge a battery we must provide a certain minimum amount of external energy of a certain quality to re-form the chemical combinations in the battery plates. In the case of the battery, the input energy can be in the form of electricity. Some of this low-entropy electrical energy is lost as it is converted into high-entropy heat in the current-carrying wires. In freezing ice, we increase order by decreasing the entropy of the water in the ice cube trays through removal of heat. The removed heat is transferred into a substance that is at a lower temperature, increasing its entropy and disorder. The net change in entropy is positive. For ice cubes in a freezer, we also supply to the motor low-entropy electrical energy, which ultimately is degraded to heat. In life processes, highly ordered structures are built from simpler structures of various chemicals, but to accomplish this living entities take in relatively low-entropy energy – sunlight and chemical energy – and release high-entropy heat and other wastes. The entropy of the overall system again increases.

Figure 1.2 illustrates a heat transfer process from the entropy point of view. During the heat transfer process, the net entropy increases, with the increase in entropy of the cold body more than offsetting the decrease in entropy of the hot body. This must occur to avoid violating the SLT.

More generally, processes can occur only in the direction of increased overall entropy or disorder. This implies that the entire universe is becoming more chaotic every day.

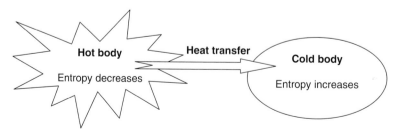

Fig. 1.2. Illustration of entropy increase and decrease for cold and hot bodies during heat transfer.

Another way of explaining this consequence of the SLT is to state that all energy transfers or conversions are irreversible. Absent external energy inputs, such processes occur spontaneously in the direction of increasing entropy. In a power plant, for example, although some of the losses can be reduced, they cannot be entirely eliminated. Entropy must increase. Usual mechanisms for low-entropy energy to be converted to high-entropy heat are irreversibilities like friction or electrical resistance or leakage of high-temperature, low-entropy heat to a lower-temperature region and its subsequent degradation.

1.3.4. Carnot's contribution

Another statement of the SLT was developed more than one hundred years ago. One of the most brilliant contributions was made by a young French physicist, Sadi Carnot, in the 19th century. Carnot, studying early steam engines, was able to abstract from the pumping pistons and spinning wheels that the conversion of heat to mechanical work requires a difference of temperature. The purpose of a heat engine, as he described it, is to take heat from a high-temperature source, convert some of it to mechanical work, and then reject the rest of the heat to a lower-temperature heat reservoir. Carnot described heat engines using a simple analogy to waterwheels. The energy available for conversion in a waterwheel is the gravitational energy contained in water as it flows from some height (behind a dam or from a mountain lake) down through the wheel. The amount of energy available depends on the difference in height – the 'head' as it is called – between the source and the pool below the wheel. The energy available to a heat engine depends on the 'temperature head.' Just as a high dam can provide more energy than a low one, a large temperature difference can provide more energy to be converted by a heat engine than can a small temperature difference. In the example of a heat engine, the high-temperature reservoir is the hot steam produced in the power plant furnace.

For a steam turbine and condenser assembly, the low-temperature reservoir to which the device rejects the unconverted energy is the condenser cooling water. The important temperature difference is thus the difference in temperature between the incoming steam, usually about 700°C, and the water in the condenser, which is typically between environmental conditions (around 0–25°C) and the boiling temperature of water (100°C). The 'temperature head' in this example would therefore be 600–700°C. Carnot's explanation of heat engines led to the second law. Once energy is in the form of heat, it cannot be converted entirely to mechanical energy. Some heat will always be exhausted.

1.3.5. The second law of thermodynamics

Although a spontaneous process can proceed only in a definite direction, the FLT gives no information about direction; it merely states that when one form of energy is converted to another, the quantities of energy involved are conserved regardless of feasibility of the process. Thus, processes can be envisioned that do not violate the FLT but do violate the SLT, e.g., transfer of a certain quantity of heat from a low-temperature body to a high-temperature body, without the input of an adequate external energy form like work. However, such a process is impossible, emphasizing that the FLT is itself inadequate for explaining energy processes.

The SLT establishes the difference in the quality of different forms of energy and explains why some processes can spontaneously occur while other cannot. The SLT is usually expressed as an inequality, stating that the total entropy after a process is equal to or greater than that before. The equality only holds for ideal or reversible processes. The SLT has been confirmed experimentally.

The SLT defines the fundamental quantity entropy as a randomized energy state unavailable for direct conversion to work. It also states that all spontaneous processes, both physical and chemical, proceed to maximize entropy, i.e., to become more randomized and to convert energy to a less available form. A direct consequence of fundamental importance is the implication that at thermodynamic equilibrium the entropy of a system is at a relative maximum; i.e., no further increase in disorder is possible without changing the thermodynamic state of the system by some external means (such as adding heat). A corollary of the SLT is the statement that the sum of the entropy changes of a system and that of its surroundings must always be positive. In other words, the universe (the sum of all systems and surroundings) is constrained to become forever more disordered and to proceed toward thermodynamic equilibrium with some absolute maximum value of entropy. From a biological standpoint this is intuitively reasonable since, unless gradients in concentration and temperature are forcibly maintained by the consumption of energy, organisms proceed spontaneously toward the biological equivalent of equilibrium-death.

The SLT is general. However, when intermolecular forces are long range, as in the case of particles interacting through gravitation, there are difficulties because our classification into extensive variables (proportional to size) and intensive variables (independent of size) does not apply. The total energy is no longer proportional to size. Fortunately gravitational forces are very weak compared to short-range intermolecular forces. It is only on the astrophysical scale that this problem becomes important. The generality of the SLT provides a powerful means to understand the thermodynamic aspects of real systems through the use of ideal systems. A classic example is Planck's analysis of radiation in thermodynamic equilibrium with matter (blackbody radiation) in which Planck considered idealized simple harmonic oscillators interacting with radiation. Planck considered simple harmonic oscillators not merely because they are good approximations of molecules but because the properties of radiation in thermal equilibrium with matter are universal, regardless of the particular nature of the matter with which the radiation interacts. The conclusions one arrives at using idealized oscillators and the laws of thermodynamics must also be valid for all other forms of matter, however complex.

What makes this statement of the SLT valuable as a guide to formulating energy policy is the relationship between entropy and the usefulness of energy. Energy is most useful to us when it is available to do work or we can get it to flow from one substance to another, e.g., to warm a house. Useful energy thus must have low entropy so that the SLT will allow transfer or conversions to occur spontaneously.

1.3.6. SLT statements

Although there are various formulations of the SLT, two are particularly well known:

1. *Clausius statement*: It is impossible for heat to move of itself from a lower-temperature reservoir to a higher-temperature reservoir. That is, heat transfer can only occur spontaneously in the direction of temperature decrease. For example, we cannot construct a refrigerator that operates without any work input.
2. *Kelvin–Planck statement*: It is impossible for a system to receive a given amount of heat from a high-temperature reservoir and to provide an equal amount of work output. While a system converting work to an equivalent energy transfer as heat is possible, a device converting heat to an equivalent energy transfer as work is impossible. Alternatively, a heat engine cannot have a thermal efficiency of 100%.

1.3.7. The Clausius inequality

The Clausius inequality provides a mathematical statement of the second law, which is a precursor to second law statements involving entropy. German physicist RJE Clausius, one of the founders of thermodynamics, stated

$$\oint (\delta Q/T) \leq 0 \tag{1.6}$$

where the integral symbol $\oint$ shows the integration should be done for the entire system. The cyclic integral of $\delta Q/T$ is always less than or equal to zero. The system undergoes only reversible processes (or cycles) if the cyclic integral equals zero, and irreversible processes (or cycles) if it is less than zero.

Equation (1.6) can be expressed without the inequality as

$$S_{gen} = -\oint (\delta Q/T) \tag{1.7}$$

where

$$S_{gen} = \Delta S_{total} = \Delta S_{sys} + \Delta S_{surr}$$

The quantity S_{gen} is the entropy generation associated with a process or cycle, due to irreversibilities. The following are cases for values of S_{gen}:

- $S_{gen} = 0$ for a reversible process
- $S_{gen} > 0$ for an irreversible process
- $S_{gen} < 0$ for no process (i.e., negative values for S_{gen} are not possible)

Consequently, one can write for a reversible process,

$$\Delta S_{sys} = (Q/T)_{rev} \quad \text{and} \quad \Delta S_{surr} = -(Q/T)_{rev} \tag{1.8}$$

For an irreversible process,

$$\Delta S_{sys} > (Q/T)_{surr} \tag{1.9}$$

due to entropy generation within the system as a result of internal irreversibilities. Hence, although the change in entropy of the system and its surroundings may individually increase, decrease or remain constant, the total entropy change or the total entropy generation cannot be less than zero for any process.

1.3.8. Useful relationships

It is helpful to list some common relations for a process involving a pure substance and assuming the absence of electricity, magnetism, solid distortion effects and surface tension. The following four equations apply, subject to the noted restrictions:

- $\delta q = du + \delta w$ (a FLT statement applicable to any simple compressible closed system)
- $\delta q = du + p\,dv$ (a FLT statement restricted to reversible processes for a closed system)
- $T\,ds = du + \delta w$ (a combined statement of the FLT and SLT, with $Tds = \delta q$)
- $T\,ds = du + p\,dv$ (a combined statement of the FLT and SLT valid for all processes between equilibrium states)

1.4. Exergy

A very important class of problems in engineering thermodynamics concerns systems or substances that can be modeled as being in equilibrium or stable equilibrium, but that are not in mutual stable equilibrium with the surroundings. For example, within the earth there are reserves of fuels that are not in mutual stable equilibrium with the atmosphere and the sea. The requirements of mutual chemical equilibrium are not met. Any system at a temperature above or below that of the environment is not in mutual stable equilibrium with the environment. In this case the requirements of mutual thermal equilibrium are not met. Any lack of mutual stable equilibrium between a system and the environment can be used to produce shaft work.

With the SLT, the maximum work that can be produced can be determined. Exergy is a useful quantity that stems from the SLT, and helps in analyzing energy and other systems and processes.

1.4.1. The quantity exergy

The exergy of a system is defined as the maximum shaft work that can be done by the composite of the system and a specified reference environment. The reference environment is assumed to be infinite, in equilibrium, and to enclose all other systems. Typically, the environment is specified by stating its temperature, pressure and chemical composition. Exergy is not simply a thermodynamic property, but rather is a property of both a system and the reference environment.

The term exergy comes from the Greek words *ex* and *ergon*, meaning from and work. The exergy of a system can be increased if exergy is input to it (e.g., work is done on it). The following are some terms found in the literature that are equivalent or nearly equivalent to exergy: available energy, essergy, utilizable energy, available energy and availability.

1.4.2. Exergy analysis

Exergy has the characteristic that it is conserved only when all processes occuring in a system and the environment are reversible. Exergy is destroyed whenever an irreversible process occurs. When an exergy analysis is performed on a plant such as a power station, a chemical processing plant or a refrigeration facility, the thermodynamic imperfections can be quantified as exergy destructions, which represent losses in energy quality or usefulness (e.g., wasted shaft work or wasted potential for the production of shaft work). Like energy, exergy can be transferred or transported across the boundary of a system. For each type of energy transfer or transport there is a corresponding exergy transfer or transport.

Exergy analysis takes into account the different thermodynamic values of different energy forms and quantities, e.g., work and heat. The exergy transfer associated with shaft work is equal to the shaft work. The exergy transfer associated with heat transfer, however, depends on the temperature at which it occurs in relation to the temperature of the environment.

1.4.3. Characteristics of exergy

Some important characteristics of exergy are described and illustrated:

- A system in complete equilibrium with its environment does not have any exergy. No difference appears in temperature, pressure, concentration, etc. so there is no driving force for any process.
- The exergy of a system increases the more it deviates from the environment. For instance, a specified quantity of hot water has a higher exergy content during the winter than on a hot summer day. A block of ice carries little exergy in winter while it can have significant exergy in summer.
- When energy loses its quality, exergy is destroyed. Exergy is the part of energy which is useful and therefore has economic value and is worth managing carefully.
- Exergy by definition depends not just on the state of a system or flow, but also on the state of the environment.
- Exergy efficiencies are a measure of approach to ideality (or reversibility). This is not necessarily true for energy efficiencies, which are often misleading.
- Exergy can generally be considered a valuable resource. There are both energy or non-energy resources and exergy is observed to be a measure of value for both:
- Energy forms with high exergy contents are typically more valued and useful than energy forms with low exergy. Fossil fuels, for instance, have high energy and energy contents. Waste heat at a near environmental condition, on the other hand, has little exergy, even though it may contain much energy, and thus is of limited value. Solar energy, which is thermal radiation emitted at the temperature of the sun (approximately 5800 K), contains much energy and exergy.
- A concentrated mineral deposit 'contrasts' with the environment and thus has exergy. This contrast and exergy increase with the concentration of the mineral. When the mineral is mined the exergy content of the mineral is retained, and if it is enriched or purified the exergy content increases. A poor quality mineral deposit contains less exergy and can accordingly be utilized only through a larger input of external exergy. Today this substitution of exergy often comes from exergy forms such as coal and oil. When a concentrated mineral is dispersed the exergy content is decreased.

An engineer designing a system makes trade-offs among competing factors. The engineer is expected to aim for the highest reasonable technical efficiency at the lowest reasonable cost under the prevailing technical, economic and legal conditions, and also accounting for ethical, ecological and social consequences and objectives. Exergy analysis is a tool that can facilitate this work. Exergy methods provide unique insights into the types, locations and causes of losses and can thereby help identify possible improvements. For instance, Exergetic Life Cycle Assessment (ExLCA) is suggested as a method to better meet environmental objectives as studied in detail in Chapter 19.

Before discussing in detail linkages between energy and exergy, and the relations between exergy and both the environment and sustainable development, some key points that highlight the importance of exergy and its utilization are provided. Specifically, exergy analysis is an effective method and tool for:

- Combining and applying the conservation of mass and conservation of energy principles together with the SLT for the design and analysis of energy systems.
- Improving the efficiency of energy and other resource use (by identifying efficiencies that always measure the approach to ideality as well as the locations, types and true magnitudes of wastes and losses).
- Revealing whether or not and by how much it is possible to design more efficient systems by reducing the inefficiencies in existing systems.
- Addressing the impact on the environment of energy and other resource utilization, and reducing or mitigating that impact.
- Identifying whether a system contributes to achieving sustainable development or is unsustainable.

1.4.4. The reference environment

Since the value of the exergy of a system or flow depends on the state of both the system or flow and a reference environment, a reference environment must be specified prior to the performance of an exergy analysis.

The environment is often modeled as a reference environment similar to the actual environment in which a system or flow exists. This ability to tailor the reference environment to match the actual local environment is often an advantage of exergy analysis.

Some, however, consider the need to select a reference environment a difficulty of exergy analysis. To circumvent this perceived difficulty, some suggest that a 'standard environment' be defined with a specified chemical composition, temperature and pressure. A possible chemical standard environment for global use could, for instance, be based on a standard atmosphere, a standard sea and a layer of the earth's crust. The definition of such a reference environment is usually problematic, however, as these systems are not in equilibrium with each other.

In accounting for local conditions, a reference environment can vary spatially and temporally. The need to account for spatial dependence is clear if one considers an air conditioning and heating system operating in the different climates across the earth. In addition, an aircraft or rocket experiences different environmental conditions as it ascends through the atmosphere. The importance of accounting for temporal dependence is highlighted by considering a technology like a seasonal thermal energy storage unit, in which heating or cooling capacity can be stored from one season where it is available in the environment to another season when it is unavailable but in demand.

1.4.5. Exergy vs. energy

Energy analysis is the traditional method of assessing the way energy is used in an operation involving the physical or chemical processing of materials and the transfer and/or conversion of energy. This usually entails performing energy balances, which are based on the FLT, and evaluating energy efficiencies. This balance is employed to determine and reduce waste exergy emissions like heat losses and sometimes to enhance waste and heat recovery.

However, an energy balance provides no information on the degradation of energy or resources during a process and does not quantify the usefulness or quality of the various energy and material streams flowing through a system and exiting as products and wastes.

The exergy method of analysis overcomes the limitations of the FLT. The concept of exergy is based on both the FLT and the SLT. Exergy analysis clearly indicates the locations of energy degradation in a process and can therefore lead to improved operation or technology. Exergy analysis can also quantify the quality of heat in a waste stream. A main aim of exergy analysis is to identify meaningful (exergy) efficiencies and the causes and true magnitudes of exergy losses. Table 1.1 presents a general comparison of energy and exergy.

It is important to distinguish between exergy and energy in order to avoid confusion with traditional energy-based methods of thermal system analysis and design. Energy flows into and out of a system with mass flows, heat transfers and

Table 1.1. Comparison of energy and exergy.

Energy	Exergy
Dependent on properties of only a matter or energy flow, and independent of environment properties	Dependent on properties of both a matter or energy flow and the environment
Has values different from zero when in equilibrium with the environment (including being equal to mc^2 in accordance with Einstein's equation)	Equal to zero when in the dead state by virtue of being in complete equilibrium with the environment
Conserved for all processes, based on the FLT	Conserved for reversible processes and not conserved for real processes (where it is partly or completely destroyed due to irreversibilities), based on the SLT
Can be neither destroyed nor produced	Can be neither destroyed nor produced in a reversible process, but is always destroyed (consumed) in an irreversible process
Appears in many forms (e.g., kinetic energy, potential energy, work, heat) and is measured in that form	Appears in many forms (e.g., kinetic exergy, potential exergy, work, thermal exergy), and is measured on the basis of work or ability to produce work
A measure of quantity only	A measure of quantity and quality

work interactions (e.g., work associated with shafts and piston rods). Energy is conserved, in line with the FLT. Exergy, although similar in some respects, is different. It loosely represents a quantitative measure of the usefulness or quality of an energy or material substance. More rigorously, exergy is a measure of the ability to do work (or the work potential) of the great variety of streams (mass, heat, work) that flow through a system. A key attribute of exergy is that it makes it possible to compare on a common basis interactions (inputs, outputs) that are quite different in a physical sense. Another benefit is that by accounting for all the exergy streams of the system it is possible to determine the extent to which the system destroys exergy. The destroyed exergy is proportional to the generated entropy. Exergy is always destroyed in real processes, partially or totally, in line with the SLT. The destroyed exergy, or the generated entropy, is responsible for the less-than-ideal efficiencies of systems or processes.

1.4.6. Exergy efficiencies

The *exergy efficiency* is an efficiency based on the SLT. In this section, we describe the use of exergy efficiencies in assessing the utilization efficiency of energy and other resources.

Engineers make frequent use of efficiencies to gauge the performance of devices and processes. Many of these expressions are based on energy, and are thus FLT-based. Also useful are measures of performance that take into account limitations imposed by the second law. Efficiencies of this type are SLT-based efficiencies.

To illustrate the idea of a performance parameter based on the SLT and to contrast it with an analogous energy-based efficiency, consider a control volume at steady-state for which energy and exergy balances can be written, respectively, as

$$\text{(Energy in)} = \text{(Energy output in product)} + \text{(Energy emitted with waste)} \tag{1.10}$$

$$\text{(Exergy in)} = \text{(Exergy output in product)} + \text{(Exergy emitted with waste)} + \text{(Exergy destruction)} \tag{1.11}$$

In these equations, the term product might refer to shaft work, electricity, a certain heat transfer, one or more particular exit streams, or some combination of these. The latter two terms in the exergy balance (Eq. (1.11)) combine to constitute the exergy losses. Losses include such emissions to the surroundings as waste heat and stack gases. The exergy destruction term in the exergy balance is caused by internal irreversibilities.

From energy or exergy viewpoints, a gauge of how effectively the input is converted to the product is the ratio of product to input. That is, the energy efficiency η can be written as

$$\eta = \text{Energy output in product/Energy input} = 1 - [\text{Energy loss/Energy input}] \tag{1.12}$$

and the exergy efficiency ψ as

$$\psi = \text{Exergy output in product/Exergy input} = 1 - [\text{Exergy loss/Exergy input}]$$

$$= 1 - [(\text{Exergy waste emission} + \text{Exergy destruction})/\text{Exergy input}] \tag{1.13}$$

The exergy efficiency ψ frequently gives a finer understanding of performance than the energy efficiency η. In evaluating η, the same weight is assigned to energy whether it is shaft work or a stream of low-temperature fluid. Also, the energy efficiency centers attention on reducing energy emissions to improve efficiency. The parameter ψ weights energy flows by accounting for each in terms of exergy. It stresses that both waste emissions (or external irreversibilities) and internal irreversibilities need to be dealt with to improve performance. In many cases it is the irreversibilities that are more significant and more difficult to address.

Efficiency expressions each define a class of efficiencies because judgment has to be made about what is the product, what is counted as a loss and what is the input. Different decisions about these lead to different efficiency expressions within the class.

Other SLT-based efficiency expressions also appear in the literature. One of these is evaluated as the ratio of the sum of the exergy exiting to the sum of the exergy entering. Another class of second law efficiencies is composed of task efficiencies.

1.4.7. Solar exergy and the earth

Most energy in the thin top layer of the earth's surface, where life is found, derives from the sun.

Sunlight, rich in exergy, is incident on the earth. Much is reflected by the atmosphere, while some is absorbed by atmospheric constituents or reaches the surface of the earth where it is absorbed. Most of the absorbed solar radiation is converted to thermal energy which is emitted at the temperature of the earth's surface and atmosphere, and leaves the earth as thermal radiation (heat) with no exergy relative to the earth. Thus, while almost all the energy input to the earth with solar energy is re-emitted to space as thermal energy, the exergy associated with solar radiation is delivered to the earth.

The net exergy absorbed by the earth is gradually destroyed but during this destruction it manages to drive the earth's water and wind systems and to support life. Green plants absorb exergy from the sunlight and convert it via photosynthesis into chemical exergy. The chemical exergy then passes through different food chains in ecosystems, from micro-organisms to people. There exists no material waste.

1.5. Illustrative examples

We provide four illustrative examples to highlight the concepts discussed in this chapter and their importance and to demonstrate their application in engineering settings. The examples cover entropy generation during heat transfer processes, entropy generation in a wall due to heat transfer and sensible energy storage. Examples 1–3 are adapted from examples of Cengel and Boles (2006).

1.5.1. Illustrative example 1

A heat source at 800 K loses 2000 kJ of heat to a sink at (a) 500 K and (b) 750 K. Determine which heat transfer process is more irreversible.

Solution: A sketch of the reservoirs is shown in Fig. 1.3. Both cases involve heat transfer through a finite temperature difference and are therefore irreversible. The magnitude of the irreversibility associated with each process can be determined by calculating the total entropy change for each case. The total entropy change for a heat transfer process involving two reservoirs (a source and a sink) is the sum of the entropy changes of each reservoir since the two reservoirs form an adiabatic system.

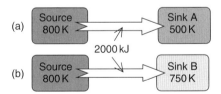

Fig. 1.3. Schematic for the example on entropy generation during heat transfer.

Or do they? The problem statement gives the impression that the two reservoirs are in direct contact during the heat transfer process. But this cannot be the case since the temperature at a point can have only one value, and thus it cannot be 800 K on one side of the point of contact and 500 K on the other side. In other words, the temperature function cannot have a discontinuity. Therefore, it is reasonable to assume that the two reservoirs are separated by a partition through which the temperature drops from 800 K on one side to 500 K (or 750 K) on the other. Therefore, the entropy change of the partition should also be considered when evaluating the total entropy change for this process. However, considering that entropy is a property and the values of properties depend on the state of a system, we can argue that the entropy change of the partition is zero since the partition appears to have undergone a *steady* process and thus experienced no change in its properties at any point. We base this argument on the fact that the temperature on both sides of the partition and thus throughout remain constant during this process. Therefore, we are justified to assume that $\Delta S_{\text{partition}} = 0$ since the entropy (as well as the energy) content of the partition remains constant during the process.

Since each reservoir undergoes an internally reversible, isothermal process, the entropy change for each reservoir can be determined from $\Delta S = Q/T$ where T is the constant absolute temperature of the system and Q is the heat transfer for the internally reversible process.

(a) For the heat transfer process to a sink at 500 K:

$$\Delta S_{source} = Q_{source}/T_{source} = -2000\,kJ/800\,K = -2.5\,kJ/K$$

$$\Delta S_{sink} = Q_{sink}/T_{sink} = 2000\,kJ/500\,K = 4.0\,kJ/K$$

and

$$S_{gen} = \Delta S_{total} = \Delta S_{source} + \Delta S_{sink} = (-2.5 + 4.0)\,kJ/K = 1.5\,kJ/K$$

Therefore, 1.5 kJ/K of entropy is generated during this process. Noting that both reservoirs undergo internally reversible processes, the entire entropy generation occurs in the partition.

(b) Repeating the calculations in part (a) for a sink temperature of 750 K, we obtain

$$\Delta S_{source} = -2.5\,kJ/K$$

$$\Delta S_{sink} = 2.7\,kJ/K$$

and

$$S_{gen} = \Delta S_{total} = (-2.5 + 2.7)\,kJ/K = 0.2\,kJ/K$$

The total entropy change for the process in part (b) is smaller, and therefore it is less irreversible. This is expected since the process in (b) involves a smaller temperature difference and thus a smaller irreversibility.

Discussion: The irreversibilities associated with both processes can be eliminated by operating a Carnot heat engine between the source and the sink. For this case it can be shown that $\Delta S_{total} = 0$.

1.5.2. Illustrative example 2

Consider steady heat transfer through a 5 m × 6 m brick wall of a house of thickness 30 cm. On a day when the temperature of the outdoors is 0°C, the house is maintained at 27°C. The temperatures of the inner and outer surfaces of the brick wall are measured to be 20°C and 5°C, respectively, and the rate of heat transfer through the wall is 1035 W. Determine the rate of entropy generation in the wall, and the rate of total entropy generation associated with this heat transfer process.

Solution: We first take the *wall* as the system (Fig. 1.4). This is a *closed system* since no mass crosses the system boundary during the process. We note that the entropy change of the wall is zero during this process since the state and thus the entropy of the wall do not change anywhere in the wall. Heat and entropy enter from one side of the wall and leave from the other side.

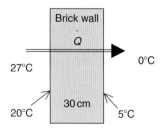

Fig. 1.4. Schematic for the example on entropy generation in a wall due to heat transfer.

Assumptions: (i) The process is steady, and thus the rate of heat transfer through the wall is constant. (ii) Heat transfer through the wall is one-dimensional.

Analysis: The rate form of the entropy balance for the wall simplifies to

$$\left(\dot{S}_{in} - \dot{S}_{out}\right) + \dot{S}_{gen} = \Delta\dot{S}_{system} = 0$$

$$(\dot{Q}/T)_{in} - (\dot{Q}/T)_{out} + \dot{S}_{gen} = 0 \Rightarrow (1035\,\text{W}/293\,\text{K}) - (1035\,\text{W}/278\,\text{K}) + \dot{S}_{gen} = 0$$

$$\Rightarrow \dot{S}_{gen} = 0.191\,\text{W}/\text{K}$$

Therefore, the rate of entropy generation in the wall is 0.191 W/K.

Note that entropy transfer by heat at any location is Q/T at that location, and the direction of entropy transfer is the same as the direction of heat transfer.

To determine the rate of total entropy generation during this heat transfer process, we extend the system to include the regions on both sides of the wall that experience a temperature change. Then one side of the system boundary becomes room temperature while the other side becomes the temperature of the outdoors. The entropy balance for this *extended system* (system and its immediate surroundings) will be the same as that given above, except the two boundary temperatures will be 300 and 273 K instead of 293 and 278 K, respectively. Then the rate of total entropy generation becomes

$$(1035\,\text{W}/300\,\text{K}) - (1035\,\text{W}/273\,\text{K}) + \dot{S}_{gen,total} = 0 \Rightarrow \dot{S}_{gen,total} = 0.341\,\text{W}/\text{K}$$

Discussion: Note that the entropy change of this extended system is also zero since the state of air does not change at any point during the process. The difference between the two entropy generation rates is 0.150 W/K, and represents the entropy generation rate in the air layers on both sides of the wall. The entropy generation in this case is entirely due to irreversible heat transfer across a finite temperature difference.

1.5.3. Illustrative example 3

Consider a frictionless piston–cylinder device containing a saturated liquid–vapor mixture of water at 100°C. During a constant-pressure process, 600 kJ of heat is transferred to the surrounding air at 25°C. As a result, part of the water vapor contained in the cylinder condenses. Determine (a) the entropy change of the water and (b) the total entropy generation during this heat transfer process.

Solution: We first take the *water in the cylinder* as the system (Fig. 1.5). This is a *closed system* since no mass crosses the system boundary during the process. Note that the pressure and thus the temperature of water in the cylinder remain constant during the process. Also, the entropy of the system decreases during the process because of heat loss.

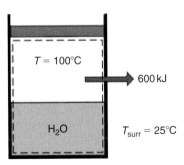

Fig. 1.5. Schematic for the example on entropy generation associated with heat transfer from a piston–cylinder device.

Assumptions: (i) There are no irreversibilities involved within the system boundaries, and thus the process is internally reversible. (ii) The water temperature remains constant at 100°C everywhere, including at the boundaries.

Analysis: (a) Noting that water undergoes an internally reversible isothermal process, its entropy change can be determined from

$$\Delta S_{\text{system}} = Q/T_{\text{system}} = -600 \, \text{kJ}/(100 + 273) \, \text{K} = -1.61 \, \text{kJ/K}$$

(b) To determine the total entropy generation during this process, we consider the *extended system*, which includes the water, the piston–cylinder device, and the region immediately outside the system that experiences a temperature change so that the entire boundary of the extended system is at the surrounding temperature of 25°C. The entropy balance for this *extended system* (system and its immediate surroundings) yields

$$S_{\text{in}} - S_{\text{out}} + S_{\text{gen}} = \Delta S_{\text{system}} \Rightarrow -Q_{\text{out}}/T_{\text{b}} + S_{\text{gen}} = \Delta S_{\text{system}}$$

or

$$S_{\text{gen}} = Q_{\text{out}}/T_{\text{b}} + \Delta S_{\text{system}} = 600 \, \text{kJ}/(25 + 273)\text{K} + (-1.61 \, \text{kJ/K}) = 0.40 \, \text{kJ/K}$$

The entropy generation in this case is entirely due to irreversible heat transfer through a finite temperature difference.

Note that the entropy change of this extended system is equivalent to the entropy change of the water since the piston–cylinder device and the immediate surroundings do not experience any change of state at any point, and thus any change in any property, including entropy.

Discussion: For illustration, consider the reverse process (i.e., the transfer of 600 kJ of heat from the surrounding air at 25°C to saturated water at 100°C) and see if the increase of entropy principle can detect the impossibility of this process. This time, heat transfer is to the water (heat is gained instead of lost), and thus the entropy change of water is 1.61 kJ/K. Also, the entropy transfer at the boundary of the extended system has the same magnitude but opposite direction. This process results in an entropy generation of −0.4 kJ/K. A negative entropy generation indicates that the reverse process is *impossible*.

To complete the discussion, consider the case where the surrounding air temperature is a differential amount below 100°C (say 99.999...9°C) instead of being 25°C. This time, heat transfer from the saturated water to the surrounding air occurs through a differential temperature difference rendering this process *reversible*. It can be shown that $S_{\text{gen}} = 0$ for this process.

Remember that reversible processes are idealized, and they can be approached but never reached in reality.

Further discussion on entropy generation associated with heat transfer:
In the example above it is determined that 0.4 kJ/K of entropy is generated during the heat transfer process, but it is not clear exactly where the entropy generation takes place, and how. To pinpoint the location of entropy generation, we need to be more precise about the description of the system, its surroundings, and the system boundary.

In the example, we assumed both the system and the surrounding air to be isothermal at 100°C and 25°C, respectively. This assumption is reasonable if both fluids are well mixed. The inner surface of the wall must also be at 100°C while the outer surface is at 25°C since two bodies in physical contact must have the same temperature at the point of contact. Considering that entropy transfer with heat transfer Q through a surface at constant temperature T is Q/T, the entropy transfer from the water to the wall is $Q/T_{\text{sys}} = 1.61 \, \text{kJ/K}$. Likewise, the entropy transfer from the outer surface of the wall into the surrounding air is $Q/T_{\text{surr}} = 2.01 \, \text{kJ/K}$. Clearly, entropy in the amount of $(2.01 - 1.61) = 0.4 \, \text{kJ/K}$ is generated in the wall, as illustrated in Fig. 1.6b.

Identifying the location of entropy generation enables us to determine whether a process is internally reversible. A process is internally reversible if no entropy is generated within the system boundaries. Therefore, the heat transfer process discussed in the example above is internally reversible if the inner surface of the wall is taken as the system boundary, and thus the system excludes the container wall. If the system boundary is taken to be the outer surface of the container wall, then the process is no longer internally reversible since the wall, which is the site of entropy generation, is now part of the system.

For thin walls, it is tempting to ignore the mass of the wall and to regard the wall as the boundary between the system and the surroundings. This seemingly harmless choice hides the site of entropy generation and is a source of confusion. The temperature in this case drops suddenly from T_{sys} to T_{surr} at the boundary surface (see Fig. 1.6a), and confusion arises as to which temperature to use in the relation Q/T for entropy transfer at the boundary.

Note that if the system and the surrounding air are not isothermal as a result of insufficient mixing, then part of the entropy generation occurs in both the system and the surrounding air in the vicinity of the wall, as shown in Fig. 1.6c.

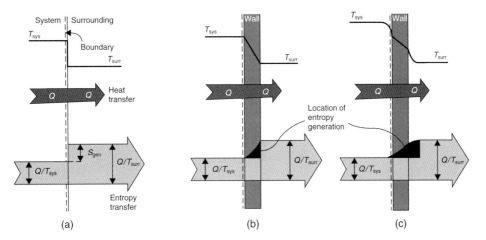

Fig. 1.6. Schematic representation of entropy generation during a heat transfer process through a finite temperature difference.

1.5.4. Illustrative example 4

Consider two sensible thermal energy storage systems, X and Y, in an environment at a temperature of 25°C. Each storage receives a quantity of heat from a stream of 500 kg of water, which is cooled from 80°C to 30°C. The recovery operation and the storage duration for the two storages differ. Determine (a) the energy recovery, loss and efficiency for each storage and (b) the corresponding exergy parameters. Compare the storages.

Solution:
Assumption: The thermal storage system is assumed to be comprised of three processes: charging, storing and discharging (Fig. 1.7). This simple model is used in order to distinguish energy and exergy concepts clearly, and highlight the importance of exergy as a tool for practical thermodynamic systems.

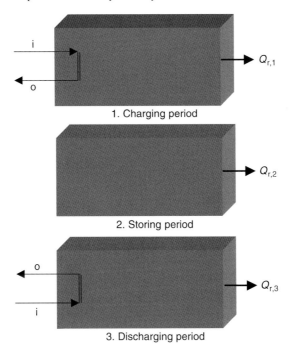

Fig. 1.7. Schematic of three processes in a storage cycle for a sensible thermal energy storage system.

(a) Both sensible thermal storage systems X and Y receive a quantity of heat from a stream of 500 kg of water, which is cooled from 80°C to 30°C. Therefore, the heat input to each storage during the charging period is

$$Q_i = m_i c_p \Delta T = 500 \times 4.186 \times (80 - 30) = 104{,}650 \, kJ$$

For system X:
After one day, 94,185 kJ of heat is recovered during the discharging period from storage system X by a stream of 4500 kg of water being heated from 30°C to 35°C. That is,

$$Q_o = m_o c_p \Delta T = 4500 \times 4.186 \times (35 - 30) = 94{,}185 \, kJ$$

Therefore, the energy efficiency for sensible thermal storage X becomes

$$\eta_x = Q_o/Q_i = 94{,}185/104{,}650 = 0.90$$

The heat rejection to the surroundings during storage is

$$Q_r = Q_i - Q_o = 104{,}650 - 94{,}185 = 10{,}465 \, kJ$$

For system Y:
The heat recovered during discharging, the energy efficiency and the heat rejection to the surroundings can be evaluated for storage system Y in a similar way. Storage system Y stores the heat for 90 days, after which a quantity of heat is recovered during the discharging period by heating a stream of 500 kg of water from 30°C to 75°C. The heat recovered during discharging for storage Y is

$$Q_o = m_o c_p \Delta T = 500 \times 4.186 \times (75 - 30) = 94{,}185 \, kJ$$

which is the same as that for storage X. Thus, the energy efficiency for storage Y is the same as that for storage X, i.e. $\eta_Y = 0.90$, and the heat rejection to the surroundings for storage Y is the same as that for storage X, i.e. 10,465 kJ.

It is useful to note that the ability to store sensible heat in a tank or container strongly depends on the value for the storage material of ρc_p.

Both storage systems have the same energy efficiency, but storage system Y, which stores the heat for 90 days rather than 1 day, and which returns the heat at the much more useful temperature of 75°C rather than 35°C, provides considerably better performance. It is clear that a more perceptive comparative measure than that provided by the energy efficiency of the storage is needed if the true usefulness of a sensible thermal storage is to be assessed and a rational basis for the optimization of its economic value established. An efficiency defined simply as the percentage of the total energy stored in a system which can be recovered ignores the quality of the recovered energy, and so cannot provide a measure of ideal performance as mentioned earlier. The exergy efficiency provides such a measure of the effectiveness of a sensible thermal storage system and thus is advantageous.

(b) An exergy analysis is conducted for sensible thermal energy storage systems X and Y. The exergy change during the charging period can be obtained as

$$\Delta Ex_i = m_i c_p[(T_1 - T_2) - T_0(\ln(T_1/T_2)] = 500 \times 4.186 \times [(353 - 303) - 298 \times (\ln(353/303)] = 9387 \, kJ$$

For system X:
The exergy change during the discharging period for storage system X can be evaluated as follows:

$$\Delta Ex_o = m_o c_p[(T_1 - T_2) - T_0(\ln(T_1/T_2)] = 4500 \times 4.186 \times [(308 - 303) - 298 \times (\ln(308/303)] = 2310 \, kJ$$

Therefore, the exergy efficiency for storage X is

$$\psi_x = \Delta Ex_o/\Delta Ex_i = 2310/9387 = 0.25$$

For system Y:

Since heat is recovered from the storage after 90 days by a stream of 500 kg of water entering at 30°C and leaving at 75°C, the exergy change and efficiency of storage system Y can be obtained as follows:

$$\Delta Ex_o = m_o c_p[(T_1 - T_2) - T_0(\ln(T_1/T_2)] = 500 \times 4.186 \times [(348 - 303) - 298 \times (\ln(348/303)] = 7820 \text{ kJ}$$

$$\psi_y = \Delta Ex_o / \Delta Ex_i = 7820/9387 = 0.83$$

The energy efficiencies for both storage systems are determined to be the same at 90% despite having two different storage periods and different heat recovery temperatures. This situation clearly indicates that the FLT is not sufficient to distinguish these two sensible thermal energy storage systems. This deficiency highlights the advantage of exergy methods. The different behaviors of the two storage systems are more apparent using exergy analysis, as different exergy efficiencies are obtained (25% for system X and 83% for system Y). Thermal storage system Y is more effective than system X, because it returns heat at a higher temperature and thus with greater exergy content, despite having a longer storage period.

This example illustrates in a practical manner some of the more abstract concepts discussed in this chapter and highlights the importance of understanding and considering exergy, rather than or together with energy. Exergy is demonstrated to be a more effective tool for analyzing and comparing the performance of sensible thermal energy storage systems. Specifically, exergy analysis

- Provides a clearer, more meaningful and useful accounting of efficiencies and losses in a thermal energy storage system, by using the conservation of mass and energy principles together with the second law for design and analysis.
- Reflects more correctly the thermodynamic and economic values of thermal energy storage systems.
- Reveals whether or not and by how much it is possible to design more efficient sensible thermal energy storage systems by reducing inefficiencies.

1.6. Closing remarks

Despite the existence of the Zeroth and third laws of thermodynamics, thermodynamics is very much a science of energy and exergy (including entropy). The FLT refers to energy conservation and treats all energy forms equally, thus not identifying losses of energy quality or work potential and possible improvements in the use of resources. For example, energy alone cannot identify the losses in an adiabatic throttling process. However, the SLT involves exergy and entropy concepts, and considers irreversibilities and the consequent non-conservation of exergy and entropy.

During the past several decades, exergy-related studies have received increasing attention from various disciplines ranging from mechanical and chemical engineering, to environmental engineering and ecology and so on. As a consequence, the international exergy community has expanded significantly in recent years.

Problems

1.1 Define the macroscopic and microscopic forms of energy and explain their use in thermodynamic analyses.

1.2 Define the following forms of energy and explain their differences: internal energy, thermal energy, heat, heat transfer, sensible energy, latent energy, chemical energy, nuclear energy, flow energy, flow work and enthalpy.

1.3 Give some actual engineering examples where a high-quality energy form is used to satisfy a low-quality energy application. Explain how the same applications can be accomplished using lower-quality energy sources.

1.4 The most common application of geothermal energy is power generation. Some other uses are process heating, district heating, district cooling, greenhouse heating and heating for fish farming. From a thermodynamic point of view and considering the quality of energy, explain which of these uses you recommend most.

1.5 Some people claim that solar energy is in fact the source of all other renewable energy sources such as geothermal, wind, hydro and biomass. Evaluate this claim, providing appropriate justification.

1.6 Solar energy is renewable, free and available in many parts of the world. Despite these advantages, there are few commercial solar steam power plants currently operating. Explain why this is the case and if you think there will be more of these power plants in the future.

1.7 Can the entropy of a system decrease during a process? If so, does this violate the increase of entropy principle?

1.8 Consider a system undergoing a process that takes it from one state to another. Does the entropy change of the system depend on the process path followed during the process? Explain.

1.9 There are two classical statements of the second law of thermodynamics (the Kelvin–Plank and Clausius statements). Can there be other valid statements? If so, why are these classical definitions repeated in almost every thermodynamics textbook?

1.10 Can you determine if a process is possible by performing an entropy analysis? Explain.

1.11 The first law of thermodynamics is also known as the conservation of energy principle and leads to energy balances. Is there a corresponding entropy balance or exergy balance? Express general energy, entropy and exergy balances using words.

1.12 Provide a statement for the second law of thermodynamics using exergy.

1.13 What is the relationship between entropy generation and irreversibility?

1.14 Which is a more effective method for improving the efficiency of a system: reducing entropy generation or reducing exergy destruction due to irreversibilities?

1.15 Does an exergy analysis replace an energy analysis? Describe any advantages of exergy analysis over energy analysis.

1.16 Can you perform an exergy analysis without an energy analysis? Explain.

1.17 Does the reference environment have any effect on the results of an exergy analysis?

1.18 Do you recommend using standard atmospheric conditions (1 atm, 25°C) or actual atmospheric conditions as reference environment in an exergy analysis? Explain.

1.19 Some people claim that, for a fossil-fuel power plant, energy analysis is superior to exergy analysis because the heat input is directly related to the amount of fuel burned and consequently the cost of fuel to generate a certain amount of power. Evaluate this claim.

1.20 The energy efficiency of a cogeneration plant involving power and process heat outputs may be expressed as the sum of the power and process heat divided by the heat input. This is also sometimes known as the utilization efficiency. As a senior or graduate student of engineering does it seem counter-intuitive to add power and heat, or like you are adding apples and oranges. Can you get around this awkward situation by using exergy efficiency? How?

1.21 Explain why friction and heat transfer across a finite temperature difference cause a process to be irreversible.

1.22 Noting that heat transfer does not occur without a temperature difference and that heat transfer across a finite temperature difference is irreversible, is there such a thing as reversible heat transfer? Explain.

Chapter 2

EXERGY AND ENERGY ANALYSES

2.1. Introduction

Exergy analysis is a thermodynamic analysis technique based on the second law of thermodynamics which provides an alternative and illuminating means of assessing and comparing processes and systems rationally and meaningfully. In particular, exergy analysis yields efficiencies which provide a true measure of how nearly actual performance approaches the ideal, and identifies more clearly than energy analysis the causes and locations of thermodynamic losses. Consequently, exergy analysis can assist in improving and optimizing designs.

Increasing application and recognition of the usefulness of exergy methods by those in industry, government and academia has been observed in recent years. Exergy has also become increasingly used internationally. The present authors, for instance, have examined exergy analysis methodologies and applied them to industrial systems (e.g., Rosen and Horazak, 1995; Rosen and Scott, 1998; Rosen and Dincer, 2003a, b, 2004b; Rosen and Etele, 2004; Rosen et al., 2005), thermal energy storage (e.g., Dincer and Rosen, 2002; Rosen et al., 2004), countries (e.g., Rosen, 1992; Rosen and Dincer, 1997b) and environmental impact assessments (e.g., Crane et al., 1992; Rosen and Dincer, 1997a, 1999; Gaggioli, 1998; Gunnewiek and Rosen, 1998).

In this chapter, theoretical and practical aspects of thermodynamics most relevant to energy and exergy analyses are described. This section reviews fundamental principles and such related issues as reference-environment selection, efficiency definition and material properties acquisition are discussed. General implications of exergy analysis results are discussed, and a step-by-step procedure for energy and exergy analyses is given.

2.2. Why energy and exergy analyses?

Thermodynamics permits the behavior, performance and efficiency to be described for systems for the conversion of energy from one form to another. Conventional thermodynamic analysis is based primarily on the first law of thermodynamics, which states the principle of conservation of energy. An energy analysis of an energy-conversion system is essentially an accounting of the energies entering and exiting. The exiting energy can be broken down into products and wastes. Efficiencies are often evaluated as ratios of energy quantities, and are often used to assess and compare various systems. Power plants, heaters, refrigerators and thermal storages, for example, are often compared based on energy efficiencies or energy-based measures of merit.

However, energy efficiencies are often misleading in that they do not always provide a measure of how nearly the performance of a system approaches ideality. Further, the thermodynamic losses which occur within a system (i.e., those factors which cause performance to deviate from ideality) often are not accurately identified and assessed with energy analysis. The results of energy analysis can indicate the main inefficiencies to be within the wrong sections of the system, and a state of technological efficiency different than actually exists.

Exergy analysis permits many of the shortcomings of energy analysis to be overcome. Exergy analysis is based on the second law of thermodynamics, and is useful in identifying the causes, locations and magnitudes of process inefficiencies. The exergy associated with an energy quantity is a quantitative assessment of its usefulness or quality. Exergy analysis acknowledges that, although energy cannot be created or destroyed, it can be degraded in quality, eventually reaching a state in which it is in complete equilibrium with the surroundings and hence of no further use for performing tasks.

For energy storage systems, for example, exergy analysis allows one to determine the maximum potential associated with the incoming energy. This maximum is retained and recovered only if the energy undergoes processes in a reversible manner. Losses in the potential for exergy recovery occur in the real world because actual processes are always irreversible.

The exergy flow rate of a flowing commodity is the maximum rate that work may be obtained from it as it passes reversibly to the environmental state, exchanging heat and materials only with the surroundings. In essence, exergy analysis states the theoretical limitations imposed on a system, clearly pointing out that no real system can conserve exergy and that only a portion of the input exergy can be recovered. Also, exergy analysis quantitatively specifies practical limitations by providing losses in a form in which they are a direct measure of lost exergy.

2.3. Nomenclature

Although a relatively standard terminology and nomenclature has evolved for conventional classical thermodynamics, there is at present no generally agreed on terminology and nomenclature for exergy analysis. A diversity of symbols and names exist for basic and derived quantities (Kotas et al., 1987; Lucca, 1990). For example, exergy is often called available energy, availability, work capability, essergy, etc.; and exergy consumption is often called irreversibility, lost work, dissipated work, dissipation, etc. The exergy-analysis nomenclature used here follows that proposed by Kotas et al. (1987) as a standard exergy-analysis nomenclature. For the reader unfamiliar with exergy, a glossary of selected exergy terminology is included (see Appendix A).

2.4. Balances for mass, energy and entropy

2.4.1. Conceptual balances

A general balance for a quantity in a system may be written as

$$\text{Input} + \text{Generation} - \text{Output} - \text{Consumption} = \text{Accumulation} \tag{2.1}$$

Input and output refer respectively to quantities entering and exiting through system boundaries. Generation and consumption refer respectively to quantities produced and consumed within the system. Accumulation refers to build up (either positive or negative) of the quantity within the system.

Versions of the general balance equation above may be written for mass, energy, entropy and exergy. Mass and energy, being subject to conservation laws (neglecting nuclear reactions), can be neither generated nor consumed. Consequently, the general balance (Eq. (2.1)) written for each of these quantities becomes

$$\text{Mass input} - \text{Mass output} = \text{Mass accumulation} \tag{2.2}$$

$$\text{Energy input} - \text{Energy output} = \text{Energy accumulation} \tag{2.3}$$

Before giving the balance equation for exergy, it is useful to examine that for entropy:

$$\text{Entropy input} + \text{Entropy generation} - \text{Entropy output} = \text{Entropy accumulation} \tag{2.4}$$

Entropy is created during a process due to irreversibilities, but cannot be consumed.

These balances describe what is happening in a system between two instants of time. For a complete cyclic process where the initial and final states of the system are identical, the accumulation terms in all the balances are zero.

2.4.2. Detailed balances

Two types of systems are normally considered: open (flow) and closed (non-flow). In general, open systems have mass, heat and work interactions, and closed systems have heat and work interactions. Mass flow into, heat transfer into and work transfer out of the system are defined to be positive. Mathematical formulations of the principles of mass and energy conservation and entropy non-conservation can be written for any system, following the general physical interpretations in Eqs. (2.2) through (2.4).

Consider a non-steady flow process in a time interval t_1 to t_2. Balances of mass, energy and entropy, respectively, can be written for a control volume as

$$\sum_i m_i - \sum_e m_e = m_2 - m_1 \tag{2.5}$$

$$\sum_i (e + Pv)_i m_i - \sum_e (e + Pv)_e m_e + \sum_r (Q_r)_{1,2} - (W')_{1,2} = E_2 - E_1 \tag{2.6}$$

$$\sum_i s_i m_i - \sum_e s_e m_e + \sum_r (Q_r/T_r)_{1,2} + \prod_{1,2} = S_2 - S_1 \tag{2.7}$$

Here, m_i and m_e denote respectively the amounts of mass input across port i and exiting across port e; $(Q_r)_{1,2}$ denotes the amount of heat transferred into the control volume across region r on the control surface; $(W')_{1,2}$ denotes the amount of work transferred out of the control volume; $\prod_{1,2}$ denotes the amount of entropy created in the control volume (and is also referred to as S_{gen} in this text); m_1, E_1 and S_1 denote respectively the amounts of mass, energy and entropy in the control volume at time t_1 and m_2, E_2 and S_2 denote respectively the same quantities at time t_2; and e, s, P, T and v denote specific energy, specific entropy, absolute pressure, absolute temperature and specific volume, respectively. The total work W' done by a system excludes flow work, and can be written as

$$W' = W + W_x \tag{2.8}$$

where W is the work done by a system due to change in its volume and W_x is the shaft work done by the system. The term 'shaft work' includes all forms of work that can be used to raise a weight (i.e., mechanical work, electrical work, etc.) but excludes work done by a system due to change in its volume. The specific energy e is given by

$$e = u + ke + pe \tag{2.9}$$

where u, ke and pe denote respectively specific internal, kinetic and potential (due to conservative force fields) energies. For irreversible processes $\prod_{1,2} > 0$, and for reversible processes $\prod_{1,2} = 0$.

The left sides of Eqs. (2.5) through (2.7) represent the net amounts of mass, energy and entropy transferred into (and in the case of entropy created within) the control volume, while the right sides represent the amounts of these quantities accumulated within the control volume.

For the mass flow m_j across port j,

$$m_j = \int_{t_1}^{t_2} \left[\int_j (\rho V_n dA)_j \right] dt \tag{2.10}$$

Here, ρ is the density of matter crossing an area element dA on the control surface in time interval t_1 to t_2 and V_n is the velocity component of the matter flow normal to dA. The integration is performed over port j on the control surface. One-dimensional flow (i.e., flow in which the velocity and other intensive properties do not vary with position across the port) is often assumed. Then the previous equation becomes

$$m_j = \int_{t_1}^{t_2} (\rho V_n A)_j dt \tag{2.11}$$

It has been assumed that heat transfers occur at discrete regions on the control surface and the temperature across these regions is constant. If the temperature varies across a region of heat transfer,

$$(Q_r)_{1,2} = \int_{t_1}^{t_2} \left[\int_r (q \, dA)_r \right] dt \tag{2.12}$$

and

$$(Q_r/T_r)_{1,2} = \int_{t_1}^{t_2} \left[\int_r (q/T)_r dA_r \right] dt \tag{2.13}$$

where T_r is the temperature at the point on the control surface where the heat flux is q_r. The integral is performed over the surface area of region A_r.

The quantities of mass, energy and entropy in the control volume (denoted by m, E and S) on the right sides of Eqs. (2.5) through (2.7), respectively, are given more generally by

$$m = \int \rho \, dV \tag{2.14}$$

$$E = \int \rho e \, dV \tag{2.15}$$

$$S = \int \rho s \, s \, dV \tag{2.16}$$

where the integrals are over the control volume.

For a closed system, $m_i = m_e = 0$ and Eqs. (2.5) through (2.7) become

$$0 = m_2 - m_1 \tag{2.17}$$

$$\sum_r (Q_r)_{1,2} - (W')_{1,2} = E_2 - E_1 \tag{2.18}$$

$$\sum_r (Q_r/T_r)_{1,2} + \prod_{1,2} = S_2 - S_1 \tag{2.19}$$

2.5. Exergy of systems and flows

Several quantities related to the conceptual exergy balance are described here, following the presentations by Moran (1989) and Kotas (1995).

2.5.1. Exergy of a closed system

The exergy $Ex_{\text{non-flow}}$ of a closed system of mass m, or the non-flow exergy, can be expressed as

$$Ex_{\text{non-flow}} = Ex_{\text{ph}} + Ex_{\text{o}} + Ex_{\text{kin}} + Ex_{\text{pot}} \tag{2.20}$$

where

$$Ex_{\text{pot}} = \text{PE} \tag{2.21}$$

$$Ex_{\text{kin}} = \text{KE} \tag{2.22}$$

$$Ex_{\text{o}} = \sum_i (\mu_{io} - \mu_{ioo})N_i \tag{2.23}$$

$$Ex_{\text{non-flow,ph}} = (U - U_{\text{o}}) + P_{\text{o}}(V - V_{\text{o}}) - T_{\text{o}}(S - S_{\text{o}}) \tag{2.24}$$

where the system has a temperature T, pressure P, chemical potential μ_i for species i, entropy S, energy E, volume V and number of moles N_i of species i. The system is within a conceptual environment in an equilibrium state with intensive properties T_{o}, P_{o} and μ_{ioo}. The quantity μ_{io} denotes the value of μ at the environmental state (i.e., at T_{o} and P_{o}). The terms on the right side of Eq. (2.20) represent respectively physical, chemical, kinetic and potential components of the non-flow exergy of the system.

The exergy Ex is a property of the system and conceptual environment, combining the intensive and extensive properties of the system with the intensive properties of the environment.

Physical non-flow exergy is the maximum work obtainable from a system as it is brought to the environmental state (i.e., to thermal and mechanical equilibrium with the environment), and chemical non-flow exergy is the maximum work obtainable from a system as it is brought from the environmental state to the dead state (i.e., to complete equilibrium with the environment).

2.5.2. Exergy of flows

Exergy of a matter flow

The exergy of a flowing stream of matter Ex_{flow} is the sum of non-flow exergy and the exergy associated with the flow work of the stream (with reference to P_o), i.e.,

$$Ex_{flow} = Ex_{non\text{-}flow} + (P - P_o)V \tag{2.25}$$

Alternatively, Ex_{flow} can be expressed following Eq. (2.20) in terms of physical, chemical, kinetic and potential components:

$$Ex_{flow} = Ex_{ph} + Ex_o + Ex_{kin} + Ex_{pot} \tag{2.26}$$

where

$$Ex_{pot} = PE \tag{2.27}$$

$$Ex_{kin} = KE \tag{2.28}$$

$$Ex_o = \sum_i (\mu_{io} - \mu_{ioo})N_i \tag{2.29}$$

$$Ex_{flow,ph} = (H - H_o) - T_o(S - S_o) \tag{2.30}$$

Exergy of thermal energy

Consider a control mass, initially at the dead state, being heated or cooled at constant volume in an interaction with some other system. The heat transfer experienced by the control mass is Q. The flow of exergy associated with the heat transfer Q is denoted by Ex_Q, and can be expressed as

$$Ex_Q = \int_i^f (1 - T_o/T)\delta Q \tag{2.31}$$

where δQ is an incremental heat transfer, and the integral is from the initial state (i) to the final state (f). This 'thermal exergy' is the minimum work required by the combined system of the control mass and the environment in bringing the control mass to the final state from the dead state.

Often the dimensionless quantity in parentheses in this expression is called the 'exergetic temperature factor' and denoted τ:

$$\tau = 1 - T_o/T \tag{2.32}$$

The relation between τ and the temperature ratio T/T_o is illustrated in Fig. 2.1.

If the temperature T of the control mass is constant, the thermal exergy transfer associated with a heat transfer is

$$Ex_Q = (1 - T_o/T)Q = \tau Q \tag{2.33}$$

For heat transfer across a region r on a control surface for which the temperature may vary,

$$Ex_Q = \int_r [q_r(1 - T_o/T_r)dA_r] \tag{2.34}$$

where q_r is the heat flow per unit area at a region on the control surface at which the temperature is T_r.

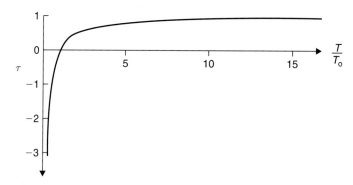

Fig. 2.1. The relation between the exergetic temperature factor τ and the absolute temperature ratio T/T_o. The factor τ is equal to zero when $T = T_o$. For heat transfer at above-environment temperatures (i.e., $T > T_o$), $0 < \tau \leq 1$. For heat transfer at sub-environment temperatures (i.e., $T < T_o$), $\tau < 0$, implying that exergy and energy flow in opposite directions in such cases. Note that the magnitude of the exergy flow exceeds that of the energy flow when $\tau < -1$, which corresponds to $T < T_o/2$.

Exergy of work

Equation (2.8) separates total work W' into two components W_x and W. The exergy associated with shaft work Ex_W is by definition W_x.

    The exergy transfer associated with work done by a system due to volume change is the net usable work due to the volume change, and is denoted by W_{NET}. Thus for a process in time interval t_1 to t_2,

$$(W_{NET})_{1,2} = W_{1,2} - P_o(V_2 - V_1) \tag{2.35}$$

where $W_{1,2}$ is the work done by the system due to volume change $(V_2 - V_1)$. The term $P_o(V_2 - V_1)$ is the displacement work necessary to change the volume against the constant pressure P_o exerted by the environment.

Exergy of electricity

As for shaft work, the exergy associated with electricity is equal to the energy.

2.6. Exergy consumption

For a process occuring in a system, the difference between the total exergy flows into and out of the system, less the exergy accumulation in the system, is the exergy consumption I, expressible as

$$I = T_o S_{gen} \tag{2.36}$$

Equation (2.36) points out that exergy consumption is proportional to entropy creation, and is known as the Gouy–Stodola relation.

2.7. Exergy balance

By combining the conservation law for energy and non-conservation law for entropy, the exergy balance can be obtained:

$$\text{Exergy input} - \text{Exergy output} - \text{Exergy consumption} = \text{Exergy accumulation} \tag{2.37}$$

Exergy is consumed due to irreversibilities. Exergy consumption is proportional to entropy creation. Equations (2.4) and (2.37) demonstrate an important main difference between energy and exergy: energy is conserved while exergy, a measure of energy quality or work potential, can be consumed.

An analogous balance to those given in Eqs. (2.5) through (2.7) can be written for exergy, following the physical interpretation of Eq. (2.37). For a non-steady flow process during time interval t_1 to t_2,

$$\sum_i ex_i m_i - \sum_e ex_e m_e + \sum_r (Ex_{Q_r})_{1,2} - (Ex_W)_{1,2} - (W_{\text{NET}})_{1,2} - I_{1,2} = Ex_2 - Ex_1 \tag{2.38}$$

where $(W_{\text{NET}})_{1,2}$ is given by Eq. (2.35) and

$$(Ex_{Q_r})_{1,2} = \int_{t_1}^{t_2} \left[\int_r (1 - T_o/T_r) q_r dA_r \right] dt \tag{2.39}$$

$$I_{1,2} = T_o S_{\text{gen},1,2} \tag{2.40}$$

$$Ex = \int \rho \xi \, dV \tag{2.41}$$

Here, I and S_{gen} respectively denote exergy consumption and entropy creation, Ex denotes exergy, and the integral for Ex is performed over the control volume. The first two terms on the left side of Eq. (2.38) represent the net input of exergy associated with matter, the third term the net input of exergy associated with heat, the fourth and fifth terms the net input of exergy associated with work, and the sixth term the exergy consumption. The right side of Eq. (2.38) represents the accumulation of exergy.

For a closed system, Eq. (2.38) simplifies to

$$\sum_r (Ex_{Q_r})_{1,2} - (Ex_W)_{1,2} - (W_{\text{NET}})_{1,2} - I_{1,2} = Ex_2 - Ex_1 \tag{2.42}$$

When volume is fixed, $(W_{\text{NET}})_{1,2} = 0$ in Eqs. (2.38) and (2.42). Also, when the initial and final states are identical, as in a complete cycle, the right sides of Eqs. (2.38) and (2.42) are zero.

2.8. Reference environment

Exergy is evaluated with respect to a reference environment, so the intensive properties of the reference environment determine the exergy of a stream or system.

2.8.1. Theoretical characteristics of the reference environment

The reference environment is in stable equilibrium, with all parts at rest relative to one another. No chemical reactions can occur between the environmental components. The reference environment acts as an infinite system, and is a sink and source for heat and materials. It experiences only internally reversible processes in which its intensive state remains unaltered (i.e., its temperature T_o, pressure P_o and the chemical potentials μ_{ioo} for each of the i components present remain constant). The exergy of the reference environment is zero. The exergy of a stream or system is zero when it is in equilibrium with the reference environment.

The natural environment does not have the theoretical characteristics of a reference environment. The natural environment is not in equilibrium, and its intensive properties exhibit spatial and temporal variations. Many chemical reactions in the natural environment are blocked because the transport mechanisms necessary to reach equilibrium are too slow at ambient conditions. Thus, the exergy of the natural environment is not zero; work could be obtained if it were to come to equilibrium. Consequently, models for the reference environment are used which try to achieve a compromise between the theoretical requirements of the reference environment and the actual behavior of the natural environment.

2.8.2. Models for the reference environment

Several classes of reference-environment models are described below:

- *Natural-environment-subsystem models*: An important class of reference-environment models is the natural-environment-subsystem type. These models attempt to simulate realistically subsystems of the natural environment. One such model consisting of saturated moist air and liquid water in phase equilibrium was proposed by Baehr and Schmidt (1963). An extension of the above model which allowed sulfur-containing materials to be analyzed was proposed by Gaggioli and Petit (1977) and Rodriguez (1980). The temperature and pressure of this reference environment (see Table 2.1) are normally taken to be 25°C and 1 atm, respectively, and the chemical composition is taken to consist of air saturated with water vapor, and the following condensed phases at 25°C and 1 atm: water (H_2O), gypsum ($CaSO_4 \cdot 2H_2O$) and limestone ($CaCO_3$). The stable configurations of C, O and N respectively are taken to be those of CO_2, O_2 and N_2 as they exist in air saturated with liquid water at T_o and P_o; of hydrogen is taken to be in the liquid phase of water saturated with air at T_o and P_o; and of S and Ca respectively are taken to be those of $CaSO_4 \cdot 2H_2O$ and $CaCO_3$ at T_o and P_o.

Analyses often use the natural-environment-subsystem model described in Table 2.1, but with a temperature modified to reflect the approximate mean ambient temperature of the location of the system or process for the time period under consideration.

Table 2.1. A reference-environment model.

Temperature	$T_o = 298.15\,K$		
Pressure	$P_o = 1\,atm$		
Composition	(i)	Atmospheric air saturated with H_2O at T_o and P_o, having the following composition:	
		Air constituents	Mole fraction
		N_2	0.7567
		O_2	0.2035
		H_2O	0.0303
		Ar	0.0091
		CO_2	0.0003
		H_2	0.0001
	(ii)	The following condensed phases at T_o and P_o:	
		Water (H_2O)	
		Limestone ($CaCO_3$)	
		Gypsum ($CaSO_4 \cdot 2H_2O$)	

Source: Adapted from Gaggioli and Petit (1977).

- *Reference-substance models*: Here, a 'reference substance' is selected and assigned zero exergy for every chemical element. One such model in which the reference substances were selected as the most valueless substances found in abundance in the natural environment was proposed by Szargut (1967). The criteria for selecting such reference substances is consistent with the notion of simulating the natural environment, but is primarily economic in nature, and is vague and arbitrary with respect to the selection of reference substances. Part of this environment is the composition of moist air, including N_2, O_2, CO_2, H_2O and the noble gases; gypsum (for sulfur) and limestone (for calcium).

  Another model in this class, in which reference substances are selected arbitrarily, was proposed by Sussman (1980, 1981). This model is not similar to the natural environment. Consequently absolute exergies evaluated with this model do not relate to the natural environment, and cannot be used rationally to evaluate efficiencies. Since exergy-consumption values are independent of the choice of reference substances, they can be rationally used in analyses.
- *Equilibrium models*: A model in which all the materials present in the atmosphere, oceans and a layer of the crust of the earth are pooled together and an equilibrium composition is calculated for a given temperature was proposed by Ahrendts (1980). The selection of the thickness of crust considered is subjective and is intended to include all materials accessible to technical processes. Ahrendts considered thicknesses varying from 1 to 1000 m, and a temperature of 25°C. For all thicknesses, Ahrendts found that the model differed significantly from the natural environment. Exergy values obtained using these environments are significantly dependent on the thickness of

crust considered, and represent the absolute maximum amount of work obtainable from a material. Since there is no technical process available which can obtain this work from materials, Ahrendts' equilibrium model does not give meaningful exergy values when applied to the analysis of real processes.

- *Constrained-equilibrium models*: Ahrendts (1980) also proposed a modified version of his equilibrium environment in which the calculation of an equilibrium composition excludes the possibility of the formation of nitric acid (HNO_3) and its compounds. That is, all chemical reactions in which these substances are formed are in constrained equilibrium, and all other reactions are in unconstrained equilibrium. When a thickness of crust of 1 m and temperature of 25°C were used, the model was similar to the natural environment.
- *Process-dependent models*: A model which contains only components that participate in the process being examined in a stable equilibrium composition at the temperature and total pressure of the natural environment was proposed by Bosnjakovic (1963). This model is dependent on the process examined, and is not general. Exergies evaluated for a specific process-dependent model are relevant only to the process; they cannot rationally be compared with exergies evaluated for other process-dependent models.

Many researchers have examined the characteristics of and models for reference environments (e.g., Ahrendts, 1980; Wepfer and Gaggioli, 1980; Sussman, 1981), and the sensitivities of exergy values to different reference-environment models (Rosen and Dincer, 2004a).

2.9. Efficiencies and other measures of merit

Efficiency has always been an important consideration in decision making regarding resource utilization. Efficiency is defined as 'the ability to produce a desired effect without waste of, or with minimum use of, energy, time, resources, etc.,' and is used by people to mean the effectiveness with which something is used to produce something else, or the degree to which the ideal is approached in performing a task.

For general engineering systems, non-dimensional ratios of quantities are typically used to determine efficiencies. Ratios of energy are conventionally used to determine efficiencies of engineering systems whose primary purpose is the transformation of energy. These efficiencies are based on the first law of thermodynamics. A process has maximum efficiency according to the first law if energy input equals recoverable energy output (i.e., if no 'energy losses' occur). However, efficiencies determined using energy are misleading because in general they are not measures of 'an approach to an ideal.'

To determine more meaningful efficiencies, a quantity is required for which ratios can be established which do provide a measure of an approach to an ideal. Thus, the second law must be involved, as this law states that maximum efficiency is attained (i.e., ideality is achieved) for a reversible process. However, the second law must be quantified before efficiencies can be defined.

The 'increase of entropy principle,' which states that entropy is created due to irreversibilities, quantifies the second law. From the viewpoint of entropy, maximum efficiency is attained for a process in which entropy is conserved. Entropy is created for non-ideal processes. The magnitude of entropy creation is a measure of the non-ideality or irreversibility of a process. In general, however, ratios of entropy do not provide a measure of an approach to an ideal.

A quantity which has been discussed in the context of meaningful measures of efficiency is negentropy (Hafele, 1981). Negentropy is defined such that the negentropy consumption due to irreversibilities is equal to the entropy creation due to irreversibilities. As a consequence of the 'increase of entropy principle,' maximum efficiency is attained from the viewpoint of negentropy for a process in which negentropy is conserved. Negentropy is consumed for non-ideal processes. Negentropy is a measure of order. Consumptions of negentropy are therefore equivalent to degradations of order. Since the abstract property of order is what is valued and useful, it is logical to attempt to use negentropy in developing efficiencies. However, general efficiencies cannot be determined based on negentropy because its absolute magnitude is not defined.

Negentropy can be further quantified through the ability to perform work. Then, maximum efficiency is attainable only if, at the completion of a process, the sum of all energy involved has an ability to do work equal to the sum before the process occurred. Exergy is a measure of the ability to perform work, and from the viewpoint of exergy, maximum efficiency is attained for a process in which exergy is conserved. Efficiencies determined using ratios of exergy do provide a measure of an approach to an ideal. Exergy efficiencies are often more intuitively rational than energy efficiencies because efficiencies between 0% and 100% are always obtained. Measures which can be greater than 100% when energy is considered, such as coefficient of performance, normally are between 0% and 100% when exergy is considered. In fact, some researchers (e.g., Gaggioli, 1983) call exergy efficiencies 'real' or 'true' efficiencies, while calling energy efficiencies 'approximations to real' efficiencies.

Energy (η) and exergy (ψ) efficiencies are often written for steady-state processes occurring in systems as

$$\eta = \frac{\text{Energy in product outputs}}{\text{Energy in inputs}} = 1 - \frac{\text{Energy loss}}{\text{Energy in inputs}} \tag{2.43}$$

$$\psi = \frac{\text{Exergy in product outputs}}{\text{Exergy in inputs}} = \frac{\text{Exergy loss plus consumption}}{\text{Exergy in inputs}} \tag{2.44}$$

Two other common exergy-based efficiencies for steady-state devices are as follows:

$$\text{Rational efficiency} = \frac{\text{Total exergy output}}{\text{Total exergy input}} = 1 - \frac{\text{Exergy consumption}}{\text{Total exergy input}} \tag{2.45}$$

$$\text{Task efficiency} = \frac{\text{Theoretical minimum exergy input required}}{\text{Actual exergy input}} \tag{2.46}$$

Exergy efficiencies often give more illuminating insights into process performance than energy efficiencies because (i) they weigh energy flows according to their exergy contents and (ii) they separate inefficiencies into those associated with effluent losses and those due to irreversibilities. In general, exergy efficiencies provide a measure of potential for improvement.

2.10. Procedure for energy and exergy analyses

A simple procedure for performing energy and exergy analyses involves the following steps:

- Subdivide the process under consideration into as many sections as desired, depending on the depth of detail and understanding desired from the analysis.
- Perform conventional mass and energy balances on the process, and determine all basic quantities (e.g., work, heat) and properties (e.g., temperature, pressure).
- Based on the nature of the process, the acceptable degree of analysis complexity and accuracy, and the questions for which answers are sought, select a reference-environment model.
- Evaluate energy and exergy values, relative to the selected reference-environment model.
- Perform exergy balances, including the determination of exergy consumptions.
- Select efficiency definitions, depending on the measures of merit desired, and evaluate values for the efficiencies.
- Interpret the results, and draw appropriate conclusions and recommendations, relating to such issues as design changes, retrofit plant modifications, etc.

2.11. Energy and exergy properties

Many material properties are needed for energy and exergy analyses of processes. Sources of conventional property data are abundant for many substances (e.g., steam, air and combustion gases and chemical substances).

Energy values of heat and work flows are absolute, while the energy values of material flows are relative. Enthalpies are evaluated relative to a reference level. Since energy analyses are typically concerned only with energy differences, the reference level used for enthalpy calculations can be arbitrary. For the determination of some energy efficiencies, however, the enthalpies must be evaluated relative to specific reference levels (e.g., for energy-conversion processes, the reference level is often selected so that the enthalpy of a material equals its higher heating value (HHV).

If, however, the results from energy and exergy analyses are to be compared, it is necessary to specify reference levels for enthalpy calculations such that the enthalpy of a compound is evaluated relative to the stable components of the reference environment. Thus, a compound which exists as a stable component of the reference environment is defined to have an enthalpy of zero at T_o and P_o. Enthalpies calculated with respect to such conditions are referred to as 'base enthalpies' (Rodriguez, 1980). The base enthalpy is similar to the enthalpy of formation. While the latter is the enthalpy of a compound (at T_o and P_o) relative to the elements (at T_o and P_o) from which it would be formed, the former is the enthalpy of a component (at T_o and P_o) relative to the stable components of the environment (at T_o and P_o). For many environment models, the base enthalpies of material fuels are equal to their HHVs.

Base enthalpies for many substances, corresponding to the reference-environment model in Table 2.1, are listed in Table 2.2.

It is necessary for chemical exergy values to be determined for exergy analysis. Many researchers have developed methods for evaluating chemical exergies and tabulated values (e.g., Szargut, 1967; Rodriguez, 1980; Sussman, 1980). Included are methods for evaluating the chemical exergies of solids, liquids and gases. For complex materials (e.g., coal, tar, ash), approximation methods have been developed. By considering environmental air and gaseous process streams as ideal gas mixtures, chemical exergy can be calculated for gaseous streams using component chemical exergy values (i.e., values of $(\mu_{io} - \mu_{ioo})$ listed in Table 2.2).

Table 2.2. Base enthalpy and chemical exergy values of selected species.

Species	Specific base enthalpy (kJ/g-mol)	Specific chemical exergy* (kJ/g-mol)
Ammonia (NH_3)	382.585	$2.478907 \ln y + 337.861$
Carbon (graphite) (C)	393.505	410.535
Carbon dioxide (CO_2)	0.000	$2.478907 \ln y + 20.108$
Carbon monoxide (CO)	282.964	$2.478907 \ln y + 275.224$
Ethane (C_2H_6)	1564.080	$2.478907 \ln y + 1484.952$
Hydrogen (H_2)	285.851	$2.478907 \ln y + 235.153$
Methane (CH_4)	890.359	$2.478907 \ln y + 830.212$
Nitrogen (N_2)	0.000	$2.478907 \ln y + 0.693$
Oxygen (O_2)	0.000	$2.478907 \ln y + 3.948$
Sulfur (rhombic) (S)	636.052	608.967
Sulfur dioxide (SO_2)	339.155	$2.478907 \ln y + 295.736$
Water (H_2O)	44.001	$2.478907 \ln y + 8.595$

* y represents the molal fraction for each of the respective species.
Source: Compiled from data in Rodriguez (1980) and Gaggioli and Petit (1977).

2.12. Implications of results of exergy analyses

The results of exergy analyses of processes and systems have direct implications on application decisions and on research and development (R&D) directions.

Further, exergy analyses more than energy analyses provide insights into the 'best' directions for R&D effort. Here, 'best' is loosely taken to mean 'most promising for significant efficiency gains.' There are two main reasons for this statement:

1. Exergy losses represent true losses of the potential that exists to generate the desired product from the given driving input. This is not true in general for energy losses. Thus, if the objective is to increase efficiency, focusing on exergy losses permits R&D to focus on reducing losses that will affect the objective.
2. Exergy efficiencies always provide a measure of how nearly the operation of a system approaches the ideal or theoretical upper limit. This is not in general true for energy efficiencies. By focusing R&D effort on those plant sections or processes with the lowest exergy efficiencies, the effort is being directed to those areas which inherently have the largest margins for efficiency improvement. By focusing on energy efficiencies, on the other hand, one can expend R&D effort on topics for which little margins for improvement, even theoretically, exist.

Exergy analysis results typically suggest that R&D efforts should concentrate more on internal rather than external exergy losses, based on thermodynamic considerations, with a higher priority for the processes having larger exergy losses. Although this statement suggests focusing on those areas for which margins for improvement are greatest, it does not indicate that R&D should not be devoted to those processes having low exergy losses, as simple and cost-effective ways to increase efficiency by reducing small exergy losses should certainly be considered when identified.

More generally, it is noted that application and R&D allocation decisions should not be based exclusively on the results of energy and exergy analyses, even though these results provide useful information to assist in such decision making. Other factors must be considered also, such as economics, environmental impact, safety, and social and political implications.

2.13. Closing remarks

This chapter has covered theoretical and practical aspects of thermodynamics that are of most relevance to energy and exergy analyses of systems. The chapter discusses fundamental principles and such related issues as reference-environment selection, efficiency definition and acquisition of material properties. General implications of exergy analysis results are elaborated, and a step-by-step procedure for both energy and exergy analyses is given.

Problems

2.1 Explain why exergy analysis has become a major topic in many thermodynamics courses and why increasing numbers of people use it in the analysis of energy systems.

2.2 Exergy analysis allows the determination of the upper limits of system efficiency and quantification of the causes of degradation of system performance. Can similar results be obtained using an energy analysis?

2.3 Define the following terms and explain, where appropriate, their differences: exergy, available energy, availability, work capability, essergy, exergy consumption, irreversibility, lost work, dissipated work, dissipation, exergy destruction and recovered exergy.

2.4 Carry out research to find out who invented the word 'exergy' and when. Why is the term 'exergy' more commonly used than 'availability'?

2.5 Investigate the literature to identify the various symbols used for exergy, closed system exergy, flow exergy and irreversibility. Are there any standard or commonly accepted symbols for these terms?

2.6 What is the difference between closed system exergy and flow exergy? Express flow exergy in terms of closed system exergy.

2.7 The exergies of kinetic and potential energy are equal to the kinetic and potential energy, respectively. Consequently, some people argue that an exergy analysis involving a wind turbine or a hydroelectric power plant is meaningless. Do you agree? Explain.

2.8 Is there any relationship between the exergy of heat transfer and work output for a Carnot heat engine? Explain.

2.9 Is this statement always correct: the exergy of work is equal to work? Explain.

2.10 Write mass, energy, entropy and exergy balances for the following devices: (a) an adiabatic steam turbine, (b) an air compressor with heat loss from the air to the surroundings, (c) an adiabatic nozzle and (d) a diffuser with heat loss to the surroundings.

2.11 Why have several classes of reference-environment models been proposed?

2.12 Some researchers argue that 'efficiency should be used only after it is clearly defined.' Do you agree? Explain.

2.13 What is the difference between energy and exergy efficiency? Define both for an adiabatic turbine.

2.14 Some researchers consider the 'isentropic efficiency' a type of 'first-law adiabatic efficiency' even though isentropic is a term associated with the second law of thermodynamics. What is your opinion?

2.15 One person claims that the exergy efficiency of a system is always greater than its energy efficiency while another person claims the opposite. Which person is correct? Explain with examples.

2.16 Define each of the following efficiencies for a compressor and explain under which conditions they should be used: (a) isentropic efficiency, (b) isothermal efficiency and (c) exergy efficiency.

2.17 An engineer wants to express the performance of a hydraulic turbine using an isentropic efficiency, in terms of enthalpies. Is this reasonable? Explain with examples. Can you recommend better alternatives?

2.18 How do you define and express the exergy of a fuel? Is there any relationship between the exergy of a fuel and its lower or higher heating value?

2.19 A student claims that the thermal and exergy efficiencies of a fossil-fuel power plant are very close to each other. Do you agree? Explain.

2.20 How can one use the results of an exergy analysis to improve system efficiency?

2.21 Can exergy analysis be used in the design of a thermal system? Explain.

2.22 Can exergy analysis be used in the optimization of a thermal system? Explain.

2.23 An exergy analysis of the components of a system indicates that the exergy efficiency of component A is much smaller than that of component B. Does this mean that the priority for investing resources should be improving the performance of component A rather than component B?

Chapter 3

EXERGY, ENVIRONMENT AND SUSTAINABLE DEVELOPMENT

3.1. Introduction

The relationship between energy and economics was a prime concern in 1970s. At that time, the linkage between energy and the environment did not receive much attention. As environmental concerns, such as acid rain, ozone depletion and global climate change, became major issues in the 1980s, the link between energy utilization and the environment became more recognized. Since then, there has been increasing attention on this connection, as it has become more clear that energy production, transformation, transport and use all impact the earth's environment, and that environmental impacts are associated with the thermal, chemical and nuclear emissions which are a necessary consequence of the processes that provide benefits to humanity. Simultaneously concerns have been expressed about the non-sustainable nature of human activities, and extensive efforts have begun to be devoted toward developing methods for achieving sustainable development.

The relation between sustainable development and the use of resources, particularly energy resources, is of great significance to societies. Attaining sustainable development requires that sustainable energy resources be used, and is assisted if resources are used efficiently. Exergy methods are important since they are useful for improving efficiency. The relations between exergy and both energy and the environment makes it clear that exergy is directly related to sustainable development.

That these topics are connected can be seen relatively straightforwardly. For instance, the environmental impact of emissions can be reduced by increasing the efficiency of resource utilization. As this measure helps preserve resources, it is sometimes referred to as '*energy conservation.*' However, increasing efficiency has sustainability implications as it lengthens the lives of existing resource reserves, but generally entails greater use of materials, labor and more complex devices. Depending on the situation and the players involved, the additional cost may be justified by the added security associated with a decreased dependence on energy resources, by the reduced environmental impact and by the social stability obtained through increased productive employment.

Many suggest that mitigating the environmental impact of energy resource utilization and achieving increased resource utilization efficiency are best addressed by considering exergy. By extension, since these topics are critical elements in achieving sustainable development, exergy also appears to provide the basis for developing comprehensive methodologies for sustainability. The exergy of an energy form or a substance is a measure of its usefulness or quality or potential to cause change. The latter point suggests that exergy may be, or provide the basis for, an effective measure of the potential of a substance or energy form to impact the environment. In practice, the authors feel that a thorough understanding of exergy and the insights it can provide into the efficiency, environmental impact and sustainability of energy systems are required for the engineer or scientist working in the area of energy systems and the environment. Further, as energy policies increasingly play an important role in addressing sustainability issues and a broad range of local, regional and global environmental concerns, policy makers also need to appreciate the exergy concept and its ties to these concerns. The need to understand the linkages between exergy and energy, sustainable development and environmental impact has become increasingly significant.

Despite the fact that many studies appeared during the past two decades concerning the close relationship between energy and the environment, there has only recently been an increasing number of works on the linkage between the exergy and the environment (e.g., Reistad, 1970; Szargut, 1980; Wepfer and Gaggioli, 1980; Crane et al., 1992; Rosen and Dincer, 1997a; 2001; Dincer and Rosen, 1999a; Sciubba, 1999). In this chapter, which extends ideas presented in our earlier works, we consider exergy as the confluence of energy, environment and sustainable development, as illustrated in Fig. 3.1. The basis for this treatment is the interdisciplinary character of exergy and its relation to each of these disciplines. The primary objective of this chapter is to present a unified exergy-based structure that provides useful

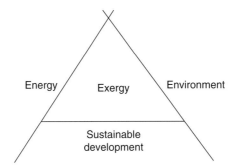

Fig. 3.1. Interdisciplinary triangle covered by the field of exergy analysis.

insights and direction to those involved in exergy, environment and sustainable development for analyzing and addressing appropriately each of these areas using exergy concepts.

3.2. Exergy and environmental problems

3.2.1. Environmental concerns

Environmental problems, issues and concerns span a continuously growing range of pollutants, hazards and ecosystem degradation factors that affect areas ranging from local through regional to global. Some of these concerns arise from observable, chronic effects on, for instance, human health, while others stem from actual or perceived environmental risks such as possible accidental releases of hazardous materials.

Many environmental issues are caused by or related to the production, transformation and use of energy. For example, 11 major areas of environmental concern in which energy plays a significant role have been identified, namely major environmental accidents, water pollution, maritime pollution, land use and siting impact, radiation and radioactivity, solid waste disposal, hazardous air pollutants, ambient air quality, acid deposition, stratospheric ozone depletion, and global climate change. While energy policy was concerned mainly with economic considerations in the 1970s and early 1980s, environmental-impact control, though clean fuels and energy technologies as well as energy efficiency, have received increasing attention over the last couple of decades.

Environmental problems are often complex and constantly evolving. Generally, our ability to identify and quantify scientifically the sources, causes and effects of potentially harmful substances has greatly advanced. Throughout the 1970s most environmental analyses and legal instruments of control focused on conventional pollutants (e.g., SO_x, NO_x, CO and particulates). Recently, environmental control efforts have been extended to (i) hazardous air pollutants, which are usually toxic chemical substances that are harmful in small doses and (ii) globally significant pollutants such as CO_2. Developments in industrial processes and systems often lead to new environmental problems. For instance, major increases in recent decades in the transport of industrial goods and people by car have led to increases in road traffic which in turn have enlarged the attention paid to the effects and sources of NO_x and volatile organic compound (VOC) emissions.

Other important aspects of environmental impact are the effects of industrial devices on the esthetics and ecology of the planet. The relatively low costs of fossil fuels has made humanity increasingly dependent on them and caused significant pollution, endangering many biological systems and reducing the planet's ecological diversity. Researchers and others can play a vital role in our planet's evolution by guiding the development of industrial society, in part by using exergy as a tool to reduce energy consumption and environmental degradation.

In the past two decades the risks and reality of environmental degradation have become apparent. The environmental impact of human activities has grown due to increasing world population, energy consumption, industrial activity, etc. Details on pollutants and their impacts on the environment and humans may be found in Dincer (1998). The most internationally significant environmental issues are usually considered to be acid precipitation, stratospheric ozone depletion and global climate change, which are the focus of this section.

Global climate change

Global climate change, including global warming, refers to the warming contribution of the earth of increased atmospheric concentration of CO_2 and other greenhouse gases. In Table 3.1, the contributions of various greenhouse gases to the

Table 3.1. Contributions of selected substances to global climate change.

Substance	ARIRR[a]	Atmospheric concentration (ppm)		AGR[b] (%)	SGEHA[c] (%)	SGEIHA[d] (%)
		Pre-industrial	1990s			
CO_2	1	275	346	0.4	71	50 ± 5
CH_4	25	0.75	1.65	1	8	15 ± 5
N_2O	250	0.25	0.35	0.2	18	9 ± 2
R-11	17,500	0	0.00023	5	1	13 ± 3
R-12	20,000	0	0.00040	5	2	13 ± 3

[a] Ability to retain infrared radiation relative to CO_2.
[b] Annual growth rate.
[c] Share in the greenhouse effect due to human activities.
[d] Share in the greenhouse effect increase due to human activities.
Source: Dincer and Rosen (1999b).

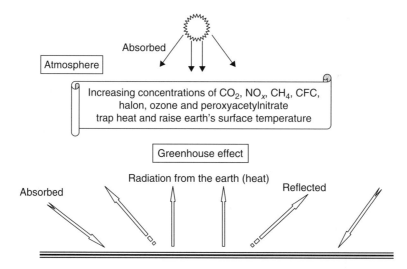

Fig. 3.2. Processes involved in the greenhouse effect.

processes involved in global climate change are summarized. CO_2 emissions account for about 50% of the anthropogenic greenhouse effect. Other gases such as CH_4, CFCs, halons, N_2O, ground ozone and peroxyacetylnitrate, produced by industrial and domestic activities, also contribute to raising the earth's temperature (Fig. 3.2).

Global climate change is associated with increasing atmospheric concentrations of greenhouse gases, which trap heat radiated from the earth's surface, thereby raising the surface temperature of the earth. The earth's surface temperature has increased about 0.6°C over the last century, and as a consequence the sea level has risen by perhaps 20 cm. The role of various greenhouse gases is summarized in Dincer and Rosen (1999b).

Humankind contributes to the increase in atmospheric concentrations of greenhouse gases. CO_2 releases from fossil fuel combustion, methane emissions from human activity, chlorofluorocarbon releases and deforestation all contribute to the greenhouse effect. Most scientists and researchers agree that emissions of greenhouse gases have led to global warming and that if atmospheric concentrations of greenhouse gases continue to increase, as present trends in fossil fuel consumption suggest, the earth's temperature may increase this century by 2–4°C. If this prediction is realized, the sea level could rise from 30 to 60 cm by 2100, leading to flooding of coastal settlements, displacement of fertile zones for agriculture and food production toward higher latitudes, reduced fresh water for irrigation and other uses, and other consequences that could jeopardize populations. The magnitude of the greenhouse effect now and in the future is debated, but most agree that greenhouse gas emissions are to some extent harmful to the environment.

Most efforts to control global climate change must consider the costs of reducing carbon emissions. Achieving a balance between economic development and emissions abatement requires policies aimed at improving the efficiency of energy use, encouraging energy conservation and renewable energy use, and facilitating fuel switching (particularly to hydrogen), and increasing access to advanced technologies.

Stratospheric ozone depletion

Ozone in the stratosphere (altitudes of 12–25 km) absorbs ultraviolet (UV) radiation (wavelengths 240–320 nm) and infrared radiation. The regional depletion of the stratospheric ozone layer, which has been shown to be caused by emissions of CFCs, halons (chlorinated and brominated organic compounds) and nitrogen oxides (NO_x) (Fig. 3.3), can lead to increased levels of damaging UV radiation reaching the ground, causing increased rates of skin cancer, eye damage and other harm to biological species.

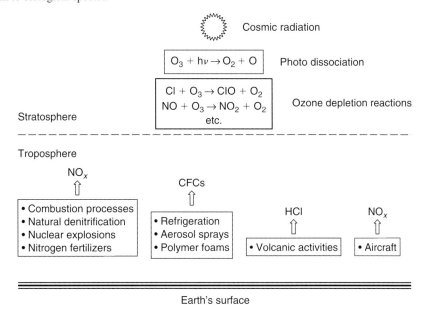

Fig. 3.3. Processes involved in stratospheric ozone depletion.

Many activities lead to stratospheric ozone depletion. CFCs, which are used in air conditioning and refrigerating equipment as refrigerants and in foam insulation as blowing agents, and NO_x emissions from fossil fuel and biomass combustion, natural denitrification, nitrogen fertilizers and aircrafts, are the most significant contributors to ozone depletion. In 1987 an international landmark protocol was signed in Montreal to reduce the production of CFCs and halons, and commitments for further reductions and eventually banning were undertaken subsequently (e.g., the 1990 London Conference). Researchers have studied the chemical and physical phenomena associated with ozone depletion, mapped ozone losses in the stratosphere, and investigated the causes and impacts of the problem.

Alternative technologies that do not use CFCs have increased substantially and may allow for a total ban of CFCs. More time will be needed in developing countries, some of which have invested heavily in CFC-related technologies.

Acid precipitation

Acid rain (acid precipitation) is the result of emissions from combustion of fossil fuels from stationary devices, such as smelters for non-ferrous ores and industrial boilers, and transportation vehicles. The emissions are transported through the atmosphere and deposited via precipitation on the earth. The acid precipitation from one country may fall on other countries, where it exhibits its damaging effects on the ecology of water systems and forests, infrastructure and historical and cultural artifacts.

Acid rain is mainly attributable to emissions of SO_2 and NO_x which react with water and oxygen in the atmosphere to form such substances as sulfuric and nitric acids (Fig. 3.4). In the atmosphere, these substances react to form acids, which are sometimes deposited on ecosystems that are vulnerable to excessive acidity. The control of acid precipitation requires control of SO_2 and NO_x emissions. These pollutants cause local concerns related to health and contribute to

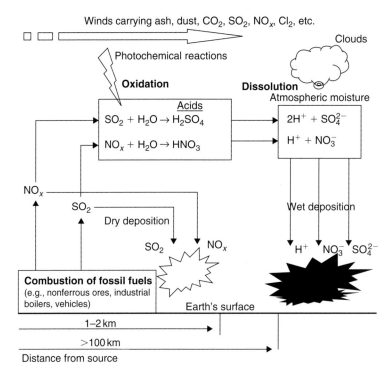

Fig. 3.4. Processes involved in the formation and transport of acid precipitation. Note that acid-formation reactions shown in box are shown only to illustrate reactants and products, and thus exclude coefficients.

the regional and trans-boundary problem of acid precipitation. Attention has also begun focusing on other contributing substances such as VOCs, chlorides, ozone and trace metals, which may participate in chemical transformations in the atmosphere that result in acid precipitation and other pollutants.

The impacts of acid precipitation are as follows:

- acidification of lakes, streams and ground waters;
- damage to forests, crops and plants due to the toxicity of excessive acid concentration;
- harm to fish and aquatic life;
- deterioration of materials (e.g., buildings, metal structures and fabrics);
- alterations of the physical and optical properties of clouds due to the influence of sulfate aerosols.

Many energy-related activities lead to acid precipitation. Electric power generation, residential heating and industrial energy use account for about 80% of SO_2 emissions. Sour gas treatment releases H_2S that reacts to form SO_2 when exposed to air. Most NO_x emissions are from fossil fuel combustion in stationary devices and road transport. VOCs from various sources contribute to acid precipitation. The largest contributors to acid precipitation have been the USA, China and the countries from the former Soviet Union.

Some gaseous pollutants are listed in Table 3.2, along with their environmental impacts.

3.2.2. Potential solutions to environmental problems

Potential solutions to current environmental problems including pollutant emissions have recently evolved, including:

- recycling;
- process change and sectoral modification;
- acceleration of forestation;
- application of carbon and/or fuel taxes;
- materials substitution;
- promoting public transport;

Table 3.2. Impacts on the environment of selected gaseous pollutants.

Pollutants	Greenhouse effect	Stratospheric ozone depletion	Acid precipitation	Smog
Carbon monoxide (CO)	•	•	•	•
Carbon dioxide (CO_2)	+	+/−	•	•
Methane (CH_4)	+	+/−	•	•
Nitric oxide (NO) and nitrogen dioxide (NO_2)	•	+/−	+	+
Nitrous oxide (N_2O)	+	+/−	•	•
Sulfur dioxide (SO_2)	−	+	•	•
Chlorofluorocarbons (CFCs)	+	+	•	•
Ozone (O_3)	+	•	•	+

Note: + denotes a contributing effect, − denotes that the substance exhibits an impact which varies with conditions and chemistry and may not be a general contributor and • denotes no impact.
Source: Speight (1996).

- changing lifestyles;
- increasing public awareness of energy-related environmental problems;
- increased education and training;
- policy integration.

Potential solutions to energy-related environmental concerns have also evolved. These include:

- use of renewable and advanced energy technologies;
- energy conservation and increasing the efficiency of energy utilization;
- application of cogeneration, trigeneration, and district heating and cooling;
- use of alternative energy forms and sources for transport;
- energy source switching from fossil fuels to environmentally benign energy forms;
- use of clean coal technologies;
- use of energy storage;
- optimum monitoring and evaluation of energy indicators.

Among the potential solutions listed above, some of the most important are the use of renewable and advanced energy technologies. An important step in moving toward the implementation of such technologies is to identify and remove barriers. Several barriers have in the past been identified to the development and introduction of cleaner energy processes, devices and products. The barriers can also affect the financing of efforts to augment the supply of renewable and advanced energy technologies.

Some of the barriers faced by many renewable and advanced energy technologies include:

- technical constraints;
- financial constraints;
- limited information and knowledge of options;
- lack of necessary infrastructure for recycling, recovery and re-use of materials and products;
- lack of facilities;
- lack of expertise within industry and research organizations, and/or lack of coordinated expertise;
- poorly coordinated and/or ambiguous national aims related to energy and the environment;
- uncertainties in government regulations and standards;
- lack of adequate organizational structures;
- lack of differentiated electrical rates to encourage off-peak electricity use;
- mismanagement of human resources;
- lack of societal acceptability of new renewable and advanced energy technologies;
- absence of, or limited consumer demand for, renewable and advanced energy products and processes.

Establishing useful methods for promoting renewable and advanced energy technologies requires analysis and clarification about how to combine environmental objectives, social and economic systems, and technical development. It is important to employ tools that encourage technological development and diffusion, and to align government policies in energy and environment.

3.2.3. Energy and environmental impact

Energy resources are required to supply the basic human needs of food, water and shelter, and to improve the quality of life. The United Nations indicates that the energy sector must be addressed in any broad atmosphere-protection strategy through programs in two major areas: increasing energy efficiency and shifting to environmentally sound energy systems. The major areas investigated promote: (i) the energy transition; (ii) increased energy efficiency and, consequently, increased exergy efficiency; (iii) renewable energy sources and (iv) sustainable transportation systems. It was reported that (i) a major energy efficiency program would provide an important means of reducing CO_2 emissions and (ii) the activities should be accompanied by measures to reduce the fossil fuel component of the energy mix and to develop alternative energy sources. These ideas have been reflected in many recent studies concentrating on the provision of energy services with the lowest reasonable environmental impact and cost and the highest reasonable energy security.

Waste heat emissions to the environment are also of concern, as irresponsible management of waste heat can significantly increase the temperature of portions of the environment, resulting in thermal pollution. If not carefully controlled so that local temperature increases are kept within safe and desirable levels, thermal pollution can disrupt marine life and ecological balances in lakes and rivers.

Measures to increase energy efficiency can reduce environmental impact by reducing energy losses. Within the scope of exergy methods, as discussed in the next section, such activities lead to increased exergy efficiency and reduced exergy losses (both waste exergy emissions and internal exergy consumptions). In practice, potential efficiency improvements can be identified by means of modeling and computer simulation. Increased efficiency can help achieve energy security in an environmentally acceptable way by reducing the emissions that might otherwise occur. Increased efficiency also reduces the requirement for new facilities for the production, transportation, transformation and distribution of the various energy forms, and the associated environmental impact of these additional facilities. To control pollution, efficiency improvement actions often need to be supported by pollution amelioration technologies or fuel substitution. The most significant measures for environmental protection are usually those undertaken at the regional or national levels, rather than by individual projects.

3.2.4. Thermodynamics and the environment

People have long been intrigued by the implications of the laws of thermodynamics on the environment. One myth speaks of Ouroboros, a serpent like creature which survived and regenerated itself by eating only its own tail. By neither taking from nor adding to its environment, this creature was said to be completely environmentally benign and selfsufficient. It is useful to examine this creature in light of the thermodynamic principles recognized today. Assuming that Ouroboros was an isolated system (i.e., it received no energy from the sun or the environment, and emitted no energy during any process), Ouroboros' existence would have violated neither the conservation law for mass nor the first law of thermodynamics (which states energy is conserved). However, unless it was a reversible creature, Ouroboros' existence would have violated the second law (which states that exergy is reduced for all real processes), since Ouroboros would have had to obtain exergy externally to regenerate the tail it ate into an equally ordered part of its body (or it would ultimately have dissipated itself to an unordered lump of mass). Thus, Ouroboros would have to have had an impact on its environment.

Besides demonstrating that, within the limits imposed by the laws of thermodynamics, all real processes must have some impact on the environment, this example is intended to illustrate the following key point: the second law is instrumental in providing insights into environmental impact (e.g., Hafele, 1981; Edgerton, 1992; Rosen and Dincer, 1997a). Today, the principles demonstrated through this example remain relevant, and technologies are sought having Ouroboros' characteristics of being environmentally benign and self-sufficient (e.g., University of Minnesota researchers built an 'energy-conserving' house called Ouroboros (Markovich, 1978)). The importance of the second law in understanding environmental impact implies that exergy, which is based on the second law, has an important role to play in this field.

The most appropriate link between the second law and environmental impact has been suggested to be exergy (Rosen and Dincer, 1997a), in part because it is a measure of the departure of the state of a system from that of the environment.

The magnitude of the exergy of a system depends on the states of both the system and the environment. This departure is zero only when the system is in equilibrium with its environment. The concept of exergy analysis as it applies to the environment is discussed in detail elsewhere (Rosen and Dincer, 1997).

An understanding of the relations between exergy and the environment may reveal the underlying fundamental patterns and forces affecting changes in the environment, and help researchers deal better with environmental damage. Tribus and McIrvine (1971) suggest that performing exergy analyses of the natural processes occurring on the earth could form a foundation for ecologically sound planning because it would indicate the disturbance caused by large-scale changes. Three relationships between exergy and environmental impact (Rosen and Dincer, 1997a) are discussed below:

- *Order destruction and chaos creation*: The destruction of order, or the creation of chaos, is a form of environmental damage. Entropy is fundamentally a measure of chaos, and exergy of order. A system of high entropy is more chaotic or disordered than one of the low entropy, and relative to the same environment, the exergy of an ordered system is greater than that of a chaotic one. For example, a field with papers scattered about has higher entropy and lower exergy than the field with the papers neatly piled. The exergy difference of the two systems is a measure of (i) the exergy (and order) destroyed when the wind scatters the stack of papers and (ii) the minimum work required to convert the chaotic system to the ordered one (i.e., to collect the scattered papers). In reality, more than this minimum work, which only applies if a reversible clean-up process is employed, is required. The observations that people are bothered by a landscape polluted with papers chaotically scattered about, but value the order of a clean field with the papers neatly piled at the side, suggests that, on a more abstract level, ideas relating exergy and order in the environment may involve human values (Hafele, 1981) and that human values may in part be based on exergy and order.

- *Resource degradation*: The degradation of resources found in nature is a form of environmental damage. Kestin (1980) defines a resource as a material, found in nature or created artificially, which is in a state of disequilibrium with the environment, and notes that resources have exergy as a consequence of this disequilibrium. Two main characteristics of resources are valued:

  (i) *Composition (e.g., metal ores)*: Many processes exist to increase the value of such resources by purifying them, which increases their exergy. Note that purification is accomplished at the expense of consuming at least an equivalent amount of exergy elsewhere (e.g., using coal to drive metal ore refining).

  (ii) *Reactivity (e.g., fuels)*: Their potential to cause change, or 'drive' a task or process.

  Two principal general approaches exist to reduce the environmental impact associated with resource degradation:

  (i) *Increased efficiency*: Increased efficiency preserves exergy by reducing the exergy necessary for a process, and therefore reduces environmental damage. Increased efficiency also usually reduces exergy emissions which, as discussed in the next section, also play a role in environmental damage.

  (ii) *Using external exergy resources (e.g., solar energy)*: The earth is an open system subject to a net influx of exergy from the sun. It is the exergy (or order states) delivered with solar radiation that is valued; all the energy received from the sun is ultimately radiated out to the universe. Environmental damage can be reduced by taking advantage of the openness of the earth and utilizing solar radiation (instead of degrading resources found in nature to supply exergy demands). This would not be possible if the earth was a closed system, as it would eventually become more and more degraded or 'entropic.'

- *Waste exergy emissions*: The exergy associated with waste emissions can be viewed as a potential for environmental damage in that the exergy of the wastes, as a consequence of not being in stable equilibrium with the environment, represents a potential to cause change. When emitted to the environment, this exergy represents a potential to change the environment. Usually, emitted exergy causes a change which is damaging to the environment, such as the deaths of fish and plants in some lakes due to the release of specific substances in stack gases as they react and come to equilibrium with the environment, although in some cases the change may be perceived to be beneficial (e.g., the increased growth rate of fish and plants near the cooling-water outlets from thermal power plants). Further, exergy emissions to the environment can interfere with the net input of exergy via solar radiation to the earth (e.g., emissions of CO_2 and other greenhouse gases from many processes appear to cause changes to the atmospheric CO_2 concentration, affecting the receiving and re-radiating of solar radiation by the earth). The relation between waste exergy emissions and environmental damage has been recognized by several researchers (e.g., Reistad, 1970). By considering the economic value of exergy in fuels, Reistad developed an air-pollution rating that he felt was preferable to the mainly empirical ratings then in use, in which the air-pollution cost for a fuel was estimated as either (i) the cost to remove the pollutant or (ii) the cost to society of the pollution in the form of a tax which should be levied if pollutants are not removed from effluent streams.

Although the previous two points indicate simultaneously that exergy in the environment in the form of resources is of value while exergy in the environment in the form of emissions is harmful due to its potential to cause environmental damage, confusion can be avoided by considering whether or not the exergy is constrained (see Fig. 3.5). Most resources found in the environment are constrained and by virtue of their exergy are of value, while unconstrained emissions of exergy are free to impact in an uncontrolled manner on the environment. To elaborate further on this point, consider a scenario in which emissions to the environment are constrained (e.g., by separating sulfur from stack gases). This action yields two potential benefits: the potential for environmental damage is restrained from entering the environment, and the now-constrained emission potentially becomes a valued commodity (i.e., a source of exergy).

The decrease in the environmental impact of a process, in terms of several measures, as the process exergy efficiency increases is illustrated approximately in Fig. 3.6.

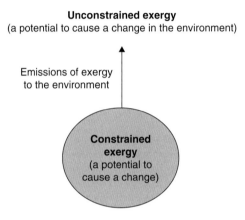

Fig. 3.5. Comparison of constrained and unconstrained exergy illustrating that exergy constrained in a system represents a resource, while exergy emitted to the environment becomes unconstrained and represents a driving potential for environmental damage.

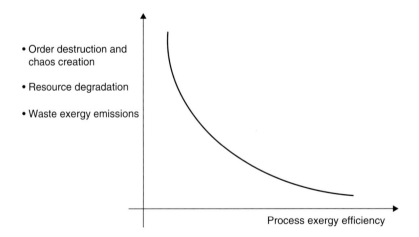

Fig. 3.6. Qualitative illustration of the relation between the exergy efficiency of a process and the associated environmental impact in terms of order destruction and chaos creation, or resource degradation, or waste exergy emissions.

3.3. Exergy and sustainable development

Energy resources are needed for societal development, and sustainable development requires a supply of energy resources that is sustainably available at reasonable cost and causes no or minimal negative societal impacts. Clearly, energy resources such as fossil fuels are finite and thus lack the characteristics needed for sustainability, while others such as renewable energy sources are sustainable over the relatively long term. Environmental concerns are also a major factor in sustainable development, as activities which degrade the environment are not sustainable. Since much environmental impact is associated with energy use, sustainable development requires the use of energy resources which cause as little environmental impact as reasonably possible. Clearly, limitations on sustainable development due to environmental emissions can be in part overcome through increased efficiency, as this usually leads to less environmental impact for the same services or products.

The diversity of energy choices is but one reason why exergy plays a key role in the context of sustainable development. Many factors contribute to achieving sustainable development. For example, for development to be sustainable:

- it must satisfy the needs and aspirations of society;
- it must be environmentally and ecologically benign;
- sufficient resources (natural and human) must be available.

The second point reinforces the importance of environmental concerns in sustainable development. Clearly, activities which continually degrade the environment are not sustainable over time, while those that have no or little negative impact on the environment are more likely to contribute to sustainable development (provided, of course, that they satisfy the other conditions for sustainable development).

3.3.1. Sustainable development

The term *sustainable development* was introduced in 1980, popularized in the 1987 report of the World Commission on Environment and Development (the Brundtland Commission), and given a global mission status by the UN Conference on Environment and Development in Rio de Janeiro in 1992.

The Brundtland Commission defined *sustainable development* as 'development that meets the needs of the present without compromising the ability of future generations to meet their own needs.' The Commission noted that its definition contains two key concepts: *needs*, meaning 'in particular the essential needs of the world's poor,' and *limitations*, meaning 'limitations imposed by the state of technology and social organization on the environment's ability to meet present and future needs' (OECD, 1996).

The Brundtland Commission's definition was thus not only about sustainability and its various aspects but also about equity, equity among present inhabitants of the planet and equity among generations. *Sustainable development* for the Brundtland Commission includes environmental, social and economic factors, but considers remediation of current social and economic problems an initial priority. The chief tools cited for remediation are 'more rapid economic growth in both industrial and developing countries, freer market access for the products of developing countries, greater technology transfer, and significantly larger capital flows, both concessional and commercial.' Such growth was said to be compatible with recognized environmental constraints, but the extent of the compatibility was not explored.

An enhanced definition of global sustainable development is presented in the Encyclopedia of Life Support Systems (EOLSS, 1998): '*the wise use of resources through critical attention to policy, social, economic, technological, and ecological management of natural and human engineered capital so as to promote innovations that assure a higher degree of human needs fulfillment, or life support, across all regions of the world, while at the same time ensuring intergenerational equity.*'

3.3.2. Sustainability and its need

The world is changing rapidly due in part to the increasing wealth and size of the population. A growing need exists for more efficient and sustainable production processes. As our world increasingly strives for a more sustainable society, we must overcome some major problems, for example, increasing population, lack of and inequitable distribution of wealth, insufficient food production and energy supply, and increasing environmental impact.

Sustainability has been called a key to solving current ecological, economic and developmental problems. Sustainability has been broadly discussed since it was brought to public attention by the Brundtland Report and has since been developed into a blueprint for reconciling economic and ecological necessities. Many have contributed to make this concept scientifically acceptable so that it can be utilized as a yard stick for strategic planning. Two features that make sustainability useful for strategic planning are its inherent long-term view and its ability to accommodate changing conditions.

Sustainability and *sustainable development* became fashionable terms in the 1990s, a legacy of concerns about the environment expressed during the 1970s and 1980s. The media often refer to the same concepts via such terms as sustainable architecture, sustainable food production, sustainable future, sustainable community, sustainable economic development, sustainable policy, etc.

Some key component requirements for sustainable development (see Fig. 3.7) are societal, economic, environmental and technological sustainability. Some topics within each of these component areas are listed in the figure.

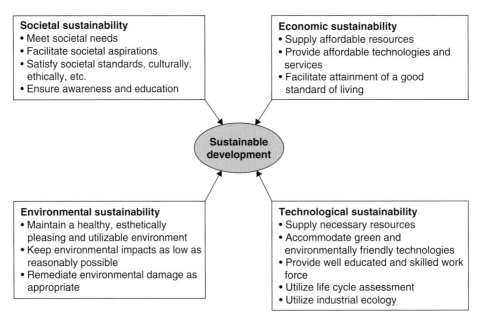

Fig. 3.7. Some key requirements of sustainable development.

3.3.3. Dimensions of sustainability

The kinds of techno-economic changes envisaged by many as necessary for long-term sustainability usually include sharp reductions in the use of fossil fuels to minimize the danger of global climate change. Alternatives to using fossil fuels include use of nuclear power, large-scale photovoltaics, intensive biomass cultivation and large-scale hydroelectric projects (in applicable regions), as well as major changes in patterns of energy consumption and conservation. Again, there are disputes over which of these energy alternatives is the most desirable, feasible, etc. However, understanding future energy patterns, from both supply and demand perspectives, is critical (van Schijndel et al., 1998).

The ecological criterion for sustainability acknowledges the likelihood that some important functions of the natural environment cannot be replaced within any realistic time frame, if ever, by human technology, however sophisticated. Some examples include the need for arable land, water and a benign climate for agriculture, the role of reducing bacteria in recycling nutrient elements in the biosphere, and the protection provided by the stratospheric ozone layer. The ecological criterion for long-term sustainability implicitly allows for some technological intervention. For example, methods of artificially accelerating tree growth may compensate for some net decrease in the land area devoted to forests. But, absent any plausible technological fixes, sustainability does not allow for major climate change, widespread desertification, deforestation of the tropics, accumulation of toxic heavy metals and non-biodegradable halogenated organics in soils and sediments, sharp reductions in biodiversity, etc.

3.3.4. Environmental limits and geographic scope

The report of the Brundtland Commission stimulated debate not only about the environmental impacts of industrialization and the legacy of present activities for coming generations, but also about what might be the physical or ecological limits to economic growth. From this perspective, sustainability can be defined in terms of carrying capacity of the ecosystem, and described with input–output models of energy and resource consumption. Sustainability then becomes an economic state where the demands placed on the environment by people and commerce can be met without reducing the capacity of the environment to provide for future generations. Some (e.g., Dincer, 2002) have expressed this idea in simple terms as an economic rule for a restorative economy as: '*Leave the world better than you found it, take no more than you need, try not to harm life or the environment, make amends if you do.*'

Sustainability-related limits on society's material and energy throughputs might be set as follows (OECD, 1996):

- The rates of use of renewable resources should not exceed their rates of regeneration.
- The rates of use of non-renewable resources should not exceed the rates at which renewable substitutes are developed.
- The rates of pollutant emissions should not exceed the corresponding assimilative capacity of the environment.

Sustainability, or alternatively unsustainability, must also be considered in terms of geographic scope. Some activities may be globally unsustainable. For example, they may result in climate change or depletion of the stratospheric ozone layer, and so affect several geographic regions, if not the whole world. Other activities may be regionally unsustainable, perhaps by producing and dispersing tropospheric ozone or acidifying gases that can kill vegetation and cause famine in one region but not in other parts of the world. Still other activities may be locally unsustainable, perhaps because they lead to hazardous ambient levels of CO locally or because the noise they produce makes habitation impossible. Overall, sustainability appears to be more a global than a regional or local concern. If an environmental impact exceeds the carrying capacity of the planet, for instance, then life is threatened, but it is beyond the carrying capacity of one area, then that area may become uninhabitable but life can most likely continue elsewhere.

3.3.5. Environmental, social and economic components of sustainability

The focus of this discussion on physical limits does not ignore the social and economic aspects of sustainability. Some may consider a way of life may not worth sustaining under certain circumstances, such as extreme oppression or deprivation. In fact, oppression or deprivation can interfere with efforts to make human activity environmentally benign. Nonetheless, if ecosystems are irreparably altered by human activity, then subsequent human existence may become not merely unpalatable, but infeasible. Thus the environmental component of sustainability is essential.

The heterogeneity of the environmental, social and economic aspects of sustainability should also be recognized. Environmental and social considerations often refer to *ends*, the former having perhaps more to do with the welfare of future generations and the latter with the welfare of present people. Rather than an end, economic considerations can perhaps more helpfully be seen as a *means* to the various ends implied by environmental and social sustainability.

3.3.6. Industrial ecology and resource conservation

In the field of industrial ecology, Connelly and Koshland (2001a) have recently stated that the several processes that de-link consumption from depletion in evolving biological ecosystems can be used as resource conservation *strategies* for de-linking consumption from depletion in immature industrial systems. These processes include waste cascading, resource cycling, increasing exergy efficiency and renewable exergy use.

Connelly and Koshland (2001a, b) demonstrate that the relation between these strategies and an exergy-based definition they propose for ecosystem evolution follows directly from first- and second-law principles. They discuss the four conservation strategies in the context of a simple, hypothetical industrial ecosystem consisting initially of two solvent consumption processes and the chain of industrial processes required to deliver solvent to these two processes. One solvent consuming process is assumed to require lower-purity feedstock than the other. All solvent feedstocks are derived from non-renewable, fossil sources, and all solvent leaving the two consumptive processes is emitted to the atmosphere. This is a linear process that takes in useful energy and materials and releases waste energy and material. The four conservation strategies are each described below:

- *Waste cascading*: Waste cascading may be described in thermodynamic terms as using outputs from one or more consumptive processes as inputs to other consumptive processes requiring equal or lower exergy. Waste cascading

reduces resource consumption in two ways: by reducing the rate of exergy loss caused by the dissipation of potentially usable wastes in the environment, and by reducing the need to refine virgin resources. In our hypothetical
industrial ecosystem, cascading allows used (i.e., partially consumed) solvent from the first process to be used in
the second solvent consumption process, eliminating solvent emissions from the first process and the need to refine
and supply pure solvent to the second process. The solvent consumption rate in the two processes is unchanged,
but the rate of resource depletion associated with these processes is reduced. Although waste cascading reduces
demand for other resources and hence is an important resource conservation strategy, cascading does not avoid
emissions of waste exergy entirely. Thus, cascading cannot form a resource cycle. Losses associated with the
upgrade and supply of solvent to the top of the cascade, the consumption of resources in the two processes
constituting the cascade, and dissipation of waste solvent released from the bottom of the cascade cannot be
avoided. Cascading can thus reduce the linkage between consumption and depletion, but it cannot fully de-link
the two.

- *Resource cycling*: To reduce emissions from the bottom of a waste cascade (or at the outlet of a single consumptive process) and return this bottom waste to the top of a resource cascade, the exergy removed from a
  resource during consumption must be returned to it. This process of exergy loss through consumption followed
  by exergy return through transfer is the basis of resource cycling. Adding a solvent recycling process and its
  associated chain of industrial processes to the hypothetical system reduces depletion both by eliminating exergy
  loss from the dissipation of released solvents and by substituting a post-consumption upgrade path for a virgin
  resource upgrade path. An activated carbon solvent separation system, for example, will generally be far less
  exergy intensive than the fossil-based manufacture of virgin solvent. Cycling cannot, however, eliminate depletion. In accordance with the second law, all exergy transfers in real (irreversible) processes must be accompanied
  by exergy loss (i.e., total exergy must always decrease). Hence, in any real cycling process, the overall resource
  depletion rate will exceed the rate of exergy loss in the consumptive process whose wastes are being cycled. In
  the above example, the two solvent consumption processes and the exergy removed from non-renewed resources
  for the purpose of upgrading the solvent would contribute to resource depletion in the case of complete solvent
  cycling.
- *Increasing exergy efficiency*: One way to reduce the resource depletion associated with cycling is to reduce the
  losses that accompany the transfer of exergy to consumed resources by increasing the efficiency of exergy transfer
  between resources (i.e., increasing the fraction of exergy removed from one resource that is transferred to another).
  Exergy efficiency may be defined as

$$\text{Exergy efficiency} = \text{Exergy output/exergy input}$$

where

$$\text{Exergy loss} = \text{Exergy input} - \text{exergy output}$$

Compared to energy efficiency, exergy efficiency may be thought of as a more meaningful measure of efficiency
that accounts for quantity *and* quality aspects of energy flows. Unlike energy efficiency, exergy efficiency provides an *absolute* measure of efficiency that accounts for first- and second-law limitations. In the current example,
increasing exergy efficiency in the case of complete cycling would involve increasing the efficiency of the solvent
upgrade process. Although technological and economic limitations to efficiency gains prevent exergy efficiency
from approaching unity, many industrial processes today operate at very low efficiencies, and it is widely recognized that large margins for efficiency improvement often remain. However, even if exergy efficiency could
be brought to 100%, the resource depletion associated with solvent consumption and upgrade in the example
would still not be eliminated. Recycling with a 100% exergy efficient upgrade process would result in a depletion
rate equal to the consumption rate of the two solvent consumption processes. To fully de-link consumption from
depletion, it is necessary to use resources that supply exergy without being depleted.

- *Renewable exergy use*: To fully de-link consumption from depletion, the exergy used to upgrade consumption
  products must be derived from renewable exergy sources (i.e., sources such as electricity generated directly
  or indirectly from solar radiation or sources such as sustainably harvested biomass feedstocks). In the solvent
  cycling example, using a sustainably harvested biomass fuel as the exergy source for the solvent upgrade process
  could in theory create a solvent cycling system in which a closed solvent cycle is driven entirely by renewable
  exergy inputs. In this situation, the depletion rate becomes independent of the exergy efficiency of the solvent
  upgrade process.

3.3.7. Energy and sustainable development

The relation between sustainable development and the use of resources, particularly energy resources, is of great significance to societies (e.g., Goldemberg et al., 1988; MacRae, 1992). A supply of energy resources is generally agreed to be a necessary, but not sufficient, requirement for development within a society. Societies, such as countries or regions, that undergo significant industrial and economic development almost always have access to a supply of energy resources. In some countries (e.g., Canada) energy resources are available domestically, while in others (e.g., Japan) they must be imported.

For development that is sustainable over long periods of time, there are further conditions that must be met. Principally, such societies must have access to and utilize energy resources that are sustainable in a broad sense, i.e., that are obtainable in a secure and reliable manner, safely utilizable to satisfy the energy services for which they are intended with minimal negative environmental, health and societal impacts, and usable at reasonable costs.

An important implication of the above statements is that sustainable development requires not just that sustainable energy resources be used, but that the resources be used efficiently. Exergy methods are essential in evaluating and improving efficiency. Through efficient utilization, society maximizes the benefits it derives from its resources, while minimizing the negative impacts (such as environmental damage) associated with their use. This implication acknowledges that most energy resources are to some degree finite, so that greater efficiency in utilization allows such resources to contribute to development over a longer period of time, i.e., to make development more sustainable. Even if one or more energy resources eventually become inexpensive and widely available, increases in efficiency will likely remain sought to reduce the associated environmental impacts, and the resource requirements (energy, material, etc.) to create and maintain systems to harvest the energy.

3.3.8. Energy and environmental sustainability

The environmental aspects of energy use merit further consideration, as a large portion of the environmental impact in a society is associated with energy resource utilization. Ideally, a society seeking sustainable development utilizes only energy resources which cause no environmental impact. Such a condition can be attained or nearly attained by using energy resources in ways that cause little or no wastes to be emitted into the environment, and/or that produce only waste emissions that have no or minimal negative impact on the environment. This latter condition is usually met when relatively inert emissions that do not react in the environment are released, or when the waste emissions are in or nearly in equilibrium (thermal, mechanical and chemical) with the environment, i.e., when the waste exergy emissions are minimal.

In reality, however, all resource use leads to some degree of environmental impact. A direct relation exists between exergy efficiency (and sometimes energy efficiency) and environmental impact, in that through increased efficiency, a fixed level of services can be satisfied with less energy resources and, in most instances, reduced levels of related waste emissions. Therefore, it follows that the limitations imposed on sustainable development by environmental emissions and their negative impacts can be in part overcome through increased efficiency, i.e., increased efficiency can make development more sustainable.

3.3.9. Exergy and sustainability

Exergy can be considered the confluence of energy, environment and sustainable development, as shown in Fig. 3.1, which illustrates the interdisciplinary character of exergy and its central focus among these disciplines.

Exergy methods can be used to improve sustainability. For example, in a recent study on thermodynamics and sustainable development, Cornelissen (1997) points out that one important element in obtaining sustainable development is the use of exergy analysis. By noting that energy can never be 'lost', as it is conserved according to the first law of thermodynamics, while exergy can be lost due to internal irreversibilities, that study suggests that exergy losses, particularly due to the use of non-renewable energy forms, should be minimized to obtain sustainable development. Further, the study shows that environmental effects associated with emissions and resource depletion can be expressed in terms of one exergy-based indicator, which is founded on physical principles.

Sustainable development also includes economic viability. Thus, the methods relating exergy and economics also reinforce the link between exergy and sustainable development. The objectives of most existing analysis techniques integrating exergy and economics include the determination of (i) the appropriate allocation of economic resources so as to optimize the design and operation of a system and/or (ii) the economic feasibility and profitability of a system. Exergy-based economic analysis methods are referred to by such names as thermoeconomics, second-law costing, cost accounting and exergoeconomics.

Figure 3.8 illustratively presents the relation between exergy and sustainability and environmental impact. There, sustainability is seen to increase and environmental impact to decrease as the process exergy efficiency increases. The two limiting efficiency cases are significant. First, as exergy efficiency approaches 100%, environmental impact approaches zero, since exergy is only converted from one form to another without loss, either through internal consumptions or waste emissions. Also sustainability approaches infinity because the process approaches reversibility. Second, as exergy efficiency approaches 0%, sustainability approaches zero because exergy-containing resources are used but nothing is accomplished. Also, environmental impact approaches infinity because, to provide a fixed service, an ever increasing quantity of resources must be used and a correspondingly increasing amount of exergy-containing wastes are emitted.

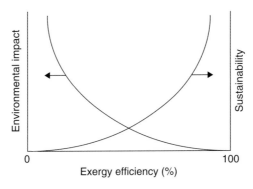

Fig. 3.8. Qualitative illustration of the relation between the environmental impact and sustainability of a process, and its exergy efficiency.

Some important contributions, that can be derived from exergy methods, for increasing the sustainability of development which is non-sustainable are presented in Fig. 3.9. Development typical of most modern processes, which are generally non-sustainable, is shown at the bottom of the figure. A future in which development is sustainable is shown at the top of the figure, while some key exergy-based contributions toward making development more sustainable are shown, and include increased exergy efficiency, reduction of exergy-based environmental degradation and use of sustainable exergy resources.

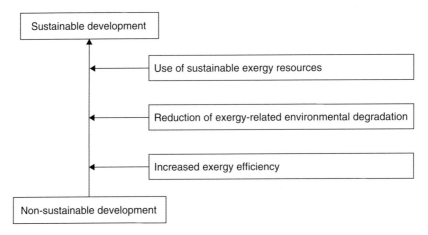

Fig. 3.9. Some key contributions of exergy methods to increasing the sustainability of non-sustainable systems and processes.

3.3.10. Exergetic aspects of sustainable processes

Figure 3.10 outlines a typical industrial process, with its throughputs of materials and energy. Cleaner production of materials, goods and services is one of the tools for sustainable development. Such production entails the efficient use of resources and the corresponding production of only small amounts of waste. Clean production often also involves the use of renewable resources. Yet the quality of the products remains important. This does not mean that cleaner production is necessarily contradictory to the economic approach of minimizing costs and maximizing profits. The challenge is often to generate win–win situations, such as those that, by minimizing the use of resources and the corresponding emissions, also decrease the costs of a given process.

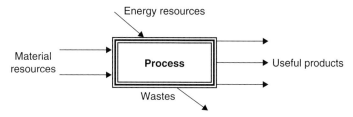

Fig. 3.10. Model of an industrial process.

As mentioned earlier, life cycle assessment (LCA) aims to improve or to optimize processes so that they consume fewer resources and produces less emissions and wastes. Common routes for achieving this often include end-of-pipe treatment such as wastewater treatment plants, filters and scrubbers. These provide only partial solutions, as they do not decrease the environmental load, but rather shift it from one phase and location to another (e.g., water or air to soil). In many cases, however, expensive end-of-pipe treatment solutions are unavoidable. Exergy analysis appears to be a significant tool for improving processes by changing their characteristics, rather than simply via end-of-pipe fixes. Thus exergy methods can help achieve more sustainable processes.

As a basic example, consider the conversion of mechanical work to heat ideally, i.e., with 100% efficiency. Heat has a lower exergy, or quality than work. Therefore, heat cannot be converted to work with a 100% energy efficiency. But, the conversion can be in theory achieved with a 100% exergy efficiency. Thus exergy analysis helps identify the upper limit for efficiency improvements.

Some examples of the difference between energy and exergy are shown in Table 3.3. Hot water and steam with the same enthalpy have different exergy, the value for steam being higher than for hot water. Fuels like natural gas and gasoline have exergetic values comparable to their net heating values. Work and electricity have the same exergy and energy. Exergy is calculated in Table 3.3 as the product of energy and a quality factor.

Table 3.3. Energy and exergy values of various energy forms.

Source	Energy (J)	Exergy (J)	Exergy/energy ratio*
Water at 80°C	100	16	0.16
Steam at 120°C	100	24	0.24
Natural gas	100	99	0.99
Electricity/work	100	100	1.00

* For heat, the exergy/energy ratio is the exergetic temperature factor $\tau = 1 - T_0/T_s$, where T_0 is the absolute temperature of the environment and T_s is the absolute temperature of the stream. Calculations can often be simplified as Exergy = Energy × exergy/energy ratio.
Modified from: Van Schijndel et al. (1998).

3.3.11. Renewables and tools for sustainable development

Renewable energy resources are often sustainable. Most energy supplies on earth derive from the sun, which continually warms us and supports plant growth via photosynthesis. Solar energy heats the land and sea differentially and so causes winds and consequently waves. Solar energy also drives evaporation, which leads to rain and in turn hydropower. Tides are the result of the gravitational pull of the moon and sun, and geothermal heat is the result of radioactive decay within the earth.

Many diverse energy-related problems and challenges which relate to renewable energy are faced today. Some examples follow (Dincer, 2000):

- Growing energy demand. The annual population growth rate is currently around 2% worldwide and higher in many countries. By 2050, world population is expected to double and economic development is expected to continue improving standards of living in many countries. Consequently, global demand for energy services is expected to increase by up to 10 times by 2050 and primary energy demand by 1.5–3 times.
- Excessive dependence on specific energy forms. Society is extremely dependent on access to specific types of energy currencies. The effect of the multi-day blackout of 2003 in Ontario and several northeastern U.S. states illustrated the dependency on electricity supply, as access was lost or curtailed to computers, elevators, air conditioners, lights and health care. Developed societies would come to a virtual standstill without energy resources.
- Energy-related environmental impacts. Continued degradation of the environment by people, most agree, will have a negative impact on the future, and energy processes lead to many environmental problems, including global climate change, acid precipitation, stratospheric ozone depletion, emissions of a wide range of pollutants including radioactive and toxic substances, and loss of forests and arable land.
- The dominance of non-sustainable and non-renewable energy resources. Limited use is made today of renewable energy resources and corresponding technologies, even though such resources and technologies provides a potential solution to current and future energy resource shortages. By considering engineering practicality, reliability, applicability, economics and public acceptability, appropriate uses for sustainable and renewable energy resources can be found. Of course, financial and other resources should not always be dedicated to renewable energy resources, as excessively extravagant or impractical plans are often best avoided.
- Energy pricing that does not reflect actual costs. Many energy resource prices have increased over the last couple of decades, in part to account for environmental costs, yet many suggest that energy prices still do not reflect actual societal costs.
- Global disparity in energy use. Wealthy industrialized economies which contain 25% of the world's population use 75% of the world's energy supply.

These and other energy-related issues need to be resolved if humanity and society are to develop sustainably in the future. Renewable energy resources appear to provide one component of an effective sustainable solution, and can contribute over the long term to achieving sustainable solutions to today's energy problems.

Attributes, benefits and drawbacks of renewables

The attributes of renewable energy technologies (e.g., modularity, flexibility, low operating costs) differ considerably from those for traditional, fossil fuel-based energy technologies (e.g., large capital investments, long implementation lead times, operating cost uncertainties regarding future fuel costs). Renewable energy technologies can provide cost-effective and environmentally beneficial alternatives to conventional energy systems. Some of the benefits that make energy conversion systems based on renewable energy attractive follow:

- They are relatively independent of the cost of oil and other fossil fuels, which are projected to rise significantly over time. Thus, cost estimates can be made reliably for renewable energy systems and they can help reduce the depletion of the world's non-renewable energy resources.
- Implementation is relatively straightforward.
- They normally do not cause excessive environmental degradation and so can help resolve environmental problems. Widespread use of renewable energy systems would certainly reduce pollution levels.
- They are often advantageous in developing countries. The market demand for renewable energy technologies in developing nations will likely grow as they seek better standards of living.

Renewable energy resources have some characteristics that lead to problematic but often solvable technical and economic challenges:

- generally diffuse,
- not fully accessible,
- sometimes intermittent,
- regionally variable.

The overall benefits of renewable energy technologies are often not well understood, leading to such technologies often being assessed as less cost effective than traditional technologies. For renewable energy technologies to be assessed comprehensively, all of their benefits must be considered. For example, many renewable energy technologies can provide, with short lead times, small incremental capacity additions to existing energy systems. Such power generation units usually provide more flexibility in incremental supply than large devices like nuclear power stations.

The role of renewables in sustainable development

Renewable energy has an important role to play in meeting future energy needs in both rural and urban areas (Hui, 1997). The development and utilization of renewable energy should be given a high priority, especially in the light of increased awareness of the adverse environmental impacts of fossil-based generation. The need for sustainable energy development is increasing rapidly in the world. Widespread use of renewable energy is important for achieving sustainability in the energy sectors in both developing and industrialized countries.

Renewable energy resources and technologies are a key component of sustainable development for three main reasons:

1. They generally cause less environmental impact than other energy sources. The variety of renewable energy resources provides a flexible array of options for their use.
2. They cannot be depleted. If used carefully in appropriate applications, renewable energy resources can provide a reliable and sustainable supply of energy almost indefinitely. In contrast, fossil fuel and uranium resources are diminished by extraction and consumption.
3. They favor system decentralization and local solutions that are somewhat independent of the national network, thus enhancing the flexibility of the system and providing economic benefits to small isolated populations. Also, the small scale of the equipment often reduces the time required from initial design to operation, providing greater adaptability in responding to unpredictable growth and/or changes in energy demand.

Not all renewable energy resources are inherently clean in that they cause no burden on the environment in terms of waste emissions, resource extraction or other environmental disruptions. Nevertheless, the use of renewable energy resources almost certainly can provide a cleaner and more sustainable energy system than increased controls on conventional energy systems.

To seize the opportunities, countries should establish renewable energy markets and gradually develop experience with renewable technologies. The barriers and constraints to the diffusion of renewables should be removed. The legal, administrative and financing infrastructure should be established to facilitate planning and application of renewable energy projects. Government could play a useful role in promoting renewable energy technologies by initiating surveys and studies to establish their potential in both urban and rural areas. Figure 3.11 shows the major considerations for developing renewable energy technologies.

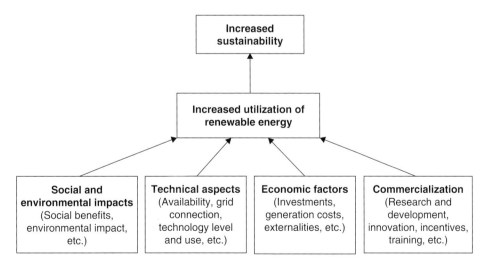

Fig. 3.11. Major considerations involved in the development of renewable energy technologies for sustainable development.

As existing energy utilities often play a key role in determining the adoption and contribution of renewable energy technologies, the utility structure and the strategy for integrating renewables should be reviewed and studied. Utility regulations should be framed to reflect the varying costs over the networks, increase competitiveness and facilitate access of independent renewable energy production. A major challenge for renewables is to get them into a reliable market at a price which is competitive with energy derived from fossil fuels, without disrupting local economies. Since the use of renewable energy often involves awareness of perceived needs and sometimes a change of lifestyle and design, it is essential to develop effective information exchange, education and training programs. Knowledge of renewable energy technologies should be strengthened by establishing education and training programs. Energy research and development and demonstration projects should be encouraged to improve information and raise public awareness. The technology transfer and development process should be institutionalized through international exchanges and networking.

To overcome obstacles in initial implementation, programs should be designed to stimulate a renewable energy market so that options can be exploited by industries as soon as they become cost effective. Financial incentives should be provided to reduce up-front investment commitments and to encourage design innovation.

Tools for environmental impact and sustainability

An energy system is normally designed to work under various conditions to meet different expectations (e.g., load, environment and social expectations). Table 3.4 lists some available environmental tools, and detailed descriptions of each tool follow (e.g., Lundin et al., 2000; Tangsubkul et al., 2002):

- *Life cycle assessment (LCA)*: LCA is an analytical tool used to assess the environmental burden of products at the various stages in a product's life cycle. In other words, LCA examines such products 'from cradle-to-grave'. The term 'product' is used in this context to mean both physical goods as well as services. LCA can be applied to help design an energy system and its subsystems to meet sustainability criteria through every stage of the life cycle. LCA, as an environmental accounting tool, is very important.
- *Environmental impact assessment (EIA)*: EIA is an environmental tool used in assessing the potential environmental impact of a proposed activity. The derived information can assist in making a decision on whether or not the proposed activity will pose any adverse environmental impacts. The EIA process assesses the level of impacts and provides recommendations to minimize such impacts on the environment.
- *Ecological footprints*: Ecological footprints analysis is an accounting tool enabling the estimation of resource consumption and waste assimilation requirements of a defined human population or economy in terms of corresponding productive land use.
- *Sustainable process index (SPI)*: SPI is a measure of the sustainability of a process producing goods. The unit of measure is m^2 of land. It is calculated from the total land area required to supply raw materials, process energy (solar derived), provide infrastructure and production facilities, and dispose of wastes.
- *Material flux analysis (MFA)*: MFA is a materials accounting tool that can be used to track the movement of elements of concern through a specified system boundary. The tool can be adapted further to perform a comparative study of alternatives for achieving environmentally sound options.
- *Risk assessment*: Risk assessment can estimate the likelihood of potential impacts and the degree of uncertainty in both the impact and the likelihood it will occur. Once management has been informed about the level of risk involved in an activity, the decision of whether such a risk is acceptable can be subsequently made.
- *Exergy analysis*: As discussed throughout this book, exergy is the quality of a flow of energy or matter that represents the useful part of the energy or matter. The conversion of energy in a process usually is driven by the consumption of energy quality. It is found that using the exergy concept to estimate the consumption of physical resources can improve the quality of the data necessary for LCA.

Table 3.4. Selected methods and tools for environmental assessment and improvement.

Risk tools	Environmental tools	Thermodynamic tools	Sustainability tools
Risk assessment	Environmental performance indicators	Exergy analysis	Life cycle assessment
	Environmental impact assessment	Material flux analysis	Sustainable process index
	Ecological footprints		Industrial ecology

Ecologically and economically conscious process engineering

Numerous efforts have been made to develop and promote ecologically and economically sustainable engineering. When applying ecologically and economically sustainable engineering, industrial and ecological systems are treated as networks of energy flows.

Ecosystems convert sunlight to natural resources, while industrial systems convert natural resources to economic goods and services. Thus, all products and services can be considered as transformed and stored forms of solar energy. An energy flow chart for a typical industrial system that includes ecological and economic inputs is shown in Fig. 3.12. Traditional economic analysis only accounts for economic inputs and outputs, since industry does not pay money to nature for its products and services. LCA focuses mainly on the waste streams, and their impact, while systems ecology ignores wastes and their impacts. Figure 3.12 can be used for assessing the economic and environmental viability of products and processes.

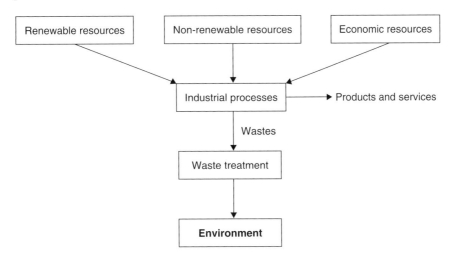

Fig. 3.12. Flow diagram for an industrial process that includes resource and economic inputs.

The thermodynamic approach to LCA and design accounts for economic and ecological inputs and services, and the impact of emissions. This approach is related to exergy. Exergy analysis is popular for improving the thermodynamic efficiency of industrial processes. However, it ignores ecological inputs and the impact of emissions. These shortcomings of exergy analysis have been overcome by combining it with life cycle impact assessment and emergy analysis. Emergy analysis is a popular approach for analyzing and modeling ecosystems. The resulting approach bridges systems ecology with systems engineering. Applications of this approach to LCA and process design are being developed.

3.3.12. Exergy as a common sustainability quantifier for process factors

Exergy has several qualities that make it suitable as a common quantifier of the sustainability of a process (Sciubba, 2001):

- Exergy is an extensive property whose value is uniquely determined by the parameters of both the system and the reference environment.
- If a flow undergoes any combination of work, heat and chemical interactions with other systems, the change in its exergy expresses not only the *quantity* of the energetic exchanges but also the *quality*.
- Provided a chemical reference state is selected that is reflective of the actual typical chemical environment on earth, the chemical portion of the exergy of a substance can be evaluated. The exergy of a substance such as a mineral ore or of a fossil fuel is known when the composition and the thermodynamic conditions of the substance and the environment at the extraction site are known. The chemical exergy of a substance is zero when it is in equilibrium with the environment, and increases as its state deviates from the environment state. For a mineral, for example, the exergy of the raw ore is either zero (if the ore is of the same composition as the environmental material) or higher if the ore is somewhat concentrated or purified.

- The *value* of a product of a process, expressed in terms of 'resource use consumption,' may be obtained by adding to the exergy of the original inputs all the contributions due to the different streams that were used in the process.
- If a process effluent stream is required to have no impact on the environment, the stream must be brought to a state of thermodynamic equilibrium with the reference state before being discharged into the environment. The minimum amount of work required to perform this task is by definition the exergy of the stream. For this reason, many suggest that the exergy of an effluent is a correct measure of its potential environmental cost.

Some researchers (e.g., Sciubba, 2001) propose that an 'invested exergy' value be attached to a process product, defined as the sum of the cumulative exergy content of the product and the 'recycling exergy' necessary to allow the process to have zero impact on the environment. They further suggest the following, for any process:

- A proper portion of the invested exergy plus the exergy of a stream under consideration can be assigned to the stream, thereby allowing the process to be 'charged' with the physical and invested exergy of its effluents.
- If a feasible formulation exists to convert the remaining 'non-energetic externalities' (labor and capital) into exergetic terms, their equivalent input in any process could be added to the exergy and invested exergy of each stream. The exergy flow equivalent to labor can perhaps be estimated by assigning a resource value to the work hour, computed as the ratio of the yearly total exergetic input in a society or region to the total number of work hours generated in the same period of time. Similarly, the exergy flow equivalent to a capital flow can perhaps be estimated by assigning a resource value to the monetary unit, computed as the ratio between the yearly total exergetic input in a society or region and the total monetary circulation (perhaps in terms of gross domestic product, or total retail sales, or a different financial measure) in the same period of time.

In summary, we consider sustainable development here to involve four key factors (see Fig. 3.13): environmental, economic, social and resource/energy. The connections in Fig. 3.13 illustrate that these factors are interrelated.

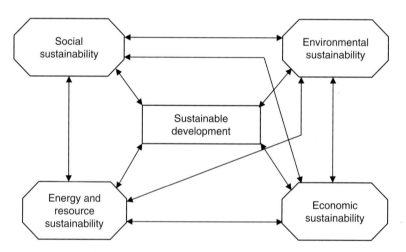

Fig. 3.13. Factors impacting sustainable development and their interdependences.

3.4. Illustrative example

The ideas discussed in this chapter are demonstrated for the process of electricity generation using a coal-fired steam power plant. The plant considered is the Nanticoke generating station, which is examined in detail in Section 11.6. Individual units of the station each have net electrical outputs of approximately 500 MW. A single unit (Fig. 3.14) consists of four main sections (Rosen and Dincer, 1997a):

1. *Steam generators*: Pulverized coal-fired natural circulation steam generators produce primary and reheat steam. Regenerative air preheaters are used and flue gas exits through chimneys.

2. *Turbine generators and transformers*: Primary steam from the steam generators passes through turbine generators, which are connected to a transformer. Steam exhausted from the high-pressure cylinder is reheated and extraction steam from several points on the turbines preheats feed water.
3. *Condensers*: Cooling water condenses the steam exhausted from the turbines.
4. *Preheaters and pumps*: The temperature and pressure of the condensate are increased.

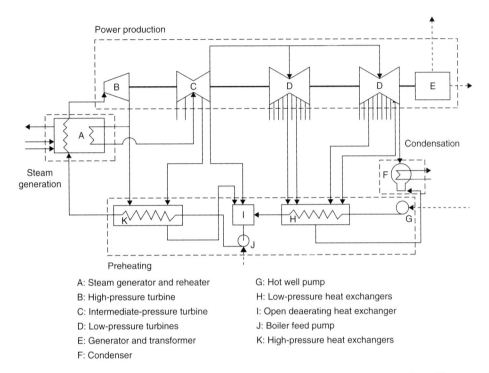

A: Steam generator and reheater
B: High-pressure turbine
C: Intermediate-pressure turbine
D: Low-pressure turbines
E: Generator and transformer
F: Condenser

G: Hot well pump
H: Low-pressure heat exchangers
I: Open deaerating heat exchanger
J: Boiler feed pump
K: High-pressure heat exchangers

Fig. 3.14. Breakdown of a unit in the coal-fired electrical generating station into four main sections. The external inputs are coal and air, and the output is stack gas and solid waste for unit A. The external outputs for unit E are electricity and waste heat. Electricity is input to units G and J, and cooling water enters and exits unit F.

3.4.1. Implications regarding exergy and energy

Energy and exergy analyses of the station have been performed (see Section 11.6). Overall balances of exergy and energy for the station are illustrated in Fig. 3.15, where the rectangle in the center of each diagram represents the station. The main findings follow:

- For the overall plant, the energy efficiency, defined as the ratio of net electrical energy output to coal energy input, is found to be about 37%, and the corresponding exergy efficiency 36%.
- In the steam generators, the energy and exergy efficiencies are evaluated, considering the increase in energy or exergy of the water as the product. The steam generators appear significantly more efficient on an energy basis (95%) than on an exergy basis (50%). Physically, this discrepancy implies that, although most of the input energy is transferred to the preheated water, the energy is degraded as it is transferred. Most of the exergy losses in the steam generators are associated with internal consumptions (mainly due to combustion and heat transfer).
- In the condensers, a large quantity of energy enters (about 775 MW for each unit), of which close to 100% is rejected, and a small quantity of exergy enters (about 54 MW for each unit), of which about 25% is rejected and 75% internally consumed.
- In other plant devices, energy losses are found to be small (about 10 MW total), and exergy losses are found to be moderately small (about 150 MW total). The exergy losses are almost completely associated with internal consumptions.

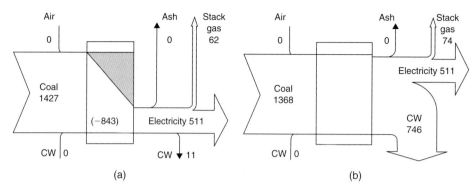

Fig. 3.15. Overall energy and exergy balances for the coal-fired electrical generating station. The rectangle in the center of each diagram represents the station. Widths of flow lines are proportional to the relative magnitudes of the represented quantities. CW denotes cooling water. (a) Exergy balance showing flow rates (positive values) and consumption rate (negative value, denoted by hatched region) of exergy (in MW); (b) Energy balance showing flow rates of energy (in MW).

3.4.2. Implications regarding exergy and the environment

In this example of a conventional coal-fired electrical generating station, each of the relationships between exergy and environmental impact described in Section 3.2.4 is demonstrated:

- Waste exergy is emitted from the plant with waste stack gas, solid combustor wastes, and the waste heat released to the atmosphere and the lake from which condenser cooling water is obtained. The exergy of these emissions represents a potential to impact on the environment. Societal concern already exists regarding emissions of harmful chemical constituents in stack gases and thermal pollution in local water bodies of water, but the exergy-based insights into environmental-impact potential of these phenomena are not yet well understood or recognized.
- Coal, a finite resource, is degraded as it drives the electricity generation process. Although a degree of resource degradation cannot be avoided for any real process, increased exergy efficiency can reduce the amount of degradation, for the same services or products. In the extreme, if the process in the example were made thermodynamically ideal by increasing the exergy efficiency from 37% to 100%, coal use and the related emissions would each decrease by over 60%.
- Order destruction occurs during the exergy consuming conversion of coal to less ordered stack gases and solid wastes, and chaos creation occurs as wastes are emitted to the environment, allowing the products of combustion to move and interact without constraints throughout the environment.

3.4.3. Implications regarding exergy and sustainable development

The exergy-related implications discussed in this section assist in achieving sustainable development by providing insights into efficiency improvement and environmental-impact reduction. These insights, combined with economics and other factors, can assist in improving the sustainability of (i) the electricity generation process considered and (ii) the broader provision of electricity and electrical-related services in regions.

3.5. Closing remarks

This chapter discusses the relations between exergy and energy, environmental impact and sustainable development. Three main relations between exergy and environmental impact are extensively discussed in terms of order destruction and chaos creation, resource degradation and waste exergy emissions. The potential usefulness of exergy analysis in addressing and solving energy-related sustainable development and environmental problems is shown to be substantial. In addition, thermodynamic principles, particularly the concepts encompassing exergy, are shown to have a significant role to play in evaluating energy and environmental technologies. Some key points, which will likely be useful to scientists and engineers as well as decision and policy makers, can be drawn from this chapter:

- Moving toward sustainable development requires that environmental problems be resolved. These problems cover a range of pollutants, hazards and types of ecosystem degradation, and extend over various geographic areas. Some of the most significant environmental problems are acid precipitation, stratospheric ozone depletion and global climate change, with the latter being potentially the most significant.
- Sustainable development requires a sustainable supply of energy resources that, in the long term, is sustainably available at reasonable cost and can be utilized for all required tasks without causing negative societal impacts. Energy sources such as sunlight, wind and falling water are generally considered renewable and therefore sustainable over the relatively longer term.
- Assessments of the sustainability of processes and systems, and efforts to improve sustainability, should be based in part on thermodynamic principles, and especially the insights revealed through exergy analysis.
- For societies to attain or try to attain sustainable development, effort should be devoted to developing renewable energy resources and technologies. Renewable energy technologies can provide environmentally responsible and sustainable alternatives to conventional energy systems, as well as more flexibility and decentralization.
- To realize the energy, exergy, economic and environmental benefits of renewable energy sources, an integrated set of activities should be conducted including research and development, technology assessment, standards development and technology transfer. These can be aimed at improving efficiency, facilitating the substitution of renewable energy and other environmentally benign energy currencies for more harmful ones, and improving the performance characteristics of renewable energy technologies.

Through the example in Section 3.4, the authors have attempted to provide some practical illustrations of the more abstract concepts discussed in this chapter, particularly by highlighting the importance of understanding and considering the relations of exergy to energy and environmental impact, especially for sustainable development challenges and problems. Such an enhanced understanding of sustainability and environmental issues relating to energy resource use is needed both to allow the problems to be better addressed and to help develop solutions that are beneficial for the economy and society.

Problems

3.1   What is the relationship between energy efficiency, environment and sustainability?

3.2   What is the relationship between the use of renewable energy sources, environment and sustainability?

3.3   What is the relationship between exergy, environment and sustainable development?

3.4   Name two policies that promote the better use of energy so as to support sustainable development and explain how these policies can relate to exergy.

3.5   Name several major environmental concerns and explain how the exergy concept can help reduce or mitigate them.

3.6   Is there any way of eliminating emissions of gases that cause global warming while using fossil fuels?

3.7   Describe the concept of 'greenhouse gas emissions trading' and discuss its implications.

3.8   Identify ten potential solutions to environmental problems. Identify and describe three of the most effective solutions.

3.9   List the barriers to the development of renewable and advanced energy technologies. Discuss how these barriers can be overcome.

3.10 Is it possible to achieve sustainable development as defined by The Brundtland Commission? Discuss if this definition is realistic.

3.11 Does a process with a high exergy efficiency necessarily cause little environmental impact and lead to a high level of sustainability? Does a process with a low exergy efficiency necessarily cause much environmental impact and lead to a low level of sustainability? Discuss with examples.

3.12 What is industrial ecology and how is it, or can it be, related to exergy and sustainable development?

3.13 What is life cycle assessment and how is it, or can it be, related to environment and sustainable development?

3.14 What is exergetic life cycle assessment and how is it, or can it be, related to environment and sustainable development?

3.15 Identify a fossil-fuel-fired electrical power station in your area and conduct a general exergy analysis using actual plant data following the illustrative example given in Section 3.4. Discuss the results.

3.16 Identify a fossil-fuel-fired electrical power station in your area and determine the amount of emissions it generated in the last year. Determine the number of cars that emit the same amount of emissions annually.

Chapter 4

APPLICATIONS OF EXERGY IN INDUSTRY

4.1. Introduction

Many researchers and practicing engineers refer to exergy methods as powerful tools for analyzing, assessing, designing, improving and optimizing systems and processes. It is not surprising, therefore, that exergy methods are used in some industries.

Others have also noticed an increase in exergy use by industry. For instance, Tadeusz Kotas wrote in the preface to his 1995 book on exergy (Kotas, 1995), 'ever since the . . . early 70s, there has been a steady growth in the interest in exergy analysis . . . This increase manifests itself in . . . the more widespread use of exergy analysis in industry.' Also, Bejan (2001) recently wrote, 'As the new century begins, we are witnessing revolutionary changes in the way thermodynamics is . . . practiced. The methods of exergy analysis . . . are the most visible and established forms of this change.'

A few examples can help illustrate this point. Some electrical generation companies utilize exergy methods to design better stations, and to improve efficiency and avoid performance deterioration in existing stations. Also, some cogeneration (or combined heat and power) facilities use exergy methods both to improve efficiency and to resolve economic costing and pricing issues.

Given that exergy analysis is first and foremost a technical tool for guiding efficiency-improvement efforts in engineering and related fields, one would expect industry to wholeheartedly embrace the use of exergy. However, exergy analysis is used only sparingly by many industries, and not at all by others. Clearly, therefore, exergy is not generally and widely accepted by industry at present.

There have been several initiatives to improve this situation. One of the key successes has been the formation in 2006 of a new group entitled 'Exergy Analysis for Sustainable Buildings,' within the American Society for Heating, Refrigerating and Air-Conditioning Engineers (ASHRAE). That group has the following mission and strategic goals:

Mission:
Exergy Analysis for Sustainable Buildings is concerned with all exergy aspects of energy and power utilization of systems and equipment for comfort and service, assessment of their impact on the environment, and development of analysis techniques, methodologies and solutions for environmentally safer, sustainable low-exergy buildings.

Strategic goals:

- Make 'Exergy analysis' a primary tool for design, analysis and performance improvement of building HVAC systems for better environment and sustainability.
- Develop simple to understand, easy to apply yet very effective and comprehensive analysis packages.
- Develop exergy as a common eco-engineering metric.
- Develop new exergy policies and strategies to complement energy policies and strategies for better global sustainability and future.
- Develop robust, seamless and easy to understand definitions and simplified equations, charts, tables, etc. for green buildings that are easy to understand by every discipline involved (architects, builders, decision makers, etc.).
- Develop and maintain products and services to meet the needs of ASHRAE members and the engineering community at large; develop guides, standards, handbook chapters, organize professional development courses (PDC), e-learning course material, maintain a very strong web site, organize symposia, forums, seminars, publish technical bulletins, co-operate with other organizations.
- Create a culture of exergetic innovation, resilience and flexibility within ASHRAE that recognizes and responds to technological and ecological needs of HVAC and building industry.

- Develop exergy-related design and evaluation parameters, algorithms to be used in various certification and evaluation codes like LEED. Develop equipment rating system similar to EER (e.g., exergy efficiency (EE), exergetic improvement ratio (EIR)).
- Close co-ordination and co-operation with other ASHRAE Technical Committees, groups and other national and international associations: identify institutions within and outside ASHRAE and select the group liaisons to these institutions.
- Develop a common and inter-disciplinary exergy definitions and nomenclature library.
- Identify different exergy analysis parameters like embodied, operating and other parameters in exergy analysis assigned to this group, like the energy/exergy required to produce and assemble the materials the building is made of (upstream) and exergy destruction (downstream) and environmental impact.
- Optimize the use of energy and exergy analysis for the next generation designs and optimize volunteer's time.

This activity provides a strong indication that industries and engineers are now more keen than ever to improve their performance by utilizing exergy methods.

4.2. Questions surrounding industry's use of exergy

Industry's relatively limited use of exergy methods at present leads to several pertinent questions:

- Why are exergy methods not more widely used by industry?
- What can be done to increase industry's use – or even acceptance – of exergy?
- Is industry's minimal use of exergy appropriate?
- Should steps be taken to make exergy methods more widely used by industry?

In addition, questions arise due to the observation that the use of exergy methods appears to vary geographically. For instance, more companies in Europe than in North America seem to utilize exergy methods to enhance and maintain plant performance. Perhaps this situation is attributable to the fact that European companies often take a longer-term view of efficiency measures than their counterparts in North America, and so exergy methods are more naturally considered in Europe. But perhaps there are other reasons that merit investigation.

To answer better these questions, it is useful to examine some of the advantages and disadvantages of exergy methods.

4.3. Advantages and benefits of using exergy

The benefits of exergy analysis are numerous, especially compared to energy analysis. Usually the benefits are clearly identifiable and sometimes they are remarkable. Some of the more significant ones follow below:

- Efficiencies based on exergy, unlike those based on energy, are always measures of the approach to true ideality, and therefore provide more meaningful information when assessing the performance of energy systems. Also, exergy losses clearly identify the locations, causes and sources of deviations from ideality in a system.
- In complex systems with multiple products (e.g., cogeneration and trigeneration plants), exergy methods can help evaluate the thermodynamic values of the product energy forms, even though they normally exhibit radically different characteristics.
- Exergy-based methods have evolved that can help in design-related activities. For example, some methods (e.g., exergoeconomics and thermoeconomics) can be used to improve economic evaluations. Other methods (e.g., environomics) can assist in environmental assessments.
- Exergy can improve understanding of terms like energy conservation and energy crisis, facilitating better responses to problems.

In addition, exergy methods can help in optimization activities. Berg (1980) noted this advantage when examining the different degrees of use of exergy in industry. He wrote, 'In some industries, particularly electric utilities, use of second law analysis in various forms has been a long standing practice in design. In other industries, the more direct techniques of second law analysis were not widely used; other less direct and less exacting techniques were used instead. Even though the approach to optimization in the latter cases was slower, and ultimately less perfect, the approach was nevertheless made.' Consequently, many design applications of exergy analysis have occurred that aim to evaluate, compare, improve or optimize energy systems.

Some of the advantages of exergy methods are described in the following subsections.

4.3.1. Understanding thermodynamic efficiencies and losses through exergy

Decisions regarding resource utilization and technical design have traditionally been based on conventional parameters like performance, economics and health and safety. In recent decades, new concerns like environmental damage and scarcity of resources have increased the considerations in decision making. But always, efficiencies and losses have been important. We hear references to energy efficiencies and energy losses all the time, whether we are dealing with companies, government, the public or others. People have developed a sense of comfort when dealing with such terms as energy efficiencies and losses, perhaps through repetitious use and exposure.

Yet numerous problems are associated with the meaning of energy efficiencies and losses. For instance, efficiencies based on energy can often be non-intuitive or even misleading, in part because energy efficiencies do not necessarily provide a measure of how nearly a process approaches ideality. Also, losses of energy can be large in quantity, when they are in fact not that significant thermodynamically due to the low quality or usefulness of the energy that is lost.

Exergy efficiencies do provide measures of approach to ideality, and exergy losses do provide measures of the deviation from ideality.

This situation, where confusion and lack of clarity exist regarding measures as important as efficiency and loss, is problematic. In general, such clarity can be achieved through the use of efficiency and loss measures that are based on exergy. Consequently, several actions are needed:

- Be clear about what is meant when we discuss thermodynamic efficiencies and losses.
- Ensure that the measures we use for efficiencies and losses are meaningful.
- Utilize efficiency and loss measures based on exergy as much as possible.
- Where energy-based measures are used, indicate clearly the meaning and proper interpretation of the values, as well as any limitations associated with them.

To further explain, it is instructive to consider the basic question of what is intended when we use terms like efficiency and loss.

4.3.2. Efficiency

To understand what we mean – or intend to mean – when we cite an efficiency, it is helpful to consider definitions. Efficiency is defined in one dictionary as 'the ability to produce a desired effect without waste of, or with minimum use of energy, time, resources, etc.' Efficiency is used by people to mean the effectiveness with which something is used to produce something else, or the degree to which the ideal is approached in performing a task.

For engineering systems, non-dimensional ratios of quantities are typically used to determine efficiencies. For engineering systems whose primary purpose is the transformation of energy, ratios of energy are conventionally used to determine efficiencies. A process has maximum efficiency according to such energy-based measures if energy input equals product energy output (i.e., if no 'energy losses' occur). However, efficiencies determined using energy are misleading because in general they are not measures of 'an approach to an ideal.'

To determine more meaningful efficiencies, a quantity is required for which ratios can be established which do provide a measure of an approach to an ideal. The second law of thermodynamics must be involved in obtaining a measure of an approach to an ideal. This law states that maximum efficiency is attained (i.e., ideality is achieved) for a reversible process. However, the second law must be quantified before efficiencies can be defined. Some approaches follow:

- The 'increase of entropy principle' quantifies the second law, stating that the entropy creation due to irreversibilities is zero for ideal processes and positive for real ones. From the viewpoint of entropy, maximum efficiency is attained for a process in which the entropy creation due to irreversibilities is zero. The magnitude of the entropy creation due to irreversibilities is a measure of the non-ideality or irreversibility of a process. In general, however, ratios of entropy do not provide a measure of an approach to an ideal.
- A quantity which has been discussed in the context of meaningful measures of efficiency is negentropy (e.g., see Chapter 21 of the report (Hafele, 1981) for a major study carried out by the International Institute of Applied Systems Analysis). Negentropy is defined such that the negentropy consumption due to irreversibilities equals the entropy creation due to irreversibilities. From the viewpoint of negentropy, maximum efficiency is attained for a process in which negentropy is conserved. Negentropy is consumed for non-ideal processes. Furthermore, negentropy is a measure of order. Consumptions of negentropy are therefore equivalent to degradations of order. Since the abstract property of order is what is valued and useful, it is logical to attempt to use negentropy in

developing efficiencies. However, general efficiencies cannot be determined based on negentropy because its absolute magnitude is not defined.

- Order and negentropy can be further quantified through the ability to perform work. Then, maximum efficiency is attainable only if, at the completion of the process, the sum of all energy involved has an ability to do work equal to the sum before the process occurred. Such measures are based on both the first and second laws.
- Exergy is defined as the maximum work that can be produced by a stream or system in a specified environment. Exergy is a quantitative measure of the 'quality' or 'usefulness' of an amount of energy. From the viewpoint of exergy, maximum efficiency is attained for a process in which exergy is conserved. Efficiencies determined using ratios of exergy do provide a measure of an approach to an ideal. Also, exergy efficiencies quantify the potential for improvement.

Other researchers have also indicated support for the use of exergy efficiencies. For example, Gaggioli and Petit (1977) refer to exergy efficiencies as 'real' or 'true' efficiencies, while referring to energy efficiencies as 'approximations to real' efficiencies. Exergy efficiencies are often more intuitively rational than energy efficiencies, because efficiencies between 0% and 100% are always obtained. Measures of merit that can be greater than 100%, such as coefficient of performance (COP), normally are between 0% and 100% when exergy is considered.

Of course, other exergy-based measures of efficiency than the one described earlier can be defined. The different definitions simply answer different questions. For instance, Kotas (1995) prefers the use of the rational efficiency, citing features that render it 'particularly suitable as a criterion of the degree of thermodynamic perfection of a process.' The key point: efficiencies based on exergy are normally meaningful and useful.

Sometimes confusion can arise when using exergy efficiencies, in part because several exist. For instance, different values can be obtained when evaluating a turbine using different exergy efficiencies (and using entropy-related efficiencies like isentropic efficiency, which are indirectly related to exergy efficiencies). These must be well understood before they are used. What is important though is that, unlike energy efficiencies, exergy-based efficiencies are reasonable in that they provide measures of approach to ideality.

4.3.3. Loss

Losses occur when the efficiency of a device or process deviates from the efficiency that would occur if the device or process were ideal. The value of a loss is a measure of this deviation from ideality.

Energy losses are not necessarily indicative of a deviation from ideality. For instance, some processes lose heat to the surroundings, but if this heat is emitted at the temperature of the surroundings the loss does not lead to an irreversibility. Conversely, some processes have no energy losses, such as the combustion of fuel in air in an isolated vessel, yet the process is highly irreversible and therefore non-ideal.

Exergy losses, on the other hand, do provide quantitative measures of deviations from ideality. In addition, exergy losses allow the location, type and cause of a loss, or inefficiency, to be clearly identified. This information is critical for efforts to increase exergy efficiency. As stated by Moran and Shapiro (2005), 'exergy analysis is particularly suited for furthering the goal of more efficient energy use, since it enables the locations, types, and true magnitudes of waste and loss to be determined.'

An additional insight obtained through exergy losses relates to the fact that they can be divided into two types: the losses associated with waste exergy effluents and the losses associated with internal irreversibilities in a system or process (i.e., exergy consumptions).

4.3.4. Examples

Many examples can be used to illustrate how use of exergy clarifies measures of thermodynamic efficiency and loss. Only a few are presented here. However, more clear illustrations are needed, both to educate and convince people of the benefits of exergy methods.

Consider a Carnot heat engine operating between a heat source at a temperature of 600 K and a heat sink at 300 K. The energy efficiency of this device is 50% (i.e., $1 - 300/600 = 0.5$). Yet a Carnot engine is ideal. Clearly, the energy efficiency is misleading as it indicates that a significant margin for improvement exists when in fact there is none. The exergy efficiency of this device is 100%, properly indicating its ideal nature in a straightforward and clear manner.

Consider next an electrical resistance space heater. Almost all of the electricity that enters the unit is dissipated to heat within the space. Thus the energy efficiency is nearly 100% and there are almost no energy losses. Yet the exergy efficiency of such a device is typically less than 10%, indicating that the same space heating can in theory be achieved using one-tenth of the electricity. In reality, some of these maximum savings in electricity use can be attained using a heat pump. The use of even a relatively inefficient heat pump can reduce the electricity used to achieve the same space heating by two thirds. Clearly the use of energy efficiencies and losses is quite misleading for electrical heating.

Finally, consider a buried thermal energy storage tank. A hot medium flows through a heat exchanger within the storage and heat is transferred into the storage. After a period of time, a cold fluid is run through the heat exchanger and heat is transferred from the storage into the cold fluid. The amount of heat thus recovered depends on how much heat has escaped from the storage into the surrounding soil, and how long the recovery fluid is passed through the heat exchanger. But a problem arises in evaluating the energy efficiency of this storage because the energy efficiency can be increased simply by lengthening the time that the recovery fluid is circulated. What is neglected here is the fact that the temperature at which the heat is recovered is continually decreasing toward the ambient soil temperature as the fluid circulates. Thus although the energy recovered increases as the recovery fluid continues to circulate, the exergy recovered hardly increases at all after a certain time, reflecting the fact that recovering heat at near-environmental temperatures does not make a storage more efficient thermodynamically.

4.3.5. Discussion

The points raised here are practical, since efforts to improve efficiency are guided by what we perceive to be efficiencies and losses. If that is wrong, then all efforts may be in vain. When we allow ourselves to be guided by energy efficiencies and losses, in particular, we may be striving for the wrong goal or even be trying to achieve the unachievable.

The ramifications of such errors can vary. They can be relatively small when a company engineer wastes time trying to improve the efficiency of an already nearly ideal device. But the ramifications can also be very large in many situations, such as when a company or government invests millions of dollars on research and development to improve efficiencies of technologies that are – perhaps – not as in need of improvement as others that deviate excessively from ideality. Consequently, exergy-based efficiencies are required to address energy problems effectively and to prioritize efficiency-improvement efforts appropriately.

4.4. Understanding energy conservation through exergy

Energy conservation, although widely used, is an odd term. It is prone to be confusing and is often misleading. Energy conservation is nothing more than a statement of the principle of conservation of energy, which is embodied in the first law of thermodynamics. Yet the term energy conservation normally means something much different when it is used by lay people – as well as many technical people.

Exergy can help us understand this dual set of views about energy conservation in a rational and meaningful way. Further, exergy can help clarify this confusion by preserving the appropriate use of the term energy conservation as a statement of a scientific principle, while giving proper understanding to the meaning implied by most people when they discuss energy conservation. In fact, it can be argued that the meaning in the latter case is better expressed through the term *exergy conservation*.

4.4.1. What do we mean by energy conservation?

To non-thermodynamicists, many meanings are expressed by the term energy conservation, meanings usually related to solving problems regarding energy resources or technologies. For example, energy conservation can mean:

- Increasing efficiencies of devices and processes so they use less energy resources to provide the same levels of services or products, thereby preserving the energy resources. Increasing efficiency can be accomplished either by incremental improvements to existing devices or processes, or by major design alterations.
- Reducing energy requirements by reconsidering what the energy is being used for, in hopes of finding ways to satisfy the overall objective(s) while using less energy resources. In the electrical sector of an economy, this concept involves reducing electrical energy demands of users and is sometimes referred to as 'demand side management.'
- Changing lifestyles so that we need and use less energy resources (e.g., substituting use of more mass transit and bicycles for automobile use). In the extreme, some suggest we 'return to the past' and radically curtail our use of energy resources by retreating from the highly energy-intensive lives adopted over the last few centuries. These ideas are usually equated to accepting lower standards of living.

- Substituting alternate energy resources and forms for ones we deem precious and wish to preserve. This interpretation of energy conservation may, involve switching heating systems from natural gas to a renewable energy resource like solar energy.

As noted earlier, those familiar with thermodynamics regard the term energy conservation simply as a statement of a scientific principle or law. So how do we reconcile these two radically different interpretations and understandings of energy conservation? How can energy conservation be the essence of a scientific principle or law, while it simultaneously reflects a wide range of objectives for solving energy-related problems?

4.4.2. Exergy conservation

Exergy is the key to providing simple, meaningful and practical answers to the above questions. Exergy is based on the first and second laws of thermodynamics. It is the second law that defines an ideal or perfect process or device as one that is reversible. This idea can be clearly grasped because energy is conserved in any system – ideal or otherwise – while exergy is conserved for only an ideal or perfect process or device. Exergy is not conserved for real processes or devices. de Nevers and Seader (1980) put it another way when they elegantly write: 'Energy is conserved . . . in all of our most wasteful uses of fuels and electricity.' Thus, if one aims for thermodynamic perfection, exergy conservation is a logical and meaningful target that is fully consistent with the objective. Energy conservation is not and, in fact, is utterly meaningless in this regard. We of course can never in reality achieve the ideality associated with exergy conservation, but knowing of its hypothetical existence certainly provides a clear upper limit for conservation efforts.

These ideas are consistent with statements of Tsatsaronis and Valero (1989) who wrote '. . . energy analysis generally fails to identify waste or the effective use of fuels and resources. For instance, the first law does not recognize any waste in an adiabatic throttling process – one of the worst processes from the thermodynamic viewpoint.' They go on to state 'exergy analysis . . . calculates the useful energy associated with a thermodynamic systems . . . [and] identifies and evaluates the inefficiencies of an exergy system.'

Of course, we never aim for thermodynamic perfection in the real world. Too many other factors come into play, like economics, convenience, reliability, safety, etc. Thus decision making about how far we take efforts to shift the actual level of performance nearer to the ideal, i.e., to conserve exergy, involves complex trade-offs among competing factors. What is critical is that, although other factors temper conservation goals, it is exergy – or commodities and resources that have high exergy contents – we seek to preserve when we speak of energy conservation. Exergy is what we value because it, not energy, consistently represents the potential to drive processes and devices that deliver services or products. In fact, it seems that *exergy conservation* is what lay people mean when they say energy conservation.

This is important because we need to be clear about what we say and mean. If we confuse ourselves by using energy conservation not just to describe a basic scientific conservation principle, but also to describe efforts to solve energy-related problems, we cannot effectively address those problems.

By accepting that it is exergy conservation to which we aspire, in concert with other objectives, we can effectively address important energy-related problems in society like security of supplies of useful energy resources, and resolving shortages of useful energy resources. In addition, this understanding of exergy conservation provides us with the underpinning needed to develop useful and meaningful measures of efficiency.

4.4.3. Examples

Others have in the past also noted the misleading aspects of energy conservation, while recognizing the need to focus on exergy instead. Some interesting examples follow below:

- After the first 'energy crisis' in the early 1970s, Keenan et al. (1973) wrote '. . . energy, rather than being consumed in any process, is always conserved. When opportunities for fuel conservation are to be assessed, it becomes necessary to use a measure other than energy.' They went on to discuss what is now commonly called exergy as the preferred measure.
- Around the same time, Berg (1974), then an engineer with the U.S. Federal Power Commission, wrote 'National efforts to conserve energy resources could be much enhanced by the adoption of [exergy] to measure the effectiveness of energy utilization.' He also noted that 'the first law of thermodynamics guarantees that energy can be neither created nor destroyed; thus it would hardly seem necessary to have a national policy addressed to its conservation.'
- Gaggioli (1983) closed the preface in the book of contributed chapters on exergy, by stating: 'Exergy analyses not only avoid many misconceptions resulting from energy analyses but also point out the way to economic energy

conservation.' Although he clearly used the term energy conservation with the lay meaning (probably to make the statement clearer to lay readers), Gaggioli reinforced the need to consider exergy in conservation efforts and activities.

- Recently, Scott (2000), a proponent of the use of exergy methods, wrote a particularly articulate article on the confusing nature of the term energy conservation, highlighting dangers associated with its use.

Clarity of thought and soundness of understanding are needed to address issues successfully. This idea applies to energy conservation as well as to other issues. The lack of clarity and understanding that often seems to be present when energy conservation is discussed ought to be resolved. Exergy provides the means to this resolution, which is needed if energy-related issues are to be addressed effectively.

4.5. Disadvantages and drawbacks of using exergy

Those in industry who choose not to utilize exergy often do so for several reasons, some of which are described below:

- Exergy methods are considered too cumbersome or complex by some users. For example, the need to choose a reference environment in exergy analysis is considered by some to render the technique too challenging.
- The results of exergy analyses are regarded by some as difficult to interpret, understand and utilize.
- Many potential users are simply unfamiliar with exergy, being educated about energy and therefore more comfortable with it.
- Perhaps most importantly, some practicing engineers have simply not found exergy methods to lead to tangible, direct results.

It is important to consider these reasons carefully, as they may provide insights regarding what actions are needed to improve the situation.

Certainly, there are varied opinions on the subject of industry's use of exergy. Many views related to exergy and industry were expressed at a 1996 panel session on The Second Law in Engineering Education, at the American Society of Mechanical Engineers' International Mechanical Engineering Congress and Exposition (El-Sayed et al., 1997). The panelists were from academia and industry. Some panelists explained the need for education in exergy methods for the benefit of industry, while some felt otherwise, questioning even the level to which the second law of thermodynamics should be taught. This point is relevant because industry feels the impacts of thermodynamics education.

4.6. Possible measures to increase applications of exergy in industry

Exergy methods are useful and can be extremely beneficial to industry and others. The concerns about exergy are in reality barriers that can – and in fact should – be overcome to increase industry's adoption of exergy. The use of exergy can benefit not only industry, but also society (e.g., through a cleaner environment). These benefits are too great to be passed by.

Although industry's grounds for often not using exergy are not well founded, it is true that perception is often reality. So, if industry is to adopt – or be convinced to adopt – exergy methods on a more widespread basis, several actions by exergy proponents are needed.

Some examples of actions that are necessary or would be beneficial follow. This list is by no means intended to be exhaustive, as many other suggestions can be made:

- Practitioners must be educated about exergy methods and their applications, through college and university programs, continuing-education courses and on-the-job training.
- Concerted efforts must be made to point out clearly and unambiguously to industry the benefits of using exergy methods.
- These efforts should be supplemented by case studies and 'demonstration projects' where exergy has been applied beneficially, and promotion activities. In particular, we need clear and understandable success stories about exergy applications.

The last point above is certainly not new. When asked how to increase industry's use of exergy analysis, in the mid-1980s, James Funk in 1985, a thermodynamicist with much experience in exergy who has focused on hydrogen production methods, replied that grand demonstrations of exergy analysis are required to increase awareness of it.

The launching of *International Journal of Exergy*, which evolved from *Exergy – An International Journal*, will almost certainly help increase industry's acceptance and use of exergy. By providing a focal point for reports of exergy research and applications, this journal provides an excellent conduit through which advances in exergy methods and their uses can be clearly conveyed to potential users in industry. Of course, it is also important that articles on new applications of exergy appear in journals focused on applications (e.g., journals on energy technologies and resources, like nuclear, solar and hydrogen energy). But the focused outlet provided by the *International Journal of Exergy* is essential, especially for research on the intricacies of exergy methods. This need became most clear to me on learning that some articles on exergy methods were being rejected by applications-oriented journals, almost solely because the articles were deemed outside the scope of the journals.

4.7. Closing remarks

Through measures such as those outlined in here and others, the broad potential of exergy can come to be fully realized by industry in the future. For the direct benefit of industry in particular, and society in general, it is critical that the potential benefits of exergy be exploited.

Problems

4.1   Identify the reasons for the reluctance to use exergy by many in industry.

4.2   Some claim that the main reason exergy is not used to a great extent by industry is that the results of an exergy analysis are difficult to interpret. Do you agree? Explain.

4.3   Is exergy analysis more useful for power plants or for cogeneration plants? Explain.

4.4   Is exergy analysis more useful for a steam power plant or a hydroelectric power plant? Explain.

4.5   The COP of refrigeration and heat pump systems can be greater than 1. Can the exergy efficiency of such systems be greater than 1? Explain.

4.6   Provide two different expressions for the exergy efficiency of refrigeration and heat pump systems. Incorporate COP in one expression.

4.7   Give an expression for the exergy efficiency of a turbine that includes exergy loss as one of the terms.

4.8   A student calculates the exergy efficiency of an adiabatic steam turbine to be greater than 100%. Is this result reasonable? Explain.

4.9   Does a process with an exergy efficiency greater than 100% necessarily violate the second law of thermodynamics? Explain.

4.10   Does the use of exergy efficiency for evaluating a hydraulic or wind turbine provide any advantage compared to using energy efficiency?

4.11   The exergy efficiency of an electric resistance heater is typically less than 10%. Explain how the same heating can be accomplished using a reversible heat pump. Consider 5 kW of heating with an outdoor air temperature of $2°C$ and an indoor temperature of $24°C$. Draw a schematic of this heat pump unit and calculate the COP and the exergy efficiency of this heat pump.

4.12   Repeat the previous problem considering a ground source heat pump with a ground temperature of $13°C$.

4.13   We are continuously encouraged to conserve or save energy even though according to the first law of thermodynamics energy is conserved anyway. Do you favor replacing the term 'energy conservation' with 'exergy conservation' to avoid this confusion?

Chapter 5

EXERGY IN POLICY DEVELOPMENT AND EDUCATION

5.1. Introduction

It is important for the public to have a basic understanding, appreciation and awareness of many technical issues. Such understanding and awareness fosters healthy public debate about problems and possible solutions, often helps guide how public funds are spent and facilitates policy development.

Energy issues are no exception. Yet, the public's understanding of energy issues is often confused. In large part, this situation is attributable to the public having next to no understanding of exergy. It is explained why such an understanding is necessary in this chapter.

It is easier for the public to be educated about and aware of exergy if students are adequately educated about exergy in appropriate venues (university and college thermodynamics courses, primary and secondary schools, etc.).

Consequently, this chapter deals with education and awareness of exergy, first by focusing on the public and then by dwelling on the education of thermodynamicists as well as other technical people. The objective is to demonstrate that exergy has a place and role to play in policy development regarding energy-related education and awareness.

Exergy can play a key role in developing appropriate and beneficial energy-related policies relating to education and awareness. Two main areas where exergy can have an impact on policies are discussed in this chapter: public education and awareness and student education. The former is more general, but is supported by the latter. Regarding public education and awareness about exergy, it appears that the public is often confused when it discusses energy, and needs to be better educated about exergy if energy issues and problems are to be addressed appropriately. Regarding the education of students about exergy, it appears that the coverage of exergy in thermodynamics education is often insufficient and inappropriate. Better coverage of exergy is needed to improve thermodynamics education and to make it more interesting to students, and a basic level of 'exergy literacy' is needed among engineers and scientists – particularly those involved in decision making.

5.2. Exergy methods for analysis and design

Some features of exergy follow:

- When energy quality decreases, exergy is destroyed. Exergy is the 'part' of energy that is useful to society and has economic value and is thus worth managing carefully.
- A system has no exergy when it is in complete equilibrium with its environment. Then, no differences appear between the system and the environment in temperature, pressure or constituent concentrations.
- The exergy of a system increases as the deviation of its state from that of the environment increases. For instance, hot water has a higher exergy content in winter than on a hot summer day, while a block of ice contains little exergy in winter but a significant quantity in summer.

Two simple examples are used to illustrate the attributes of exergy:

- Consider an adiabatic system containing fuel and air at ambient conditions. The fuel and air react to form a mixture of hot combustion gases. During the combustion process, the energy in the system remains fixed because it is adiabatic. But the exergy content declines as combustion proceeds due to the irreversibilities associated with the conversion of the high-quality energy of fuel to the lower-quality energy of combustion gases. The different behaviors of energy and exergy during this process are illustrated qualitatively in Fig. 5.1.

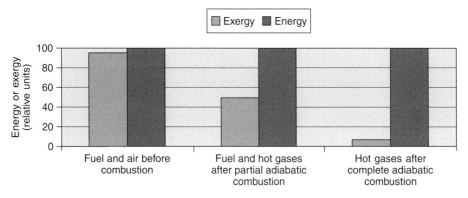

Fig. 5.1. Qualitative comparison of energy and exergy during combustion.

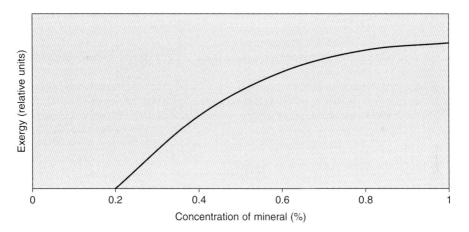

Fig. 5.2. Qualitative variation of the exergy of a mineral with concentration.

- A mineral deposit 'contrasts' with the environment of Earth, and is thus a carrier of exergy. This contrast increases with the concentration of the mineral, as shown in Fig. 5.2. When the mineral is mined, its exergy content is low or zero (depending on the concentration of the mineral in the environmental deposit) while the exergy content increases if its concentration is enriched. A poorer mineral deposit contains less exergy than a concentrated one. Conversely, when a concentrated mineral is dispersed in the environment, its exergy content decreases.

Exergy analysis is a methodology for the analysis, design and improvement of energy and other systems. The exergy method is useful for improving the efficiency of energy resource use.

Exergy has many other implications on and links with other disciplines. A link exists between exergy and environmental impact and sustainability. Energy production, transformation, transport and use impact on the Earth's environment. The exergy of a quantity of energy or a substance can be viewed as a measure of its usefulness, quality or potential to cause change. Exergy appears to be an effective measure of the potential of a substance to impact the environment. This link between exergy and environmental impact is particularly significant since energy and environment policies are likely to play an increasingly prominent role in the future in a broad range of local, regional and global environmental concerns. The tie between exergy and the environment has implications regarding environmental impact and has been investigated previously by several researchers, including the authors. This has been presented in detail in Chapter 3.

Exergy is a useful concept in economics. In macroeconomics, exergy offers a way to reduce resource depletion and environmental destruction, by such means as exergy taxes or rebates. In microeconomics, exergy has been combined beneficially with cost–benefit analysis to improve designs. By minimizing life cycle cost (LCC), we find the 'best' system given prevailing economic conditions and, by minimizing exergy losses, we also minimize environmental effects.

Finally, exergy has been proposed as an important consideration in policy making related to energy. The present work extends this work, by focusing specifically on education and awareness.

5.3. The role and place for exergy in energy-related education and awareness policies

Before considering understanding and awareness by the public of exergy, and its role and place in energy-related education and awareness policies, it is informative to consider the public's understanding and awareness of the more conventional quantity energy.

5.3.1. Public understanding and awareness of energy

The typical lay person hears of energy and energy issues daily, and is generally comfortable with receiving that energy-related information and feels he/she follows it. He/she even understands it, or at least thinks he/she does.

This sense of comfort and understanding exists despite all of the problems associated with energy. For instance, consider the following:

- Efficiencies based on energy can often be non-meaningful or even misleading, because energy efficiency is not a consistent measure of how nearly a process approaches ideality. For instance, the energy efficiency of electric space heating is high (nearly 100%) even though the process is far from ideal. The fact that the same space heat can be delivered by an electric heat pump using much less electricity than the electric space heater corroborates this observation.
- Losses of energy can be large in quantity, when they are in fact not that significant thermodynamically due to the low quality or usefulness of the energy that is lost. For example, the waste heat exiting a power plant via cooling water has a lot of energy, but little exergy (because its state is near to that of the environment).

5.3.2. Public understanding and awareness of exergy

An understanding of exergy, similar to that which exists for energy, is almost entirely non-existent in lay members of the public. This lack of understanding exists despite the fact that exergy overcomes many of the deficiencies described above of energy methods.

Worse still, the public is often confused when it refers to energy. To those who deal with exergy, it often seems that members of the public actually mean exergy when they say energy. For example, two respected exergy researchers, Wepfer and Gaggioli (1980), begin an article with 'Exergy … is synonymous with what the layman calls "energy." It is exergy, not energy, that is the resource of value, and it is this commodity, that "fuels" processes, which the layman is willing to pay for.'

These points illustrate why it is essential that the public develop – or be helped to develop – a basic understanding of exergy. The level of understanding needed by the public about exergy should at least be comparable to that for energy.

To help illustrate the above contentions, some examples follow below of the problems associated with a lack of knowledge of exergy by the public:

- One example of the confusion exhibited by the public when they speak of energy is the well-used term *energy conservation*. When the public says energy conservation, it is usually referring to an objective of efforts to solve energy problems. Yet the term energy conservation is meaningless in that regard, in that it simply states the first law of thermodynamics. Exergy, however, is not conserved and it appears that what the public is really interested in conserving is exergy, the potential to drive processes and systems that deliver services or products.
- Another example of confusion in the public surrounds the drive for increased *energy efficiency*. Energy efficiencies do not necessarily provide a measure of how nearly a process approaches ideality, yet that is what the public means when it says energy efficiency. Exergy efficiencies do provide measures of approach to ideality, and so it appears that the public means increased exergy efficiency when it discusses increased energy efficiency.
- A third example of the problems that can develop when the public does not have a knowledge of exergy, but retains only a confused understanding of energy, relates to the *energy crisis* of the 1970s, when oil scarcities existed due to reductions in oil production. Most of the energy that was available to the public before the crisis was available during it. For instance, huge amounts of solar energy continued to stream into the Earth every day. Waste thermal energy was continually emitted from facilities and buildings. The commodity for which there was a crisis, therefore, appeared to be exergy, not energy. That is, energy forms capable of delivering a wide range of

energy services (like oil), which have high exergies, were in short supply. Of course, there were also other issues related to the energy crisis, particularly the shortage of reasonably inexpensive and widely available resources. But, the key point here is that the crisis was about exergy, not energy, yet the public referred to the situation as an *energy crisis*.

- A fourth example of public confusion about energy relates to the oft-pronounced need for *energy security*. If it were simply energy for which we desire a secure supply, there would be no real problem. We have energy in abundance available in our environment, and even when we use energy we still have equivalent quantities of energy left over because our use is really only energy conversion or transformation. However, we are not concerned about ensuring secure supplies of energy, but rather of only those resources that are useful to us, that can be used to provide a wide range of energy services, that can satisfy all our energy-related needs and desires. That is, we are concerned with having secure supplies of exergy, or what might be called *exergy security*.

The lack of clarity regarding the points raised in these four examples has been discussed in more detail previously, focusing on scientists, engineers and other technical readers. This discussion, however, is intended to raise these points in a different context, and emphasize that this lack of clarity extends to the public, where the problems caused are different, but perhaps just as or more important.

5.3.3. Extending the public's need to understand and be aware of exergy to government and the media

By extension of the above arguments, government officials require a rudimentary understanding of exergy to improve – or at least complement – their understanding of energy issues. This understanding can help guide the development of rational energy policies. Government, being a type of reflection of the public, will be far less prone to use exergy methods, even when they can be beneficial, if it feels that the public does not understand exergy even in the most simple way and therefore will not appreciate government efforts.

The importance of such government involvement should not be understated and has been investigated by researchers. For example, Wall (1993) alludes to the importance of exergy in relation to democracy in an article that dealing with the dilemmas of modern society.

Similarly, members of the media – including the press, television and radio – need to be informed, at least at a basic level, about exergy and its roles. In a sense, the media are also a reflection of the public. If the media have an appreciation of exergy, they can help ensure lay members of the public have an understanding about exergy. Educating via television, in particular, can be an especially powerful method for increasing public awareness about exergy.

However, for the press and media to run exergy-related articles, it requires that the public have a rudimentary understanding of and interest in exergy matters. Otherwise, the press and media tend to neglect exergy-related topics for fear of boring or confusing the public. A first step to resolving the reluctance of the press to write about exergy is education.

The next section of this article focuses on educating students about exergy, which is one manner of directly and indirectly educating the public, in the long run, about exergy.

5.4. The role and place for exergy in education policies

Thermodynamics education is often thought of as a mature discipline, yet it remains is the subject of continual debate. The emergence of exergy methods as important elements and tools of thermodynamics has provided additional subject matter for debate, especially regarding the role and place for exergy in curricula.

5.4.1. Education about exergy

The impact of exergy on the teaching of thermodynamics has been and continues to be significant. Developments in this area abound. For instance, Bejan (2001) noted in an editorial in the inaugural issue of *Exergy, An International Journal*, 'As the new century begins, we are witnessing revolutionary changes in the way thermodynamics is taught.' Further on, he observed that 'the methods of exergy analysis … are the most visible and established forms of this change.'

One point of contention is whether present coverage of exergy in thermodynamics education is sufficient and appropriate. Views on this issue are often not in agreement.

Exergy, where it forms part of the curriculum, is normally taught at the college and university levels. However, many feel that it should be covered in primary and/or secondary education levels. That point is also disputed. Most of the remainder of this section focuses on college and university levels, since exergy is normally taught at the post-secondary level.

In some ways and at some schools, present coverage of exergy is sufficient. Some evidence to support this claim follows:

- Several articles have appeared in the engineering and education literature on teaching exergy analysis. For instance, Cengel (1996) proposed 'a "physical" or "intuitive" approach … as an alternative to the current "formula-based" approach to learning thermodynamics' and incorporated exergy into the approach. Dunbar and Lior (1992) recommended an exergy-based approach to teaching energy systems. They noted that the approach highlights 'important conclusions from exergy analysis, not obtainable from the conventional energy analysis.' In addition, they felt that 'the approach evoked the intellectual curiosity of students and increased their interest in the course.'
- Most texts on thermodynamics have over the last two decades incorporated sections or chapters on exergy methods. Even in 1988, while commenting on the increased attention being paid to exergy analysis, Bejan (1988) pointed out that, 'every new undergraduate engineering thermodynamics textbook has at least one chapter on the subject.'
- Several excellent texts devoted to exergy analysis have been published, including some particularly useful ones, e.g., by Kotas (1995), Edgerton (1982) and Szargut et al. (1988).

Such materials have made it easier to expand the coverage of exergy in thermodynamics courses. Yet in general room exists for improvement in the area of exergy coverage in thermodynamics education, and efforts should be made to achieve these improvements. Three points related to improving thermodynamics education through better coverage of exergy are addressed in the following three subsections.

5.4.2. The need for exergy literacy in scientists and engineers

We need to ensure that our education systems provide all students who study thermodynamics with a good grounding in exergy. For exergy methods to become more widely used and beneficially exploited, those who study and work in technical fields – particularly where thermodynamics is applied – should have a basic understanding of exergy. In addition, technical managers and decision makers require at least an appreciation of what exergy is and how it is used, if they are to make proper decisions on matters where exergy is, or should be, considered.

Some may suggest that we do not, for practical purposes, need such an understanding of exergy among technical personnel. Some examples help refute such suggestions:

- Research proposals have at times not been funded, in large part because exergy methods formed part of the approach. On reading the comments of reviewers of such proposals, it is sometimes evident that the reviewers have read *energy* in place of *exergy* throughout the proposals – rendering them meaningless. This result can occur despite the fact that great pains are often taken within proposals to emphasize the need to use exergy methods. What perhaps is and remains most disconcerting is that, in many of these instances, the reviewers have technical backgrounds.
- Government officials or company managers have admitted to exergy researchers and practitioners that in many instances they chose not to use exergy on a project or activity, not because an exergy approach was unsuitable but rather because they did not understand it or had never heard of it. This situation is problematic because decision making, when it is based on avoiding topics about which one is ignorant, cannot be productive.

We consequently feel in general that a strong need exists to improve the 'exergy literacy' of engineers and scientists.

5.4.3. Understanding the second law through exergy

The second law of thermodynamics often makes students of thermodynamics fearful. Introducing the concept of entropy usually only increases their trepidation. Even students who pass courses on thermodynamics and ultimately graduate often retain fears of the second law and entropy and feel they do not really understand these topics. Ahrendts (1980), for example, begins one of his articles on exergy methods with 'Thermodynamics is not a very popular science, because the concepts in thermodynamics do not conform to the unsophisticated human experience.' Focusing on the second law, he continues 'Traditional formulations of the second law prevent a simple understanding of energy conversions, because the application of the entropy concept to those processes is often looked upon as a miracle.'

Others also have agreed with these concerns and developed different approaches to teaching the second law. One example is a thermodynamics text by Dixon (1975), the preface of which states 'entropy is [not] the most significant or useful aspect of the Second Law' and 'the Second Law has to do with the concept of degradation of energy; that is, with loss of useful work potential.' Dixon introduces the second law through the concept of degradation of energy, claiming 'degradation … because it is a work term, is an easily grasped concept.' By focusing on degradation of energy rather than the abstract property entropy, Dixon feels his book results in 'a clearer physical meaning for entropy.'

The approach described above, although instructive for some, is still somewhat abstract and vague. The approach comes close to teaching the second law through exergy, and would likely be clearer if it did so. Exergy provides a perspective of the second law and entropy that is much clearer and more intuitively understandable to students. The definition of exergy, which states that exergy is the maximum work that can be obtained from a system or flow within a reference environment, is usually much more straightforward than the definition of entropy. Also, characteristics of exergy – such as it being a non-conserved quantity – are normally easier for students to grasp than the concepts surrounding entropy and its characteristics.

Thus, we feel that exergy should in general form a central component of thermodynamics courses. More specifically, exergy should play a significant role in dealing with and teaching the second law. Such an approach would likely make the second law more understandable and practical and less fear inspiring. Some readers will point out correctly that there exist some additional complexities when dealing with exergy rather than entropy (e.g., a reference environment must be introduced and defined). Nevertheless, we believe that the overall benefits of approaching the second law through exergy outweigh the difficulties.

5.4.4. Exergy's place in a curriculum

A challenging issue is where exergy should be covered in a curriculum. In engineering programs, for example, exergy is sometimes covered lightly in thermodynamics courses at the undergraduate level. Sometimes exergy is covered separately, as either a core or an elective undergraduate course, while in some schools exergy is only covered at the graduate level. In the latter case, the rationale often provided is that students need a firm grounding in traditional thermodynamics before they deal with exergy. Those who support including exergy as a part of the undergraduate curriculum, on the other hand, claim this approach is necessary because exergy forms a critical and important part of basic thermodynamics. Further support for this argument is added by earlier statements in this article about exergy providing a preferable approach to dealing with and teaching the second law.

Although there is likely some merit in each of the rationales for different placements of exergy in curricula, we nevertheless believe that some coverage of exergy should be required in all undergraduate courses in engineering thermodynamics. Beyond such a core of exergy material, however, there are multiple ways in which additional exergy material can be incorporated into engineering curricula – at undergraduate or graduate levels.

Exergy methods can also be incorporated into courses that apply to thermodynamics. In *Thermal Design and Optimization*, for instance, Bejan et al. (1996) feature a substantial amount of material on exergy and related methods. They explain in the preface that they include exergy in the text 'because an increasing number of engineers and engineering managers worldwide agree that it has considerable merit and are advocating its use.' They state further that their aim in featuring exergy and related methods is 'to contribute to the education of the next generation of thermal system designers and to the background of currently active designers who feel the need for more effective design methods.' These authors clearly regard exergy as a critical component in thermal design education, whose importance will only increase in the future.

Much room for debate exists about the place for exergy in a curriculum. Perhaps no single answer exists, and each approach will retain its proponents. Some engineering thermodynamics curricula, to their detriment, do not cover exergy at all, but these are likely in the minority. The range of views on how exergy can and should be incorporated into a thermodynamics curriculum is sufficiently diverse that a consensus on the best approach is almost certainly not possible in the near term. It is even questionable whether a consensus can be reached in the long term, after exergy methods mature. Two points help illustrate this view:

- A panel session on The Second Law in Engineering Education was held within the Symposium on Thermodynamics and the Design, Analysis and Improvement of Energy Systems, at the 1996 International Mechanical

Engineering Congress and Exposition of the American Society of Mechanical Engineers (El-Sayed et al., 1997). Among the topics covered were whether or not undergraduate engineers need to be educated in the second law of thermodynamics and how much depth is required in this area. The panelists included representatives of academia and industry, as well as respected thermodynamicists and exergy proponents. Although several opinions were expressed on the need to include exergy methods in the teaching of the second law of thermodynamics, the views expressed varied greatly – emphasizing the difficulties in reaching a consensus on the best approach for incorporating exergy into a thermodynamics curriculum.

- Consideration of the nomenclature and terminology of exergy analysis reveals that a diversity of names and symbols presently exist for the same quantities in exergy analysis. This situation persists despite efforts to reach a common nomenclature and terminology. This weakness demonstrates how difficult it is to reach a consensus just on the relatively narrow topic of nomenclature and terminology, not to mention reaching a consensus on the broader topic of incorporating exergy into a thermodynamics curriculum.

5.5. Closing remarks

The key elements of this chapter are awareness, understanding and education as they relate to exergy and its role in policy making. The relation between these key elements is illustrated in Fig. 5.3. There, it is shown that an understanding and awareness of exergy requires, for all people, education. The types of education that are appropriate for technical persons such as engineers and scientists are shown to be different from those that are appropriate for non-technical persons such as members of the public, government or media. But, in a general sense, the factors involved in raising awareness and understanding are similar conceptually, and differ mainly in depth and rigor of treatment.

The arguments presented in this chapter demonstrate that the public is often confused when it discusses energy, and a need exists to improve public understanding and awareness of exergy. Such understanding and awareness is essential if we are to better address the energy issues and problems of today and tomorrow. Thus, exergy can play a key role in developing appropriate and beneficial energy-related policies, but exploiting the potential of exergy requires appropriate support for public education and awareness about exergy.

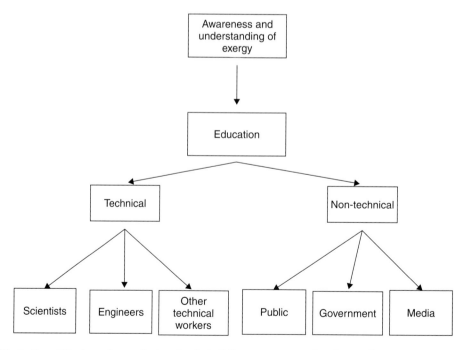

Fig. 5.3. Illustration of the importance of education in building awareness and understanding of exergy among different categories of people.

In support of the need for public understanding and awareness of exergy, exergy should take on a prominent place in thermodynamics courses. Beyond elucidating the concepts of the second law and entropy, such an approach can help ensure a rudimentary understanding of exergy in all technical personnel. An approach based on exergy could make the second law more interesting, appealing and practical and less daunting and confusing. Then, it may be easier to improve general understanding of exergy in the scientific and engineering communities, as well as the general public, by ensuring that a basic level of 'exergy literacy' exists among engineers and scientists – particularly those involved in decision making. Education policies that support inclusion of exergy in relevant curricula, at all appropriate education levels, should be considered.

Problems

5.1 Conduct research to determine for thermodynamics textbooks written 30 years ago if exergy is covered and to what extent.

5.2 Almost all undergraduate thermodynamics textbooks have at least one chapter devoted to exergy. If you were the instructor of a single thermodynamics course and you could only cover some of the chapters, would you cover exergy? Provide reasons for your answer.

5.3 Do you favor changing the way thermodynamics is taught by adopting an exergy-based approach?

5.4 Explain how a better understanding of exergy by engineering students can help improve public awareness of exergy.

5.5 Explain how exergy is a useful tool in policy making related to energy. Provide examples.

5.6 Is it realistic to expect that the level of understanding of the public about exergy should be comparable to that for energy? Explain.

5.7 Explain how a better understanding and appreciation of exergy by the public and by policy makers can help foster more efficient use of energy sources and minimizing their effects on the environment.

Chapter 6

EXERGY ANALYSIS OF PSYCHROMETRIC PROCESSES

6.1. Basic psychrometric concepts

Psychrometrics is the science of air and water vapor and deals with the properties of moist air. A thorough understanding of psychrometrics is of great significance, particularly to the HVAC community. It plays a key role, not only in heating, cooling and humidification processes and the resulting comfort of building occupants, but also in building insulation, roofing properties, and the stability, deformation and fire resistance of building materials. Hence, a good understanding of the main concepts and principles of psychrometrics are important.

Definitions of some of the most common terms in psychrometrics follow (Dincer, 2003):

Dry air: Atmospheric air generally contains a number of constituents, water vapor and miscellaneous components (e.g., smoke, pollen, gaseous pollutants, etc.). Dry air refers to air without the water vapor and miscellaneous components.
Moist air: Moist air is the basic medium in psychrometrics and is defined as a binary or two-component mixture of dry air and water vapor. The amount of water vapor in moist air varies from nearly zero (dry air) to a maximum of 0.020 kg water vapor/kg dry air under atmospheric conditions, depending on the temperature and pressure.
Saturated air: Saturated air is a saturated mixture of air and water vapor mixture, where the vapor is at the saturation temperature and pressure.
Dew point temperature: The dew point temperature is defined as the temperature of moist air saturated at the same pressure and with the same humidity ratio as that of a given sample of moist air (i.e., temperature at state 2 in Fig. 6.1). When moist air is cooled at constant pressure (i.e., process 1–2), the temperature reaches the dew point temperature when water vapor begins to condense.

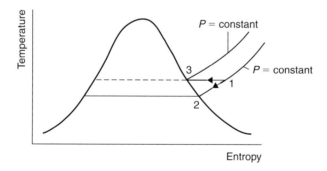

Fig. 6.1. Representation of dew point temperature on a T–s diagram.

Relative humidity: The relative humidity is defined as the ratio of the mole fraction of water vapor in a mixture to the mole fraction of water vapor in a saturated mixture at the same temperature and pressure. The relative humidity ϕ is normally based on the mole fraction equation since water vapor is considered to be an ideal gas:

$$\phi = \frac{P_v}{P_s} = \frac{\rho_v}{\rho_s} = \frac{v_s}{v_v} \qquad (6.1)$$

where P_v is the partial pressure of the vapor and P_s is the saturation pressure of vapor at the same temperature, which can be taken directly from a saturated water table. The total pressure is $P = P_a + P_v$. According to Fig. 6.1, $\phi = P_1/P_3$.

Humidity ratio: The humidity ratio of moist air (or the *mixing ratio*) is defined as the ratio of the mass of water vapor to the mass dry air contained in the mixture at the same temperature and pressure:

$$\omega = \frac{m_v}{m_a} = 0.622\frac{P_v}{P_a} \tag{6.2}$$

where $m_v = P_v V/R_v T = P_v V M_v/\check{R}T$ and $m_a = P_a V/R_a T = P_a V M_a/\check{R}T$ since both water vapor and air, as well as their mixture, are treated as ideal gases.

With the expressions for relative humidity and humidity ratio in terms of pressure ratios in Eqs. (6.1) and (6.2), the following expression can be derived:

$$\phi = \frac{\omega P_a}{0.622P_s} \tag{6.3}$$

Degree of saturation: The degree of saturation is defined as the ratio of actual humidity ratio to the humidity ratio of a saturated mixture at the same temperature and pressure.

Dry-bulb and wet-bulb temperatures: Dry-bulb and wet-bulb thermometers have been traditionally used to measure the specific humidity of moist air. The dry-bulb temperature is the temperature measured by a dry-bulb thermometer directly. The bulb of the wet-bulb thermometer is covered with a wick which is saturated with water. When the wick is subjected to an air flow (Fig. 6.2), some of the water in the wick evaporates into the surrounding air, resulting in a lower temperature than that obtained with a dry-bulb thermometer. The wet-bulb temperature is dependent on the moisture content of air, and can thus be used in conjunction with the dry-bulb temperature to determine the humidity of air. In the past, the wick conventionally was boiled in distilled water and allowed to dry before being used for wet-bulb temperature measurements. Nowadays, electronic devices are preferred for measuring the humidity of air due to their simplicity and accuracy.

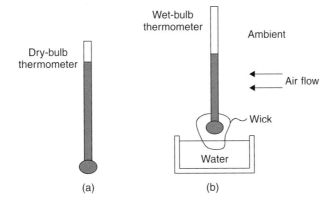

Fig. 6.2. Illustration of (a) dry-bulb and (b) wet-bulb thermometers.

Adiabatic saturation process: An adiabatic saturation process is an adiabatic process in which an air and water vapor mixture with a relative humidity less than 100% has liquid water added. Some of the water evaporates into the mixture bringing it to saturation, i.e., 100% relative humidity. The temperature of the mixture exiting an adiabatic saturation process is the *adiabatic saturation temperature* (Fig. 6.3).

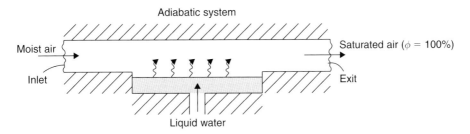

Fig. 6.3. An adiabatic saturation process.

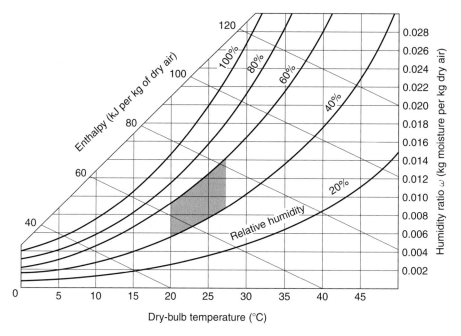

Fig. 6.4. Psychrometric chart.

Air-conditioning processes are usually shown on a psychrometric chart, which was developed in the early 1900s by a German engineer named Richard Mollier. A psychrometric chart (Fig. 6.4) shows the properties of moist air in terms of dry-bulb temperature, wet-bulb temperature, relative humidity, humidity ratio and enthalpy. Three of these properties are sufficient to identify a state of the moist air.

6.2. Balance equations for air-conditioning processes

In analyzing air-conditioning processes, we need to address three important balances: a mass balance (i.e., the continuity equation), an energy balance (i.e., the first law of thermodynamics) and an exergy balance (i.e., the second law of thermodynamics). Air-conditioning processes are essentially steady-flow processes, so general mass, energy and exergy balances may be written as shown below.

Mass balance for dry air:

$$\sum_{\text{in}} \dot{m}_a = \sum_{\text{out}} \dot{m}_a \tag{6.4}$$

Mass balance for water:

$$\sum_{\text{in}} \dot{m}_w = \sum_{\text{out}} \dot{m}_w \quad \text{or} \quad \sum_{\text{in}} \dot{m}_a \omega = \sum_{\text{out}} \dot{m}_a \omega \tag{6.5}$$

Energy balance (assuming negligible kinetic and potential energies and work):

$$\dot{Q}_{\text{in}} + \sum_{\text{in}} \dot{m}h = \dot{Q}_{\text{out}} + \sum_{\text{out}} \dot{m}h \tag{6.6}$$

Exergy balance (assuming negligible kinetic and potential energies and work):

$$\sum_{\text{in}} \dot{Ex}_{\dot{Q}} + \sum_{\text{in}} \dot{m}(ex) - \sum_{\text{out}} \dot{Ex}_{\dot{Q}} - \sum_{\text{out}} \dot{m}(ex) - \dot{Ex}_{\text{dest}} = 0 \tag{6.7}$$

$$\sum_{\text{in}} \dot{Q}\left(1 - \frac{T_0}{T}\right) + \sum_{\text{in}} \dot{m}(ex) - \sum_{\text{out}} \dot{Q}\left(1 - \frac{T_0}{T}\right) - \sum_{\text{out}} \dot{m}(ex) - \dot{Ex}_{\text{dest}} = 0$$

where

$$ex = h - h_0 - T_0(s - s_0) \tag{6.8}$$

Considering a dry air and water vapor mixture as an ideal gas, the total flow exergy of humid air per kg dry air may be expressed as (Wepfer et al., 1979)

$$ex = (c_{p,a} + \omega c_{p,v})T_0 \left(\frac{T}{T_0} - 1 - \ln \frac{T}{T_0} \right) + (1 + \tilde{\omega})R_a T_0 \ln \frac{P}{P_0} + R_a T_0 \left[(1 + \tilde{\omega}) \ln \frac{1 + \tilde{\omega}_0}{1 + \tilde{\omega}} + \tilde{\omega} \ln \frac{\tilde{\omega}}{\tilde{\omega}_0} \right] \tag{6.9}$$

where the last term is the specific chemical exergy. The proportionality between specific humidity ratio ω and specific humidity ratio on a molal basis $\tilde{\omega}$ is given by

$$\tilde{\omega} = 1.608\omega \tag{6.10}$$

where the humidity ratio is

$$\omega = m_v/m_a \tag{6.11}$$

The exergy efficiency of an overall air-conditioning process, or the subprocesses comprising it, may be written as

$$\psi = \frac{\dot{Ex}_{out}}{\dot{Ex}_{in}} = 1 - \frac{\dot{Ex}_{dest}}{\dot{Ex}_{in}} \tag{6.12}$$

where $\dot{Ex}_{out}$ and $\dot{Ex}_{in}$ respectively are the exergy output and input during the process or subprocess, and $\dot{Ex}_{dest}$ is the exergy destruction.

Common air-conditioning processes are shown on the psychrometric chart in Fig. 6.5. Figure 6.5a exhibits cooling and heating processes. During these processes, dry-bulb temperature decreases or increases, and only a change in sensible heat occurs. There is no latent heat change due to the constant humidity ratio of the air. Figure 6.5b shows a heating and humidification process. Air is first heated with a heater (process 1–2) and then humidified (process 2–3) by an injection of water. In the cooling and dehumidification process shown in Fig. 6.5c, air is cooled at constant humidity ratio until it is saturated (process 1–2). Further cooling of air (process 2–3) results in dehumidification. Figure 6.5d exhibits a process of adiabatic humidification (i.e., evaporative cooling) at a constant wet-bulb temperature. This process occurs during spray humidification. Figure 6.5e represents a process in which two air streams (at states 1 and 2, respectively) are mixed (state 3).

Balances for each of these air-conditioning processes are given below. The state numbers refer to Fig. 6.5 and the state of water is represented by the subscript 'w'.

Simple heating or cooling
Dry air mass balance:

$$\dot{m}_{a1} = \dot{m}_{a2} \tag{6.13a}$$

Water mass balance:

$$\dot{m}_{w1} = \dot{m}_{w2}$$

Energy balance:

$$\dot{Q}_{in} + \dot{m}_{a1}h_1 = \dot{m}_{a2}h_2 \text{ (heating)} \tag{6.13b}$$

$$\dot{m}_{a1}h_1 = \dot{m}_{a2}h_2 + \dot{Q}_{out} \text{ (cooling)} \tag{6.13c}$$

Exergy balance:

$$\dot{Q}_{in} \left(1 - \frac{T_0}{T} \right) + \dot{m}_{a1}(ex)_1 - \dot{m}_{a2}(ex)_2 - \dot{Ex}_{dest} = 0 \text{ (heating)} \tag{6.13d}$$

$$\dot{m}_{a1}(ex)_1 - \dot{m}_{a2}(ex)_2 - \dot{Q}_{out} \left(1 - \frac{T_0}{T} \right) - \dot{Ex}_{dest} = 0 \text{ (cooling)} \tag{6.13e}$$

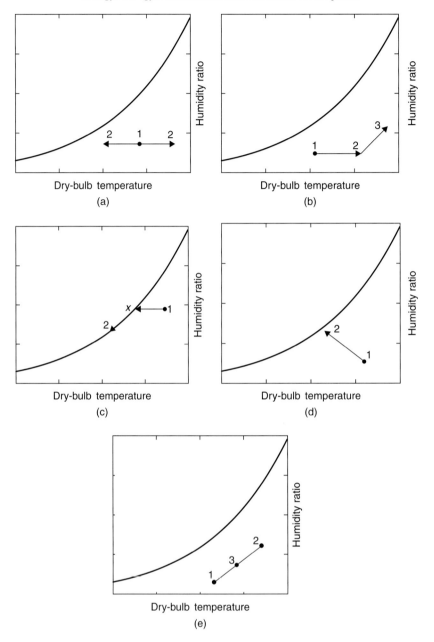

Fig. 6.5. Depiction of common air-conditioning processes on a psychrometric chart. (a) Simple heating or cooling. (b) Heating with humidification. (c) Cooling with dehumidification. (d) Evaporative cooling. (e) Adiabatic mixing of air streams.

Heating with humidification
Dry air mass balance:

$$\dot{m}_{a1} = \dot{m}_{a2} = \dot{m}_{a3} \tag{6.14a}$$

Water mass balance:

$$\dot{m}_{w1} = \dot{m}_{w2} \tag{6.14b}$$

$$\dot{m}_{w2} + \dot{m}_w = \dot{m}_{w3} \longrightarrow \dot{m}_{a2}\omega_2 + \dot{m}_w = \dot{m}_{a3}\omega_3 \tag{6.14c}$$

Energy balance:

$$\dot{Q}_{in} + \dot{m}_{a1}h_1 = \dot{m}_{a2}h_2 \tag{6.14d}$$

$$\dot{m}_{a2}h_2 + \dot{m}_w h_w = \dot{m}_{a3}h_3 \tag{6.14e}$$

Exergy balance:

$$\dot{Q}_{in}\left(1 - \frac{T_0}{T}\right) + \dot{m}_{a1}(ex)_1 - \dot{m}_{a2}(ex)_2 - \dot{Ex}_{dest} = 0 \tag{6.14f}$$

$$\dot{m}_{a2}(ex)_2 + \dot{m}_w(ex)_w - \dot{m}_{a3}(ex)_3 - \dot{Ex}_{dest} = 0 \tag{6.14g}$$

Cooling with dehumidification
Dry air mass balance:

$$\dot{m}_{a1} = \dot{m}_{a2} \tag{6.15a}$$

Water mass balance:

$$\dot{m}_{w1} = \dot{m}_{w2} + \dot{m}_w \longrightarrow \dot{m}_{a1}\omega_1 = \dot{m}_{a2}\omega_2 + \dot{m}_w \tag{6.15b}$$

Energy balance:

$$\dot{m}_{a1}h_1 = \dot{Q}_{out} + \dot{m}_{a2}h_2 + \dot{m}_w h_w \tag{6.15c}$$

Exergy balance:

$$\dot{m}_{a1}(ex)_1 - \dot{Q}_{out}\left(1 - \frac{T_0}{T}\right) - \dot{m}_{a2}(ex)_2 - \dot{m}_w(ex)_w - \dot{Ex}_{dest} = 0 \tag{6.15d}$$

Evaporative cooling
Dry air mass balance:

$$\dot{m}_{a1} = \dot{m}_{a2} \tag{6.16a}$$

Water mass balance:

$$\dot{m}_{w1} + \dot{m}_w = \dot{m}_{w2} \longrightarrow \dot{m}_{a1}\omega_1 + \dot{m}_w = \dot{m}_{a2}\omega_2 \tag{6.16b}$$

Energy balance:

$$\dot{m}_{a1}h_1 = \dot{m}_{a2}h_2 \longrightarrow h_1 = h_2 \tag{6.16c}$$

$$\dot{m}_{a1}h_1 + \dot{m}_w h_w = \dot{m}_{a2}h_2 \tag{6.16d}$$

Exergy balance:

$$\dot{m}_{a1}(ex)_1 + \dot{m}_w(ex)_w - \dot{m}_{a2}(ex)_2 - \dot{Ex}_{dest} = 0 \tag{6.16e}$$

Adiabatic mixing of air streams
Dry air mass balance:

$$\dot{m}_{a1} + \dot{m}_{a2} = \dot{m}_{a3} \tag{6.17a}$$

Water mass balance:

$$\dot{m}_{w1} + \dot{m}_{w2} = \dot{m}_{w3} \longrightarrow \dot{m}_{a1}\omega_1 + \dot{m}_{a2}\omega_2 = \dot{m}_{a3}\omega_3 \tag{6.17b}$$

Energy balance:

$$\dot{m}_{a1}h_1 + \dot{m}_{a2}h_2 = \dot{m}_{a3}h_3 \tag{6.17c}$$

Exergy balance:

$$\dot{m}_{a1}(ex)_1 + \dot{m}_{a2}(ex)_2 - \dot{m}_{a3}(ex)_3 - \dot{Ex}_{\text{dest}} = 0 \tag{6.17d}$$

6.3. Case study: exergy analysis of an open-cycle desiccant cooling system

In this section, an air-conditioning case study is presented involving energy and exergy analyses of an experimental desiccant cooling unit, adapted from Kanoglu et al. (2004).

6.3.1. Introduction

Desiccant cooling systems are heat-driven cooling units and they can be used as an alternative to the conventional vapor compression and absorption cooling systems. The operation of a desiccant cooling system is based on the use of a rotary dehumidifier (desiccant wheel) in which air is dehumidified. The resulting dry air is somewhat cooled in a sensible heat exchanger (rotary regenerator), and then further cooled by an evaporative cooler. The resulting cool air is directed into a room. The system may be operated in a closed cycle or more commonly in an open cycle in ventilation or recirculation modes. A heat supply is needed to regenerate the desiccant. Low-grade heat at a temperature of about 60–95°C is sufficient for regeneration, so renewable energies such as solar and geothermal heat as well as waste heat from conventional fossil-fuel systems may be used. The system is simple and thermal coefficient of performance (COP) is usually satisfactory.

Despite numerous theoretical and experimental studies of the first law aspects of desiccant cooling systems, few investigations have been performed of the second law aspects. Lavan et al. (1982) present a general second law analysis of desiccant cooling systems, without details on operation of the system and its components, and introduce an equivalent Carnot temperature concept for evaluating the reversible COP. The reversible COP depends on the operating parameters for the desiccant dehumidifier. Van Den Bulck et al. (1988) and Shen and Worek (1996) focus on only the desiccant dehumidifier. The latter consider the recirculation mode of the system operation and attempt to optimize the number of transfer units and the regeneration temperature of a dehumidifier based on the first and second laws. Maclaine-cross (1985) proposed a cycle with reversible components, which has an infinite COP. More recently, Pons and Kodama (2000a) evaluated the internal and external entropy generation in a desiccant cooling system operating in ventilation mode, and developed formulations for the Carnot COP of the system. In the second part of the study, Kodama et al. (2000b) applied the formulation to an experimental unit and investigated the effects of varying certain operating parameters on the entropy generation. This case study presents an exergy analysis of a desiccant cooling system and its components.

Exergy analysis has been used for work consuming refrigeration cycles. Such analyses often aim to minimize the work required for a given refrigeration task. For heat-driven cooling systems such as desiccant cooling units, exergy analysis may be used to determine the reversible COP and the sites of exergy losses that account for the difference between the reversible COP and actual COP.

In this study, a procedure for the energy and exergy analyses of open-cycle desiccant cooling systems is described and applied to an experimental working unit. First, the operation and design of the experimental unit are described. Then, energy-based system performance parameters are presented, and an exergy analysis is given in which exergy destruction and exergy efficiency relations for the system and its components are derived. Finally, the energy and exergy formulations are applied to the experimental unit using typical operating data.

6.3.2. Operation and design of experimental system

A simple schematic of the experimental system operating in ventilation mode and its representation on a psychrometric chart are shown in Fig. 6.6. Air enters the desiccant wheel, where it is dehumidified and heated by the returning air in the regeneration line. The hot dry air is then cooled in a rotary regenerator, which is basically a counter-flow sensible heat exchanger. The process air is further cooled in an evaporative cooler (EC1) before being routed to the room. An equal flow of air is withdrawn from the room for regeneration. The regeneration air is cooled in an evaporative cooler (EC2) and then preheated in the rotary regenerator by the warmer air in the process line. A heating unit supplies external heat to the regeneration air before it passes through the desiccant wheel, where the desiccant material picks up moisture

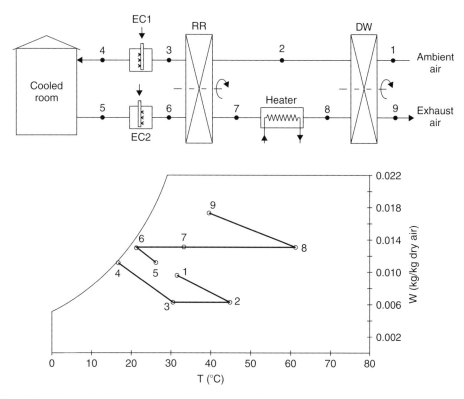

Fig. 6.6. Simplified schematic of an experimental desiccant cooling system in ventilation mode, and its psychrometric chart for typical operation as specified in Table 6.1. DW: desiccant wheel; RR: rotary regenerator; EC1: evaporative cooler 1 and EC2: evaporative cooler 2.

from the process air and transports it to the hot regenerated air. The air leaving the desiccant wheel is exhausted to the environment.

The components of the experimental unit were designed and constructed at University of Gaziantep, Turkey. The desiccant used in the desiccant wheel is natural zeolite found in Balikesir, Turkey. Natural zeolite has a density of 1450 kg/m³, a thermal conductivity of 0.60 W/m°C, and a specific heat of 1.11 kJ/kg°C. Although natural zeolite is not the best adsorbent of all possible desiccant materials, this shortcoming can be compensated for by using more of the desiccant material for a given air rate. Natural zeolite was used in part to investigate its suitability in desiccant cooling systems.

The desiccant bed consists of 12 equally shaped sections along the circumference. The zeolites are irregular in shape, 4–6 mm in size, and packed freely in the desiccant bed. The desiccant wheel has inner and outer diameters of 11 and 55 cm, respectively, and a length of 25 cm. The wheel is mounted in a sheet housing containing iron profiles. The frame is constructed using 0.8 mm sheet iron separated into 12 equal sections along the circumference. A corrosion resistant aluminum sheet with a thickness of 0.5 mm is used along the circumference to provide complete sealing. The rotor and air duct connections are well fitted using rubber gaskets. The rotary regenerator is constructed of a light honeycomb structure made of copper foil sheets with a center hole to accommodate the rotating shaft. The rotor is mounted on a reinforced iron frame divided into two sections, and is allowed to rotate on two bearings. The rotor is 50 cm in diameter and has a length of 25 cm. Copper is used as the matrix material with a thickness of 0.075 mm. The compactness of the matrix, defined as the ratio of heat transfer area to the volume, is approximately 3600 m²/m³. The housing and sealing of the rotary regenerator are similar to those for the desiccant wheel. Although aluminum is commonly used as a matrix material in rotary regenerators, copper is used here because in part it has more favorable thermal properties (e.g., the thermal conductivity of aluminum is 237 W/m°C and of copper is 401 W/m°C),

The evaporative coolers are constructed to maximize the contact area between water droplets and the air stream. Resistance heaters are used to supply regeneration heat to the unit.

6.3.3. Energy analysis

Desiccant cooling units are heat-driven systems, and the COP is defined as

$$\text{COP} = \frac{\dot{Q}_{\text{cool}}}{\dot{Q}_{\text{regen}}} = \frac{\dot{m}_a(h_5 - h_4)}{\dot{m}_a(h_8 - h_7)} \tag{6.18}$$

where $\dot{Q}_{\text{cool}}$ is the rate of heat removal from the cooled room, $\dot{Q}_{\text{regen}}$ is the rate of regeneration heat supply to the unit, $\dot{m}_a$ is the mass flow rate of air and h is the specific enthalpy of moist air. The numerical subscripts refer to state points in Fig. 6.6 throughout this chapter. Since the mass flow rates are equal in the process and regeneration lines, the effectiveness of the rotary regenerator may be expressed as

$$\varepsilon_{\text{RR}} = \frac{T_2 - T_3}{T_2 - T_6} \tag{6.19}$$

where T is the temperature of moist air. The effectiveness of the desiccant wheel may be expressed similarly:

$$\varepsilon_{\text{DW},1} = \frac{T_2 - T_1}{T_8 - T_1} \tag{6.20}$$

Another effectiveness relation for the desiccant wheel may be defined based on the dehumidification performance of the wheel with respect to the regeneration heat input, as follows:

$$\varepsilon_{\text{DW},2} = \frac{(\omega_1 - \omega_2)h_{\text{fg}}}{h_8 - h_7} \tag{6.21}$$

where the numerator represents the latent heat and the denominator, the regeneration heat. Also, w denotes the specific humidity (humidity ratio) and h_{fg}, the enthalpy of vaporization for water. A third and perhaps better performance expression for the desiccant wheel is an effectiveness based on specific humidity, which is given by Van Den Bulck et al. (1988) as

$$\varepsilon_{\text{DW},3} = \frac{\omega_1 - \omega_2}{\omega_1 - \omega_{2,\text{ideal}}} \tag{6.22}$$

Here, $\omega_{2,\text{ideal}}$ is the specific humidity at the wheel exit in the ideal case and may be taken to be zero since an ideal wheel completely dehumidifies the air.

The effectiveness relations for the evaporative coolers are

$$\varepsilon_{\text{EC1}} = \frac{T_3 - T_4}{T_3 - T_{\text{wb},3}} \tag{6.23a}$$

$$\varepsilon_{\text{EC2}} = \frac{T_5 - T_6}{T_5 - T_{\text{wb},5}} \tag{6.23b}$$

where T_{wb} is the wet-bulb temperature of the moist air.

A mass balance on the two evaporative coolers gives

$$\dot{m}_{\text{w1}} = \dot{m}_a(\omega_4 - \omega_3) \tag{6.24a}$$

$$\dot{m}_{\text{w2}} = \dot{m}_a(\omega_6 - \omega_5) \tag{6.24b}$$

where $\dot{m}_{\text{w1}}$ and $\dot{m}_{\text{w2}}$ are the rates of moisture addition to the air in the evaporative coolers in the process and regeneration lines, respectively.

6.3.4. Exergy analysis

The maximum COP of a heat-driven cooling system can be determined by assuming that the entire cycle is reversible. The cooling system would be reversible if the heat from the heat source were transferred to a Carnot heat engine, and the work output of this engine is supplied to a Carnot refrigerator to remove heat from the cooled space (Cengel and Boles, 2006). The expressions for the work output from the Carnot heat engine, the cooling load of the Carnot refrigerator and the Carnot COP of this reversible system are respectively

$$q_{\text{in}} = \frac{w_{\text{out}}}{\eta_{\text{th,C}}} \tag{6.25}$$

$$q_{\text{L}} = \text{COP}_{\text{R,C}} w_{\text{out}} \tag{6.26}$$

$$\text{COP}_{\text{C}} = \frac{q_{\text{L}}}{q_{\text{in}}} = \eta_{\text{th,C}} \text{COP}_{\text{R,C}} = \left(1 - \frac{T_{\text{ambient}}}{T_{\text{source}}}\right) \left(\frac{T_{\text{space}}}{T_{\text{ambient}} - T_{\text{space}}}\right) \tag{6.27}$$

where $\eta_{\text{th,C}}$ is the thermal efficiency of the Carnot heat engine, $\text{COP}_{\text{R,C}}$ is the COP of the Carnot refrigerator, and T_{ambient}, T_{space} and T_{source} are the temperatures of the ambient environment, the cooled space and the heat source, respectively. In the ideal system considered in Fig. 6.6, the temperatures of the ambient environment and the cooled space are T_1 and T_5, respectively, and the temperature of the heat source may be taken to be the regeneration temperature T_8. Then, Eq. (6.27) may be written as

$$\text{COP}_{\text{C}} = \left(1 - \frac{T_1}{T_8}\right) \left(\frac{T_5}{T_1 - T_5}\right) \tag{6.28}$$

The Carnot COP relation in Eq. (6.27) is the upper limit for any heat-driven cooling system which operates in a closed cycle (e.g., an absorption chiller). The open-cycle desiccant cooling system in both ventilation and recirculation modes, however, involves mass transfer between the ambient environment and the room. Water is added in evaporative coolers and to the process air in the room. All of this added water is heated and evaporated, and returned to the ambient environment during system operation. Lavan et al. (1982) and Pons and Kodama (2000a) investigated the impact on performance measures of the open nature of the cycle with varying approaches. We follow the approach of Lavan et al. (1982), which is based on using equivalent Carnot temperatures for the evaporator, the condenser and the heat source. The reversible COP of open desiccant cooling systems is expressed using this approach as

$$\text{COP}_{\text{rev}} = \left(1 - \frac{T_{\text{c}}}{T_{\text{s}}}\right) \left(\frac{T_{\text{e}}}{T_{\text{c}} - T_{\text{e}}}\right) \tag{6.29}$$

where T_{s}, T_{e} and T_{c} are the equivalent temperatures of the heat source, the evaporator and the condenser, respectively. From Fig. 6.6,

$$T_{\text{s}} = \frac{h_7 - h_8}{s_7 - s_8} \tag{6.30}$$

$$T_{\text{e}} = \frac{\dot{m}_{\text{a}} h_4 - \dot{m}_{\text{a}} h_5 + \dot{m}_{\text{w3}} h_{\text{w}}}{\dot{m}_{\text{a}} s_4 - \dot{m}_{\text{a}} s_5 + \dot{m}_{\text{w3}} s_{\text{w}}} \tag{6.31}$$

$$T_{\text{c}} = \frac{\dot{m}_{\text{a}} h_9 - \dot{m}_{\text{a}} h_1 - (\dot{m}_{\text{w1}} + \dot{m}_{\text{w2}} + \dot{m}_{\text{w3}}) h_{\text{w}}}{\dot{m}_{\text{a}} s_9 - \dot{m}_{\text{a}} s_1 - (\dot{m}_{\text{w1}} + \dot{m}_{\text{w2}} + \dot{m}_{\text{w3}}) s_{\text{w}}} \tag{6.32}$$

where $\dot{m}_{\text{w3}}$ is the rate of moisture addition to the process air in the cooled room, and h_{w} and s_{w} are the specific enthalpy and specific entropy of liquid water, respectively. It is clear that the reversible COP is a function of the operating conditions.

There is no commonly accepted definition for the exergy efficiency of a desiccant cooling system. Two definitions are considered here. The first is the ratio of actual COP to reversible COP under the same operating conditions:

$$\psi_{\text{DCS,1}} = \frac{\text{COP}}{\text{COP}_{\text{rev}}} \tag{6.33}$$

The second exergy efficiency for the system can be expressed as

$$\psi_{\text{DCS,2}} = \frac{\dot{Ex}_{\text{cool}}}{\dot{Ex}_{\text{heat}}} \tag{6.34}$$

where $\dot{Ex}_{cool}$ is the exergy difference between states 1 and 4 and corresponds to the minimum work input (reversible work) required to cool the moist air from the ambient state to the air inlet state to the room. That is,

$$\dot{Ex}_{cool} = \dot{m}_a[h_1 - h_4 - T_1(s_1 - s_4)] \tag{6.35}$$

Also, $\dot{Ex}_{heat}$ is the exergy increase of the air stream due to heat transfer in the regenerator:

$$\dot{Ex}_{heat} = \dot{m}_a[h_8 - h_7 - T_1(s_8 - s_7)] \tag{6.36}$$

This quantity may be viewed as the exergy transfer to the air stream by heat assuming that heat transfer is reversible. If the system were a work-driven cooling unit, we would use the actual work input in the denominator of Eq. (6.34), since the exergy efficiency of a vapor-compression refrigeration system is the ratio of minimum work and actual work for a given cooling task.

We now consider the exergy destruction and exergy efficiency relations for the individual components of the system. Neglecting kinetic and potential energies, the exergy rate for a flow stream is given by

$$\dot{Ex} = \dot{m}_a(ex) = \dot{m}_a[h - h_0 - T_0(s - s_0)] \tag{6.37}$$

where ex is the specific flow exergy and state 0 represents the dead state, which is the ambient state (state 1) in our system in Fig. 6.6. The rate of exergy destruction in any component may be obtained from

$$\dot{Ex}_{dest} = T_0\dot{S}_{gen} \tag{6.38}$$

where $\dot{S}_{gen}$ is the rate of entropy generation in the component, obtainable from an entropy balance on the component. The exergy destructions for the desiccant wheel, rotary regenerator and evaporative coolers may consequently be determined as

$$\dot{Ex}_{dest,DW} = T_0\dot{m}_a(s_2 + s_9 - s_1 - s_8) \tag{6.39}$$

$$\dot{Ex}_{dest,RR} = T_0\dot{m}_a(s_3 + s_7 - s_2 - s_6) \tag{6.40}$$

$$\dot{Ex}_{dest,EC1} = T_0(\dot{m}_a s_4 - \dot{m}_a s_3 - \dot{m}_{w1} s_w) \tag{6.41}$$

$$\dot{Ex}_{dest,EC2} = T_0(\dot{m}_a s_6 - \dot{m}_a s_5 - \dot{m}_{w2} s_w) \tag{6.42}$$

The components in the experimental system are well insulated, so no entropy transfer by heat appears in these entropy generation expressions.

The exergy efficiency may be defined for the rotary regenerator and the desiccant wheel as the ratio of the exergy increase of the cold stream to the exergy decrease of the hot stream. That is,

$$\psi_{DW} = \frac{h_2 - h_1 - T_0(s_2 - s_1)}{h_8 - h_9 - T_0(s_8 - s_9)} \tag{6.43}$$

$$\psi_{RR} = \frac{h_7 - h_6 - T_0(s_7 - s_6)}{h_3 - h_4 - T_0(s_3 - s_4)} \tag{6.44}$$

The general exergy efficiency is sometimes expressed as

$$\psi = \frac{\dot{Ex}_{out}}{\dot{Ex}_{in}} = 1 - \frac{\dot{Ex}_{dest}}{\dot{Ex}_{in}} \tag{6.45}$$

where $\dot{Ex}_{out}$ and $\dot{Ex}_{in}$ are the exergy recovered and the exergy input for the component, respectively. Using this definition for evaporative coolers, we obtain

$$\psi_{EC1} = 1 - \frac{\dot{E}_{dest,EC1}}{\dot{Ex}_3} \tag{6.46a}$$

$$\psi_{EC2} = 1 - \frac{\dot{E}_{dest,EC2}}{\dot{E}x_5} \tag{6.46b}$$

The regeneration heat is supplied to the unit by a heating source, e.g., a gas burner, a solar collector, geothermal heat. The heating source must be known in order to determine the exergy destruction and the exergy efficiency for the heating system. Assuming that the heat source is at a constant temperature, which may be ideally taken to be the air temperature at state 8 in Fig. 6.6, the exergy destruction and the exergy efficiency for the heating system can be expressed as

$$\dot{E}_{dest,HS} = T_1 \left(\dot{m}_a s_8 - \dot{m}_a s_7 - \frac{\dot{Q}_{regen}}{T_8} \right) \tag{6.47}$$

$$\psi_{HS} = \frac{h_8 - h_7 - T_0(s_8 - s_7)}{h_8 - h_7 - T_0(h_8 - h_7)/T_8} \tag{6.48}$$

Equation (6.48) can be viewed as the ratio of the exergy supplied to the regeneration air to the exergy released by the heat source.

6.3.5. Results and discussion

The energy and exergy formulations are applied to the experimental desiccant cooling system described earlier. Table 6.1 lists the measured dry- and wet-bulb temperatures and calculated properties for the system during typical operation. Properties of moist air and water are obtained from an equation solver with built-in thermodynamic functions for many substances (Klein, 2006).

Table 6.1. Measured and calculated state properties of the system shown in Fig. 6.6.[a]

State	$T(°C)$	$T_{wb}(°C)$	ω (kg water/kg dry air)	ϕ	h (kJ/kg dry air)	s (kJ/K kg dry air)
1	31.5	19.7	0.00950	0.329	56.01	5.803
2	43.5	21.0	0.00630	0.115	60.00	5.813
3	30.2	16.7	0.00630	0.237	46.48	5.769
4	17.3	16.7	0.01162	0.940	46.85	5.772
5[b]	26.7	19.8	0.01162	0.530	56.50	5.805
6	20.4	19.8	0.01427	0.950	56.72	5.806
7	33.7	23.7	0.01427	0.435	70.43	5.852
8	60.8	30.1	0.01427	0.110	98.41	5.940
9	49.8	29.4	0.01747	0.227	95.39	5.933

[a] The mass flow rate of air is 400 kg/h.
[b] The temperature at state 5 is the ARI (1998) value for indoors.

In Table 6.1, effectiveness, exergy efficiency and exergy destruction values are given for the system and its components. The rotary regenerator has a low effectiveness (57.5%) and exergy efficiency (38.7%). The evaporative coolers have high effectivenesses (95.3% and 91.8%) and low exergy efficiencies (14.7% and 58.3%). One reason for the lower exergy efficiency for evaporative cooler 1 (EC1) compared to evaporative cooler 2 (EC2) is that there is a greater rate of water evaporation in EC1, resulting in higher irreversibility. The experimental system uses electricity as the heat source for convenience and ease of control, but an actual system would likely use a different heat source. For simplicity, we assume here an ideal heat source at a constant temperature (taken to be equal to the temperature at state 8 in Fig. 6.6). The exergy efficiency of the heating system is found to be 53.7% for this case. All three effectiveness values for the desiccant wheel are low, particularly the third one (33.7%), indicating poor dehumidification performance. This poor performance may be largely attributable to natural zeolite not being the best desiccant and the internal design and construction of the wheel. Dehumidification effectiveness and system COP are directly related since the desiccant wheel performance has the greatest effect on system performance. The exergy efficiency (76.1%) as appears quite high, given that Van Den Bulck et al. (1988) report that the maximum exergy efficiency for desiccant dehumidifiers is about 85%.

Exergy destructions in absolute terms and as a percentage of total exergy destruction are given in the last two columns of Table 6.2. The desiccant wheel is responsible for the greatest portion of the total exergy destruction (33.8%) followed by the heating system (31.2%). The rotary regenerator and evaporative coolers account for the remaining exergy destructions. These results are in agreement with those of Kodama et al. (2000b), who found that the desiccant wheel and heating system account for the majority of the entropy generation for most operating conditions of their experimental system.

Table 6.2. Selected energy and exergy performance data for the system and its components.

	Effectiveness ε (%)	Exergy efficiency ψ (%)	Exergy destruction rate $\dot{Ex}_{\text{dest}}$ (kW)	Exergy destruction rate (% of total)
Rotary regenerator	57.5	38.7	0.07075	17.5
Evaporative cooler 1	95.3	14.7	0.05817	14.4
Evaporative cooler 2	91.8	58.3	0.01272	3.1
Heating system	–	53.7	0.1261	31.2
Desiccant wheel	40.9 (Eq. (6.20)) 27.4 (Eq. (6.21)) 33.7 (Eq. (6.22))	76.1	0.1369	33.8
System		11.1 (Eq. (6.33)) 3.3 (Eq. (6.34))	0.40464	100

Van Den Bulck et al. (1988) identify the causes of irreversibility for the dehumidifier as the mixing of process and regeneration air streams, the transfer of energy and mass across finite temperature differences, and vapor pressure differences between the desiccant matrix and the regeneration air stream. Similarly, the adiabatic humidification process in the evaporative coolers involves irreversibilities caused by concentration difference and mass transfer across finite temperature differences. Heat transfer across a finite temperature difference and the mixing of air streams are the primary causes of irreversibility in the rotary regenerator. The causes of irreversibility for the heating system depend on the method of heat input. For the ideal heat source considered in the analysis, the irreversibility is due to the heat transfer across a finite temperature difference whose maximum value is $(T_8 - T_7)$.

The exergy efficiency of the system is evaluated to be 11.1% with Eq. (6.33) and 3.3% with Eq. (6.34). We find the former value more meaningful since it compares the actual and reversible COPs of the system. To approach the reversible COP, the exergy destructions in components of the system must be reduced. Efforts to reduce exergy destructions should focus initially on the highest exergy destruction sites. Significant increases in the exergy efficiency, and thus the COP, of desiccant cooling systems may be achieved by minimizing exergy destructions in the desiccant wheel, the heating system and the rotary regenerator, and developing less irreversible processes as alternatives to the inherently irreversible evaporative cooling process. Maclaine-cross (1985) attempted this by replacing evaporative coolers with a reversible wet surface heat exchanger. The present exergy efficiency of only 11.1% indicates a high potential for improvement and is typical of desiccant cooling units.

Other performance data for the experimental system are given in Table 6.3. The Carnot COP is greater than the reversible COP since the definition for the reversible COP provides a more realistic upper limit for the system performance by considering the open nature of the cycle. In other words, the reversible COP definition excludes the external irreversibilities resulting from the open nature of the cycle because they cannot be eliminated. Also listed in Table 6.3 are the Carnot equivalent temperatures calculated from Eqs. (6.30) through (6.32), the cooling load and the regeneration heat supply to the unit. Note that the equivalent temperatures for the open cycle are lower than the corresponding temperatures for a heat-driven closed cycle.

The actual COP of the system is 0.345. The COPs of other actual and experimental desiccant cooling units are reported to be between 0.5 and 0.8 (e.g., Krishna and Murthy, 1989; Kodama et al., 2000b). Noting that the rotary regenerator

Table 6.3. Additional performance data
for the system.

COP values	
COP	0.345
COP_{rev}	3.112
COP_C	5.472
Heat flow rates (kW)	
$\dot{Q}_{cool}$	1.072
$\dot{Q}_{regen}$	3.109
Temperatures (°C)	
T_c	27.5
T_e	21.8
T_s	46.6

and the evaporative coolers have somewhat satisfactory performance, the present low COP is mostly due to inadequate performance of the desiccant wheel. This is indicated by the low dehumidification effectiveness (33.7%) and high exergy destruction (33.8% of the total) of the desiccant wheel. Desiccant material, internal geometry (i.e., how the desiccant is distributed within the dehumidifier matrix), regeneration temperature, the ratio of regeneration air to process air and wheel rotational speed are parameters affecting wheel performance. Optimization of these parameters can improve the COP of the unit.

The irreversibilities due to the heating system, although significant, cannot be eliminated in this system since they are caused by the temperature difference. The exergy destruction in the rotary regenerator is also significant in degrading the overall system performance and it can be reduced by better design and operation. Perhaps the operating parameter that affects the regenerator performance most is the rotational speed. Evaporative coolers are inherently irreversible and there is little that can be done to reduce their irreversibilities.

The effects of measurement inaccuracies in the dry- and wet-bulb temperatures on the results are small. The thermo-couples used to measure the temperatures have an estimated inaccuracy of ±0.5°C. A 0.5°C change in the regeneration dry-bulb temperature (state 8) changes the actual COP by 1.8%, the reversible COP by 4.1%, the exergy efficiency of the desiccant wheel by 1.5% and the exergy destruction in the wheel by 1.5%. Inaccuracies in the wet-bulb temperature have smaller effects on the results. A change of 0.5°C in the regeneration wet-bulb temperature changes the same results by under 0.5%.

6.4. Closing remarks

Exergy analysis has been applied to psychrometric processes, and a case study has been considered. The results are clearer and more meaningful than those obtained by energy analysis, and help indicate potential modifications to improve efficiency.

Problems

6.1 Explain the importance of using exergy analysis for assessing and designing air-conditioning processes.
6.2 Explain how exergy analysis can help improve the performance of air-conditioning processes.
6.3 Consider the following processes, and evaluate them from an exergetic point of view and assess the degree of energy degradation in each: (a) heating a room with resistance heaters, (b) heating a room with a natural gas furnace,

(c) air-conditioning a room using an electric chiller unit in summer and (d) air-conditioning a room using an evaporative cooler in summer.

6.4 Why are air-conditioning processes typically highly irreversible? Explain the causes of exergy destructions in such processes and propose methods for reducing or minimizing the destructions.

6.5 Using the process diagrams and balance equations provided in this chapter, write exergy efficiency expressions for the following air-conditioning processes: (a) simple heating and cooling, (b) heating with humidification, (c) cooling with dehumidification, (d) evaporative cooling and (e) adiabatic mixing of airstreams.

6.6 Is the adiabatic mixing of two airstreams with the same pressure and different temperatures reversible? How about mixing of two airstreams with the same pressure and same temperature?

6.7 Is the adiabatic humidification of air with water at the same temperature as the air a reversible process? How about adiabatic humidification of air with water at a different temperature from the air?

6.8 Consider natural environmental processes like rain, snow, wind, etc. Are these processes reversible or irreversible? If your answer is irreversible, do you worry about the negative effects of the exergy destructions?

6.9 What is the difference between an evaporative cooling process and a desiccant cooling process? Describe climatic conditions for which each of these cooling systems is more appropriate.

6.10 Air enters an evaporative cooler at 40°C and 20% relative humidity. If the effectiveness of the cooler is 90%, determine the exergy destruction and the exergy efficiency of this process.

6.11 Ambient air at 1 atm, 5°C and 35% relative humidity is heated to 20°C in the heating section of an air processing device at a rate of 80 m^3/min and then humidified to 25°C and 70% relative humidity. Determine the total rate of exergy destruction and the exergy efficiency of the entire process. Assume heat is supplied by a heat source at 100°C.

Chapter 7

EXERGY ANALYSIS OF HEAT PUMP SYSTEMS

7.1. Introduction

The principle governing the operation of the heat pump was discovered before the start of the 1900s and is the basis of all refrigeration. The idea of using a heat engine in a reverse mode as a heat pump was proposed by Lord Kelvin in the 19th century, but it was only in the 20th century that practical machines came into common use, mainly for refrigeration. Beginning in the 1970s, air-source heat pumps came into common use. They have the advantage of being combustion free, and thus do not generate indoor pollutants like carbon monoxide. Heat pumps are also installation cost competitive with a central combustion furnace/central air conditioner combinations. Hence, heat pumps now routinely provide central air conditioning as well as heating.

Today, heat pumps are widely used not only for air conditioning and heating, but also for cooling, producing hot water and preheating feed water in various types of facilities including office buildings, computer centers, public buildings, restaurants, hotels, district heating and cooling systems and industrial plants.

Efficient energy use, including waste heat recovery and the application of renewable energy can reduce carbon dioxide emissions and global warming. A heat pump system can contribute to this objective, normally delivering more thermal energy than the electrical energy required to operate it.

A significant portion of global energy consumption is attributable to domestic heating and cooling. Heat pumps are advantageous and widely used in many applications due to their high utilization efficiencies compared to conventional heating and cooling systems. There are two common types of heat pumps: air-source heat pumps and ground-source (or geothermal) heat pumps.

A heat pump is essentially a heat engine operating in reverse and can be defined as a device that moves heat from a region of low temperature to a region of higher temperature. The residential air-to-air heat pump, the type most commonly in use, extracts heat from low temperature outside air and delivers this heat indoors. To accomplish this and avoid violating the second law of thermodynamics, work is done on the heat-pump working fluid (e.g., a refrigerant).

Four different energy-based criteria are commonly used to describe the efficiency of a heat pump (Dincer, 2003). For each of these criteria, the higher is the value the higher is the efficiency of the system. Heat-pump efficiency is determined by comparing the amount of energy delivered by the heat pump to the amount of energy it consumes. Note that efficiency measures are usually based on laboratory tests and do not necessarily measure how a heat pump performs in actual use.

- *Coefficient of performance*
  The coefficient of performance (COP) is the most common measure of heat-pump efficiency. The COP is the ratio of the heat output of a heat pump to its electrical energy input, expressible as

  $$COP = \text{Heat output/electrical energy input}$$

  For example, air-source heat pumps generally have COPs ranging from 2 to 4, implying that they deliver 2–4 times more energy than they consume. Water- and ground-source heat pumps normally have COPs of 3–5. The COP of an air-source heat pumps decreases as the outside temperature drops. Therefore, two COP ratings are usually given for a system: one at 8.3°C (47°F) and the other at −9.4°C (17°F). When comparing COPs, one must be sure the ratings are based on the same outside air temperature to avoid inconsistencies. COPs for ground- and water-source heat pumps do not vary as widely because ground and water temperatures are more constant than air temperatures.

While comparing COPs can be informative, it does not provide a complete picture. When the outside temperature drops below 4.4°C (40°F), the outdoor coils of a heat pump must be defrosted periodically. The outdoor coil temperature can be below freezing when a heat pump is in the heating cycle. Under these conditions, moisture in the air freezes on the surface of the cold coil. Eventually frost can build up sufficiently to keep air from passing over the coil, causing it to lose efficiency. When the coil efficiency is reduced enough to appreciably affect system capacity, the frost must be eliminated. To defrost the coils, the heat pump reverses its cycle and moves heat from the house to the outdoor coil to melt the ice. This process reduces the average COP significantly.

Some heat pump units have an energy-saving feature that allows the unit to defrost only when necessary. Others enter a defrost cycle at set intervals whenever the unit is in the heating mode.

Another factor that lowers the overall efficiency of air-to-air heat pumps is their inability to provide sufficient heat during the coldest days of winter. This weakness causes a back-up heating system to be required. The back-up is often provided by electric resistance heating, which has a COP of only one. When the temperature drops to the $-3.8°C$ to $-1.1°C$ range, or a different system-specific balance point, this electric resistance heating engages and overall system efficiency decreases.

- *Primary energy ratio*

Heat pumps may be driven electrically or by engines (e.g., internal combustion engines or gas motors). Unless electricity is derived from an alternative source (e.g., hydro, wind, solar), heat pumps also utilize primary energy sources upstream or onsite, as in the case of a natural gas motor. When comparing heat pump systems driven by different energy sources it is appropriate to use the primary energy ratio (PER) as the ratio of useful heat delivered to primary energy input. The PER is related to the COP as follows:

$$PER = \eta\,COP$$

where η is the efficiency with which the primary energy input is converted to work in the shaft of the compressor.

Due to the high COP of heat pumps, their PER values can be high relative to those for conventional fossil fuel fired systems. In the case of an electrically driven compressor where the electricity is generated in a coal power plant, the efficiency η may be as low as 25%. The PER equation indicates that gas engine-driven heat pumps are very attractive from a PER point of view since values for η (up to 75%) can be obtained. However, heat recovery systems tend to be judged on their potential financial savings rather than their potential energy savings.

- *Energy efficiency ratio*

The energy efficiency ratio (EER) is used for evaluating the efficiency of a heat pump in the cooling cycle. EER is defined as the ratio of cooling capacity provided to electricity consumed as follows:

$$EER = Cooling\ capacity/electrical\ energy\ input$$

The same rating system is used for air conditioners, allowing for straightforward comparisons of different units. In practice, EER ratings higher than 10 are desirable.

- *Heating season performance factor*

A heat pump's performance varies depending on the weather and how much supplementary heat is required. Therefore, a more realistic performance measure, especially for air-to-air heat pumps, is evaluated on a seasonal basis. One such measure is referred to as the heating season performance factor (HSPF) for the heating cycle. An industry standard test for overall heating efficiency provides an HSPF rating. Such laboratory testing attempts to take into account the reductions in efficiency caused by defrosting, temperature fluctuations, supplemental heat, fans and on/off cycling. The HSPF is estimated as the seasonal heating output divided by the seasonal power consumption, as follows:

$$HSPF = Total\ seasonal\ heating\ output/total\ electrical\ energy\ input$$

The HSPF can be thought of as the 'average COP' for the entire heating system. To estimate the average COP, one divides the HSPF by 3.4. Hence, an HSPF of 6.8 corresponds roughly with an average COP of 2. HSPFs of 5–7 are considered good. The higher the HSPF, the more efficient the heat pump on a seasonal basis.

Most utility-sponsored heat pump incentive programs require that heat pumps have an HSPF of at least 6.8. Many heat pumps meet this requirement, and some have HSPF ratings above 9. More efficient heat pumps are generally more expensive, so financial assessments must also account for the annual energy savings along with the added cost.

- *Seasonal energy efficiency ratio*

  As noted above, a heat pump's performance varies depending on the weather and the amount of supplementary heat required, so a more realistic efficiency measure can be obtained on a seasonal basis. The seasonal energy efficiency ratio (SEER) for the cooling cycle is such a measure. The SEER is the ratio of the total cooling of the heat pump to the total electrical energy input during the same period, i.e.,

$$\text{SEER} = \text{Total seasonal cooling output/total electrical energy input}$$

  The SEER rates the seasonal cooling performance of the heat pump. The SEER for a unit varies depending on where it is located. SEER values of 8–10 are considered good. The higher the SEER, the more efficiently the heat pump cools. The SEER compares the heat removed from a house or structure being cooled and the energy used by the heat pump, including fans. The SEER is usually noticeably higher than the HSPF since defrosting is not needed and there is usually no need for expensive supplemental heat during conditions when air conditioning is used.

Exergy-based measures of efficiency exist, based on the above measures or on other definitions. As with other technologies, these exergy-based measures offer advantages over energy-based measures.

In this chapter, energy and exergy analyses of an air-source heat pump system are presented. Exergy losses for each component of the system are identified, while the potential for efficiency improvements is described.

7.2. System description

A schematic of an air/water heat pump system considered is shown in Fig. 7.1. The system consists of two separate circuits: (1) a heat pump circuit and (2) a heat distribution circuit (water circuit). The refrigerant circuit consists of a

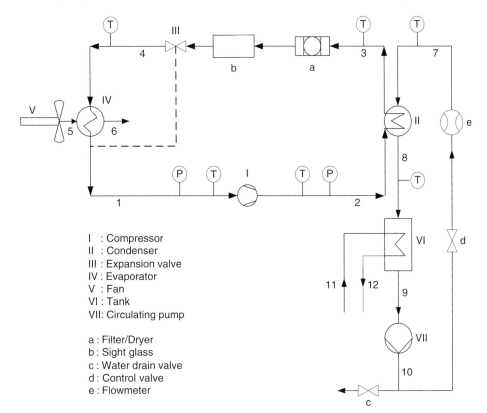

I   : Compressor
II : Condenser
III : Expansion valve
IV : Evaporator
V : Fan
VI : Tank
VII: Circulating pump

a : Filter/Dryer
b : Sight glass
c : Water drain valve
d : Control valve
e : Flowmeter

Fig. 7.1. Schematic of an air/water heat pump system.

compressor, a condenser, an expansion valve and an evaporator. The refrigerant is R-134a. The heat distribution circuit consists of a storage tank and a circulating pump.

Device I in Fig. 7.1 is a fully hermetically sealed reciprocating piston compressor. The condenser (II) is of a coaxial pipe cluster heat exchanger construction that works on the counter-flow principle. The refrigerant expands in an expansion valve (III). The evaporator (IV) is of a finned tube construction and has a large surface area. The refrigerant flows through the evaporator and draws heat from the ambient air over the large surface area. The heat transfer is enhanced by two fans that draw air through the fins.

During the operation period assessed, the control valve is adjusted so that the flow rate in the hot water circuit is approximately 0.020 m³/h. After the pressures on the suction and delivery sides of the working medium circuit have stabilized, data are recorded, including compressor power hot water flow rate, and pressure and temperatures at various points of the unit.

7.3. General analysis

Mass, energy and exergy balances are employed to determine the heat input, the rate of exergy destruction, and energy and exergy efficiencies. Steady-state, steady-flow processes are assumed. A general mass balance can be expressed in rate form as

$$\sum \dot{m}_{in} = \sum \dot{m}_{out} \tag{7.1}$$

where $\dot{m}$ is the mass flow rate, and the subscript 'in' stands for inlet and 'out' for outlet. Energy and exergy balances can be written as

$$\dot{E}_{in} = \dot{E}_{out} \tag{7.2}$$

$$\dot{Ex}_{in} - \dot{Ex}_{out} = \dot{Ex}_{dest} \tag{7.3}$$

The specific flow exergy of the refrigerant or water is evaluated as

$$ex_{r,w} = (h - h_0) - T_0(s - s_0) \tag{7.4}$$

where h is enthalpy, s is entropy and the subscript zero indicates properties at the reference (dead) state (i.e., at P_0 and T_0). The total flow exergy of air is determined as (Wepfer et al., 1979)

$$ex_{air} = (C_{p,a} + \omega C_{p,v})T_0[(T/T_0) - 1 - \ln(T/T_0)] + (1 + 1.6078\omega)R_a T_0 \ln(P/P_0) \\ + R_a T_0\{(1 + 1.6078\omega)\ln[(1 + 1.6078\omega_0)/(1 + 1.6078\omega)] + 1.6078\omega \ln(\omega/\omega_0)\} \tag{7.5}$$

where the specific humidity ratio is

$$\omega = \dot{m}_v/\dot{m}_a \tag{7.6}$$

The exergy rate is determined as

$$\dot{Ex} = \dot{m}(ex) \tag{7.7}$$

The exergy destructions in the heat exchanger (condenser or evaporator) and circulating pump respectively are evaluated as

$$\dot{Ex}_{dest,HE} = \dot{Ex}_{in} - \dot{Ex}_{out} \tag{7.8}$$

$$\dot{Ex}_{dest,pump} = \dot{W}_{pump} - (\dot{Ex}_{out} - \dot{Ex}_{in}) \tag{7.9}$$

where $\dot{W}_{pump}$ is the work rate of the pump.

The energy-based efficiency measure of the heat pump unit (COP$_{HP}$) and the overall heat pump system (COP$_{sys}$) can be defined as follows:

$$COP_{HP} = \frac{\dot{Q}_{cond}}{\dot{W}_{comp}} \tag{7.10a}$$

or, in terms of electrical input,

$$COP_{HP} = \frac{\dot{Q}_{cond}}{\dot{W}_{comp,elec}} \tag{7.10b}$$

and

$$COP_{sys} = \frac{\dot{Q}_{cond}}{\dot{W}_{comp} + \dot{W}_{pump} + \dot{W}_{fans}} \tag{7.11a}$$

or, in terms of electrical input,

$$COP_{sys} = \frac{\dot{Q}_{cond}}{\dot{W}_{comp,elec} + \dot{W}_{pump,elec} + \dot{W}_{fans,elec}} \tag{7.11b}$$

Here,

$$\dot{W}_{comp,elec} = \dot{W}_{comp}/(\eta_{comp,elec}\eta_{comp,mech}) \tag{7.12a}$$

$$\dot{W}_{pump,elec} = \dot{W}_{pump}/(\eta_{pump,elec}\eta_{pump,mech}) \tag{7.12b}$$

$$\dot{W}_{fans,elec} = \dot{W}_{fans}/(\eta_{fan,elec}\eta_{fan,mech}) \tag{7.12c}$$

The exergy efficiency can be expressed as the ratio of total exergy output to total exergy input:

$$\psi = \frac{\dot{Ex}_{out}}{\dot{Ex}_{in}} \tag{7.13}$$

where 'out' refers to 'net output' or 'product' or 'desired value,' and 'in' refers to 'driving input' or 'fuel.' The exergy efficiency of the heat exchanger (condenser or evaporator) is determined as the increase in the exergy of the cold stream divided by the decrease in the exergy of the hot stream, on a rate basis, as follows:

$$\psi_{HE} = \frac{\dot{Ex}_{cold,out} - \dot{Ex}_{cold,in}}{\dot{Ex}_{hot,in} - \dot{Ex}_{hot,out}} = \frac{\dot{m}_{cold}(ex_{cold,out} - ex_{cold,in})}{\dot{m}_{hot}(ex_{hot,in} - ex_{hot,out})} \tag{7.14}$$

Van Gool's (1997) improvement potential on a rate basis, denoted $\dot{IP}$, is expressible as

$$\dot{IP} = (1 - \psi)(\dot{Ex}_{in} - \dot{Ex}_{out}) \tag{7.15}$$

The relative irreversibility, RI, is evaluated as (Szargut et al., 2002)

$$RI = \frac{\dot{Ex}_{dest,i}}{\dot{Ex}_{dest,tot}} = \frac{\dot{I}_i}{\dot{I}_{tot}} \tag{7.16}$$

where the subscript 'i' denotes the ith device.

7.4. System exergy analysis

The following assumptions are made during the energy and exergy analyses:

(a) All processes are steady-state and steady-flow with negligible potential and kinetic energy effects and no chemical reactions.
(b) Heat transfer to the system and work transfer from the system are positive.
(c) Air behaves as an ideal gas with a constant specific heat.
(d) Heat transfer and refrigerant pressure drops in the tubing connecting the components are negligible since their lengths are short.
(e) The compressor mechanical ($\eta_{comp,mech}$) and the compressor motor electrical ($\eta_{comp,elec}$) efficiencies are 68% and 69%, respectively. These values are based on actual data in which the power input to the compressor is 0.149 kW.

(f) The circulating pump mechanical ($\eta_{\text{pump,mech}}$) and the circulating pump motor electrical ($\eta_{\text{pump,elec}}$) efficiencies are 82% and 88%, respectively. These values are based on an electric power of 0.050 kW obtained from the pump characteristic curve (Grundfos, 2006).

(g) The fan mechanical ($\eta_{\text{fan,mech}}$) and the fan motor electrical ($\eta_{\text{fan,elec}}$) efficiencies are 40% and 80%, respectively. These values are based on fan characteristic data (Ebmpapst, 2006) and the proposed efficiency values for a small propeller fan (Nagano et al., 2003).

Mass and energy balances as well as exergy destructions obtained from exergy balances for each of the heat pump components illustrated in Fig. 7.1 can be expressed as follows:

Compressor (I):

$$\dot{m}_1 = \dot{m}_{2,s} = \dot{m}_{\text{act,s}} = \dot{m}_r \tag{7.17a}$$

$$\dot{W}_{\text{comp}} = \dot{m}_r(h_{2,\text{act}} - h_1) \tag{7.17b}$$

$$\dot{Ex}_{\text{dest,comp}} = \dot{m}_r(ex_1 - ex_{2,\text{act}}) + \dot{W}_{\text{comp}} \tag{7.17c}$$

where heat interactions with the environment are neglected.

Condenser (II):

$$\dot{m}_2 = \dot{m}_3 = \dot{m}_r; \quad \dot{m}_7 = \dot{m}_8 = \dot{m}_w \tag{7.18a}$$

$$\dot{Q}_{\text{cond}} = \dot{m}_r(h_{2,\text{act}} - h_3); \quad \dot{Q}_{\text{cond}} = \dot{m}_w C_{p,w}(T_8 - T_7) \tag{7.18b}$$

$$\dot{Ex}_{\text{dest,cond}} = \dot{m}_r(ex_{2,\text{act}} - ex_3) + \dot{m}_w(ex_7 - ex_8) \tag{7.18c}$$

Expansion (throttling) valve (III):

$$\dot{m}_3 = \dot{m}_4 = \dot{m}_r \tag{7.19a}$$

$$(h_3 = h_4) \tag{7.19b}$$

$$\dot{Ex}_{\text{dest,exp}} = \dot{m}_r(ex_3 - ex_4) \tag{7.19c}$$

Evaporator (IV):

$$\dot{m}_4 = \dot{m}_1 = \dot{m}_r \tag{7.20a}$$

$$\dot{Q}_{\text{evap}} = \dot{m}_r(h_1 - h_4); \quad \dot{Q}_{\text{evap}} = \dot{m}_{\text{air}} C_{p,\text{air}}(T_5 - T_6) \tag{7.20b}$$

$$\dot{Ex}_{\text{dest,evap}} = \dot{W}_{\text{fan}} + \dot{m}_r(ex_4 - ex_1) + \dot{m}_{\text{air}}(ex_5 - ex_6) \tag{7.20c}$$

Fan (V):

$$\dot{m}_5 = \dot{m}_{5'} = \dot{m}_{\text{air}} \tag{7.21a}$$

$$\dot{W}_{\text{fan}} = \dot{m}_{\text{air}} \left[(h_5 - h_{5'}) + \frac{V_{\text{exit}}^2}{2} \right] \tag{7.21b}$$

$$\dot{Ex}_{\text{dest,fan}} = \dot{W}_{\text{fan,elec}} + \dot{m}_{\text{air}}(ex_{5'} - ex_5) \tag{7.21c}$$

Storage tank (VI):

$$\dot{m}_8 = \dot{m}_9 = \dot{m}_w; \quad \dot{m}_{11} = \dot{m}_{12} = \dot{m}_{\text{tw}} \tag{7.22a}$$

$$\dot{Q}_{\text{st}} = \dot{m}_w C_{p,w}(T_8 - T_9); \quad \dot{Q}_{\text{tank}} = \dot{m}_{\text{tw}} C_{p,\text{tw}}(T_{12} - T_{11}) \tag{7.22b}$$

$$\dot{Ex}_{\text{dest,st}} = \dot{m}_w(ex_8 - ex_9) + \dot{m}_{\text{tw}}(ex_{11} - ex_{12}) \tag{7.22c}$$

Circulating pump (VII):

$$\dot{m}_9 = \dot{m}_{10s} = \dot{m}_{10,\text{act}} = \dot{m}_w \tag{7.23a}$$

$$\dot{W}_{\text{pump}} = \dot{m}_w(h_{10,\text{act}} - h_9) \tag{7.23b}$$

$$\dot{Ex}_{\text{dest,pump}} = \dot{m}_r(ex_9 - ex_{10,\text{act}}) + \dot{W}_{\text{pump}} \tag{7.23c}$$

where interactions with the environment are neglected.

Since the volume flow rate on the refrigerant side is not measured, COP_{act} is evaluated using Eq. (7.18b) as follows:

$$\text{COP}_{\text{act}} = \frac{\dot{m}_w C_{p,w}(T_8 - T_7)}{\dot{W}_{\text{comp,act}}} = \frac{\dot{V}_w \rho_w C_{p,w}(T_8 - T_7)}{\dot{W}_{\text{comp,act}}} \tag{7.24}$$

Exergy efficiencies of the heat pump system and its components are evaluated as follows:

Heat pump unit (I–IV):

$$\psi_{\text{HP}} = \frac{\dot{Ex}_{\text{heat}}}{\dot{W}_{\text{comp,elec}}} = \frac{\dot{Ex}_{\text{in,cond}} - \dot{Ex}_{\text{out,cond}}}{\dot{W}_{\text{comp,elec}}} \tag{7.25}$$

Overall heat pump system (I–VII):

$$\psi_{\text{HP,sys}} = \frac{\dot{Ex}_{\text{in,cond}} - \dot{Ex}_{\text{out,cond}}}{\dot{W}_{\text{comp,elec}} + \dot{W}_{\text{pump,elec}} + \dot{W}_{\text{fans,elec}}} \tag{7.26}$$

Compressor (I):

$$\psi_{\text{comp}} = \frac{\dot{Ex}_{2,\text{act}} - \dot{Ex}_1}{\dot{W}_{\text{comp}}} \tag{7.27}$$

Condenser (II):

$$\psi_{\text{cond}} = \frac{\dot{Ex}_8 - \dot{Ex}_7}{\dot{Ex}_{2,\text{act}} - \dot{Ex}_3} = \frac{\dot{m}_w(ex_8 - ex_7)}{\dot{m}_r(ex_{2,\text{act}} - ex_3)} \tag{7.28}$$

Expansion (throttling) valve (III):

$$\psi_{\text{exp}} = \frac{\dot{Ex}_4}{\dot{Ex}_3} = \frac{ex_4}{ex_3} \tag{7.29}$$

Evaporator (IV):

$$\psi_{\text{evap}} = \frac{\dot{Ex}_4 - \dot{Ex}_1}{\dot{Ex}_5 - \dot{Ex}_6} = \frac{\dot{m}_r(ex_4 - ex_1)}{\dot{m}_{\text{air}}(ex_5 - ex_6)} \tag{7.30}$$

Fan (V):

$$\psi_{\text{fan}} = \frac{\dot{Ex}_5 - \dot{Ex}_{5'}}{\dot{W}_{\text{fan}}} = \frac{\dot{m}_{\text{air}}(ex_5 - ex_{5'})}{\dot{W}_{\text{fan}}} \tag{7.31}$$

Storage tank (VI):

$$\psi_{\text{st}} = \frac{\dot{Ex}_{12} - \dot{Ex}_{11}}{\dot{Ex}_8 - \dot{Ex}_9} = \frac{\dot{m}_{\text{tw}}(ex_{12} - ex_{11})}{\dot{m}_w(ex_8 - ex_9)} \tag{7.32}$$

Circulating pump (VII):

$$\psi_{\text{pump}} = \frac{\dot{Ex}_{10,\text{act}} - \dot{Ex}_9}{\dot{W}_{\text{pump}}} = \frac{\dot{m}_{\text{w}}(ex_{10,\text{act}} - ex_9)}{\dot{W}_{\text{pump}}} \tag{7.33}$$

7.5. Results and discussion

Temperature, pressure and mass flow rate data for the working fluid (R-134a), water and air are given in Table 7.1 following the state numbers specified in Fig. 7.1. Exergy rates evaluated for each state are presented in Table 7.1. The reference state is taken to be the state of environment on February 4, 2006, when the temperature and the atmospheric pressure were 2.2°C and 98.80 kPa, respectively (The Weather Network, 2006). The thermodynamic properties of water, air and R-134a are found using the Engineering Equation Solver (EES) software package.

Table 7.2 presents exergy, energy and RI data for a representative unit of the heat pump system. The exergy efficiencies on a product/fuel basis for the heat pump unit and the overall system, respectively, are 72.1% and 59.8% while the corresponding COP values are 3.4 and 1.68.

It is clear from Table 7.2 that the greatest irreversibility occurs in devices I (condenser) and VII (circulating pump) for the heat pump unit and the overall system, respectively. The first irreversibility is partly due to the large degree of superheat achieved at the end of the compression process, leading to large temperature differences associated with the initial phase of heat transfer. For the heat pump unit, the compressor has the second highest irreversibility. The mechanical–electrical losses are due to imperfect electrical, mechanical and isentropic efficiencies and emphasize the need for careful selection of this equipment, since components of inferior performance can considerably reduce overall system performance. The third largest irreversibility is associated with the evaporator, and the fourth largest with the capillary tube due to the pressure drop of the refrigerant passing through it.

The component irreversibility results of the heat pump unit indicate that the greatest potential for improvement is probably in the condenser, followed by the compressor, evaporator and the expansion device. Irreversibilities in the evaporator and the condenser occur due to the temperature differences between the two heat exchanger fluids, pressure losses, flow imbalances and heat transfer with the environment. Since compressor power depends strongly on the inlet and outlet pressures, any heat exchanger improvements that reduce the temperature difference will reduce compressor power by bringing the condensing and evaporating temperatures closer together. From a design standpoint, compressor irreversibility can be reduced independently. Recent advances in the heat pump market have led to the use of scroll compressors. Replacing the reciprocating compressor by a scroll unit can increase cooling effectiveness. The only way to eliminate the throttling loss is to replace the capillary tube (the expansion device) with an isentropic turbine (an isentropic expander) and to recover some shaft work from the pressure drop.

More generally, in evaluating the efficiency of heat pump systems, the most commonly used measure is the energy (or first law) efficiency, which is modified to a COP. However, for indicating the possibilities for thermodynamic improvement, energy analysis is inadequate and exergy analysis is needed. The exergy analysis of air-source heat pump system presented in this chapter identifies improvement potential.

7.6. Concluding remarks

Comprehensive energy and exergy analyses are presented and applied for evaluating heat pump systems and their components. Actual data are utilized in the analysis. Exergy destructions in the overall heat pump system and its components are quantified.

Some concluding remarks can be drawn from the results:

- The values for COP_{HP} and COP_{sys} are found to be 3.40 and 1.68, respectively, at a dead state temperature of 2.2°C.
- The exergy efficiency values for the heat pump unit and the overall heat pump system on a product/fuel basis are 72.1% and 59.8%, respectively. The exergy efficiencies elucidate potentials for improvement.
- The largest irreversibility in the heat pump unit is associated with the condenser, followed by the compressor, the evaporator and the expansion valve.
- The results focus attention on components where the greatest potential is destroyed and quantify the extent to which modifications affect, favorably or unfavorably, the performance of the system and its components.

Table 7.1. Process data for flows in the heat pump system.

State number	Description	Fluid	Phase	Temperature, T (°C)	Pressure, P (kPa)	Specific humidity ratio, ω (kg$_{water}$/kg$_{air}$)	Specific enthalpy, h (kJ/kg)	Specific entropy, s (kJ/kg K)	Mass flow rate, $\dot{m}$ (kg/s)	Specific exergy, ex (kJ/kg)	Exergy rate, $\dot{Ex} = \dot{m}(ex)$ (kW)
0	–	Refrigerant	Dead state	2.2	98.80	–	257.4	1.041	–	0	0
0′	–	Water	Dead state	2.2	98.80	–	9.3	0.034	–	0	0
0″	–	Moist air	Dead state	2.2	98.80	0.002	–	–	–	0	0
1	Evaporator outlet/ compressor inlet	Refrigerant	Superheated vapor	2.5	307	–	252.3	0.935	0.002	24.09	0.048
2,s	Condenser inlet/ compressor outlet	Refrigerant	Superheated vapor	45.3	1011	–	256.2	0.935	0.002	27.98	0.056
2,act	Condenser inlet/ compressor outlet	Refrigerant	Superheated vapor	54.6	1011	–	287.4	0.966	0.002	50.65	0.101
3	Condenser outlet/ expansion valve inlet	Refrigerant	Compressed liquid	22.8	1011	–	83.4	0.313	0.002	26.45	0.053
4	Evaporator inlet	Refrigerant	Mixture	1.3	307	–	83.4	0.319	0.002	24.80	0.050
5	Fan air inlet to evaporator	Air	Gas	16	–	0.004	26.20	–	0.136	0.33	0.045
5′	Air inlet to fan	Air	Gas	15.9	–	0.004	26.09	–	0.136	0.23	0.032
6	Fan air outlet from evaporator	Air	Gas	14	–	0.004	24.17	–	0.136	0.27	0.037

(Continued)

Table 7.1. (Continued)

State number	Description	Fluid	Phase	Temperature, T (°C)	Pressure, P (kPa)	Specific humidity ratio, ω (kg_{water}/kg_{air})	Specific enthalpy, h (kJ/kg)	Specific entropy, s (kJ/kg K)	Mass flow rate, $\dot{m}$ (kg/s)	Specific exergy, ex (kJ/kg)	Exergy rate, $\dot{Ex} = \dot{m}(ex)$ (kW)
7	Water inlet to condenser	Water	Compressed liquid	16.9	230	–	71.1	0.252	0.011	1.77	0.020
8	Water outlet from condenser	Water	Compressed liquid	24.6	220	–	103.3	0.361	0.011	3.96	0.044
9	Water outlet from storage tank/circulating pump inlet	Water	Compressed liquid	16.0	200	–	67.3	0.239	0.011	1.55	0.017
10,s	Water outlet from tank/ circulating pump outlet	Water	Compressed liquid	16.05	240	–	69.9	0.239	0.011	4.15	0.046
10,act	Water outlet from tank/ circulating pump outlet	Water	Compressed liquid	16.9	240	–	71.0	0.251	0.011	1.96	0.022
11	Tap water inlet to tank	Water	Compressed liquid	9.1	500	–	38.7	0.138	0.024	0.76	0.018
12	Tap water outlet from tank	Water	Compressed liquid	13.1	500	–	54.9	0.196	0.024	1.00	0.024

Table 7.2. Data for devices of a representative unit in the heat pump system.

Device number	Device	Exergy destruction rate, $\dot{Ex}_{dest}$ (kW)	Utilized power (kW)	$\dot{P}$ (kW)	$\dot{F}$ (kW)	IṖ (kW)	RI(%)		Exergy efficiency, ψ (%)		COP (–)	
							On HP unit basis	On overall system basis	$\dot{P}/\dot{F}$	Using Eqs. (7.25) and (7.26)	COP$_{HP}$ Using Eqs. (7.10a) and (7.11a)	COP$_{sys}$ Using Eqs. (7.10b) and (7.11b)
I	Compressor	0.017	0.070	0.053	0.070	0.0041	31.48	15.89	75.71	–	–	–
II	Condenser	0.024	0.408	0.024	0.048	0.0120	44.44	22.43	50.00	–	–	–
III	Expansion valve	0.003	–	0.050	0.053	0.0002	5.55	2.80	94.34	–	–	–
IV	Evaporator	0.010	0.338	0.002	0.008	0.0075	18.53	9.35	25.00	–	–	–
V	Fan	0.001	0.014	0.013	0.014	0.00007	–	0.93	92.85	–	–	–
VI	Storage tank	0.021	0.389	0.006	0.027	0.0163	–	19.63	22.22	–	–	–
VII	Circulating pump	0.031	0.036	0.005	0.036	0.0267	–	28.97	13.89	–	–	–
I–IV	Heat pump unit	0.054	–	0.129	0.179	0.0151	100.00	–	72.07	16.11	5.83	3.4
I–VII	Overall system	0.107	–	0.153	0.256	0.0430	–	100.00	59.77	9.89	2.74	1.68

* IṖ denotes improvement potential rate and RI relative irreversibility.

$\psi = \dot{P}/\dot{F}$ = overall exergy efficiency, where $\dot{P}$ is the product exergy and $\dot{F}$ is the fuel exergy.

Problems

7.1 Compare air-source and ground-source heat pump systems from an exergetic point of view.

7.2 Describe the difference between a ground-source heat pump and a geothermal heat pump.

7.3 Determine the exergy efficiency of the following heat pump systems used to keep a house at 22°C and discuss the results: (a) an air-source heat pump with a COP of 1.8 that absorbs heat from outdoor air at 2°C, (b) a ground-source heat pump with a COP of 2.6 that absorbs heat from the ground at 12°C and (c) a geothermal heat pump with a COP of 3.8 that absorbs heat from underground geothermal water at 60°C.

7.4 In a heat pump system, exergy destructions occur in various components such as the compressor, the condenser, the evaporator and the expansion valve. What are the causes of exergy destructions in each of these components?

7.5 Rework the illustrative example provided in this chapter using the given input data and try to duplicate the results. If your results differ from those given in the example, discuss why. Propose methods for improving the performance of the system based on reducing or minimizing exergy destruction.

7.6 Conduct a detailed exergy analysis of a real air-conditioning system using actual operating data. Present the results using tables and figures and discuss them. The system can be (a) an air-conditioning or heat pump unit, (b) a chiller unit which produces cool water in summer and warm water in winter, (c) an absorption refrigeration system or (d) a ground-source heat pump system.

7.7 Obtain a published article on exergy analysis of heat pump systems. Using the operating data provided in the article, perform a detailed exergy analysis of the system and compare your results to those in the original article.

Chapter 8

EXERGY ANALYSIS OF DRYING PROCESSES AND SYSTEMS

8.1. Introduction

Drying is widely used in a variety of applications ranging from food drying to wood drying. The drying industry uses large amounts of energy as drying is a highly energy-intensive operation. A dryer supplies the product with more heat than is available under ambient conditions thus sufficiently increasing the vapor pressure of the moisture held within the product to enhance moisture migration from within the product and significantly decreasing the relative humidity of the drying air to increase its moisture carrying capability and to ensure a sufficiently low equilibrium moisture content.

Drying is a thermal process in which heat and moisture transfer occur simultaneously. Heat is transferred by convection from heated air to the product to raise the temperatures of both the solid and moisture that is present. Moisture transfer occurs as the moisture travels to the evaporative surface of the product and then into the circulating air as water vapor. The heat and moisture transfer rates are therefore related to the velocity and temperature of the circulating drying air.

Drying involves thermally removing volatile substances (e.g., moisture) to yield a solid product. Mechanical methods for separating a liquid from a solid are not considered drying. In this chapter, we deal with thermal drying only. When a wet solid is subjected to thermal drying, two processes occur simultaneously:

1. Transfer of energy (mostly as heat) from the surrounding environment to evaporate the surface moisture.
2. Transfer of internal moisture to the surface of the solid and its subsequent evaporation due to process 1.

The rate at which drying is accomplished is governed by the rate at which the two processes proceed. Process 1 depends strongly on external conditions such as temperature, air humidity and flow, area of exposed surface and pressure, whereas process 2 depends on the physical nature, temperature and moisture content of the solid. Surface evaporation in process 2 is controlled by the diffusion of vapor from the surface of the solid to the surrounding atmosphere through a thin film of air in contact with the surface. Excessive surface evaporation, after the initial surface moisture removal, causes a high moisture gradient from the interior to the surface, sometimes causing over-drying and excessive shrinkage. These phenomena lead to high tension within the material, resulting in cracking and warping. Excessive surface evaporation can be retarded by employing high relative humidities of the drying air while maintaining a relatively high rate of internal moisture movement by heat transfer.

An important aspect of designing drying technology is the mathematical modeling of the drying processes and equipment. Accurate modeling allows design engineers to choose the most suitable operating conditions and then size the drying equipment and drying chamber accordingly to meet the desired operating conditions. Modeling is based on a set of mathematical equations that adequately characterize the system. The solution of these equations allows process parameters to be predicted as a function of time at any point in the dryer for given initial conditions.

Drying processes are complex. Kerkhof (1994) points out that the quantitative understanding of drying processes has the following special difficulties:

(a) The physical processes are highly non-linear.
(b) There are complicated exchanges and interaction processes.
(c) The dominating phenomena depend on drying conditions, and may change within the course of drying.
(d) The transport properties inside the material are highly dependent on moisture content and temperature.

Due to these complexities, the drying industry needs appropriate analysis techniques to provide optimal solutions to drying problems.

Exergy analysis can help reduce irreversibilities and increase the efficiency for drying processes. Increased efficiency reduces the energy required by drying systems for the production, transportation, transformation and distribution of various energy forms. Exergy analysis is a powerful tool for optimizing drying conditions, and is particularly important for large-scale high-temperature drying applications in industry.

This chapter describes and illustrates the exergy analysis of a drying process, applied here to moist solids. We define exergy efficiency as a function of heat and mass transfer parameters. An illustrative example is presented to demonstrate the importance of exergy methods for the analysis and optimization of drying processes.

8.2. Exergy losses associated with drying

The main exergy losses for drying are associated with irreversibilities and are described qualitatively for three different types of industrial drying methods: air drying, drum drying and freeze drying (Dincer, 2002b).

Air drying

The following are significant sources of exergy loss for air drying:

- A sizable amount of exergy is lost with exiting air, even if it is assumed to reach the wet-bulb temperature in the drying process. At higher wet-bulb temperatures, the water present in the exiting air makes a significant contribution to the total exergy loss of the exiting air.
- The exergy exiting with the product is seen to be quite small, as might be expected, since it was shown earlier that little exergy was put into the solid products.
- The exergy loss from the walls of the dryer, due to heat rejection, is significant and needs to be taken into consideration. For example, in spray drying this amount may reach up to 25% of the total exergy input. Of course, this loss can be reduced by appropriately insulating the dryer. Another important aspect is the size of the dryer. For example, the jet-type ring dryer has a much smaller loss from its walls than an equivalent-capacity spray dryer, due to its smaller dimensions.

Drum drying

Three major sources of exergy loss can be identified, including the following:

1. Some exergy is lost from the drum due to convection of air over the drum surface, which is not very large, being of the same order as that lost with the solid products.
2. The exergy loss associated with the exhausted vapor is large when calculated on a per kilogram of water basis. However, this energy is available at a temperature only slightly above the surrounding temperature (28.9°C) and is present in a large volume of air. Therefore, it would be difficult to develop an efficient means to reclaim this exergy.
3. The steam condensate in the drum is another sizable potential loss. The saturated liquid at the drum pressure could be used in a heat exchanger at the same pressure, or it could be flashed to atmospheric pressure and then used as a heat exchange medium, though at a lower temperature.

Freeze drying

Minimizing the exergy losses in freeze drying is more significant, compared to other drying types, since the energy requirements are much higher than for the other drying processes. Some major sources of exergy losses for freeze drying follow:

- Exergy losses due to radiative heat transfer from the heating plates to the dryer walls and with the exiting products are negligible, being less than 0.1% of the exergy required to remove 1 kg of water.
- Two sizable exergy loss areas which lend themselves to energy reclamation are heat dissipated in the vacuum pumps and heat rejected to the environment by the refrigeration system condensers. The magnitude of the latter loss is almost equivalent to the exergy required to remove 1 kg of water.
- The largest portion of exergy loss occurs in the condenser of the freeze dryer refrigeration system, in which 1062 kcal/kg of water sublimed must be dissipated, probably either to cooling water or to the ambient air. Under normal refrigeration system operation, most of this heat is available at a temperature of 38°C, a fact which limits its

usefulness. However, refrigeration systems can be designed which make exergy available at higher temperature, but at a cost of requiring more energy input at the compressor. Clearly, optimization can assist in balancing factors such as these in designs.

8.3. Analysis

This section presents energy and exergy analyses of drying processes. The systems are illustrated with input and output terms in Fig. 8.1, where there are seen to be four major interactions:

1. Input of drying air to the drying chamber to dry the products.
2. Input of moist products to be dried in the chamber.
3. Output of the moist air after containing the evaporated moisture removed from the products.
4. Output of the dried products, with moisture content reduced to the desired level.

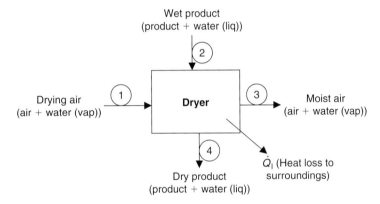

Fig. 8.1. Schematic of a drying process showing input and output terms.

8.3.1. Balances

Mass, energy and exergy balances can be written for the above system, treated as a control volume.

Mass balances

We can write mass balance equations for the dryer given above for three flows: product, dry air and water.

$$\text{Product:} \quad (\dot{m}_p)_2 = (\dot{m}_p)_4 = \dot{m}_p \tag{8.1}$$

$$\text{Air:} \quad (\dot{m}_a)_1 = (\dot{m}_a)_3 = \dot{m}_a \tag{8.2}$$

$$\text{Water:} \quad \omega_1 \dot{m}_a + (\dot{m}_w)_2 = \omega_3 \dot{m}_a + (\dot{m}_w)_4 \tag{8.3}$$

Energy balance

An energy balance can be written for the entire system, by equating input and output energy terms:

$$\dot{m}_a h_1 + \dot{m}_p (h_p)_2 + (\dot{m}_w)_2 (h_w)_2 = \dot{m}_a h_3 + \dot{m}_p (h_p)_4 + (\dot{m}_w)_4 (h_w)_4 + \dot{Q}_1 \tag{8.4}$$

where

$$h_1 = (h_a)_1 + \omega_1 (h_v)_1 = (h_a)_1 + \omega_1 (h_g)_1 \tag{8.5}$$

$$h_3 = (h_a)_3 + \omega_3 (h_v)_3 \tag{8.6}$$

The values of h_1 and h_3 can be obtained from a psychrometric chart. The heat loss rate from the chamber can be expressed as

$$\dot{Q}_l = \dot{m}_a q_l \tag{8.7}$$

Exergy balance

An exergy balance for the entire system can be written analogously to the energy balance as follows:

$$\dot{m}_a ex_1 + \dot{m}_p (ex_p)_2 + (\dot{m}_w)_2 (ex_w)_2 = \dot{m}_a ex_3 + \dot{m}_p (ex_p)_4 + (\dot{m}_w)_4 (ex_w)_4 + \dot{E}x_q + \dot{E}x_d \tag{8.8}$$

The specific exergy for the flow at point 1 can be expressed as

$$ex_1 = [(C_p)_a + \omega_1 (C_p)_v](T_1 - T_0) - T_0 \left\{ [(C_p)_a + \omega_1 (C_p)_v] \ln \left(\frac{T_1}{T_0} \right) - (R_a + \omega_1 R_v) \ln \left(\frac{P_1}{P_0} \right) \right\}$$

$$+ T_0 \left\{ (R_a + \omega_1 R_v) \ln \left(\frac{1 + 1.6078\omega^0}{1 + 1.6078\omega_1} \right) + 1.6078\omega_1 R_a \ln \left(\frac{\omega_1}{\omega^0} \right) \right\} \tag{8.9}$$

and the specific exergy at point 3 as

$$ex_3 = [(C_p)_a + \omega_3 (C_p)_v](T_3 - T_0) - T_0 \left\{ [(C_p)_a + \omega_3 (C_p)_v] \ln \left(\frac{T_3}{T_0} \right) - (R_a + \omega_3 R_v) \ln \left(\frac{P_3}{P_0} \right) \right\}$$

$$+ T_0 \left\{ (R_a + \omega_3 R_v) \ln \left(\frac{1 + 1.6078\omega^0}{1 + 1.6078\omega_3} \right) + 1.6078\omega_3 R_a \ln \left(\frac{\omega_3}{\omega^0} \right) \right\} \tag{8.10}$$

The specific exergy for the moist products can be written as

$$ex_p = [h_p(T, P) - h_p(T_0, P_0)] - T_0[s_p(T, P) - s_p(T_0, P_0)] \tag{8.11}$$

and the specific exergy for the water content as

$$ex_w = [h_f(T) - h_g(T_0)] + v_f[P - P_g(T)] - T_0[s_f(T) - s_g(T_0)] + T_0 R_v \ln \left[\frac{P_g(T_0)}{x_v^0 P_0} \right] \tag{8.12}$$

The exergy flow rate due to heat loss can be expressed as follows:

$$\dot{E}x_q = \dot{m}_a ex_q = \dot{m}_a \left(1 - \frac{T_0}{T_{av}} \right) q_l = \left(1 - \frac{T_0}{T_{av}} \right) Q_l \tag{8.13}$$

where T_{av} is the average outer surface temperature of the dryer.

Typical data for the reference environment are as follows: $T_0 = 32°C$, $P_0 = 1$ atm, $\omega^0 = 0.0153$ and $x_v^0 = 0.024$ (mole fraction of water vapor in air).

8.3.2. Exergy efficiency

We define the exergy efficiency for the drying process as the ratio of exergy use (investment) in the drying of the product to the exergy of the drying air supplied to the system. That is,

$$\psi = \frac{\text{Exergy input for evaporation of moisture in product}}{\text{Exergy of drying air supplied}}$$

or

$$\psi = \frac{(\dot{m}_w)_{ev}[(ex_w)_3 - (ex_w)_2]}{\dot{m}_a ex_1} \tag{8.14}$$

where

$$(\dot{m}_w)_{ev} = (\dot{m}_w)_2 - (\dot{m}_w)_4 \tag{8.15}$$

$$(ex_w)_3 = [h(T_3, P_{v3}) - h_g(T_0)] - T_0[s(T_3, P_{v3}) - s_g(T_0)] + T_0 R_v \ln\left[\frac{P_g(T_0)}{x_v^0 P_0}\right] \tag{8.16}$$

and

$$P_{v3} = (x_v)_3 P_3 \tag{8.17}$$

8.4. Importance of matching supply and end-use heat for drying

Various types of energy display different qualities, and these differences affect their ability to drive energy processes and be converted into other kinds of energy. For example, the quality of heat depends on the heat source temperature, since the higher is the temperature of a heat source relative to the ambient temperature, the greater is the portion of heat that can be converted to mechanical work. The surrounding air can be regarded as an infinite heat reservoir, but it normally cannot drive thermal processes.

Szargut et al. (1988) note that the capacity for doing work (or exergy) is a measure of energy quality. To be efficient, it is important to utilize energy in quantity and quality that matches the task. For heat transfer processes, this statement implies that the temperature of a heating fluid should be moderately above the temperature of the cooler substance, but not excessively so. More generally, the exergy input to a task should moderately exceed that required for the task for high efficiency.

In many applications in practice, we use fossil fuels for various low- and medium-temperature applications, especially in residential and industrial sectors. One needs to study both energy and exergy efficiencies and compare these for the thermal processes and applications. Such data may indicate that society is inefficient in its use of energy since high-quality (or high-temperature) energy sources such as fossil fuels are often used for relatively low-temperature processes like water and space heating or cooling, industrial drying, industrial steam production, etc. Exergy analysis permits a better matching of energy sources and uses so that high-quality energy is used only for tasks requiring that high quality.

8.5. Illustrative example

An exergy analysis of a dryer is performed and the effects on exergy efficiencies are examined of varying system parameters such as mass flow rate and temperature of the drying air, the quantity of products entering, the initial and final moisture contents of the product, the specific inlet exergy and humidity ratio, and the net exergy use for drying the products.

8.5.1. Approach

The procedure used here to determine the exergy efficiency of the drying process follows:

- Provide $\dot{m}_a$, $\dot{m}_p$, $(\dot{m}_w)_2$, $(\dot{m}_w)_4$ and $\omega_1 \rightarrow$ calculate ω_3
- Provide T_1, P_1, T_2, P_2, T_3, P_3, T_4 and $P_4 \rightarrow$ determine $\dot{Q}_1$
- Provide $(C_P)_a$, $(C_P)_v$, R_a, R_v, T_{av} and $(x_v)_3 \rightarrow$ determine $\dot{Ex}_d$ and ψ
- Use steam tables, the psychrometric chart and dead state properties accordingly.

The following parameters are considered to be inputs or known parameters in the procedure:

- $\dot{m}_a$, $\dot{m}_p$, $(\dot{m}_w)_2$, $(\dot{m}_w)_4$, ω_1, T_1 and T_2

Table 8.1 presents thermal data related to products and drying air. These data are used to determine the exergy efficiency change with variations in mass flow rate of air, temperature of drying air, specific exergy, specific exergy difference, moisture content of the product and humidity ratio of drying air.

8.5.2. Results

Figure 8.2 illustrates the variation of exergy efficiency with inlet air mass flow rate for several values of the product mass. Increasing mass flow rate reduces the exergy efficiency. Beyond a certain mass flow rate of air, its effect on exergy

Table 8.1. Thermal data used in the example.

(a) Thermophysical properties at different states				
	State 1	State 2	State 3	State 4
Temperature (°C)	55–100	25	25–70	50–95
RH (%)	10–35	55–85	60–95	15–30

(b) Material and reference-environment properties	
$(C_p)_a$	1.004 kJ/kg°C
$(C_p)_v$	1.872 kJ/kg°C
R_a	0.287 kJ/kg°C
R_v	0.4615 kJ/kg°C
T_{av}	50°C
$(x_v)_3$	0.055
T_o	32°C
P_o	101.3 kPa
ω_o	0.0153
$(x_v)_o$	0.024

(c) Mass flows		
$\dot{m}_a$ (kg/s)	m_p (kg)	$\dot{m}_p$ (kg/s)
0	1	0.0002778
1	5	0.0013389
1.5	10	0.0027778
2	15	0.0041667
2.5	20	0.0055556

efficiency decreases. This observation occurs because increasing mass flow rate increases the exergy input to the system, which in turn lowers the exergy efficiency, based on Eq. (8.14). In addition, increasing the mass of product considerably influences the exergy efficiency, i.e., exergy efficiency increases with increasing product mass due to more of the input exergy being utilized in the drying operation rather than flowing through the dryer unchanged.

Figure 8.3 shows the variation of exergy efficiency with inlet drying air temperature for several values of product mass. The behavior of the curves is similar to those shown in Fig. 8.2. Increasing drying air temperature reduces the exergy efficiency, since exergy efficiency is inversely proportional to the exergy rate of drying air. However, the exergy efficiency increases considerably with increasing product mass.

Figure 8.4 exhibits the variation of exergy efficiency of the dryer with specific exergy content of the input drying air for various product amounts. As expected, the exergy efficiency decreases with increasing specific exergy at point 1 for the same amount of products since more exergy loss occurs in the system.

Figure 8.5 shows the variation of dryer exergy efficiency with the difference in specific evaporation exergies of water content for different drying air flow rates. Increasing the specific exergy difference decreases the exergy efficiency. For the same magnitude of the specific exergy difference, a greater mass flow rate of drying air results in a lower exergy efficiency, due to the fact that higher mass flow rates of drying air consume more energy and hence cause greater exergy losses.

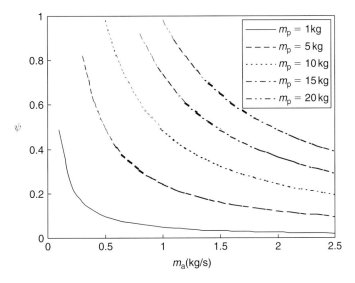

Fig. 8.2. Variation of process exergy efficiency with mass flow rate of drying air for different product weights.

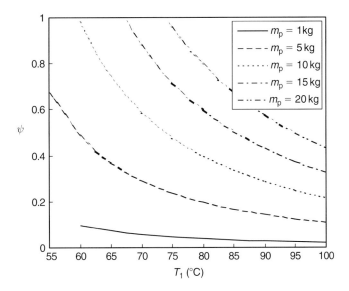

Fig. 8.3. Variation of process exergy efficiency with temperature of drying air for different product weights.

Figure 8.6 shows the exergy efficiency variation with product mass as the mass flow rate of drying air is varied. The exergy efficiency increases linearly with product mass. The exergy efficiency also increases as the mass flow rate of drying air decreases, as expected since the exergy efficiency is inversely proportional to the mass flow rate of drying air. The linear increase of exergy efficiency with product mass indicates that the ratio of the specific exergy difference between the product and the exergy exiting to the exergy of the drying air remains constant for a given product mass.

Figure 8.7 depicts the exergy efficiency variation with the moisture content of the incoming products as the mass flow rate of evaporated water varies. The exergy efficiency increases with increasing moisture content of the products. This effect is more pronounced as the evaporation rate increases. In this case, the energy utilized for drying the product increases when the moisture content of the products increases. Consequently, for given air inlet conditions, the energy utilization in the system is enhanced.

Figure 8.8 exhibits the variation of exergy efficiency of the drying process with the humidity ratio of drying air entering the dryer at different drying air flow rates. A linear relationship is observed between exergy efficiency and

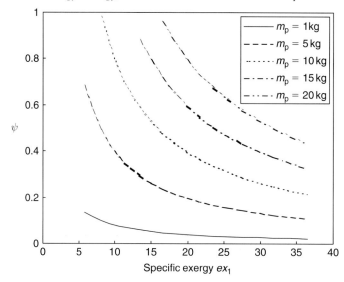

Fig. 8.4. Variation of process exergy efficiency with specific exergy of inlet drying air for different product weights.

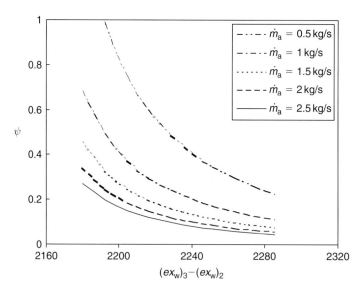

Fig. 8.5. Variation of process exergy efficiency with specific exergy difference of products for different mass flow rates of air.

humidity ratio. Interestingly we note that exergy efficiency varies little (decreasing slightly) with increasing humidity ratio of the drying air.

8.5.3. Discussion

The example demonstrates the usefulness of exergy analysis in assessments of drying systems, and should be useful in optimizing the designs of drying systems and their components, and identifying appropriate applications and optimal configurations for drying systems. Some of the advantages of exergy analysis of drying systems are that it provides:

- A better accounting of the loss of availability of heat in drying.
- More meaningful and useful information than energy analysis on drying efficiency.

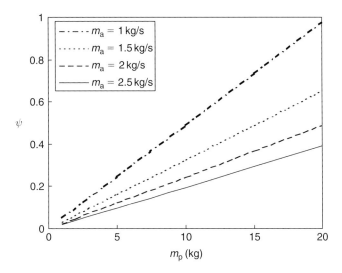

Fig. 8.6. Variation of process exergy efficiency with product weight for different mass flow rates of air.

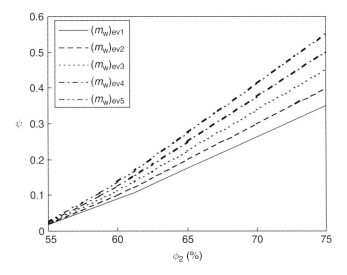

Fig. 8.7. Variation of process exergy efficiency with product moisture content for different moisture evaporation rates.

- A good understanding whether or not and by how much it is possible to design more efficient drying systems by reducing the inefficiencies in the existing units.

Exergy analysis can help reduce the irreversibilities in a drying system and thus increase the exergy efficiency of the system. Increased efficiency also reduces the energy requirement for drying facilities regarding production, transportation, transformation and distribution of energy forms, each of which involve some environmental impact. Exergy analysis is thus useful for determining optimum drying conditions.

When addressing environmental issues for drying systems, it is important to understand the relations between exergy and environmental impact. Enhanced understanding of the environmental problems relating to energy can assist efforts to improve the environmental performance of the drying industry. The relationships between exergy and environmental impact described earlier (Chapter 3) are demonstrated for high-temperature industrial drying systems driven by fossil fuels:

- Order destruction and chaos creation are observed in industrial drying through the degradation of fossil fuel to stack gases and solid wastes, and the unconstrained emission of wastes to the environment.

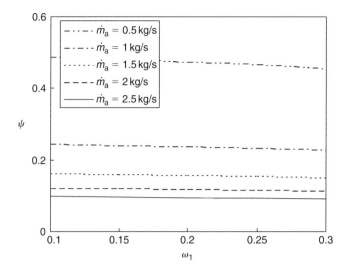

Fig. 8.8. Variation of process exergy efficiency with humidity ratio of drying air for different mass flow rates of air.

- In driving the drying process, a finite resource, fossil fuel, is degraded. Increased process efficiency can reduce this degradation for the same products. If the process considered here were made thermodynamically ideal, the exergy efficiency would increase from about 20% to 100%, and fossil fuel use as well as related emissions would decrease by more than 50%.
- Waste exergy emissions, which represent a potential to impact on the environment, from the drying plant occur with combustion wastes and the waste heat and moist air released to the atmosphere.

8.6. Energy analysis of fluidized bed drying of moist particles

An important and common operation for many products, drying involves the removal of moisture from a wet solid by bringing this moisture into a gaseous state. Usually, water is the liquid evaporated and air is used as the purge gas. Fluidized beds can be advantageously used for drying. Gas–solid fluidization is a process in which the solid phase, under fluidization conditions created by a rapidly flowing gas, assumes a 'fluid-like' state. Fluidized bed drying is carried out in a bed fluidized by the drying medium.

To improve or optimize dryer performance, operating conditions, the material being dried and the drying fluid must be correctly specified. Operating conditions influence the quality of the dried product and include gas velocity, inlet and outlet gas temperatures, feed temperature, and start-up and shutdown parameters.

The energy used in drying is significant and therefore often represents a reducible process cost. Exergy analysis can be used to identify operating conditions in which potential savings can be made. The goal in drying is to use a minimum amount of exergy for maximum moisture removal so as to achieve the desired final conditions of the product.

In this section, energy and exergy analyses are performed of fluidized bed drying and used to optimize the input and output conditions. The effects on energy and exergy efficiencies of hydrodynamic and thermodynamic conditions such as inlet air temperature, fluidization velocity and initial moisture content are analyzed. Two materials are considered: wheat and corn. This section also demonstrates how exergy analysis helps better understand fluidized bed drying and determine effective process improvements.

8.6.1. Fluidized bed drying

Drying involves simultaneous heat and mass transfer. Heat, necessary for evaporation, is supplied to the particles of the material and moisture vapor is transferred from the material into the drying medium. Heat is transported by convection from the surroundings to the particle surfaces and from there, by conduction, further into the particle. Moisture is transported in the opposite direction as a liquid or vapor; on the surface it evaporates and is convected to the surroundings.

Gas–solid fluidization is a process of contact between the two phases. Fluidizing with hot air is an attractive means for drying many moist powder and granular products. The first commercial unit was installed in the U.S. in 1948 (Becken, 1960) to dry dolomite rock. Krokida and Kiranoudis (2000) state that industrial fluidized bed dryers are the most popular family of dryers for agricultural and chemical products in dispersion or multi-dispersion states. During the past two decades various experimental and theoretical studies have been undertaken of fluidized bed drying (e.g., Mujumdar, 1995; Baker, 2000; Krokida and Kiranoudis, 2000; Langrish and Harvey, 2000; Senadeera et al., 2000;), particularly on heat, mass and fluid flow aspects.

Hydrodynamics of fluidized beds

The fluidization gas velocity dominates the behavior of fluidized beds. Fluidized bed drying retains high efficiencies at low fluidization velocities, reducing drying times and energy use. For instance, DiMattia (1993) investigated the effect of fluidization velocity on the slugging behavior of large particles (red spring wheat, long grain rice and whole peas) and found that it is not necessary to operate the bed at a high fluidization velocity.

Particle moisture content and relative humidity of the fluidizing gas can impact fluidization behavior. These effects have been investigated by Hajidavalloo (1998) for two bed materials (sand and wheat). Excessive moisture content of particles may affect the behavior of particles during fluidization.

A general correlation for minimum fluidization velocity, u_{mf}, is given by Kunii and Levenspiel (1991):

$$\frac{1.75}{\varepsilon_{mf}^3 \phi_s} \mathrm{Re}_{mf}^2 + \frac{150(1 - \varepsilon_{mf})}{\varepsilon_{mf}^3 \phi_s^2} \mathrm{Re}_{mf} = \mathrm{Ar} \tag{8.18}$$

where Re is the Reynolds number:

$$\mathrm{Re}_{mf} = \frac{d_p u_{mf} \rho_g}{\mu_g} \tag{8.19}$$

and Ar the Archimedes number:

$$\mathrm{Ar} = \frac{d_p^3 \rho_g (\rho_p - \rho_g) g}{\mu_g^2} \tag{8.20}$$

The minimum fluidization velocity depends on particle moisture content; increasing moisture content increases the minimum fluidization velocity. For wet particle fluidization, the bed pressure drop for velocities above the minimum fluidization point gradually increases with increasing gas velocity (Hajidavalloo, 1998).

At the onset of fluidization, not all particles are fluidized because of adhesive forces in the bed, and the top layers of the bed usually start fluidizing while the bottom layers are still stationary. Thus the bed pressure drop is slightly less than the pressure drop equivalent to the weight of bed material.

Increasing the gas velocity further increases the drag force exerted on the particles, which can separate more contact points between particles, thus bringing them to the fluidized state. The pressure drop increases with increasing gas velocity, as more particles need to be suspended. At a certain velocity, all particles are suspended and full fluidization occurs. At this point the pressure drop is greater than the weight-of-bed pressure drop because of the effect of adhesive forces. Further increases in gas velocity do not necessarily cause the pressure drop to increase linearly.

Material properties

Thermophysical properties (e.g., specific heat) of the particles to be dried in a fluidized bed are highly dependent on the moisture content of the particles. Many correlations for different particles are reported in the literature.

In the example considered in this section, the materials used in the experimental study of Hajidavalloo (1998), red spring wheat and shelled corn, are considered. The wheat kernel is assumed to be spherical with an average diameter of 3.66 mm, and to have a density of 1215 kg/m³ and a specific heat given by Kazarian and Hall (1965):

$$c_m = 1398.3 + 4090.2 \left(\frac{M_p}{1 + M_p} \right) \tag{8.21}$$

The corn kernel has a shape factor close to unity with an average diameter of 6.45 mm, and a density of 1260 kg/m³ and a specific heat (Kazarian and Hall, 1965) of

$$c_m = 1465.0 + 3560.0 \left(\frac{M_p}{1 + M_p} \right) \tag{8.22}$$

Moisture content data are conventionally provided on a dry basis. The normalized moisture content is calculated by dividing the weight of water by the weight of dry material, as follows:

$$M_p = \frac{W_w}{W_d} \quad \text{or} \quad M_p = \frac{W_b - W_d}{W_d} \tag{8.23}$$

8.6.2. Thermodynamic model and balances

A comprehensive thermodynamic model for a fluidized bed drying system is presented in Fig. 8.9. The fluidized bed drying system is divided into three subsystems: blower, heater and drying column. The model facilitates analyses of the effect on energy and exergy efficiencies of air temperature entering the dryer column, fluidization velocity of drying air and initial moisture content of the material. Mass, energy, entropy and exergy balances are derived for the drying column during batch fluidization.

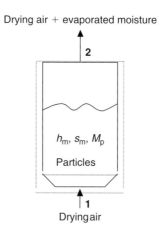

Drying air + evaporated moisture

2

h_m, s_m, M_p

Particles

1

Drying air

Fig. 8.9. Schematic of batch fluidization.

Drying in a batch fluidized bed is modeled by assuming perfect mixing of particles and isobaric behavior. In Fig. 8.9, the control volume is defined by the dashed line, and the thermodynamic state of the particle is described by enthalpy h_m, entropy s_m and moisture content M_p (uniform throughout the bed).

Mass balance for drying column

For the control volume for the drying column shown in Fig. 8.9 with a single inlet and a single exit flow, the following mass rate balance can be written:

$$\frac{dm_{cv}}{dt} = \dot{m}_{g1} - \dot{m}_{g2} \tag{8.24}$$

Here, $\dot{m}_{g1}$ and $\dot{m}_{g2}$ denote, respectively, the mass flow rate entering at (1) and exiting at (2). Similarly, a rate balance for the water in the air flowing through the dryer column leads to

$$W_d \frac{dM_p}{dt} = \dot{m}_a (X_1 - X_2) \tag{8.25}$$

where W_d is the mass of dry solid, $\dot{m}_a$ is the mass flow rate of dry air, and X_1 and X_2 respectively denote the absolute humidity of inlet and outlet air. The left side of the mass balance in Eq. (8.25) is the mass flow rate of water $\dot{m}_w$ in the air flowing out of the bed, so Eq. (8.25) can be written as

$$\dot{m}_w = \dot{m}_a(X_2 - X_1) \tag{8.26}$$

Energy balance for drying column

An energy balance is developed for the drying processes occurring in the control volume in Fig. 8.9. The main heat transfer is due to the heat of evaporation between the solid and the drying air, and there is also heat transfer with the surroundings. The energy rate balance is simplified by ignoring kinetic and potential energies. Since the mass flow rate of the dry air and the mass of dry material within the control volume remain constant with time, the energy rate balance can be expressed as

$$\frac{W_d(h_{m2} - h_{m1})}{\Delta t} = \dot{Q}_{evap} + \dot{m}_a(h_1 - h_2) - \dot{Q}_{loss} \tag{8.27}$$

The differences in specific enthalpy are given by

$$h_{m1} - h_o = c_m(T_{m1} - T_o) \tag{8.28}$$

$$h_{m2} - h_o = c_m(T_{m2} - T_o) \tag{8.29}$$

The specific enthalpy term in the energy rate balance can therefore be expressed as

$$h_{m2} - h_{m1} = c_m(T_{m2} - T_{m1}) \tag{8.30}$$

The specific enthalpy of moist air can be evaluated by adding the contribution of each component as it exits in the mixture:

$$h = h_a + X h_v \tag{8.31}$$

Entropy balance for drying column

The entropy rate balance for the control volume shown in Fig. 8.9 highlights the non-conservation of entropy and can be expressed as

$$\frac{W_d(s_{m2} - s_{m1})}{\Delta t} = \frac{\dot{Q}_{evap}}{T_m} + \dot{m}_a(s_1 - s_2) - \frac{\dot{Q}_{loss}}{T_b} + \dot{S}_{gen} \tag{8.32}$$

The specific entropies of the material are given by

$$s_{m1} - s_o = c_m \ln(T_{m1}/T_o) \tag{8.33}$$

$$s_{m2} - s_o = c_m \ln(T_{m2}/T_o) \tag{8.34}$$

so the material specific entropy term in the entropy rate balance can be expressed as

$$s_{m2} - s_{m1} = c_m \ln(T_{m2}/T_{m1}) \tag{8.35}$$

To evaluate the entropy of moist air, the contribution of each mixture component is determined at the mixture temperature and the partial pressure of the component:

$$s_{wa} = s_a - R_a \ln\frac{p_a}{p_0} + X\left(s_v - R_v \ln\frac{p_v}{p_0}\right) \tag{8.36}$$

Exergy balance for drying column

An exergy balance for the drying column can be obtained using the relevant energy and entropy balances. Multiplying the entropy balance by T_o and subtracting the resulting expression from the energy balance yields:

$$\frac{W_d(\dot{Ex}_{m2} - \dot{Ex}_{m1})}{\Delta t} = \dot{m}_a(h_1 - h_2) + \left(1 - \frac{T_0}{T_m}\right)\dot{Q}_{evap} - \left(1 - \frac{T_0}{T_b}\right)\dot{Q}_{loss} - T_0\dot{m}_a(s_1 - s_2) - T_0\dot{S}_{gen} \tag{8.37}$$

or, more simply,

$$\dot{Ex}_{m2} - \dot{Ex}_{m1} = \dot{Ex}_{da1} - \dot{Ex}_{da2} + \dot{Ex}_{evap} - \dot{Ex}_{loss} - \dot{Ex}_D \qquad (8.38)$$

where $\dot{Ex}_m$ denotes the exergy transfer rate of the material, $\dot{Ex}_{da}$ the exergy transfer rate of the drying air, $\dot{Ex}_{evap}$ the exergy evaporation rate of the dryer, $\dot{Ex}_{loss}$ the exergy loss rate to the surroundings and $\dot{Ex}_D$ the exergy destruction rate in the dryer column.

The inlet and outlet specific exergies of the material are given by

$$ex_{m1} = (h_{m1} - h_o) - T_o(s_{m1} - s_o) \qquad (8.39)$$

$$ex_{m2} = (h_{m2} - h_o) - T_o(s_{m2} - s_o) \qquad (8.40)$$

The specific exergies associated with the drying air entering and exiting the fluidized bed column are given by

$$ex_{da1} = (h_1 - h_o) - T_o(s_1 - s_o) \qquad (8.41)$$

$$ex_{da2} = (h_2 - h_o) - T_o(s_2 - s_o) \qquad (8.42)$$

where ex_{da1} and ex_{da2} are the specific exergies at the inlets and outlets, respectively; h_o and s_o denote respectively the specific enthalpy and specific entropy at the temperature of dead state (T_o); h_1 and s_1 denote respectively the specific enthalpy and the specific entropy at the temperature of drying air entering the fluidized bed column (T_{da1}); and h_2 and s_2 denote respectively the specific enthalpy and the specific entropy of drying air at the temperature of the drying air exiting the column. The potential and kinetic exergies are negligible.

The heat transfer rate due to phase change is

$$\dot{Q}_{evap} = \dot{m}_w h_{fg} \qquad (8.43)$$

where h_{fg} is latent heat of vaporization of water at the average temperature of the wet material and at atmospheric pressure. The exergy transfer rate due to evaporation in the dryer is

$$\dot{Ex}_{evap} = \left[1 - \frac{T_o}{T_m} \right] \dot{m}_w \, h_{fg} \qquad (8.44)$$

8.6.3. Efficiencies for fluidized bed drying

An energy efficiency for the dryer column can be derived using an energy balance following Giner and Calvelo (1987), who define the thermal (or energy) efficiency of the drying process as

$$\eta = \frac{\text{Energy transmitted to the solid}}{\text{Energy incorporated in the drying air}} \qquad (8.45)$$

Using an energy balance, the energy efficiency becomes

$$\eta = \frac{W_d[h_{fg}(M_{p1} - M_{p2}) + c_m(T_{m2} - T_{m1})]}{\dot{m}_{da}(h_1 - h_0)\Delta t} \qquad (8.46)$$

An exergy efficiency for the dryer column, which provides a true measure of its performance, can be derived using an exergy balance. In defining the exergy efficiency it is necessary to identify both a 'product' and a 'fuel.' Here, the product is the exergy evaporation rate and the fuel is the rate of exergy drying air entering the dryer column, and the dryer exergy efficiency is the ratio of product and fuel as outlined by Topic (1995). Then, the exergy efficiency can be expressed as

$$\psi = \frac{\dot{Ex}_{evap}}{\dot{Ex}_{da1}} \qquad (8.47)$$

8.6.4. Effects of varying process parameters

For a fluidized bed drying system, the following input parameters are useful for analyzing the efficiencies of the fluidized bed drying process:

- Temperature of drying air entering the dryer column, T_1.
- Relative humidity of drying air, RH_1.
- Velocity of drying air, u.
- Temperature of the material entering the dryer, T_{pi}.
- Initial moisture content of the material, M_{pi}.
- Weight of the material, W_b.
- Ambient temperature, T_a.

The following additional thermal parameters can also be varied for analysis purposes. In the example considered, values are obtained from Hajidavalloo (1998) and used as inputs:

- Temperature of drying air leaving the dryer column, T_2.
- Relative humidity of drying air leaving the dryer, RH_2.
- Absolute humidity of drying air leaving the dryer, X_2.
- Moisture content of the material after the drying process, M_{pf}.
- Temperature of the material after the drying process, T_{pf}.
- Drying time, Δt.

The following data are obtained from thermodynamic tables for both vapor and dry air:

- Enthalpy of dry air h_a and enthalpy of water vapor h_v entering the dryer column.
- Enthalpy of dry air h_0 and enthalpy of water vapor h_{v0} at the ambient temperature.
- Enthalpy evaporation h_{fg} at the material temperature T_m.
- Entropy of dry air s_a and entropy of water vapor s_v entering the dryer.
- Entropy of dry air s_0 and entropy of water vapor s_{v0} at the ambient temperature.

Several important points relating to the analyses are highlighted:

- The analyses are not discussed in differential form to keep the results simple and more useful to those who design and assess drying systems. A practical thermodynamic analysis is thus presented based on mass, energy, entropy and exergy balances, and is validated with experimental data from Hajidavalloo (1998).
- Spatial variations in physical and thermophysical quantities are considered negligible for simplicity. This treatment is consistent with that of Hajidavalloo (1998).
- The quantities T_0, h_0 and s_0 represent thermodynamic properties at the dead state or reference-environment conditions, which are ambient external conditions.

8.6.5. Example

An analysis is carried out for two materials: wheat and corn. These are major agricultural commodities which require extensive drying (Syahrul, 2000). The effects are examined of varying inlet air temperature, fluidization velocity and initial moisture content on both efficiency and drying rate. Experimental data of Hajidavalloo (1998) are used as input parameters and for model verification. Hajidavalloo experimentally studied various parameters, and collected various data (temperature and relative humidity of drying air and moisture content of material in the bed).

Although wheat and corn are both hygroscopic materials, the nature of their moisture diffusivity differs. The moisture diffusion coefficient of wheat is dependent only on temperature, but for corn is a function of both particle temperature and moisture content (Chu and Hustrulid, 1968). These materials also differ in size, with corn grains usually being many times larger than wheat grains. These differences can lead to different drying behaviors and different efficiencies for fluidized bed dryers that process these particles. For further details on the analysis, examples and their results see Syahrul et al. (2002a,b,c).

Results for wheat

The conditions of the inlet air and the inlet material for each drying test are given in Tables 8.2 through 8.4. The results obtained for the analysis of wheat particles are shown in Figs. 8.10 through 8.16 and include the effects of varying inlet drying air temperature, fluidization velocity and initial moisture content.

Table 8.2. Experimental conditions for investigating the effect of temperature, for wheat.

	T (°C)	M_{pi} (db)	W_b (kg)	RH (%)	u (m/s)	T_a (°C)	T_{pi} (°C)
Run 8	40.2	0.326	2.5	21.1	1.95	22.0	7.0
Run 11	65.0	0.317	2.5	18.5	1.95	22.0	6.0

Table 8.3. Experimental conditions for investigating the effect of gas velocity, for wheat.

	T (°C)	M_{pi} (db)	W_b (kg)	RH (%)	u (m/s)	T_a (°C)	T_{pi} (°C)
Run 6	49.5	0.300	2.5	13.5	1.95	18.0	6.0
Run 12	50.0	0.323	2.5	15.7	1.63	23.0	6.0

Table 8.4. Experimental conditions for investigating the effect of initial moisture content, for wheat.

	T (°C)	M_{pi} (db)	W_b (kg)	RH (%)	u (m/s)	T_a (°C)	T_{pi} (°C)
Run 2	54.5	0.409	2.54	17.0	1.91	20.5	7.0
Run 4	54.0	0.307	2.48	14.7	1.93	20.0	7.0

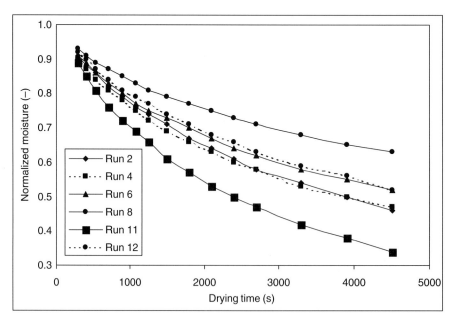

Fig. 8.10. Normalized moisture content profiles of wheat vs. drying time for different experimental runs.

It is generally observed that energy efficiencies are higher than the corresponding exergy efficiencies. Furthermore, the energy and exergy efficiencies are observed to be higher at the beginning of the drying process than at the end. The exergy of evaporation is notably high at the initial stage of the drying process due to rapid evaporation of surface moisture but decreases exponentially until the end of the process as surface moisture evaporates.

The observation that energy and exergy efficiencies for wheat drying are low at the end of drying (i.e., less than 10% for energy efficiency and 5% for exergy efficiency) can be explained by noting that surface moisture evaporates quickly

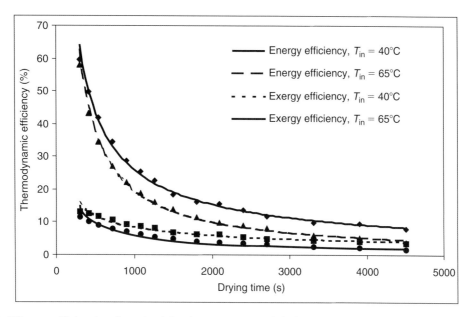

Fig. 8.11. Effect on efficiencies of varying inlet air temperature and drying time for wheat.

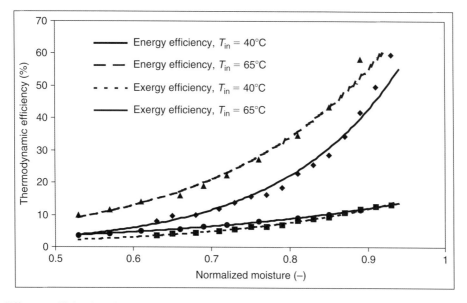

Fig. 8.12. Effect on efficiencies of varying inlet air temperature and normalized moisture for wheat.

due to high heat and mass transfer coefficients in fluidized bed systems. Thus the drying rate is high in the initial stage of the process, but low at the end when all surface moisture has evaporated and the drying front diffuses inside the material.

Figures 8.11 and 8.12 show the effects on energy and exergy efficiencies of varying inlet air temperature, drying time and normalized moisture content. The energy efficiency is found to be higher than the exergy efficiency. Furthermore, the temperature of the inlet air (the drying medium) influences the energy and exergy efficiencies non-linearly. A 25°C increase in inlet air temperature leads to an approximate increase in energy efficiency of 7% and in exergy efficiency of 1%. Higher inlet drying air temperatures also result in shorter drying times. But the inlet air temperature is limited because it can cause considerable damage to the material at high values. The final temperature of the material after long time spans approaches the temperature of the inlet drying air.

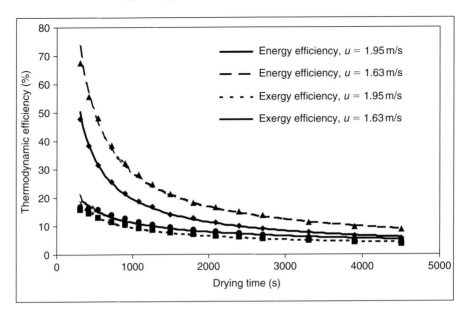

Fig. 8.13. Effect on efficiencies of varying gas velocity and drying time for wheat.

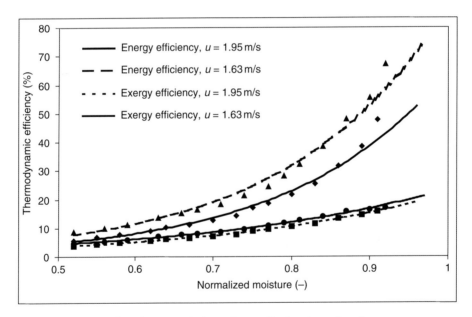

Fig. 8.14. Effect on efficiencies of varying gas velocity and normalized moisture for wheat.

Figures 8.13 and 8.14 present the effects on dryer efficiencies of varying gas velocity, drying time and normalized moisture content of the particle. A reduction of 15% in air velocity is observed to have little effect on drying time. Little change occurs in drying properties with time, which is consistent with observations of Hajidavalloo (1998). The drying rate is governed by the rate of internal moisture movement, and the influence of external variables diminishes with time, as defined by Perry et al. (1997). Figures 8.13 and 8.14 also show that a 15% reduction in air velocity roughly increases energy efficiency by 3% and exergy efficiency by 1%.

The narrow difference in velocities considered for the two test conditions is dictated by fluidization requirements. Since fluidization of wet particles requires high gas velocities, the velocity can be only somewhat reduced. A correlation based on experimental data (Hajidavalloo, 1998) shows that the minimum fluidization velocity for this investigation is

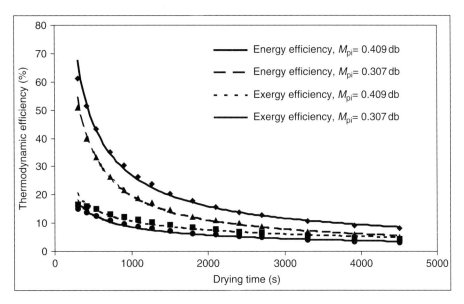

Fig. 8.15. Effect on efficiencies of varying initial moisture content and drying time for wheat.

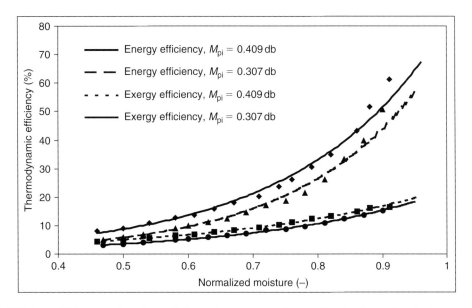

Fig. 8.16. Effect on efficiencies of varying initial moisture content and normalized moisture for wheat.

1.22 m/s. It could be advantageous to use a high air velocity at the first drying stage and then to reduce it as drying proceeds.

Figures 8.15 and 8.16 present the effects on dryer efficiencies of varying initial and normalized moisture contents and drying time. Higher energy and exergy efficiencies are observed for particles with high initial moisture contents, mainly due to the drying rate time lag. Increasing moisture content causes a time lag in the maximum drying rate in the initial stage of drying (Hajidavalloo, 1998). A greater portion of the input exergy goes toward evaporation when the material being dried has a higher moisture content. However, there are practical restrictions since the wet material must be fluidizable.

Results for corn

Corn kernels are usually many times larger than wheat kernels. The moisture diffusivity of corn is a function of temperature and particle moisture content but that of wheat is dependent only on temperature. Since mass diffusion controls drying rate, drying patterns for corn can differ from those for wheat, as can the efficiencies of the fluidized bed dryer column.

The test conditions for corn are presented in Tables 8.5 through 8.7 and the analysis results in Figs 8.17 through 8.23. The results for corn are generally similar to those for wheat, although energy and exergy efficiencies for corn are lower than for wheat. Like for wheat, energy efficiencies are higher than the corresponding exergy efficiencies, and both efficiencies are higher at the beginning of the drying process than at the end.

Table 8.5. Experimental conditions for investigating the effect of temperature, for corn.

	T (°C)	M_{pi} (db)	W_b (kg)	RH (%)	u (m/s)	T_a (°C)	T_{pi} (°C)
Run C1	50.0	0.256	2.5	15.2	2.22	17.0	7.0
Run C3	63.0	0.246	2.5	17.5	2.24	17.5	7.0

Table 8.6. Experimental conditions for investigating the effect of velocity, for corn.

	T (°C)	M_{pi} (db)	W_b (kg)	RH (%)	u (m/s)	T_a (°C)	T_{pi} (°C)
Run C1	50.0	0.256	2.5	15.2	2.22	17.0	7.0
Run C4	50.0	0.257	2.5	17.5	1.88	17.6	7.0

Table 8.7. Experimental conditions for investigating the effect of initial moisture content, for corn.

	T (°C)	M_{pi} (db)	W_b (kg)	RH (%)	u (m/s)	T_a (°C)	T_{pi} (°C)
Run C1	50.0	0.256	2.5	15.2	2.22	17.0	7.0
Run C5	50.0	0.324	2.5	17.0	2.21	18.2	6.0

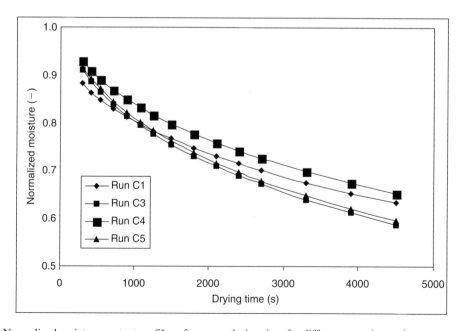

Fig. 8.17. Normalized moisture content profiles of corn vs. drying time for different experimental runs.

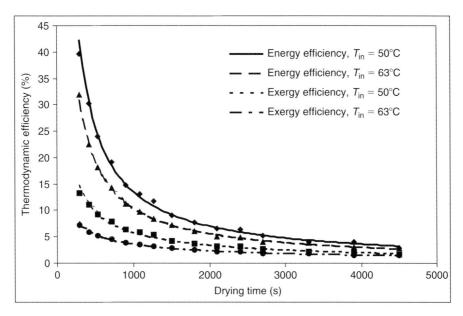

Fig. 8.18. Effect on efficiencies of varying inlet air temperature and drying time for corn.

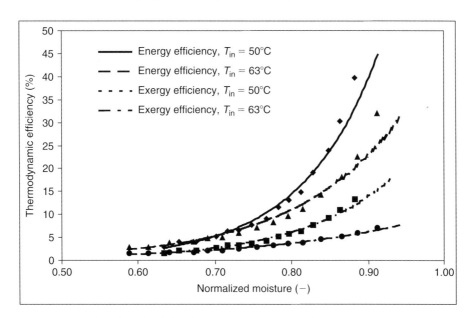

Fig. 8.19. Effect on efficiencies of varying inlet air temperature and normalized moisture for corn.

Figures 8.18 and 8.19 show the effect of inlet air temperature on efficiencies, for inlet air temperatures ranging from 50°C to 63°C. Energy and exergy efficiencies approach similar low values at the end of drying. Since the initial moisture content of the material is below the critical moisture content, the effect of external variables is of reduced importance as the drying rate is governed by the rate of internal moisture movement. Unlike for wheat, the moisture diffusion coefficient for corn is a function of temperature and moisture content. Increasing temperature does not necessarily increase drying efficiencies for corn.

The effects on efficiencies of gas velocity, drying time and normalized moisture are shown in Figs 8.20 and 8.21. Energy and exergy efficiencies are similar at the final stage of the process, but differ at the initial stage when surface moisture content is removed from the grains. Thus, it is advantageous to use a relatively low gas velocity, recognizing

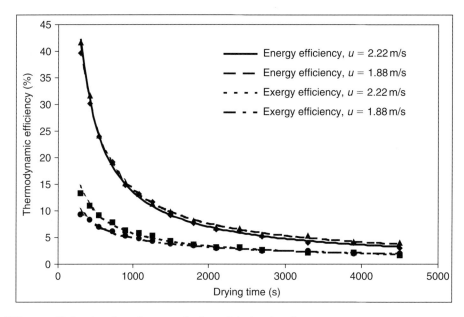

Fig. 8.20. Effect on efficiencies of varying gas velocity and drying time for corn.

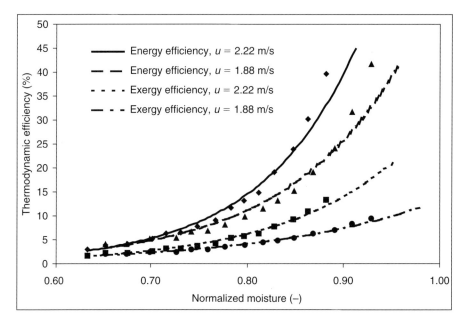

Fig. 8.21. Effect on efficiencies of varying gas velocity and normalized moisture for corn.

the practical restrictions associated with fluidization. DiMattia (1993) also found that fluidized bed dryers have high efficiencies at low fluidization velocities. Hajidavalloo's model for minimum fluidization velocity indicates for this experiment that the velocity for the onset of fluidization is 1.16 m/s.

Figures 8.22 and 8.23 show the effects on efficiencies of initial moisture content, drying time and normalized moisture. At the initial stage, the efficiencies are affected mainly by rapid evaporation but, after the surface moisture evaporates, the efficiencies decrease as the initial moisture contents of the materials decrease. A difference is observed in the efficiencies at the end of the drying process. Hajidavalloo (1998) also observed a higher drying rate for corn, which has a higher initial moisture content.

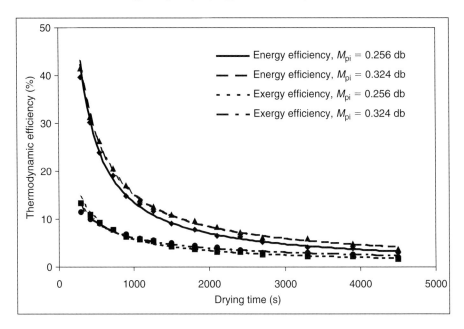

Fig. 8.22. Effect on efficiencies of varying initial moisture content and drying time for corn.

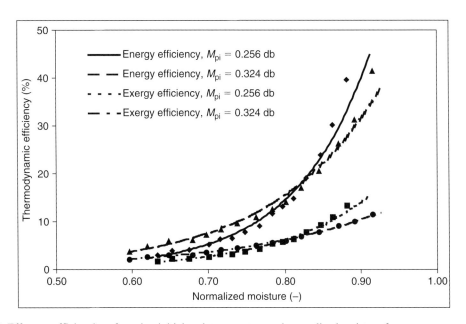

Fig. 8.23. Effect on efficiencies of varying initial moisture content and normalized moisture for corn.

Generalizations

Several generalizations can be drawn from the thermodynamic analyses of various aspects of the fluidized bed drying systems described here (Syahrul et al., 2002a,b,c):

- Energy and exergy efficiencies are higher at the beginning of the drying process than at the end since the moisture removal rate from wet particles is higher in the beginning.
- Inlet air temperature has an effect on the efficiencies of fluidized bed dryer systems, but the effect may vary with particle physical properties. For wheat particles, where the diffusion coefficient is a function only of temperature,

increasing the drying air temperature increases the efficiency non-linearly. For corn particles, where the diffusion coefficient depends on temperature and moisture content, increasing drying air temperature does not necessarily increase efficiency.

- The effect of gas velocity on energy and exergy efficiencies depends on the materials. For wheat, energy and exergy efficiencies increase with reduced air velocity. However, for corn, energy and exergy efficiencies do not exhibit any difference at the end of drying.
- The efficiencies are higher for particles with high initial moisture contents.
- Variable drying air temperature and velocity through the dryer could lead to increased efficiencies.

8.7. Concluding remarks

Exergy analyses of drying processes and systems have been presented. Exergy efficiencies are functions of heat and mass transfer parameters. An example illustrates the applicability of the method to the drying of moist solids with air, and highlights the sensitivities of the results to such parameters as drying air temperature, moisture content, humidity ratio and specific exergy, the exergy difference between inlet and outlet products, and product mass. Another example considers fluidized bed drying of moist particles. Exergy analysis is demonstrated to be a significant tool for design and optimization of drying processes.

Problems

8.1   Identify three sources of exergy loss in air drying and propose methods for reducing or minimizing them.

8.2   Identify three sources of exergy loss for drum drying and propose methods for reducing or minimizing them.

8.3   Identify three sources of exergy loss for freeze drying and propose methods for reducing or minimizing them.

8.4   High-quality energy sources capable of generating high temperatures, such as fossil fuels, are often used for relatively low-temperature processes like water and space heating or cooling, industrial drying, industrial steam production, etc. Explain how exergy analysis can pinpoint the losses in such processes and help to better match energy source to the end use.

8.5   Briefly explain the effects on the exergy efficiency of the dryer considered in the illustrative example of mass flow rate of the drying air, temperature of the drying air, the amount of products entering, the initial moisture content of the product, the final moisture content of the product, the specific inlet exergy, the humidity ratio and the net exergy use for drying the products.

8.6   Rework the illustrative example provided in Section 8.5 using the given input data and try to duplicate the results. If your results differ from those given in the example, discuss why. Propose methods for improving the performance of the system based on reducing or minimizing exergy destruction.

8.7   Obtain a published article on exergy analysis of drying systems. Using the operating data provided in the article, perform a detailed exergy analysis of the system and compare your results to those in the original article. Also, investigate the effect of varying important operating parameters on the system exergetic performance.

Chapter 9

EXERGY ANALYSIS OF THERMAL ENERGY STORAGE SYSTEMS

9.1. Introduction

Thermal energy storage (TES) generally involves the temporary storage of high- or low-temperature thermal energy for later use. Examples of TES are storage of solar energy for overnight heating, of summer heat for winter use, of winter ice for space cooling in summer and of heat or cool generated electrically during off-peak hours for use during subsequent peak demand hours. In this regard, TES is in many instances an excellent candidate to offset this mismatch between thermal energy availability and demand.

TES systems for heating or cooling capacity are often utilized in applications where the occurrence of a demand for energy and that of the economically most favorable supply of energy are not coincident. Thermal storages are used in energy conservation, industry, commercial building and solar energy systems. The storage medium can be located in storages of various types, including tanks, ponds, caverns and underground aquifers.

The storage medium in a TES can remain in a single phase (so that only sensible heat is stored) and/or undergo phase change (so that energy is stored as latent heat). Sensible TESs (e.g., liquid water systems) exhibit changes in temperature in the store as heat is added or removed. In latent TESs (e.g., liquid water/ice systems and eutectic salt systems), the storage temperature remains fixed during the phase-change portion of the storage cycle.

TES systems are used in a wide variety of applications and are designed to operate on a cyclical basis (usually daily, occasionally seasonally). TES systems achieve benefits by fulfilling one or more of the following purposes:

- *Increase generation capacity*: Demand for heating, cooling or power is seldom constant over time, and the excess generation capacity available during low-demand periods can be used to charge a TES in order to increase the effective generation capacity during high-demand periods. This process allows a smaller production unit to be installed (or to add capacity without purchasing additional units) and results in a higher load factor for the units.
- *Enable better operation of cogeneration plants*: Combined heat and power, or cogeneration, plants are generally operated to meet the demands of the connected thermal load, which often results in excess electric generation during periods of low electric use. By incorporating TES, the plant need not be operated to follow a load. Rather it can be dispatched in more advantageous ways (within some constraints).
- *Shift energy purchases to low-cost periods*: This use is the demand-side application of the first purpose listed, and allows energy consumers subject to time-of-day pricing to shift energy purchases from high- to low-cost periods.
- *Increase system reliability*: Any form of energy storage, from the uninterruptible power supply of a small personal computer to a large pumped storage, normally increases system reliability.
- *Integration with other functions*: In applications where on-site water storage is needed for fire protection, it may be feasible to incorporate thermal storage into a common storage tank. Likewise, equipment designed to solve power-quality problems may be adaptable to energy-storage purposes as well.

The most significant benefit of a TES system is often cited as its ability to reduce electric costs by using off-peak electricity to produce and store energy for daytime cooling. Indeed, TES systems successfully operate in offices, hospitals, schools, universities, airports, etc. in many countries, shifting energy consumption from periods of peak electricity rates to periods of lower rates. That benefit is accompanied by the additional benefit of lower demand charges.

Having investigated methods for TES evaluation and comparison for many years and recently combined the results (Dincer and Rosen, 2002), the authors conclude that, while many technically and economically successful thermal storages are in operation, no generally valid basis for comparing the achieved performance of one storage with that of another operating under different conditions has found broad acceptance. The energy efficiency, the ratio of the energy recovered from storage to that originally input, is conventionally used to measure TES performance. The energy efficiency, however, is inadequate because it does not take into account important factors like how nearly the performance approaches ideality, storage duration and temperatures of the supplied and recovered thermal energy and of the surroundings.

Exergy analysis provides an illuminating, rational and meaningful alternative for assessing and comparing TES systems. In particular, exergy analysis yields efficiencies which provide a true measure of how nearly actual performance approaches the ideal, and identifies more clearly than energy analysis the magnitudes, causes and locations of thermodynamic losses. Consequently, exergy analysis can assist in improving and optimizing TES designs.

Using information in the authors' recent book on TES (Dincer and Rosen, 2002), this chapter describes the application of exergy analysis to TES and demonstrates the usefulness of such analyses in providing insights into TES behavior and performance. Key thermodynamic considerations in TES evaluation are discussed, and the use of exergy in evaluating a TES system is detailed. The relation of temperature to efficiency is highlighted, and thermal stratification, cold TES and aquifer TES are considered.

9.2. Principal thermodynamic considerations in TES

Several of the principal thermodynamic considerations in TES evaluation and comparison are discussed in this section.

Energy and exergy: Energy and exergy are significant quantities in evaluating TES systems. Exergy analysis complements energy analysis and circumvents many of the difficulties associated with conventional energy-based TES methods by providing a more rational evaluation and comparison basis.

Temperature: Exergy reflects the temperature of a heat transfer and the degradation of heat quality through temperature loss. Exergy analysis applies equally well to systems for storing thermal energy at temperatures above and below the temperature of the environment because the exergy associated with such energy is always greater than or equal to zero. Energy analysis is more difficult to apply to such storage systems because efficiency definitions have to be carefully modified when cooling capacity, instead of heating capacity, is stored, or when both warm and cool reservoirs are included. Thus, exergy analysis provides for more rational evaluation of TES systems for cooling or heating capacity.

Efficiencies: The evaluation of a TES system requires a measure of performance which is rational, meaningful and practical. A more perceptive basis than energy efficiency is needed if the true usefulness of thermal storages is to be assessed, and so permit maximization of their economic benefit. Exergy efficiencies provide rational measures since they assess the approach to ideal TES performance.

Losses: With energy analysis, all losses are attributable to energy releases across system boundaries. With exergy analysis, losses are divided into two types: exergy releases from the system and internal exergy consumptions. The latter include reductions in availability of the stored heat through mixing of warm and cool fluids. The division of exergy losses allows the causes of inefficiencies to be accurately identified and improvement effort to be effectively allocated.

Stratification: Thermal stratification within a TES reduces temperature degradation. In many practical cases, a vertical cylindrical tank with a hot water inlet (outlet) at the top and a cold water inlet (outlet) at the bottom is used. The hot and cold water in the tank usually are stratified initially into two layers, with a mixing layer in between. The degree of stratification is affected by the volume and configuration of the tank, the design of the inlets and outlets, the flow rates of the entering and exiting streams, and the durations of the charging, storing and discharging periods. Increasing stratification improves TES efficiency relative to a thermally mixed-storage tank. Four primary factors degrade stored energy by reducing stratification:

1. heat leakages to or from the environment,
2. heat conduction and convection from the hot portions of the storage fluid to the colder portions,
3. vertical conduction in the tank wall,
4. mixing during charging and discharging periods (often the main cause of loss of stratification).

The effects of stratification are more clearly assessed with exergy than energy. Through carefully managing the injection, recovery and holding of heat (or cold) so that temperature degradation is minimized, better storage-cycle performance can be achieved (as measured by better thermal energy recovery and temperature retention, i.e., increased exergy efficiency).

Storage duration: Rational evaluation and comparison of TESs must account for storage duration. The length of time thermal energy is retained in a TES does not enter into expressions for efficiency, although it is clearly a dominant consideration in overall TES effectiveness. By examining the relation between storage duration and effectiveness, the authors developed an approach for comparing TESs using a time parameter.

Reference-environment temperature: Since TES evaluations based on energy and exergy are affected by the value of the reference environment temperature T_0, temporal and spatial variations of T_0 must be considered (especially for

TESs with storage periods of several months). The value of $T_o(t)$ can often be assumed to be the same as the ambient temperature variation with time, $T_{amb}(t)$, approximated on an annual basis as

$$T_{amb}(t) = \overline{T}_{amb} + \Delta T_{amb}\left[\sin\frac{2\pi t}{\text{period}} + (\text{phase shift})\right] \qquad (9.1)$$

where $\overline{T}_{amb}$ is the mean annual ambient temperature and ΔT_{amb} is the maximum temperature deviation from the annual mean. The values of the parameters in Eq. (9.1) vary spatially and the period is 1 year. For most short-term storages, a constant value of T_o can be assumed. Some possible values for T_o are the annual or seasonal mean value of the temperature of the atmosphere, or the constant temperature of soil far below the surface.

9.3. Exergy evaluation of a closed TES system

An exergy analysis of a closed tank storage with heat transfers by heat exchanger is described in this section. A complete storing cycle, as well as the charging, storing and discharging periods, are considered. Although energy is conserved in an adiabatic system, mixing of the high- and low-temperature portions of the storage medium consumes exergy, which is conserved only in reversible processes.

For the TES considered here, a closed system stores heat in a fixed amount of storage fluid, to or from which heat is transferred through a heat exchanger by means of a heat transport fluid. The TES system undergoes a complete storage cycle, with final and initial states identical. Figure 9.1 illustrates the three periods in the overall storage process considered. The TES may be stratified. Other characteristics of the considered case are:

- constant storage volume with non-adiabatic storage boundaries,
- finite charging, storing and discharging time periods,
- surroundings at constant temperature and pressure,
- negligible work interactions (e.g., pump work), and kinetic and potential energy terms.

Fig. 9.1. The three stages in a simple heat storage process: charging period (left), storing period (center) and discharging period (right).

The operation of the heat exchangers is simplified by assuming no heat losses to the environment from the charging and discharging fluids. That is, it is assumed during charging that heat removed from the charging fluid is added to the storage medium, and during discharging that heat added to the discharging fluid originates in the storage medium. This assumption is reasonable if heat losses from the charging and discharging fluids are small compared with heat losses from the storage medium. This assumption can be extended by lumping actual heat losses for the charging, storing and discharging periods. The charging and discharging fluid flows are modeled as steady and one-dimensional, with time-independent properties.

9.3.1. Analysis of the overall processes

For the cases considered (Fig. 9.1), energy and exergy balances and efficiencies are provided for the overall process.

Overall energy balance

An energy balance for the overall storage process can be written as

$$\text{Energy input} - [\text{Energy recovered} + \text{Energy loss}] = \text{Energy accumulation} \qquad (9.2)$$

or

$$(H_a - H_b) - [(H_d - H_c) + Q_l] = \Delta E \qquad (9.3)$$

where H_a, H_b, H_c and H_d are the total enthalpies of the flows at states a, b, c and d respectively; Q_l denotes the heat losses during the process and ΔE the accumulation of energy in the TES. In Eq. (9.3), $(H_a - H_b)$ represents the net heat delivered to the TES and $(H_d - H_c)$ the net heat recovered from the TES. The quantity in square brackets represents the net energy output from the system. The terms ΔE and Q_l are given by

$$\Delta E = E_f - E_i \tag{9.4}$$

$$Q_l = \sum_{j=1}^{3} Q_{l,j} \tag{9.5}$$

Here, E_i and E_f denote the initial and final energy contents of the storage and $Q_{l,j}$ denotes the heat losses during the period j, where $j = 1$, 2 and 3 correspond to the charging, storing and discharging periods, respectively. In the case of identical initial and final states, $\Delta E = 0$ and the overall energy balance simplifies.

Overall exergy balance

An overall exergy balance can be written as

Exergy input − [Exergy recovered + Exergy loss] − Exergy consumption = Exergy accumulation   (9.6)

or

$$(Ex_a - Ex_b) - [(Ex_d - Ex_c) + X_l] - I = \Delta Ex \tag{9.7}$$

Here, Ex_a, Ex_b, Ex_c and Ex_d are the exergies of the flows at states a, b, c and d respectively; and X_l denotes the exergy loss associated with Q_l; I the exergy consumption; and ΔEx the exergy accumulation. In Eq. (9.7), $(Ex_a - Ex_b)$ represents the net exergy input and $(Ex_d - Ex_c)$ the net exergy recovered. The quantity in square brackets represents the net exergy output from the system. The terms I, X_l and ΔEx are given respectively by

$$I = \sum_{j=1}^{3} I_j \tag{9.8}$$

$$X_l = \sum_{j=1}^{3} X_{l,j} \tag{9.9}$$

$$\Delta Ex = Ex_f - Ex_i \tag{9.10}$$

Here, I_1, I_2 and I_3 denote respectively the consumptions of exergy during charging, storing and discharging; $X_{l,1}$, $X_{l,2}$ and $X_{l,3}$ denote the corresponding exergy losses; and Ex_i and Ex_f denote the initial and final exergy contents of the storage. When the initial and final states are identical, $\Delta Ex = 0$.

The exergy content of the flow at the states k = a, b, c, d is evaluated as

$$Ex_k = (H_k - H_o) - T_o(S_k - S_o) \tag{9.11}$$

where Ex_k, H_k, and S_k denote the exergy, enthalpy and entropy of state k respectively, and H_o and S_o the enthalpy and the entropy at the temperature T_o and pressure P_o of the reference environment. The expression in Eq. (9.11) only includes physical exergy, as potential and kinetic exergy are assumed negligible. The chemical component of exergy is neglected because it does not contribute to the exergy flows for sensible TES systems. Thus, the exergy differences between the inlet and outlet for the charging and discharging periods are respectively:

$$Ex_a - Ex_b = (H_a - H_b) - T_o(S_a - S_b) \tag{9.12}$$

and

$$Ex_d - Ex_c = (H_d - H_c) - T_o(S_d - S_c) \tag{9.13}$$

Here it has been assumed that T_o and P_o are constant, so that H_o and S_o are constant at states a and b, and at states c and d.

For a fully mixed tank, the exergy losses associated with heat losses to the surroundings are evaluated as

$$X_{l,j} = \int_i^f \left(1 - \frac{T_o}{T_j}\right) dQ_{l,j} \quad \text{for } j = 1, 2, 3 \tag{9.14}$$

where j represents the particular period. If T_1, T_2 and T_3 are constant during the respective charging, storing and discharging periods, then $X_{i,j}$ may be written as follows:

$$X_{l,j} = \left(1 - \frac{T_o}{T_j}\right) Q_{l,j} \tag{9.15}$$

Sometimes when applying Eq. (9.15) to TES systems, T_j represents a mean temperature within the tank for period j.

Overall energy and exergy efficiencies

The energy efficiency η can be expressed as

$$\eta = \frac{\text{Energy recovered from TES during discharging}}{\text{Energy input to TES during charging}} = \frac{H_d - H_c}{H_a - H_b} = 1 - \frac{Q_l}{H_a - H_b} \tag{9.16}$$

and the exergy efficiency ψ as

$$\psi = \frac{\text{Exergy recovered from TES during discharging}}{\text{Exergy input to TES during charging}} = \frac{Ex_d - Ex_c}{Ex_a - Ex_b} = 1 - \frac{X_l + I}{Ex_a - Ex_b} \tag{9.17}$$

The efficiency expressions in Eqs. (9.16) and (9.17) do not depend on the initial energy and exergy contents of the TES.

If the TES is adiabatic, $Q_{l,j} = X_{l,j} = 0$ for all j, and the energy efficiency is fixed at unity and the exergy efficiency simplifies to

$$\psi = 1 - \frac{I}{Ex_a - Ex_b} \tag{9.18}$$

This result demonstrates that when TES boundaries are adiabatic and there are no energy losses, the exergy efficiency is less than unity due to internal irreversibilities.

9.3.2. Analysis of subprocesses

Although several energy and exergy efficiencies can be defined for charging, storing and discharging, one set of efficiencies is considered here.

Analysis of charging period

An energy balance for the charging period can be written as

$$\text{Energy input} - \text{Energy loss} = \text{Energy accumulation} \tag{9.19}$$

$$(H_a - H_b) - Q_{l,1} = \Delta E_1 \tag{9.20}$$

Here,

$$\Delta E_1 = E_{f,1} - E_{i,1} \tag{9.21}$$

and $E_{i,1} (= E_i)$ and $E_{f,1}$ denote the initial and the final energy of the TES for charging. A charging-period energy efficiency can be expressed as

$$\eta_1 = \frac{\text{Energy accumulation in TES during charging}}{\text{Energy input to TES during charging}} = \frac{\Delta E_1}{H_a - H_b} \tag{9.22}$$

An exergy balance for the charging period can be written as

$$\text{Exergy input} - \text{Exergy loss} - \text{Exergy consumption} = \text{Exergy accumulation} \tag{9.23}$$

$$(Ex_a - Ex_b) - X_{1,1} - I_1 = \Delta Ex_1 \tag{9.24}$$

Here,

$$\Delta Ex_1 = Ex_{f,1} - Ex_{i,1} \tag{9.25}$$

and $Ex_{i,1}$ $(= Ex_i)$ and $Ex_{f,1}$ denote the initial and the final exergy of the TES for charging. A charging-period exergy efficiency can be expressed as

$$\psi_1 = \frac{\text{Exergy accumulation in TES during charging}}{\text{Exergy input to TES during charging}} = \frac{\Delta Ex_1}{Ex_a - Ex_b} \tag{9.26}$$

The efficiencies in Eqs. (9.22) and (9.26) indicate the fraction of the input energy/exergy which is accumulated in the store during the charging period.

Analysis of storing period

An energy balance for the storing period can be written as

$$-\text{Energy loss} = \text{Energy accumulation} \tag{9.27}$$

$$-Q_{1,2} = \Delta E_2 \tag{9.28}$$

Here,

$$\Delta E_2 = E_{f,2} - E_{i,2} \tag{9.29}$$

and $E_{i,2}$ $(= E_{f,1})$ and $E_{f,2}$ denote the initial and final energy contents of the TES for storing. An energy efficiency for the storing period can be expressed as

$$\eta_2 = \frac{\text{Energy accumulation in TES during charging and storing}}{\text{Energy accumulation in TES during charging}} = \frac{\Delta E_1 + \Delta E_2}{\Delta E_1} \tag{9.30}$$

Using Eq. (9.28), the energy efficiency can be rewritten as

$$\eta_2 = \frac{\Delta E_1 - Q_{1,2}}{\Delta E_1} \tag{9.31}$$

An exergy balance for the storing period can be written as

$$-\text{Exergy loss} - \text{Exergy consumption} = \text{Exergy accumulation} \tag{9.32}$$

$$-X_{1,2} - I_2 = \Delta Ex_2 \tag{9.33}$$

Here,

$$\Delta Ex_2 = Ex_{f,2} - Ex_{i,2} \tag{9.34}$$

and $Ex_{i,2}$ $(= Ex_{f,1})$ and $Ex_{f,2}$ denote the initial and final exergies of the system for storing. An exergy efficiency for the storing period can be expressed as

$$\psi_2 = \frac{\text{Exergy accumulation in TES during charging and storing}}{\text{Exergy accumulation in TES during charging}} = \frac{\Delta Ex_1 + \Delta Ex_2}{\Delta Ex_1} \tag{9.35}$$

Using Eq. (9.33), the exergy efficiency can be rewritten as

$$\psi_2 = \frac{\Delta Ex_1 - (X_{1,2} + I_2)}{\Delta Ex_1} \tag{9.36}$$

The efficiencies in Eqs. (9.30) and (9.35) indicate the fraction of the energy/exergy accumulated during charging which is still retained in the store at the end of the storing period.

Analysis of discharging period

An energy balance for the discharging period can be written as

$$-[\text{Energy recovered} + \text{Energy loss}] = \text{Energy accumulation} \tag{9.37}$$

$$-[(H_d - H_c) + Q_{l,3}] = \Delta E_3 \tag{9.38}$$

Here,

$$\Delta E_3 = E_{f,3} - E_{i,3} \tag{9.39}$$

and $E_{i,3} (= E_{f,2})$ and $E_{f,3} (= E_f)$ denote the initial and final energies of the store for the discharging period. The quantity in square brackets represents the energy output during discharging. An energy efficiency for the discharging period can be defined as

$$\eta_3 = \frac{\text{Energy recovered from TES during discharging}}{\text{Energy accumulation in TES during charging and storing}} = \frac{H_d - H_c}{\Delta E_1 + \Delta E_2} \tag{9.40}$$

Using Eq. (9.28), the energy efficiency can be rewritten as

$$\eta_3 = \frac{H_d - H_c}{\Delta E_1 - Q_{1,2}} \tag{9.41}$$

An exergy balance for the discharging period can be written as follows:

$$-[\text{Exergy recovered} + \text{Exergy loss}] - \text{Exergy consumption} = \text{Exergy accumulation} \tag{9.42}$$

$$-[(Ex_d - Ex_c) + X_{l,3}] - I_3 = \Delta Ex_3 \tag{9.43}$$

Here,

$$\Delta Ex_3 = Ex_{f,3} - Ex_{i,3} \tag{9.44}$$

and $Ex_{i,3} (= Ex_{f,2})$ and $Ex_{f,3} (= Ex_f)$ denote the initial and final exergies of the store for the discharging period. The quantity in square brackets represents the exergy output during discharging. An exergy efficiency for the discharging period can be defined as

$$\psi_3 = \frac{\text{Exergy recovered from TES during discharging}}{\text{Exergy accumulation in TES during charging and storing}} = \frac{Ex_d - Ex_c}{\Delta Ex_1 + \Delta Ex_2} \tag{9.45}$$

Using Eq. (9.33), the exergy efficiency can be rewritten as

$$\psi_3 = \frac{Ex_d - Ex_c}{\Delta Ex_1 - (X_{1,2} + I_2)} \tag{9.46}$$

The efficiencies in Eqs. (9.40) and (9.45) indicate the fraction of the energy/exergy input during charging and still retained at the end of storing which is recovered during discharging.

9.3.3. Implications for subprocesses and overall process

Overall energy and exergy efficiencies can be written as the products of the energy and exergy efficiencies for charging, storing and discharging:

$$\eta = \prod_{j=1}^{3} \eta_j \tag{9.47}$$

$$\psi = \prod_{j=1}^{3} \psi_j \tag{9.48}$$

In addition it can be shown that the summations of the energy or exergy balance equations respectively for the three subprocesses give the energy or exergy balance equations for the overall process. Also, the overall changes in storage energy or exergy can be shown to be the sum of the changes during the subprocesses:

$$\sum_{j=1}^{3} \Delta E_j = E_{3,f} - E_{1,i} = E_f - E_i = \Delta E \tag{9.49}$$

$$\sum_{j=1}^{3} \Delta Ex_j = Ex_{3,f} - Ex_{1,i} = Ex_f - Ex_i = \Delta Ex \tag{9.50}$$

Note for period j that

$$\Delta E_j = E_{f,j} - E_{i,j} \tag{9.51}$$

$$\Delta Ex_j = Ex_{f,j} - Ex_{i,j} \tag{9.52}$$

and that $E_{i,1} = E_i$, $E_{f,3} = E_f$, and $E_{i,j+1} = E_{f,j}$ for $j = 1, 2$, while analogous expressions hold for the Ex terms.

This section demonstrates the application of exergy analysis to a closed TES system. Exergy analysis clearly takes into account the external and temperature losses in TES operation, and hence more correctly reflects thermodynamic behavior.

9.4. Relations between temperature and efficiency for sensible TES

Being energy based, most TES evaluation measures disregard the temperatures of the transferred heat and thus are misleading because they weight all thermal energy equally. Exergy efficiencies acknowledge that the usefulness of thermal energy depends on its quality, reflected by its temperature, and are therefore more suitable for evaluations and comparisons. This section highlights the relation between temperature and efficiency for a simple sensible TES, demonstrating that exergy analysis weights the usefulness of thermal energy appropriately, while energy analysis tends to present overly optimistic views of TES performance by neglecting the temperatures associated with thermal flows.

9.4.1. Model and analysis

Consider the overall storage process for the TES system in Fig. (9.2). Heat Q_c is injected into the system at a constant temperature T_c during a charging period. After a storing period, heat Q_d is recovered at a constant temperature T_d during a discharging period. During all periods, heat Q_l leaks to the surroundings from the system at a constant temperature T_l. For heating applications, the temperatures T_c, T_d and T_l exceed the environment temperature T_o but the discharging temperature cannot exceed the charging temperature, so the exergetic temperature factors are subject to the constraint $0 \leq \tau_d \leq \tau_c \leq 1$.

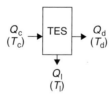

Fig. 9.2. The overall heat storage process for a general TES system. Shown are heat flows, and associated temperatures at the TES boundary (terms in parentheses).

For a process involving only heat interactions in a closed system for which the state is the same at the beginning and end, balances of energy and exergy, respectively, can be written as follows:

$$\sum_r Q_r = 0 \tag{9.53}$$

$$\sum_{r} X_r - I = 0 \tag{9.54}$$

where I denotes the exergy consumption and X_r the exergy associated with Q_r, the heat transferred into the system across region r at temperature T_r. The exergetic temperature factor τ and the temperature ratio T/T_o are compared with the temperature T in Table 9.1 for above-environmental temperatures (i.e., for $T \geq T_o$), the temperature range of interest for most heat storages.

Table 9.1. Relation between several temperature parameters for above-environment temperatures.*

Temperature ratio T/T_o	Temperature T (K)	Exergetic temperature factor τ
1.00	283	0.00
1.25	354	0.20
1.50	425	0.33
2.00	566	0.50
3.00	849	0.67
5.00	1415	0.80
10.00	2830	0.90
100.00	28,300	0.99
∞	∞	1.00

* The reference-environment temperature is $T_o = 10°C = 283$ K.

Equations (9.53) and (9.54), respectively, can be written for the modeled system as

$$Q_c = Q_d + Q_l \tag{9.55}$$

and

$$X_c = X_d + X_l + I \tag{9.56}$$

The exergy balance can be expressed as

$$Q_c \tau_c = Q_d \tau_d + Q_l \tau_l + I \tag{9.57}$$

9.4.2. Efficiencies and their dependence on temperature

The energy efficiency can be written for the modeled system as

$$\eta = \frac{Q_d}{Q_c} \tag{9.58}$$

and the exergy efficiency as

$$\psi = \frac{X_d}{X_c} = \frac{Q_d \tau_d}{Q_c \tau_c} = \frac{\tau_d}{\tau_c} \eta \tag{9.59}$$

An illuminating parameter for comparing the efficiencies is the energy-efficiency-to-exergy-efficiency ratio ψ/η. For the present TES system, this ratio can be expressed as

$$\frac{\psi}{\eta} = \frac{\tau_d}{\tau_c} \tag{9.60}$$

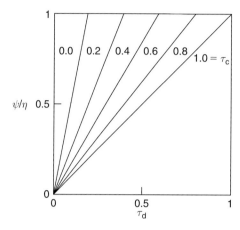

Fig. 9.3. Energy-efficiency-to-exergy-efficiency ratio, ψ/η, as a function of the discharging exergetic temperature factor τ_d, for several values of the charging exergetic temperature factor τ_c.

or

$$\frac{\psi}{\eta} = \frac{(T_d - T_o)T_c}{(T_c - T_o)T_d} \qquad (9.61)$$

The variation of the ratio ψ/η with τ_d and τ_c is illustrated in Fig. 9.3. It is seen that ψ/η varies linearly with τ_d for a given value of τ_c. Also, if the product heat is delivered at the charging temperature (i.e., $\tau_d = \tau_c$), $\psi = \eta$, while if the product heat is delivered at the temperature of the environment (i.e., $\tau_d = 0$), $\psi = 0$ regardless of the charging temperature. In the first case, there is no loss of temperature during the entire storage process, while in the second there is a complete loss of temperature. The largest deviation between values of ψ and η occurs in the second case.

The deviation between ψ and η is significant for many practical TES systems, which operate between charging temperatures as high as $T_c = 130°C$ and discharging temperatures as low as $T_d = 40°C$, and with a difference of about 30°C between charging and discharging temperatures (i.e., $T_c - T_d = 30°C$). With $T_o = 10°C$, the first condition can be shown to imply for most present day systems that $0.1 \leq \tau_d \leq \tau_c \leq 0.3$. Since it can be shown that

$$\tau_c - \tau_d = \frac{(T_c - T_d)T_o}{T_c T_d} \qquad (9.62)$$

the difference in exergetic temperature factor varies approximately between 0.06 and 0.08. Then the value of the exergy efficiency is approximately 50% to 80% of that of the energy efficiency.

Table 9.2. Values of the ratio ψ/η for a range of practical values for T_d and T_c.*

Discharging temperature, T_d (°C)	Charging temperature, T_c (°C)			
	40	70	100	130
40	1.00	0.55	0.40	0.32
70	–	1.00	0.72	0.59
100	–	–	1.00	0.81
130	–	–	–	1.00

* The reference-environment temperature is $T_o = 10°C = 283$ K.

The ratio ψ/η is illustrated in Table 9.2 for a simple TES having charging and discharging temperatures ranging between 40°C and 130°C, and a reference-environment temperature of $T_o = 10°C$. The energy and exergy efficiencies differ (with the exergy efficiency always being the lesser of the two) when $T_d < T_c$, and the difference becomes more significant as the difference between T_c and T_d increases. The efficiencies are equal only when the charging and discharging temperatures are equal (i.e., $T_d = T_c$).

Unlike the exergy efficiencies, the energy efficiencies tend to appear overly optimistic, in that they only account for losses attributable to heat leakages but ignore temperature degradation. Exergy efficiencies are more illuminating because they weight heat flows appropriately, being sensitive to the temperature at which heat is recovered relative to the temperature at which it is injected. TES energy efficiencies are good approximations to exergy efficiencies when there is little temperature degradation, as thermal energy quantities then have similar qualities. In most practical situations, however, thermal energy is injected and recovered at significantly different temperatures, making energy efficiencies poor approximations to exergy efficiencies and misleading.

9.5. Exergy analysis of thermally stratified storages

Exergy analysis recognizes differences in storage temperature, even for TESs containing equivalent energy quantities, and evaluates quantitatively losses due to degradation of storage temperature toward the environment temperature and mixing of fluids at different temperatures. These advantages of exergy over energy methods are particularly important for stratified storages since they exhibit internal spatial temperature variations. The inhibition of mixing through appropriate temperature stratification is advantageous. Through carefully managing the injection, recovery and holding of heat (or cold) to avoid stratification degradation, better storage-cycle performance can be achieved (as measured by better thermal energy recovery and temperature retention and accounted for explicitly through exergy efficiencies) (Hahne et al., 1989; Krane and Krane, 1991).

This section focuses on the energy and exergy contents of stratified storages. In the first part, several models are presented for the temperature distributions in vertically stratified thermal storages, which are sufficiently accurate, realistic and flexible for use in engineering design and analysis yet simple enough to be convenient, and which provide useful physical insights. One-dimensional gravitational temperature stratification is considered, and temperature is expressed as a function of height for each model. Expressions are derived for TES energy and exergy contents in accordance with the models. In the second part, the increase in exergy storage capacity resulting from stratification is described.

9.5.1. General stratified TES energy and exergy expressions

The energy E and exergy Ex in a TES can be found by integrating over the entire storage-fluid mass m within the TES as follows:

$$E = \int_m e \, dm \tag{9.63}$$

$$Ex = \int_m (ex) dm \tag{9.64}$$

where e denotes specific energy and ex specific exergy. For an ideal liquid, e and ex are functions only of temperature T, and can be expressed as

$$e(T) = c(T - T_o) \tag{9.65}$$

$$ex(T) = c[(T - T_o) - T_o \ln (T/T_o)] = e(T) - cT_o \ln (T/T_o) \tag{9.66}$$

Both the storage-fluid specific heat c and reference-environment temperature T_o are assumed constant.

For a TES of height H with one-dimensional vertical stratification, i.e., temperature varies only with height h, and a constant horizontal cross-sectional area, a horizontal element of mass dm can then be approximated as

$$dm = \frac{m}{H} dh \tag{9.67}$$

Since temperature is a function only of height (i.e., $T = T(h)$), the expressions for e and ex in Eqs. (9.65) and (9.66), respectively, can be written as

$$e(h) = c(T(h) - T_o) \tag{9.68}$$

$$ex(h) = e(h) - cT_o \ln (T(h)/T_o) \tag{9.69}$$

With Eq. (9.67), the expressions for E and Ex in Eqs. (9.63) and (9.64), respectively, can be written as

$$E = \frac{m}{H} \int_0^H e(h)dh \tag{9.70}$$

$$Ex = \frac{m}{H} \int_0^H ex(h)dh \tag{9.71}$$

With Eq. (9.68), the expression for E in Eq. (9.70) can be written as

$$E = mc(T_m - T_o) \tag{9.72}$$

where

$$T_m \equiv \frac{1}{H} \int_0^H T(h)dh \tag{9.73}$$

Physically, T_m represents the temperature of the TES fluid when it is fully mixed. This observation can be seen by noting that the energy of a fully mixed tank E_m at a uniform temperature T_m can be expressed, using Eq. (9.65) with constant temperature and Eq. (9.63), as

$$E_m = mc(T_m - T_o) \tag{9.74}$$

and that the energy of a fully mixed tank E_m is by the principle of conservation of energy the same as the energy of the stratified tank E:

$$E = E_m \tag{9.75}$$

With Eq. (9.69), the expression for Ex in Eq. (9.71) can be written as

$$Ex = E - mcT_o \ln (T_e/T_o) \tag{9.76}$$

where

$$T_e \equiv \exp\left[\frac{1}{H} \int_0^H \ln T(h)dh \right] \tag{9.77}$$

Physically, T_e represents the equivalent temperature of a mixed TES that has the same exergy as the stratified TES. In general, $T_e \neq T_m$, since T_e is dependent on the degree of stratification present in the TES, while T_m is independent of degree of stratification. When the TES is fully mixed, $T_e = T_m$. This can be seen by noting (with Eqs. (9.64) and (9.74)–(9.76)) that the exergy in the fully mixed TES is

$$Ex_m = E_m - mcT_o \ln (T_m/T_o) \tag{9.78}$$

The difference in TES exergy between the stratified and fully mixed (i.e., at a constant temperature T_m) cases can be expressed with Eqs. (9.76) and (9.78) as

$$Ex - Ex_m = mcT_o \ln (T_m/T_e) \tag{9.79}$$

The change given in Eq. (9.79) can be shown to be always negative. That is, the exergy consumption associated with mixing fluids at different temperatures, or the minimum work required for creating temperature differences, is always positive.

When the temperature distribution is symmetric about the center of the TES such that

$$\frac{T(h) + T(H - h)}{2} = T(H/2) \tag{9.80}$$

the mixed temperature T_m is the mean of the temperatures at the TES top and bottom.

9.5.2. Temperature-distribution models and relevant expressions

Four stratified temperature-distribution models are considered: linear (denoted by a superscript L), stepped (S), continuous-linear (C) and general three-zone (T). For each model, the temperature distribution as a function of height is given, and expressions for T_m and T_e are derived. The distributions considered are simple enough in form to permit energy and exergy values to be obtained analytically, but complex enough to be relatively realistic. The expressions developed in this section show that the exergy of a stratified storage is greater than the exergy for the same storage when it is fully mixed, even though the energy content does not change.

Linear temperature-distribution model

The linear temperature-distribution model (see Fig. 9.4) varies linearly with height h from T_b, the temperature at the bottom of the TES (i.e., at $h=0$), to T_t, the temperature at the top (i.e., at $h=H$), and can be expressed as

$$T^L(h) = \frac{T_t - T_b}{H}h + T_b \tag{9.81}$$

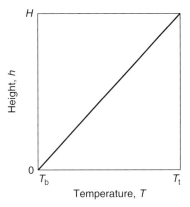

Fig. 9.4. A vertically stratified storage having a linear temperature distribution.

By substituting Eq. (9.81) into Eqs. (9.73) and (9.77), it can be shown that

$$T_m^L = \frac{T_t + T_b}{2} \tag{9.82}$$

which is the mean of the temperatures at the top and bottom of the TES, and that

$$T_e^L = \exp\left[\frac{T_t(\ln T_t - 1) - T_b(\ln T_b - 1)}{T_t - T_b}\right] \tag{9.83}$$

Stepped temperature-distribution model

The stepped temperature-distribution model (see Fig. 9.5) consists of k horizontal zones, each of which is at a constant temperature, and can be expressed as

$$T^S(h) = \begin{cases} T_1, & h_0 \le h \le h_1 \\ T_2, & h_1 < h \le h_2 \\ \dots \\ T_k, & h_{k-1} < h \le h_k \end{cases} \tag{9.84}$$

where the heights are constrained as follows:

$$0 = h_0 \le h_1 \le h_2 \dots \le h_k = H \tag{9.85}$$

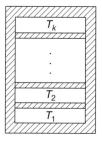

Fig. 9.5. A vertically stratified storage having a stepped temperature distribution.

It is convenient to introduce here x_j, the mass fraction for zone j:

$$x_j \equiv \frac{m_j}{m} \tag{9.86}$$

Since the TES-fluid density ρ and the horizontal TES cross-sectional area A are assumed constant, but the vertical thickness of zone j, $h_j - h_{j-1}$, can vary from zone to zone,

$$m_j = \rho V_j = \rho A(h_j - h_{j-1}) \tag{9.87}$$

and

$$m = \rho V = \rho A H \tag{9.88}$$

where V_j and V denote the volumes of zone j and of the entire TES, respectively. Substitution of Eqs. (9.87) and (9.88) into Eq. (9.86) yields

$$x_j = \frac{h_j - h_{j-1}}{H} \tag{9.89}$$

With Eqs. (9.73), (9.77), (9.84) and (9.89), it can be shown that

$$T_m^S = \sum_{j=1}^{k} x_j T_j \tag{9.90}$$

which is the weighted mean of the zone temperatures, where the weighting factor is the mass fraction of the zone, and that

$$T_e^S = \exp\left[\sum_{j=1}^{k} x_j \ln T_j\right] = \prod_{j=1}^{k} T_j^{x_j} \tag{9.91}$$

Simplified forms of the expressions for T_m and T_e can be written for the multi-zone temperature-distribution models when all zone vertical thicknesses are the same, since in this special case, the mass fractions for each of the k zones are the same (i.e., $x_j = 1/k$ for all j).

Continuous-linear temperature-distribution model

The continuous-linear temperature distribution consists of k horizontal zones, in each of which the temperature varies linearly from the bottom to the top, and can be expressed as

$$T^C(h) = \begin{cases} \phi_1^C(h) & h_0 \leq h \leq h_1 \\ \phi_2^C(h) & h_1 < h \leq h_2 \\ \quad \cdots \\ \phi_k^C(h) & h_{k-1} < h \leq h_k \end{cases} \tag{9.92}$$

where $\phi_j^C(h)$ represents the linear temperature distribution in zone j:

$$\phi_j^C(h) = \frac{T_j - T_{j-1}}{h_j - h_{j-1}} h + \frac{h_j T_{j-1} - h_{j-1} T_j}{h_j - h_{j-1}} \tag{9.93}$$

The zone height constraints in Eq. (9.85) apply here. The temperature varies continuously between zones.
   With Eqs. (9.73), (9.89), (9.92) and (9.93), it can be shown that

$$T_m^C = \sum_{j=1}^{k} x_j (T_m)_j \tag{9.94}$$

where $(T_m)_j$ is the mean temperature in zone j, i.e.,

$$(T_m)_j = \frac{T_j + T_{j-1}}{2} \tag{9.95}$$

and that

$$T_e^C = \exp\left[\sum_{j=1}^{k} x_j \ln (T_e)_j \right] = \prod_{j=1}^{k} (T_e)_j^{x_j} \tag{9.96}$$

where $(T_e)_j$ is the equivalent temperature in zone j, i.e.,

$$(T_e)_j = \begin{cases} \exp\left[\dfrac{T_j(\ln T_j - 1) - T_{j-1}(\ln T_{j-1} - 1)}{T_j - T_{j-1}} \right] & \text{if } T_j \neq T_{j-1} \\ T_j & \text{if } T_j = T_{j-1} \end{cases} \tag{9.97}$$

General three-zone temperature-distribution model

The general three-zone temperature-distribution model is a subset of the continuous-linear model in which there are only three horizontal zones (i.e., $k = 3$). The temperature varies linearly within each zone, and continuously across each zone. The temperature distribution for the general three-zone model is illustrated in Fig. 9.6, and can be expressed as follows:

$$T^T(h) = \begin{cases} \phi_1^C(h) & h_0 \leq h \leq h_1 \\ \phi_2^C(h) & h_1 < h \leq h_2 \\ \ldots \\ \phi_3^C(h) & h_2 < h \leq h_k \end{cases} \tag{9.98}$$

where $\phi_j^C(h)$ represents the temperature distribution (linear) in zone j (see Eq. (9.93)), and where the heights are constrained as in Eq. (9.85) with $k = 3$.
   Expressions for the temperatures T_m and T_e can be obtained for the general three-zone model with the expressions for T_m and T_e for the continuous-linear model with $k = 3$:

$$T_m^T = \sum_{j=1}^{3} x_j (T_m^C)_j \tag{9.99}$$

$$T_e^T = \exp\left[\sum_{j=1}^{3} x_j \ln (T_e^C)_j \right] = \prod_{j=1}^{3} (T_e^C)_j^{x_j} \tag{9.100}$$

where

$$(T_m^C)_j = \frac{T_j + T_{j-1}}{2} \tag{9.101}$$

$$(T_e^C)_j = \begin{cases} \exp\left[\dfrac{T_j(\ln T_j - 1) - T_{j-1}(\ln T_{j-1} - 1)}{T_j - T_{j-1}} \right] & \text{if } T_j \neq T_{j-1} \\ T_j & \text{if } T_j = T_{j-1} \end{cases} \tag{9.102}$$

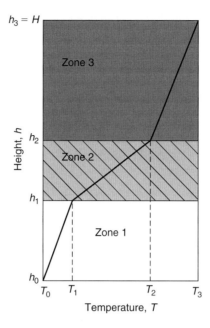

Fig. 9.6. General three-zone temperature-distribution model.

The linear temperature-distribution model is simple to utilize but not flexible enough to fit the wide range of actual temperature distributions possible, while the stepped and continuous-linear distribution models are flexible and, if the zones are made small enough, can accurately fit any actual temperature distribution. The general three-zone temperature-distribution model strives for a balance between such factors as accuracy, computational convenience and physical insight. The three-zone model simulates well the stratification distribution in many actual TES fluids, which possess lower and upper zones of slightly varying or approximately constant temperature, and a middle zone (the thermocline region) in which temperature varies substantially. The intermediate zone, which grows as thermal diffusion occurs in the tank being modeled, accounts for the irreversible effects of thermal mixing.

9.5.3. Increasing TES exergy storage capacity using stratification

The increase in exergy capacity of a thermal storage through stratification is described. For a range of realistic storage-fluid temperature profiles, the relative increase in exergy content of the stratified storage compared to the same storage when it is fully mixed is evaluated. Temperature profiles are considered having various degrees of stratification, as represented by the magnitude and sharpness of the spatial temperature variations.

Analysis

Thermal storages for heating and cooling capacity, having numerous temperature-distribution profiles, are considered. The general three-zone model is utilized to evaluate storage energy and exergy contents. For each case, the ratio is evaluated of the exergy of the stratified storage Ex to the exergy of the same storage when fully mixed Ex_m. Using Eqs. (9.72), (9.76) and (9.78) this ratio can be expressed, after simplification, as

$$\frac{Ex}{Ex_\mathrm{m}} = \frac{T_\mathrm{m}/T_\mathrm{o} - 1 - \ln\left(T_\mathrm{e}/T_\mathrm{o}\right)}{T_\mathrm{m}/T_\mathrm{o} - 1 - \ln\left(T_\mathrm{m}/T_\mathrm{o}\right)} \tag{9.103}$$

This ratio increases, from as low as unity when the storage is not stratified, to a value greater than one as the degree of stratification present increases. The ratio in Eq. (9.103) is independent of the mass m and specific heat c of the storage fluid. The ratio is also useful as an evaluation, analysis and design tool, as it permits the exergy of a stratified storage to be conveniently evaluated by multiplying the exergy of the equivalent mixed storage (a quantity straightforwardly evaluated) by the appropriate exergy ratio determined here.

Several assumptions and approximations are utilized throughout this subsection:

- Storage horizontal cross-sectional area is fixed.
- The environmental temperature T_o is fixed at 20°C, whether the case involves thermal storage for heating or cooling capacity.
- One-dimensional gravitational (i.e., vertical) temperature stratification is considered.

For simplicity, only temperature distributions which are rotationally symmetric about the center of the storage, according to Eq. (9.80), are considered. This symmetry implies that zone 2 is centered about the central horizontal axis of the storage, and that zones 1 and 3 are of equal size, i.e., $x_1 = x_3 = (1 - x_2)/2$. In the analysis, two main relevant parameters are varied realistically:

- The principal temperatures (e.g., mean, maximum, minimum).
- Temperature-distribution profiles (including changes in zone thicknesses).

Specifically, the following characterizing parameters are varied to achieve the different temperature-distribution cases considered:

- The mixed-storage temperature T_m is varied for a range of temperatures characteristic of storages for heating and cooling capacity.
- The size of zone 2, which represents the thermocline region, is allowed to vary from as little as zero to as great as the size of the overall storage, i.e., $0 \leq x_2 \leq 1$. A wide range of temperature profiles can thereby be accommodated, and two extreme cases exist: a single-zone situation with a linear temperature distribution when $x_2 = 1$, and a two-zone distribution when $x_2 = 0$.
- The maximum and minimum temperatures in the storage, which occur at the top and bottom of the storage, respectively, are permitted to vary about the mixed-storage temperature T_m by up to 15°C.

Using the zone numbering system in Fig. (9.6), and the symmetry condition introduced earlier, the following expressions can be written for the temperatures at the top and bottom of the storage, respectively:

$$T_3 = T_m + \Delta T_{st} \quad \text{and} \quad T_0 = T_m - \Delta T_{st} \tag{9.104}$$

while the the following equations can be written for the temperatures at the top and bottom of zone 2, respectively:

$$T_2 = T_m + \Delta T_{th} \quad \text{and} \quad T_1 = T_m - \Delta T_{th} \tag{9.105}$$

where the subscripts 'th' and 'st' denote thermocline region (zone 2) and overall storage, respectively, and where

$$\Delta T \equiv |T - T_m| \tag{9.106}$$

According to the last bullet above, $0 \leq \Delta T_{th} \leq \Delta T_{st} = 15°C$. Also, ΔT_{th} is the magnitude of the difference, on either side of the thermocline region (zone 2), between the temperature at the outer edge of zone 2 and T_m, while ΔT_{st} is the magnitude of the difference, on either side of the overall storage, between the temperature at the outer edge of the storage and T_m. That is,

$$\Delta T_{th} = \Delta T_1 = \Delta T_2 \quad \text{and} \quad \Delta T_{st} = \Delta T_0 = \Delta T_3 \tag{9.107}$$

where the ΔT parameters in the above equations are defined using Eq. (9.106) as follows:

$$\Delta T_j \equiv |T_j - T_m|, \quad \text{for } j = 0, 1, 2, 3 \tag{9.108}$$

Effects of varying stratification parameters

Effect of varying T_m: The variation of thermal-storage exergy with storage temperature for a mixed storage is illustrated in Fig. 9.7. For a fixed storage total heat capacity (mc), storage exergy increases, from zero when the temperature T_m is equal to the environment temperature T_o, as the temperature increases or decreases from T_o. This general trend, which is illustrated here for a mixed storage, normally holds for stratified storages since the effect on storage exergy of temperature is usually more significant than the effect of stratification.

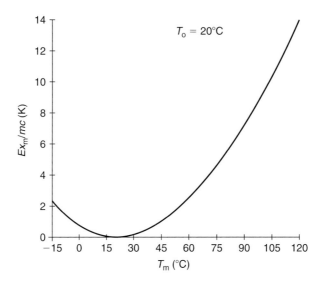

Fig. 9.7. Variation with the mixed-storage temperature T_m of the modified exergy quantity Ex_m/mc (where m and c are constant) for a mixed storage. When T_m equals the environment temperature $T_o = 20°C$, $Ex_m = 0$.

Effect of varying minimum and maximum temperatures for a linear profile: A linear temperature profile across the entire storage occurs with the three-zone model when $x_2 = 1$. Then, the upper and lower boundaries of zone 2 shift to the top and bottom of the storage, respectively, and correspondingly the temperature deviation ΔT_{th} occurs at those positions. For a linear temperature profile, the ratio Ex/Ex_m is illustrated in Figs. (9.8a), (9.8b) and (9.8c) for three temperature regimes, respectively:

- high-temperature thermal storage for heating capacity, i.e., $T_m \geq 60°C$,
- low-temperature thermal storage for heating capacity, i.e., $20°C \leq T_m \leq 60°C$,
- thermal storage for cooling capacity, i.e., $T_m \leq 20°C$.

The temperature range considered is above the environment temperature $T_o = 20°C$ for the first two cases, and below it for the third. Two key points are demonstrated in Fig. 9.8. First, for a fixed mixed-storage temperature T_m, storage exergy content increases as level of stratification increases (i.e., as ΔT_{th} increases) for all cases. Second, the percentage increase in storage exergy, relative to the mixed-storage exergy at the same T_m, is greatest when $T_m = T_o$, and decreases both as T_m increases from T_o (see Figs. 9.8a and 9.8b) and decreases from T_o (see Fig. 9.8c). The main reason for the second observation relates to the fact that the absolute magnitude of the mixed exergy for a thermal storage is small when T_m is near T_o, and larger when T_m deviates significantly from T_o (see Fig. 9.7). In the limiting case where $T_m = T_o$, the ratio Ex/Ex_m takes on the value of unity when $\Delta T_{th} = 0$ and infinity for all other values of ΔT_{th}. Hence, the relative benefits of stratification as a tool to increase the exergy-storage capacity of a thermal storage are greatest at near-environment temperatures, and less for other cases.

Effect of varying thermocline-size parameter x_2: The variation of the ratio Ex/Ex_m with the zone-2 size parameter and the temperature deviation at the zone-2 boundaries, ΔT_{th}, is illustrated in Fig. 9.9 for a series of values of the mixed-storage temperature T_m. For a fixed value of ΔT_{th} at a fixed value of T_m, the ratio Ex/Ex_m increases as the zone-2 size parameter x_2 decreases. This observation occurs because the stratification becomes less smoothly varying and more sharp and pronounced as x_2 decreases.

Effect of varying temperature-distribution profile: The temperature-distribution profile shape is varied, for a fixed value of T_m, primarily by varying values of the parameters x_2 and ΔT_{th} simultaneously. The behavior of Ex/Ex_m as x_2 and ΔT_{th} are varied for several T_m values is shown in Fig. 9.9. For all cases considered by varying these parameters at a fixed value of T_m (except for $T_m = T_o$), the ratio Ex/Ex_m increases, from a minimum value of unity at $x_2 = 1$ and $\Delta T_{th} = 0$, as x_2 decreases and ΔT_{th} increases. Physically, these observations imply that, for a fixed value of T_m, storage exergy increases as stratification becomes more pronounced, both through increasing the maximum temperature deviation from the mean storage temperature, and increasing the sharpness of temperature profile differences between storage zones.

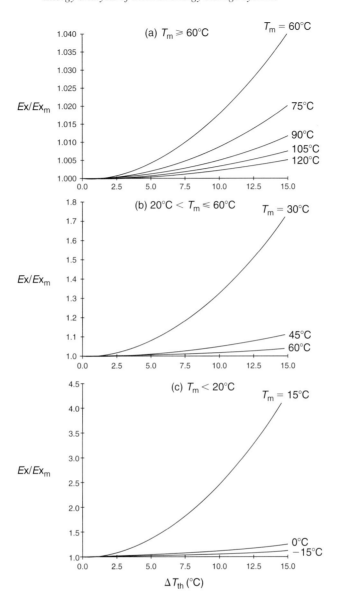

Fig. 9.8. Illustration of the variation of the ratio of the exergy values for stratified and fully mixed storages, Ex/Ex_m, for three ranges of values of the mixed-storage temperature T_m (each corresponding to a different graph).

The results clearly show that TES exergy values, unlike energy values, change due to stratification, giving a quantitative measure of the advantage provided by stratification. Also, the exergy content (or capacity) of a TES increases as the degree of stratification increases, even if the energy remains fixed. The use of stratification can therefore aid in TES analysis, design and optimization as it increases the exergy-storage capacity of a thermal storage.

9.6. Energy and exergy analyses of cold TES systems

In many countries, cold thermal energy storage (CTES) is an economically viable technology used in many thermal systems, particularly building cooling. In many CTES applications, inexpensive off-peak electricity is utilized during the night to produce with chillers a cold medium, which can be stored for use in meeting cooling needs during the day when electricity is more expensive.

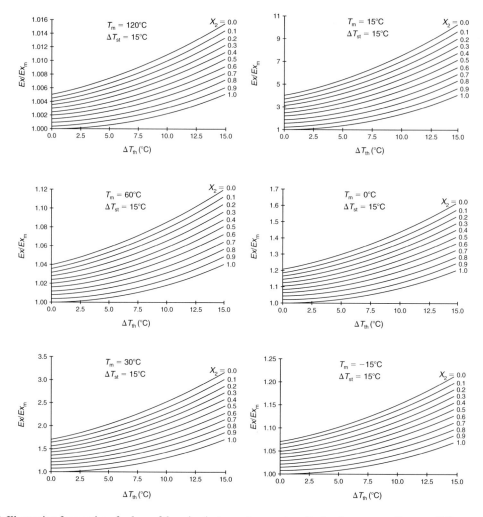

Fig. 9.9. Illustration for a series of values of the mixed-storage temperature T_m (each corresponding to a different graph) of the variation of the ratio of the exergy values for stratified and fully mixed storages, Ex/Ex_m, with temperature deviation from T_m at the upper and lower boundaries of the thermocline zone (zone 2), ΔT_{th}, and with the zone-2 mass fraction x_2. The magnitude of the temperature deviation from T_m at the top and bottom of the storage, ΔT_{st}, is 15°C for all cases.

Energy analysis is inadequate for CTES evaluation because it does not account for the temperatures at which heat (or cold) is supplied and delivered. Exergy analysis overcomes some of this inadequacy in CTES assessment. Also, exergy analysis conceptually is more direct than energy analysis since it treats cold as a valuable commodity. In this section, exergy and energy analyses are presented for CTES systems, including sensible and latent storages. Several CTES cases are considered, including storages which are homogeneous or stratified, and some which undergo phase changes. A full cycle of charging, storing and discharging is considered for each case.

9.6.1. Energy balances

Consider a cold storage consisting of a tank containing a fixed quantity of storage fluid and a heat-transfer coil through which a heat-transfer fluid is circulated. Kinetic and potential energies and pump work are neglected. An energy balance for an entire CTES cycle can be written in terms of 'cold' as follows:

$$\text{Cold input} - [\text{Cold recovered} + \text{Cold loss}] = \text{Cold accumulation} \qquad (9.109)$$

Here, 'cold input' is the heat removed from the storage fluid by the heat-transfer fluid during charging; 'cold recovered' is the heat removed from the heat-transfer fluid by the storage fluid; 'cold loss' is the heat gain from the environment to the storage fluid during charging, storing and discharging; and 'cold accumulation' is the decrease in internal energy of the storage fluid during the entire cycle. The overall energy balance for the simplified CTES system illustrated in Fig. 9.10 becomes

$$(H_b - H_a) - [(H_c - H_d) + Q_l] = -\Delta E \tag{9.110}$$

where H_a, H_b, H_c and H_d are the enthalpies of the flows at points a, b, c and d in Fig. 9.10; Q_l is the total heat gain during the charging, storing and discharging processes; and ΔE is the difference between the final and initial storage-fluid internal energies. The terms in square brackets in Eqs. (9.109) and (9.110) represent the net 'cold output' from the CTES, and $\Delta E = 0$ if the CTES undergoes a complete cycle (i.e., the initial and final storage-fluid states are identical).

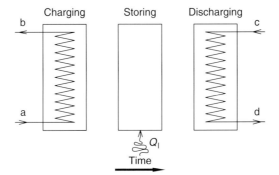

Fig. 9.10. The three processes in a general CTES system: charging (left), storing (middle) and discharging (right). The heat leakage into the system Q_l is illustrated for the storing process, but can occur in all three processes.

The energy transfer associated with the charging fluid can be expressed as

$$H_b - H_a = m_a c_a (T_b - T_a) \tag{9.111}$$

where m_a is the mass flow of heat-transfer fluid at point a (and at point b), and c_a is the specific heat of the heat-transfer fluid, which is assumed constant. A similar expression can be written for $H_c - H_b$. The energy content of a storage which is homogeneous (i.e., entirely in either the solid or the liquid phase) is

$$E = m(u - u_o) \tag{9.112}$$

which, for sensible heat interactions only, can be written as

$$E = mc(T - T_o) \tag{9.113}$$

where, for the storage fluid, c denotes the specific heat (assumed constant), m the mass, u the specific internal energy and T the temperature. Also, u_o is u evaluated at the environmental conditions.

For a mixture of solid and liquid, the energy content of the solid and liquid portions can be evaluated separately and summed as follows:

$$E = m[(1 - F)(u_s - u_o) + F(u_t - u_o)] \tag{9.114}$$

where u_s and u_t are the specific internal energies of the solid and liquid portions of the storage fluid, respectively, and F is the melted fraction (i.e., the fraction of the storage-fluid mass in the liquid phase).

For a storage fluid which is thermally stratified with a linear temperature profile in the vertical direction, the energy content can be shown with Eqs. (9.72) and (9.82) to be

$$E = mc \left(\frac{T_t + T_b}{2} - T_o \right) \tag{9.115}$$

where T_t and T_b are the storage-fluid temperatures at the top and bottom of the linearly stratified storage tank, respectively.

The change in CTES energy content from the initial (i) to the final state (f) of a process can be expressed as in Eq. (9.4).

9.6.2. Exergy balances

An exergy balance for a CTES undergoing a complete cycle of charging, storing and discharging can be written as in Eqs. (9.6) and (9.7). The exergy content of a flow of heat-transfer fluid at state k (where $k =$ a, b, c or d in Fig. 9.10) can be expressed as in Eq. (9.11). The exergy transfers associated with the charging and discharging of the storage by the heat-transfer fluid can be expressed by Eqs. (9.12) and (9.13), respectively.

The exergy loss associated with heat infiltration during the three storage periods can be expressed as in Eq. (9.15). The thermal exergy terms are negative for sub-environment temperatures, as is the case here for CTESs, indicating that the heat transfer and the accompanying exergy transfer are oppositely directed. That is, the losses associated with heat transfer are due to heat infiltration into the storage when expressed in energy terms, but due to a cold loss out of the storage when expressed in exergy terms.

The exergy content of a homogeneous storage can be shown to be

$$Ex = m[(u - u_\text{o}) - T_\text{o}(s - s_\text{o})] \tag{9.116}$$

where s is the specific entropy of the storage fluid and s_o is s evaluated at the environmental conditions. If only sensible heat interactions occur, Eq. (9.116) can be written with Eq. (9.66) as

$$Ex = mc[(T - T_\text{o}) - T_\text{o} \ln (T/T_\text{o})] \tag{9.117}$$

For a mixture of solid and liquid, the exergy content can be written as

$$Ex = m[(1 - F)[(u_\text{s} - u_\text{o}) - T_\text{o}(s_\text{s} - s_\text{o})] + F[(u_\text{t} - u_\text{o}) - T_\text{o}(s_\text{t} - s_\text{o})]] \tag{9.118}$$

where s_s and s_t are the specific entropies of the solid and liquid portions of the storage fluid, respectively.

The exergy content of a storage which is linearly stratified can be shown with Eqs. (9.76) and (9.83) to be

$$Ex = E - mcT_\text{o} \left[\frac{T_\text{t}(\ln T_\text{t} - 1) - T_\text{b}(\ln T_\text{b} - 1)}{T_\text{t} - T_\text{b}} - \ln T_\text{o} \right] \tag{9.119}$$

The change in TES exergy content can expressed as in Eq. (9.10).

9.6.3. Efficiencies

For a general CTES undergoing a cyclic operation the overall energy efficiency η can be evaluated as in Eq. (9.16), with the word energy replaced by cold for understanding. Then, following Fig. 9.10, the charging-period energy efficiencies can be expressed as in Eq. (9.22).

Energy efficiencies for the storing and discharging subprocesses can be written respectively as

$$\eta_2 = \frac{\Delta E_1 + Q_\text{l}}{\Delta E_1} \tag{9.120}$$

$$\eta_3 = \frac{H_\text{c} - H_\text{d}}{\Delta E_3} \tag{9.121}$$

where ΔE_1 and ΔE_3 are the changes in CTES energy contents during charging and discharging, respectively.

The exergy efficiency for the overall process can be expressed as in Eq. (9.17), and for the charging, storing and discharging processes, respectively, as in Eqs. (9.26), (9.35) and (9.45).

Exergy analysis provides more meaningful and useful information than energy analysis about efficiencies, losses and performance for CTES systems. The loss of low temperature is accounted for in exergy – but not in energy-based measures. Furthermore, the exergy-based information is presented in a more direct and logical manner, as exergy methods provide intuitive advantages when CTES systems are considered. Consequently, exergy analysis can assist in efforts to optimize the design of CTES systems and their components, and to identify appropriate applications and optimal configurations for CTES in general engineering systems. The application of exergy analysis to CTES systems permits mismatches in the quality of the thermal energy supply and demand to be quantified, and measures to reduce or eliminate reasonably avoidable mismatches to be identified and considered. The advantages of the exergy approach are more significant for CTES compared to heat storage due to manner in which 'cold' is treated as a resource.

9.7. Exergy analysis of aquifer TES systems

Underground aquifers are sometimes used for TES (Jenne, 1992). The storage medium in many aquifer TES (ATES) systems remains in a single phase during the storing cycle, so that temperature changes are exhibited in the store as thermal energy is added or removed.

In this section, the application of exergy analysis to ATES systems is described. For an elementary ATES model, expressions are presented for the injected and recovered quantities of energy and exergy and for efficiencies. The impact is examined of introducing a threshold temperature below which residual heat remaining in the aquifer water is not considered worth recovering. ATES exergy efficiencies are demonstrated to be more useful and meaningful than energy efficiencies because the former account for the temperatures associated with thermal energy transfers and consequently assess how nearly ATES systems approach ideal thermodynamic performance. ATES energy efficiencies do not provide a measure of approach to ideal performance and, in fact, are often misleadingly high because some of the thermal energy can be recovered at temperatures too low for useful purposes.

9.7.1. ATES model

Charging of the ATES occurs over a finite time period t_c and after a holding interval discharging occurs over a period t_d. The working fluid is water, having a constant specific heat c, and assumed incompressible. The temperature of the aquifer and its surroundings prior to heat injection is T_o, the reference-environment temperature. Only heat stored at temperatures above T_o is considered, and pump work is neglected.

During charging, heated water at a constant temperature T_c is injected at a constant mass flow rate $\dot{m}_c$ into the ATES. After a storing period, discharging occurs, during which water is extracted from the ATES at a constant mass flow rate $\dot{m}_d$. The fluid discharge temperature is taken to be a function of time, i.e., $T_d = T_d(t)$. The discharge temperature after an infinite time is taken to be the temperature of the reference-environment, i.e. $T_d(\infty) = T_o$, and the initial discharge temperature is taken to be between the charging and reference-environment temperatures, i.e. $T_o \le T_d(0) \le T_c$.

Many discharge temperature–time profiles are possible. Here, the discharge temperature is taken to decrease linearly with time from an initial value $T_d(0)$ to a final value T_o. The final temperature is reached at a time t_f and remains fixed at T_o for all subsequent times, i.e.,

$$T_d(t) = \begin{cases} T_d(0) - (T_d(0) - T_o)t/t_f & 0 \le t \le t_f \\ T_o & t_f \le t \le \infty \end{cases} \tag{9.122}$$

The simple linear discharge temperature–time profile is sufficiently realistic yet simple.

The temperature–time profiles considered in the present model for the fluid flows during the charging and discharging periods are summarized in Fig. 9.11. The two main types of thermodynamic losses that occur in ATES systems are accounted for in the model:

- *Energy losses*: Energy injected into an ATES that is not recovered is considered lost. Thus, energy losses include energy remaining in the ATES and energy injected into the ATES that is convected in a water flow or is transferred by conduction far enough from the discharge point that it is unrecoverable.
- *Mixing losses*: As heated water is pumped into an ATES, it mixes with the water already present (which is usually cooler), resulting in the recovered water being at a lower temperature than the injected water. In the present model,

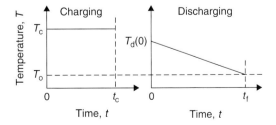

Fig. 9.11. Temperature–time profiles assumed for the charging and discharging periods in the ATES model considered.

this loss results in the discharge temperature T_d being at all times less than or equal to the charging temperature T_c, but not below the reference-environment temperature T_o (i.e., $T_o \leq T_d(t) \leq T_c$ for $0 \leq t \leq \infty$).

9.7.2. Energy and exergy analyses

The energy and exergy injected into the ATES during charging and recovered during discharging are evaluated. The energy flow associated with a flow of liquid at a constant mass flow rate $\dot{m}$, for an arbitrary period of time with T a function of t, is

$$E = \int_t \dot{E}(t)dt \tag{9.123}$$

where the integration is performed over the time period, and the energy flow rate at time t is

$$\dot{E}(t) = \dot{m}c(T(t) - T_o) \tag{9.124}$$

Here c denotes the specific heat of the liquid. Combining Eqs. (9.123) and (9.124) for constant $\dot{m}$, c and T_o,

$$E = \dot{m}c \int_t (T(t) - T_o)dt \tag{9.125}$$

The corresponding exergy flow is

$$Ex = \int_t \dot{Ex}(t)dt \tag{9.126}$$

where the exergy flow rate at time t is

$$\dot{Ex}(t) = \dot{m}c[(T(t) - T_o) - T_o \ln(T(t)/T_o)] \tag{9.127}$$

Combining Eqs. (9.126) and (9.127), and utilizing Eq. (9.125),

$$Ex = \dot{m}c \int_t [(T(t) - T_o) - T_o \ln(T(t)/T_o)]dt = E - \dot{m}cT_o \int_t \ln(T(t)/T_o)dt \tag{9.128}$$

Charging and discharging

The energy input to the ATES during charging, for a constant water injection rate $\dot{m}_c$ and over a time period beginning at zero and ending at t_c, is expressed by Eq. (9.125) with $T(t) = T_c$. That is,

$$E_c = \dot{m}_c c \int_{t=0}^{t_c} (T_c - T_o)dt = \dot{m}_c c t_c (T_c - T_o) \tag{9.129}$$

The corresponding exergy input is expressed by Eq. (9.128), with the same conditions as for E_c. Thus, after integration,

$$Ex_c = \dot{m}_c c t_c [(T_c - T_o) - T_o \ln(T_c/T_o)] = E_c - \dot{m}_c c t_c T_o \ln(T_c/T_o) \tag{9.130}$$

The energy recovered from the ATES during discharging, for a constant water recovery rate $\dot{m}_d$ and for a time period starting at zero and ending at t_d, is expressed by Eq. (9.125) with $T(t)$ as in Eq. (9.122). Thus,

$$E_d = \dot{m}_d c \int_{t=0}^{t_d} (T_d(t) - T_o)dt = \dot{m}_d c [T_d(0) - T_o]\theta(2t_f - \theta)/(2t_f) \tag{9.131}$$

where

$$\theta = \begin{cases} t_d & 0 \le t_d \le t_f \\ t_f & t_f \le t_d \le \infty \end{cases} \tag{9.132}$$

The corresponding exergy recovered is expressed by Eq. (9.128), with the same conditions as for E_d. Thus,

$$Ex_d = \dot{m}_d c \int_{t=0}^{t_d} [(T_d(t) - T_o) - T_o \ln(T_d(t)/T_o)]dt = E_d - \dot{m}_d c T_o \int_{t=0}^{t_d} \ln(T_d(t)/T_o)dt \tag{9.133}$$

Here,

$$\int_{t=0}^{t_d} \ln(T_d(t)/T_o)dt = \int_{t=0}^{t_d} \ln(at+b)dt = [(a\theta+b)/a]\ln(a\theta+b) - \theta - (b/a)\ln b \tag{9.134}$$

where

$$a = [T_o - T_d(0)]/(T_o t_f) \tag{9.135}$$

$$b = T_d(0)/T_o \tag{9.136}$$

When $t_d \ge t_f$, the expression for the integral in Eq. (9.134) reduces to

$$\int_{t=0}^{t_d} \ln(T_d(t)/T_o)dt = t_f \left[\frac{T_d(0)}{T_d(0) - T_o} \ln \frac{T_d(0)}{T_o} - 1 \right] \tag{9.137}$$

Balances and efficiencies

An ATES energy balance taken over a complete charging–discharging cycle states that the energy injected is either recovered or lost. A corresponding exergy balance states that the exergy injected is either recovered or lost, where lost exergy is associated with both waste exergy emissions and internal exergy consumptions due to irreversibilities.

If f is defined as the fraction of injected energy E_c that can be recovered if the length of the discharge period approaches infinity (i.e., water is extracted until all recoverable energy has been recovered), then

$$E_d(t_d \to \infty) = fE_c \tag{9.138}$$

It follows from the energy balance that $(1-f)E_c$ is the energy irretrievably lost from the ATES. Clearly, f varies between zero for a thermodynamically worthless ATES to unity for an ATES having no energy losses during an infinite discharge period. But mixing in the ATES can still cause exergy losses even if $f = 1$. Since E_c is given by Eq. (9.129) and $E_d(t_d \to \infty)$ by Eq. (9.131) with $\theta = t_f$, Eq. (9.138) may be rewritten as

$$\dot{m}_d c(T_d(0) - T_o)t_f/2 = f\dot{m}_c c(T_c - T_o)t_c \tag{9.139}$$

or

$$f = \frac{t_f \dot{m}_d(T_d(0) - T_o)}{2t_c \dot{m}_c(T_c - T_o)} \tag{9.140}$$

For either energy or exergy, efficiency is the fraction, taken over a complete cycle, of the quantity input during charging that is recovered during discharging, while loss is the difference between input and recovered amounts of the quantity. Hence, the energy loss as a function of the discharge time period is given by $[E_c - E_d(t_d)]$, while the corresponding exergy loss is given by $[Ex_c - Ex_d(t_d)]$. Energy losses do not reflect the temperature degradation associated with mixing, while exergy losses do.

The energy efficiency η for an ATES, as a function of the discharge time period, is given by

$$\eta(t_d) = \frac{E_d(t_d)}{E_c} = \frac{\dot{m}_d(T_d(0) - T_o)}{\dot{m}_c(T_c - T_o)} \frac{\theta(2t_f - \theta)}{2t_f t_c} \tag{9.141}$$

and the corresponding exergy efficiency ψ by

$$\psi(t_d) = Ex_d(t_d)/Ex_c \tag{9.142}$$

The energy efficiency in Eq. (9.141) simplifies when the discharge period t_d exceeds t_f, i.e., $\eta\,(t_d \geq t_f) = f$.

In practice, it is not economically feasible to continue the discharge period until as much recoverable heat as possible is recovered. As the discharge period increases, water is recovered from an ATES at ever decreasing temperatures (ultimately approaching the reference-environment temperature T_o), and the energy in the recovered water is of decreasing usefulness. Exergy analysis reflects this phenomenon, as the magnitude of the recovered exergy decreases as the recovery temperature decreases. To determine the appropriate discharge period, a threshold temperature T_t is often introduced, below which the residual energy in the aquifer water is not considered worth recovering from an ATES. For the linear temperature–time relation used here (see Eq. (9.122)), it is clear that no thermal energy could be recovered over a cycle if the threshold temperature exceeds the initial discharge temperature, while the appropriate discharge period can be evaluated using Eq. (9.122) with T_t replacing $T_d(t)$ for the case where $T_o \leq T_t \leq T_d(0)$. Thus,

$$t_d = \begin{cases} (T_d(0) - T_t)t_f/(T_d(0) - T_o) & T_o \leq T_t \leq T_d(0) \\ 0 & T_d(0) \leq T_t \end{cases} \tag{9.143}$$

In practice, a threshold temperature places an upper limit on the allowable discharge time period. Utilizing a threshold temperature usually has the effect of decreasing the difference between the corresponding energy and exergy efficiencies.

Nonetheless, ATES performance measures based on exergy are more useful and meaningful than those based on energy. Exergy efficiencies account for the temperatures associated with heat transfers to and from an ATES, as well as the quantities of heat transferred, and consequently provide a measure of how nearly ATES systems approach ideal performance. Energy efficiencies account only for quantities of energy transferred, and can often be misleadingly high, e.g., in cases where heat is recovered at temperatures too low to be useful. The use of an appropriate threshold recovery temperature can partially avoid the most misleading characteristics of ATES energy efficiencies.

9.8. Examples and case studies

9.8.1. Inappropriateness of energy efficiency for TES evaluation

A simple example demonstrates that energy efficiency is an inappropriate measure of TES performance. Consider a perfectly insulated TES containing 1000 kg of water, initially at 40°C. The ambient temperature is 20°C.

A quantity of 4200 kJ of heat is transferred to the storage through a heat exchanger from an external body of 100 kg of water cooling from 100°C to 90°C [i.e., with Eq. (9.111) (100 kg) (4.2 kJ/kg K) (100 − 90)°C = 4200 kJ]. This heat addition raises the storage temperature 1.0°C, to 41°C [i.e., (4200 kJ)/((1000 kg) (4.2 kJ/kg K)) = 1.0°C]. After a period of storage, 4200 kJ of heat is recovered from the storage through a heat exchanger which delivers it to an external body of 100 kg of water, raising the temperature of that water from 20°C to 30°C [i.e., with Eq. (9.111), $\Delta T = (4200\,\text{kJ})/((100\,\text{kg})$ (4.2 kJ/kg K)) = 10°C]. The storage is returned to its initial state at 40°C.

The energy efficiency, the ratio of the heat recovered from the storage to the heat injected, is 4200 kJ/4200 kJ = 1, or 100%. Yet the recovered heat is at only 30°C, and of little use, having been degraded. With Eq. (9.117), the exergy recovered is evaluated as (100 kg) (4.2 kJ/kg K) [(30 − 20)°C − (293 K) ln (303/293)] = 70 kJ, and the exergy supplied as (100 kg) (4.2 kJ/kg K) [(100 − 90)°C − (293 K) ln(373/363)] = 856 kJ. Thus the exergy efficiency, the ratio of the thermal exergy recovered from storage to that injected, is 70/856 = 0.082, or 8.2%, a much more meaningful expression of the achieved performance of the storage cycle.

9.8.2. Comparing thermal storages

Consider two different thermal storages, each of which undergoes a similar charging process in which heat is transferred to a closed thermal storage from a stream of 1000 kg of water which enters at 85°C and leaves at 25°C (see Fig. 9.12). Consider Cases A and B, representing two different modes of operation. For Case A, heat is recovered from the storage after 1 day by a stream of 5000 kg of water entering at 25°C and leaving at 35°C. For Case B, heat is recovered from the storage after 100 days by a stream of 1000 kg of water entering at 25°C and leaving at 75°C. In both cases the temperature

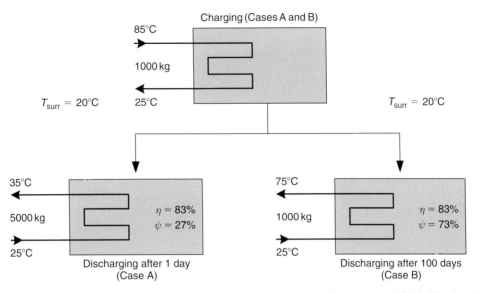

Fig. 9.12. An example in which two cases are considered. Shown are the charging process, which is identical for Cases A and B (top), the discharging process for Case A (center), and the discharging process for Case B (bottom).

of the surroundings remains constant at 20°C, and the final state of the storage is the same as the initial state. Water is taken to be an incompressible fluid having a specific heat at constant pressure of $c_p = 4.18$ kJ/kg K, and heat exchanges during charging and discharging are assumed to occur at constant pressure.

Several observations can be made from the data. First, the inlet and outlet temperatures for the charging and discharging fluids imply that a stratified temperature profile exists in the TES after charging. Second, the higher discharging fluid temperature for Case B implies that a greater degree of stratification is maintained during the storing period for Case B (or that greater internal mixing occurs for Case A). Third, the quantities of discharging fluid and the associated temperatures imply that the discharging fluid is circulated through the TES at a greater rate for Case A than for Case B.

Energy and exergy analyses of the overall processes are performed for both cases, using superscripts A and B to denote Cases A and B respectively. This example is based on the material in Section 9.3.

Energy analysis for the overall process

The net heat input to the storage during the charging period for each case is

$$H_a - H_b = m_1 c_p (T_a - T_b) = 1000 \text{ kg} \times 4.18 \text{ kJ/kg K} \times (85 - 25)\text{K} = 250{,}800 \text{ kJ}$$

For Case A, the heat recovered during the discharging period is

$$(H_d - H_c)^A = 5000 \text{ kg} \times 4.18 \text{ kJ/kg K} \times (35 - 25)\text{K} = 209{,}000 \text{ kJ}$$

The energy efficiency of storage is (see Eq. (9.16))

$$\eta^A = \frac{\text{Heat recovered}}{\text{Heat input}} = \frac{(H_d - H_c)^A}{H_a - H_b} = \frac{209{,}000 \text{ kJ}}{250{,}800 \text{ kJ}} = 0.833$$

The heat lost to the surroundings during storage is (see Eq. (9.3) with $\Delta E = 0$)

$$Q_l^A = (H_a - H_b) - (H_c - H_d)^A = 250{,}000 \text{ kJ} - 209{,}000 \text{ kJ} = 41{,}800 \text{ kJ}$$

For Case B, the heat recovered during discharging, the energy efficiency and the heat lost to the surroundings can be evaluated similarly:

$$(H_d - H_c)^B = 1000 \, \text{kg} \times 4.18 \, \text{kJ/kg K} \times (75 - 25)\text{K} = 209,000 \, \text{kJ}$$

$$\eta^B = \frac{209,000 \, \text{kJ}}{250,800 \, \text{kJ}} = 0.833$$

$$Q_l^B = 250,800 \, \text{kJ} - 209,000 \, \text{kJ} = 41,800 \, \text{kJ}$$

Exergy analysis for the overall process

The net exergy input during the charging period $(Ex_a - Ex_b)$ can be evaluated with Eq. (9.12). In that expression, the quantity $(H_a - H_b)$ represents the net energy input to the store during charging, evaluated as 250,800 kJ in the previous subsection. Noting that the difference in specific entropy can be written assuming incompressible substances having a constant specific heat as

$$s_a - s_b = c_p \ln \frac{T_a}{T_b} = 4.18 \frac{\text{kJ}}{\text{kg K}} \times \ln \frac{358 \, \text{K}}{298 \, \text{K}} = 0.7667 \frac{\text{kJ}}{\text{kg K}}$$

the quantity $T_0(S_a - S_b)$, which represents the unavailable part of the input heat, is

$$T_0(S_a - S_b) = T_0 m_1 (s_a - s_b) = 293 \, \text{K} \times 1000 \, \text{kg} \times 0.7667 \, \text{kJ/kg K} = 224,643 \, \text{kJ}$$

where m_1 denotes the mass of the transport fluid cooled during the charging period. Then, the net exergy input is

$$Ex_a - Ex_b = 250,800 \, \text{kJ} - 224,643 \, \text{kJ} = 26,1517 \, \text{kJ}$$

The net exergy output during the discharging period $(Ex_d - Ex_c)$ can be evaluated using Eq. (9.13) and, denoting the mass of the transport fluid circulated during the discharging period as m_3, in a similar three-step fashion for Cases A and B. For Case A,

$$(s_d - s_c)^A = c_p \ln \frac{T_d^A}{T_c^A} = 4.18 \frac{\text{kJ}}{\text{kg K}} \times \ln \frac{308 \, \text{K}}{298 \, \text{K}} = 0.1379 \frac{\text{kJ}}{\text{kg K}}$$

$$T_0(S_d - S_c)^A = T_0 m_3^A (s_d - s_c)^A = 293 \, \text{K} \times 5000 \, \text{kg} \times 0.1379 \, \text{kJ/kg K} = 202,023 \, \text{kJ}$$

$$(Ex_d - Ex_c)^A = 209,000 \, \text{kJ} - 202,023 \, \text{kJ} = 6977 \, \text{kJ}$$

For Case B,

$$(s_d - s_c)^B = c_p \ln \frac{T_d^B}{T_c^B} = 4.18 \frac{\text{kJ}}{\text{kg K}} \times \ln \frac{348 \, \text{K}}{298 \, \text{K}} = 0.6483 \frac{\text{kJ}}{\text{kg K}}$$

$$T_0(S_d - S_c)^B = T_0 m_3^B (s_d - s_c)^B = 293 \, \text{K} \times 1000 \, \text{kg} \times 0.6483 \, \text{kJ/kg K} = 189,950 \, \text{kJ}$$

$$(Ex_d - Ex_c)^B = 209,000 \, \text{kJ} - 189,950 \, \text{kJ} = 19,050 \, \text{kJ}$$

Thus, the exergy efficiency (see Eq. (9.17)) for Case A is

$$\psi^A = \frac{(Ex_d - Ex_c)^B}{Ex_a - Ex_b} = \frac{6977 \, \text{kJ}}{26,157 \, \text{kJ}} = 0.267$$

and for Case B is

$$\psi^B = \frac{(Ex_d - Ex_c)^B}{Ex_a - Ex_b} = \frac{19,050 \, \text{kJ}}{26,157 \, \text{kJ}} = 0.728$$

which is considerably higher than for Case A.

The exergy losses (total) can be evaluated with Eq. (9.7) (with $\Delta Ex = 0$) as the sum of the exergy loss associated with heat loss to the surroundings and the exergy loss due to internal exergy consumptions. That is,

$$(X_l + I)^A = (Ex_a - Ex_b) - (Ex_d - Ex_c)^A = 26,157 \, \text{kJ} - 6977 \, \text{kJ} = 19,180 \, \text{kJ}$$

$$(X_l + I)^B = (Ex_a - Ex_b) - (Ex_d - Ex_c)^B = 26,157 \, \text{kJ} - 19,050 \, \text{kJ} = 7107 \, \text{kJ}$$

The individual values of the two exergy loss parameters can be determined if the temperature at which heat is lost from the TES is known.

Comparison

The two cases are compared in Table 9.3. Although the same quantity of energy is discharged for Cases A and B, a greater quantity of exergy is discharged for Case B. In addition, Case B stores the energy and exergy for a greater duration of time.

Table 9.3. Comparison of the performance of a TES for two cases.

	Case A	Case B
General parameters		
Storing period (days)	1	100
Charging-fluid temperatures (in/out) (°C)	85/25	85/25
Discharging-fluid temperatures (in/out) (°C)	25/35	25/75
Energy parameters		
Energy input (kJ)	250,800	250,800
Energy recovered (kJ)	209,000	209,000
Energy loss (kJ)	41,800	41,800
Energy efficiency (%)	83.3	83.3
Exergy parameters		
Exergy input (kJ)	26,157	26,157
Exergy recovered (kJ)	6977	19,050
Exergy loss (kJ)	19,180	7107
Exergy efficiency (%)	26.7	72.8

9.8.3. Thermally stratified TES

In this example, based on material in Section 9.5, several energy and exergy quantities are determined using the general three-zone model, for a thermal storage using water as the storage fluid and having a realistic stratified temperature distribution. The use as a design tool of the material covered in Section 9.5 is also illustrated.

The actual temperature distribution considered is based on the general three-zone model and shown in Fig. 9.13, along with the general three-zone model distribution used to approximate the actual distribution. Specified general data are listed in Table 9.4.

The results of the example (see Table 9.5) demonstrate that, for the case considered, the ratio $Ex/Ex_m = 180.7/165.4 = 1.09$. This implies that the exergy of the stratified storage is about 9% greater than the exergy of the mixed storage. In effect, therefore, stratification increases the exergy storage capacity of the storage considered, relative to its mixed condition, by 9%.

The ratio Ex/Ex_m can be determined using the expressions in Section 9.5 or read from figures such as those in Fig. 9.9 (although the case here of $T_m = 60°C$, $x_2 = 0.1$ and $\Delta T_{th} = 20°C$ falls slightly outside of the range of values covered in Fig. 9.9 for the case of $T_m = 60°C$). Then, such diagrams can serve as design tools from which one can obtain a ratio that can be applied to the value of the exergy of the mixed storage to obtain the exergy of the stratified storage.

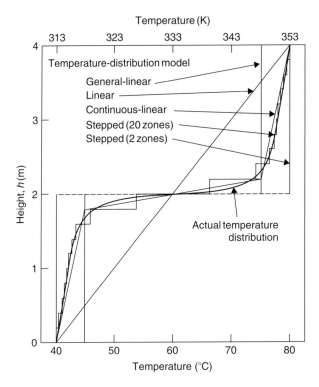

Fig. 9.13. The realistic vertically stratified temperature distribution considered in the example, and some of the temperature-distribution models used to approximate it (linear, continuous-linear, general-linear, stepped with 2 zones, and stepped with 20 zones). The shown continuous-linear distribution is equivalent to a general three-zone distribution.

Table 9.4. Specified general data for the example.

Temperatures (K)	
At TES top, $T(h=H)$	353
At TES bottom, $T(h=0)$	313
Reference environment, T_o	283
TES fluid parameters	
Height, H (m)	4
Mass, m (kg)	10,000
Specific heat, c (kJ/kg K)	4.18

9.8.4. Cold TES

Energy and exergy analyses of four different CTES cases are performed based on the material in Section 9.6. In each case, the CTES has identical initial and final states, so that the CTES operates in a cyclic manner, continuously charging, storing and discharging. The main characteristics of the cold storage cases are as follows:

- Sensible heat storage, with a fully mixed storage fluid.
- Sensible heat storage, with a linearly stratified storage fluid.

Table 9.5. Results for the stratification example.

	Temperature-distribution model						Results from numerical integration
	Linear	General-linear	Stepped			Continuous-linear*	
			$k = 200$	$k = 20$	$k = 2$		
Temperatures (K)							
T_m	333.000	333.000	333.000	333.000	333.000	333.000	333.000
T_c	332.800	332.540	332.550	332.560	332.400	332.570	332.550
Energy values (MJ)							
E	2090.000	2090.000	2090.000	2090.000	2090.000	2090.000	2090.000
E_m	2090.000	2090.000	2090.000	2090.000	2090.000	2090.000	2090.000
$E - E_m$	0.000	0.000	0.000	0.000	0.000	0.000	0.000
Exergy values (MJ)							
Ξ	172.500	181.800	181.400	181.000	186.700	180.700	181.400
Ξ_m	165.400	165.400	165.400	165.400	165.400	165.400	165.400
$\Xi - \Xi_m$	7.100	16.400	16.000	15.600	21.300	15.300	16.000
Percentage errors							
In values of T_e	+0.075	−0.030	0.000	+0.003	−0.045	+0.006	–
In values of Ξ	−4.900	+2.000	0.000	−0.200	+2.900	−0.400	–
In values of $\Xi - \Xi_m$	−55.600	+22.500	0.000	−2.500	+33.100	−4.400	–

* This case is also a general three-zone temperature-distribution model.

- Latent heat storage, with a fully mixed storage fluid.
- Combined latent and sensible heat storage, with a fully mixed storage fluid.

Assumptions and specified data

The following assumptions are made for each of the cases:

- Storage boundaries are non-adiabatic.
- Heat gain from the environment during charging and discharging is negligibly small relative to heat gain during the storing period.
- The external surface of the storage tank wall is at a temperature 2°C greater than the mean storage-fluid temperature.
- The mass flow rate of the heat-transfer fluid is controlled so as to produce constant inlet and outlet temperatures.
- Work interactions, and changes in kinetic and potential energy terms, are negligibly small.

Specified data for the four cases are presented in Table 9.6 and relate to the diagram in Fig. 9.10. In Table 9.6, T_b and T_d are the charging and discharging outlet temperatures of the heat-transfer fluid, respectively. The subscripts 1, 2 and 3 indicate the temperature of the storage fluid at the beginning of charging, storing or discharging, respectively. Also t indicates the liquid state and s indicates the solid state for the storage fluid at the phase change temperature.

In addition, for all cases, the inlet temperatures are fixed for the charging-fluid flow at $T_a = -10°C$ and for the discharging-fluid flow at $T_c = 20°C$. For cases involving latent heat changes (i.e., solidification), $F = 10\%$. The specific heat c is 4.18 kJ/kg K for both the storage and heat-transfer fluids. The phase-change temperature of the storage fluid

Table 9.6. Specified temperature data for the cases in the CTES example.

Temperature (°C)	Case			
	I	II	III	IV
T_b	4.0	15	−1	−1
T_d	11.0	11	10	10
T_1	10.5	19/2*	0 (t)	8
T_2	5.0	17/−7*	0 (s)	−8
T_3	6.0	18/−6*	0 (t&s)	0 (t&s)

* When two values are given, the storage fluid is vertically linearly stratified and the first and second values are the temperatures at the top and bottom of the storage fluid, respectively.

is 0°C. The configuration of the storage tank is cylindrical with an internal diameter of 2 m and internal height of 5 m. Environmental conditions are 20°C and 1 atm.

Results and discussion

The results for the four cases are listed in Table 9.7 and include overall and subprocess efficiencies, input and recovered cold quantities, and energy and exergy losses. The overall and subprocess energy efficiencies are identical for Cases I and II, and for Cases III and IV. In all cases the energy efficiency values are high. The different and lower exergy efficiencies for all cases indicate that energy analysis does not account for the quality of the 'cold' energy, as related to temperature, and considers only the quantity of 'cold' energy recovered.

The input and recovered quantities in Table 9.7 indicate the quantity of 'cold' energy and exergy input to and recovered from the storage. The energy values are much greater than the exergy values because, although the energy quantities involved are large, the energy is transferred at temperatures only slightly below the reference-environment temperature, and therefore is of limited usefulness.

Table 9.7. Energy and exergy quantities for the cases in the CTES example.

Period or quantity	Energy quantities				Exergy quantities			
	I	II	III	IV	I	II	III	IV
Efficiencies (%)								
Charging (1)	100	100	100	100	51	98	76	77
Storing (2)	82	82	90	90	78	85	90	85
Discharging (3)	100	100	100	100	38	24	41	25
Overall	82	82	90	90	15	20	28	17
Input, recovered and lost quantities (MJ)								
Input	361.1	361.1	5237.5	6025.9	30.9	23.2	499.8	575.1
Recovered	295.5	295.5	4713.8	5423.3	4.6	4.6	142.3	94.7
Loss (external)	65.7	65.7	523.8	602.6	2.9	2.9	36.3	48.9
Loss (internal)	–	–	–	–	23.3	15.6	321.2	431.4

The cold losses during storage, on an energy basis, are entirely due to cold losses across the storage boundary (i.e., heat infiltration). The exergy-based cold losses during storage are due to both cold losses and internal exergy losses (i.e., exergy consumptions due to irreversibilities within the storage). For the present cases, in which the exterior surface of the storage tank is assumed to be 2°C warmer than the mean storage-fluid temperature, the exergy losses include both external and internal components. Alternatively, if the heat-transfer temperature at the storage tank external surface is at the environment temperature, the external exergy losses would be zero and the total exergy losses would be entirely due to internal consumptions. If heat transfer occurs at the storage-fluid temperature, on the other hand, more of the exergy losses would be due to external losses. In all cases the total exergy losses, which are the sum of the internal and external exergy losses, remain fixed.

The four cases demonstrate that energy and exergy analyses give different results for CTES systems. Both energy and exergy analyses account for the quantity of energy transferred in storage processes. Exergy analyses take into account the loss in quality of 'cold' energy, and thus more correctly reflect the actual value of the CTES.

In addition, exergy analysis is conceptually more direct when applied to CTES systems because cold is treated as a useful commodity. With energy analysis, flows of heat rather than cold are normally considered. Thus energy analyses become convoluted and confusing as one must deal with heat flows, while accounting for the fact that cold is the useful input and product recovered for CTES systems. Exergy analysis inherently treats any quantity which is out of equilibrium with the environment (be it colder or hotter) as a valuable commodity, and thus avoids the intuitive conflict in the expressions associated with CTES energy analysis. The concept that cold is a valuable commodity is both logical and in line with one's intuition when applied to CTES systems.

9.8.5. Aquifer TES

In this case study, which is based on the material in Section 9.7, energy and exergy analyses are performed of an ATES using experimental data from the first of four short-term ATES test cycles for the Upper Cambrian Franconia-Ironton-Galesville confined aquifer. The test cycles were performed at the University of Minnesota's St. Paul campus from November 1982 to December 1983 (Hoyer et al., 1985). During the test, water was pumped from the source well, heated in a heat exchanger and returned to the aquifer through the storage well. After storage, energy was recovered by pumping the stored water through a heat exchanger and returning it to the supply well. The storage and supply wells are located 255 m apart.

For the test cycle considered here, the water temperature and volumetric flow rate vary with time during the injection and recovery processes as shown in Fig. 9.14. The storage-period duration (13 days) is also shown. Charging occurred during 5.24 days over a 17-day period. The water temperature and volumetric flow rate were approximately constant during charging, and had mean values of 89.4°C and 18.4 l/s, respectively. Discharging also occurred over 5.24 days, approximately with a constant volumetric flow rate of water and linearly decreasing temperature with time. The mean volumetric flow rate during discharging was 18.1 l/s, and the initial discharge temperature was 77°C, while the temperature after 5.24 days was 38°C. The ambient temperature was reported to be 11°C.

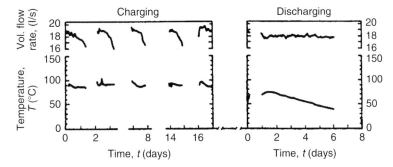

Fig. 9.14. Observed values for the temperature and volumetric flow rate of water, as a function of time during the charging and discharging periods, for the experimental test cycles used in the ATES case study.

Simplifications, analysis and results

In subsequent calculations, mean values for volumetric flow rates and charging temperature are used. Also, the specific heat and density of water are both taken to be fixed, at 4.2 kJ/kg K and 1000 kg/m^3, respectively. Since the volumetric flow rate (in l/s) is equal to the mass flow rate (in kg/s) when the density is $1000 \, \text{kg/m}^3$, $\dot{m}_c = 18.4 \, \text{kg/s}$ and $\dot{m}_d = 18.1 \, \text{kg/s}$. Also, the reference-environment temperature is fixed at the ambient temperature, i.e., $T_o = 11°C = 284 \, \text{K}$.

During charging, it can be shown using Eqs. (9.129) and (9.130), with $t_c = 5.24 \, \text{d} = 453{,}000 \, \text{s}$ and $T_c = 89.4°C = 362.4 \, \text{K}$, that

$$E_c = (18.4 \, \text{kg/s})(4.2 \, \text{kJ/kg K})(453{,}000 \, \text{s})(89.4°C - 11°C) = 2.74 \times 10^9 \, \text{kJ}$$

and

$$Ex_c = 2.74 \times 10^9 \, \text{kJ} - (18.4 \, \text{kg/s})(4.2 \, \text{kJ/kg K})(453{,}000 \, \text{s})(284 \, \text{K}) \ln (362.4 \, \text{K}/284 \, \text{K})$$

$$= 0.32 \times 10^9 \, \text{kJ}$$

During discharging, the value of the time t_f is evaluated using the linear temperature–time relation of the present model and the observations that $T_d(t = 5.24 \, \text{d}) = 38°C$ and $T_d(0) = 77°C = 350 \, \text{K}$. Then, using Eq. (9.122) with $t = 5.24 \, \text{d}$,

$$38°C = 77°C - (77°C - 11°C)(5.24 \, \text{d}/t_f)$$

which can be solved to show that $t_f = 8.87 \, \text{d}$. Thus, with the present linear model, the discharge water temperature would reach T_o if the discharge period was lengthened to almost 9 days. In reality, the rate of temperature decline would likely decrease, and the discharge temperature would asymptotically approach T_o.

The value of the fraction f can be evaluated with Eq. (9.140) as

$$f = \frac{(8.87 \, \text{d})(18.1 \, \text{kg/s})(77°C - 11°C)}{2(5.24 \, \text{d})(18.4 \, \text{kg/s})(89.4°C - 11°C)} = 0.701$$

Thus, the maximum energy efficiency achievable is approximately 70%. With these values and Eqs. (9.135) and (9.136), it can be shown that

$$a = (11°C - 77°C)/(284 \, \text{K} \times 8.87 \, \text{d}) = -0.0262 \, \text{d}^{-1} \quad \text{and} \quad b = (350 \, \text{K})/(284 \, \text{K}) = 1.232$$

Consequently, expressions dependent on discharge time period t_d can be written and plotted (see Fig. 9.15) for E_d, Ex_d, η and ψ using Eqs. (9.131)–(9.133), (9.141) and (9.142), and for the energy loss ($E_c - E_d$) and exergy loss ($Ex_c - Ex_d$).

Discussion

Both energy and exergy efficiencies in Fig. 9.15 increase from zero to maximum values as t_d increases. Further, the difference between the two efficiencies increases with increasing t_d. This latter point demonstrates that the exergy efficiency gives less weight than the energy efficiency to the energy recovered at higher t_d values since it is recovered at temperatures nearer to the reference-environment temperature.

Several other points in Fig. 9.15 are worth noting. First, for the conditions specified, all parameters level off as t_d approaches t_f, and remain constant for $t_d \geq t_f$. Second, as t_d increases toward t_f, the energy recovered increases from zero to a maximum value, while the energy loss decreases from a maximum of all the input energy to a minimum (but non-zero) value. The exergy recovery and exergy loss functions behave similarly qualitatively, but exhibit much lower magnitudes.

The difference between energy and exergy efficiencies is due to temperature differences between the charging- and discharging-fluid flows. As the discharging time increases, the deviation between these two efficiencies increases (Fig. 9.15) because the temperature of recovered heat decreases (Fig. 9.14). In this case, the energy efficiency reaches approximately 70% and the exergy efficiency of 40% by the completion of the discharge period, even though the efficiencies are both 0% when discharging commences.

To further illustrate the importance of temperature, a hypothetical modification of the present case study is considered. In the modified case, all details are as in the original case except that the temperature of the injection flow during the charging period is increased from 89.4°C to 200°C (473 K), while the duration of the charging period is decreased from its initial value of 5.24 days (453 000 s) so that the energy injected does not change. By equating the energy injected during

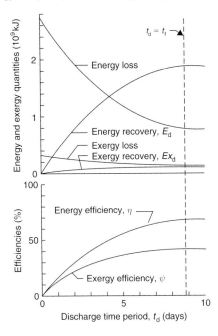

Fig. 9.15. Variation of several calculated energy and exergy quantities and efficiencies, as a function of discharge time period, for the ATES case study.

charging for the original and modified cases, the modified charging-period duration t_c' can be evaluated as a function of the new injection flow temperature T_c' as follows:

$$t_c' = t_c \frac{T_c - T_o}{T_c' - T_o} = (453{,}000\,\text{s}) \frac{(89.4°\text{C} - 11°\text{C})}{(200°\text{C} - 11°\text{C})} = 188{,}000\,\text{s}$$

The modified exergy input during charging can then be evaluated as

$$Ex_c' = 2.74 \times 10^9\,\text{kJ} - (18.4\,\text{kg/s})(4.2\,\text{kJ/kgK})(188{,}0001\,\text{s})(284\,\text{K})\ln(473\,\text{K}/284\,\text{K}) = 0.64 \times 10^9\,\text{kJ}$$

This value is double the exergy input during charging for the original case, so, since the discharging process remains unchanged in the modified case, the exergy efficiency (for any discharging time period) is half that for the original case. The altered value of exergy efficiency is entirely attributable to the new injection temperature, and occurs despite the fact that the energy efficiency remains unchanged.

If a threshold temperature is introduced and arbitrarily set at 38°C (the actual temperature at the end of the experimental discharge period of 5.24 d), then the data in Fig. 9.15 for $t_d = 5.24$ d apply and one can see that:

(i) The exergy recovered $(0.127 \times 10^9\,\text{kJ})$ is almost all (91%) of the exergy recoverable in infinite time $(0.139 \times 10^9\,\text{kJ})$, while the energy recovered $(1.60 \times 10^9\,\text{kJ})$ is not as great a portion (83%) of the ultimate energy recoverable $(1.92 \times 10^9\,\text{kJ})$.
(ii) The exergy loss $(0.19 \times 10^9\,\text{kJ})$ exceeds the exergy loss in infinite time $(0.18 \times 10^9\,\text{kJ})$ slightly (by 5.5%), while the energy loss $(1.14 \times 10^9\,\text{kJ})$ exceeds the energy loss in infinite time $(0.82 \times 10^9\,\text{kJ})$ substantially (by 39%).
(iii) The exergy efficiency (40%) has almost attained the exergy efficiency attainable in infinite time (43.5%), while the energy efficiency (58%) is still substantially below the ultimate energy efficiency attainable (70%).

To gain confidence in the model and the results, some of the quantities calculated using the linear model can be compared with the same quantities as reported in the experimental paper (Hoyer et al., 1985):

(i) The previously calculated value for the energy injection during charging of $2.74 \times 10^9\,\text{kJ}$ is 1.1% less than the reported value of $2.77 \times 10^9\,\text{kJ}$.

(ii) The energy recovered at the end of the experimental discharge period of $t_d = 5.24$ days can be evaluated with Eq. (9.131) as

$$E_d(5.24\,\text{d}) = (18.1)(4.2)(77 - 11)[5.24(2 \times 8.87 - 5.24)/(2 \times 8.87)](86{,}400\,\text{s/d}) = 1.60 \times 10^9\,\text{kJ}$$

which is 1.8% less than the reported value of 1.63×10^9 kJ.

(i) The energy efficiency at $t_d = 5.24$ d can be evaluated with Eq. (9.141) as

$$\eta(5.24\,\text{d}) = (1.60 \times 10^9\,\text{kJ})/(2.74 \times 10^9\,\text{kJ}) = 0.584$$

which is 1.0% less than the reported value of 0.59 (referred to as the 'energy recovery factor').

9.9. Concluding remarks

This chapter demonstrates that the use of exergy analysis is important for developing a sound understanding of the thermodynamic behavior of TES systems, and for rationally assessing, comparing and improving their efficiencies. Exergy analysis suggests measures to improve TES systems like:

- Reducing thermal losses (heat leakage from hot TESs and heat infiltration to cold TESs) by improving insulation levels and distributions.
- Avoiding temperature degradation by using smaller heat-exchanger temperature differences, ensuring that heat flows of appropriate temperatures are used to heat cooler flows, and increasing heat-exchanger efficiencies.
- Avoiding mixing losses by retaining and taking advantage of thermal stratification.
- Reducing pumping power by using more efficient pumps, reduced-friction heat-transfer fluids and appropriate heat recovery threshold temperatures.

The development of a standard TES evaluation methodology accounting for the thermodynamic considerations discussed in this chapter would be worthwhile. The use of exergy is important because it clearly takes into account the loss of availability and temperature of heat in storage operations, and hence more correctly reflects the thermodynamic and economic value of the storage operation. The development of better assessment methodologies will ensure effective use of energy resources by providing a basis for identifying the more productive directions for development of TES technology, and identifying the better systems without the lengthy and inefficient process of waiting for them to be sorted out by competitive economic success in the marketplace.

Problems

9.1 How are the energy and exergy efficiencies of a thermal storage system defined?

9.2 Identify the sources of exergy loss in thermal energy storage systems and propose methods for reducing or minimizing them.

9.3 What is the effect of storage temperature on the energy and exergy efficiencies of a TES system?

9.4 What is the effect of stratification on the energy and exergy efficiencies of a TES system?

9.5 Identify the operating parameters that have the greatest effects on the exergy performance of a TES system.

9.6 For which TES system are the advantages of the exergy approach more significant: cold storage or heat storage? Explain.

9.7 Is a solar water heating system using collectors a TES system? How can you express the energy and exergy efficiency of such a system? What are the causes of exergy destructions in solar water heating systems?

9.8 Obtain a published article on exergy analysis of TES systems. Using the operating data provided in the article, perform a detailed exergy analysis of the system and compare your results to those in the original article. Also, investigate the effect of varying important operating parameters on the system exergetic performance.

Chapter 10

EXERGY ANALYSIS OF RENEWABLE ENERGY SYSTEMS

Exergy analyses are performed in this chapter of several renewable energy systems including solar photovoltaic systems, solar ponds, wind turbines and geothermal district heating systems and power plants. These and other renewable energy systems are likely to play increasingly important roles in societies in the future.

10.1. Exergy analysis of solar photovoltaic systems

Solar photovoltaic (PV) technology converts sunlight directly into electrical energy. Direct current electricity is produced, which can be used in that form, converted to alternating current or stored for later use. Solar PV systems operate in an environmentally benign manner, have no moving components, and have no parts that wear out if the device is correctly protected from the environment. By operating on sunlight, PV devices are usable and acceptable to almost all inhabitants of our planet. PV systems can be sized over a wide range, so their electrical power output can be engineered for virtually any application, from low-power consumer uses like wristwatches, calculators and battery chargers to significantly energy-intensive applications such as generating power at central electric utility stations. PV systems are modular, so various incremental power capacity additions are easily accommodated, unlike for fossil or nuclear fuel plants, which require multi-megawatt plants to be economically feasible.

The solar PV cell is one of the most significant and rapidly developing renewable-energy technologies, and its potential future uses are notable. By using solar radiation, a clean energy source, PV systems are relatively benign environmentally. During the last decade, PV applications have increased in many countries and are observed throughout the residential, commercial, institutional and industrial sectors. The clean, renewable and in some instances economic features of PV systems have attracted attention from political and business decision makers and individuals. Advances in PV technology have also driven the trend to increased usage.

A PV cell is a type of photochemical energy conversion device. Others include photoelectric devices and biological photosynthesis. Such systems operate by collecting a fraction of the radiation within some range of wavelengths. In PV devices, photon energies greater than the cutoff (or band-gap) energy are dissipated as heat, while photons with wavelengths longer than the cutoff wavelength are not used.

The energy conversion factor of a solar PV system sometimes is described as the efficiency, but this usage can lead to difficulties. The efficiency of a solar PV cell can be considered as the ratio of the electricity generated to the total, or global, solar irradiation. In this definition only the electricity generated by a solar PV cell is considered. Other properties of PV systems, which may affect efficiency, such as ambient temperature, cell temperature and chemical components of the solar cell, are not directly taken into account.

The higher performance, lower cost and better reliability demonstrated by today's PV systems are leading many potential users to consider the value of these systems for particular applications. Together, these applications will likely lead industry to build larger and more cost-effective production facilities, leading to lower PV costs. Public demand for environmentally benign sources of electricity will almost certainly hasten adoption of PV. The rate of adoption will be greatly affected by the economic viability of PV with respect to competing options. Many analysts and researchers believe that it is no longer a question of if, but when and in what quantity, PV systems will gain adoption. Since direct solar radiation is intermittent at most locations, fossil fuel-based electricity generation often must supplement PV systems. Many studies have addressed this need.

This section describes solar PV systems and their components and discusses the use of exergy analysis to assess and improve solar PV systems. Exergy methods provide a physical basis for understanding, refining and predicting the variations in solar PV behavior. This section also provides and compares energy- and exergy-based solar PV efficiency definitions.

10.1.1. PV performance and efficiencies

Three PV system efficiencies can be considered: power conversion efficiency, energy efficiency and exergy efficiency. Energy (η) and exergy (ψ) efficiencies for PV systems can be evaluated based on the following definitions:

$$\eta = \text{energy in products/total energy input}$$

$$\psi = \text{exergy in products/total exergy input}$$

For solar PV cells, efficiency measures the ability to convert radiative energy to electrical energy. The electrical power output is the product of the output voltage and the current out of the PV device, taken from the current–voltage curve (I–V curve). This conversion efficiency is not a constant, even under constant solar irradiation. However, there is a maximum power output point, where the voltage value is V_m, which is slightly less than the open-circuit voltage V_{oc}, and the current value is I_m, which is slightly less than the short-circuit current I_{sc} (Fig. 10.1). In this figure, E_{GH} represents the highest energy level of the electron attainable at maximum solar irradiation conditions. It is recognized that there should be an active relational curve from I_{sc} to V_{oc} and, with this relation, E_{GH} becomes equivalent to $\int_{V=0}^{V_{oc}} I(V)\mathrm{d}V$. In addition E_L represents the low-energy content of the electron, which is the more practical energy; this energy is shown as the rectangular area in Fig. 10.1, so $E_L = I_m V_m$. The maximum power point is restricted by a 'fill factor' FF, which is the maximum power conversion efficiency of the PV device and is expressible as

$$FF = \frac{V_m I_m}{V_{oc} I_{sc}} \tag{10.1}$$

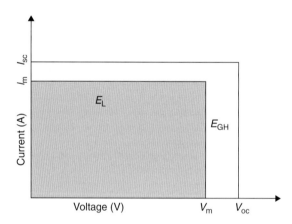

Fig. 10.1. Illustration of a general current–voltage (I–V) curve.

10.1.2. Physical exergy

The enthalpy of a PV cell with respect to the reference environment, ΔH, can be expressed as

$$\Delta H = C_p(T_{cell} - T_{amb}) \tag{10.2}$$

where C_p denotes the heat capacity, T_{amb} the ambient temperature and T_{cell} the cell temperature. The total entropy of the system relative to ambient conditions, ΔS, can be written as

$$\Delta S = \Delta S_{system} + \Delta S_{surround} \tag{10.3}$$

or

$$\Delta S = C_p \ln \left(\frac{T_{\text{cell}}}{T_{\text{amb}}} \right) - \frac{Q_{\text{loss}}}{T_{\text{cell}}} \tag{10.4}$$

where

$$Q_{\text{loss}} = C_p(T_{\text{cell}} - T_{\text{amb}}) \tag{10.5}$$

Here, Q_{loss} represents heat losses from the PV cell. With Eqs. (10.2) through (10.5), the physical exergy output for a PV cell system can be expressed as

$$Ex = E_{\text{GH}} + C_p(T_{\text{cell}} - T_{\text{amb}}) + T_{\text{amb}} \left(C_p \ln \frac{T_{\text{cell}}}{T_{\text{amb}}} - \frac{Q_{\text{loss}}}{T_{\text{cell}}} \right) \tag{10.6}$$

The first term on the right side of this equation (E_{GH}) is the generated electricity at the highest energy content of the electron. The second and third terms are the enthalpy and entropy contributions, respectively.

10.1.3. Chemical exergy

The process of PV energy conversion (Fig. 10.2) can in general be divided into two steps:

1. Electronic excitation of the absorbing component of the converter by light absorption with concomitant electronic charge creation.
2. Separation of the electronic charges.

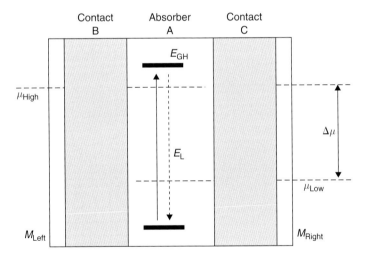

Fig. 10.2. An idealized photovoltaic converter (adapted from Bisquert et al. (2004)).

The excitation can be an electron–hole pair in a semiconductor, an electronic excitation of a molecule, or the production of excitations. In terms of the two level systems shown in Fig. 10.2, electronic excitation in the absorber promotes the system into the highest energy content with the associated electronic energy level H, simultaneously creating an electron-deficient low-energy content with associated energy level L. The electrons in these two states are separated. The departure of the populations of the states from their thermal equilibrium values implies a difference in their chemical potentials (partial-free energies) (Bisquert et al., 2004), as can be seen in Fig. 10.2. That is,

$$\Delta \mu = \mu_{\text{H}} - \mu_{\text{L}} \tag{10.7}$$

From the point of view of thermodynamics, the separation of Fermi levels arises as a result of the absorber being at a lower ambient temperature T_{amb} than the radiation 'pump' temperature T_p (i.e., the temperature of the sun). A Carnot cycle argument or statistical analysis gives the following upper limit chemical potential for the open-circuit voltage (Landsberg and Markvart, 1998; Bisquert et al., 2004):

$$\Delta\mu = \left(1 - \frac{T_{cell}}{T_p}\right)(E_{GH} - E_L) \tag{10.8}$$

where E_{GH} is the generated electricity at the highest-energy content of the electron and E_L is the available energy content of the electron (as the practical case).

Note that there is no current flow at the open-circuit voltage and that there is no voltage difference at the short-circuit current. Maximum power can be predicted to occur between these limits (Fig. 10.2). The power relations between voltage and electron charge are

$$E = qV \tag{10.9}$$

and

$$I = \frac{q}{t} \tag{10.10}$$

where V denotes circuit voltage, q electron charge, I circuit current and t time duration. The open-circuit voltage V_{oc} and short-circuit current I_{sc} represent the energy level without voltage or current, respectively.

To simplify the analysis, we take the curve for E_{GH} in Fig. 10.1 to be rectangular. Based on the Carnot cycle analogy, Eq. (10.8) then becomes

$$\Delta\mu = \left(1 - \frac{T_{cell}}{T_p}\right)[V_{oc}I_{sc} - V_m I_m] \tag{10.11}$$

This expression is used to determine the chemical exergy, following the approach presented in Fig. 10.1. As noted earlier, the efficiencies cannot be evaluated easily for some components at open-circuit voltage and short-circuit current, which are the conditions at which maximum power can be generated in a PV cell system. But from a thermodynamic perspective, the unconsidered remaining components should be extracted from the overall I–V curve. As a result, the total exergy of the PV solar cell can be formulated as

$$Ex = Ex_{physical} - (q_{sc}V_{oc} - q_L V_L)\frac{T_{cell}}{T_p} \tag{10.12}$$

where $Ex_{physical}$, q_L and V_L represent respectively the physical exergy shown in Eq. (10.6) with the excited electron charge at the low-energy content, the electron charge and the voltage.

We now define the solar cell power conversion efficiency η_{pce} as a function of E_L and S_T as follows:

$$\eta_{pce} = \frac{E_L}{S_T} = \frac{V_m I_m}{S_T} \tag{10.13}$$

where S_T represents hourly measured total solar irradiation.

The solar power conversion efficiency can also be defined in terms of the fill factor FF, based on Eq. (10.1), as follows:

$$\eta_{pce} = \frac{FF \times V_{oc}I_{sc}}{S_T} \tag{10.14}$$

The second main energy source is the solar irradiance incident on PV cells. Evaluation of the exergy efficiency of PV cells requires, therefore, the exergy of the total solar irradiation. PV cells are affected by direct and indirect components of solar irradiation, the magnitude of which depend on atmospheric effects. The exergy of solar irradiance, Ex_{solar}, can be evaluated approximately as (Bejan, 1998; Santarelli and Macagno, 2004):

$$Ex_{solar} = S_T\left(1 - \frac{T_{amb}}{T_{sun}}\right) \tag{10.15}$$

As a result of these formulations, the exergy efficiency ψ can be expressed as

$$\psi = \frac{Ex}{Ex_{solar}} \tag{10.16}$$

After substituting Eqs. (10.12) and (10.15) into Eq. (10.16) we obtain the following expression for exergy efficiency:

$$\psi = \frac{Ex_{\text{physical}} - (q_{\text{sc}} V_{\text{oc}} - q_{\text{L}} V_{\text{L}}) \dfrac{T_{\text{cell}}}{T_{\text{p}}}}{S_{\text{T}} \left(1 - \dfrac{T_{\text{amb}}}{T_{\text{sun}}}\right)} \tag{10.17}$$

The energy efficiency η depends on the generated electricity of the PV cells E_{gen} and the total energy input based on the total solar irradiation S_{T}. That is,

$$\eta = \frac{E_{\text{GH}}}{S_{\text{T}}} \tag{10.18}$$

The exergy efficiency usually gives a finer understanding of performance than the energy efficiency, and stresses that both external losses and internal irreversibilities need to be addressed to improve efficiency. In many cases, internal irreversibilities are more significant and more difficult to address than external losses.

One reason why today's commercial solar PV cells are costly is that they are inefficient. The main losses in a PV cell during electricity generation are attributable to such factors as thermalization, junction contact and recombination. These internal losses are considered in the chemical exergy part of the section. By considering the balance of energy and the heat flux absorbed and emitted by the PV cell, one can evaluate the losses due to irreversible operation of the converter. For the present analysis of PV systems, thermal exergy losses are the main external exergy losses.

10.1.4. Illustrative example

The exergy efficiency of a PV cell is evaluated based on data from a short-term test on a rack-mounted PV cell in Golden, Colorado, which is located at 105.23°W longitude and 39.71°N latitude. The test was performed from 11:00 a.m. to 5:00 p.m. on June 28, 2001 and the data measured include total solar irradiation, maximum generated power by the system, voltage, open-circuit voltage, current, short-circuit current, cell temperature and ambient temperature. The system includes two modules in series per string, and the total array nominal power rating for six strings is 631.5 W (Barker and Norton, 2003).

It can be seen that I–V curve parameters vary significantly with module temperature (Fig. 10.3). This is especially true for the current parameters I_{sc} and I_{m}, which exhibit strong linear variations with module temperature. The maximum power voltage V_{m} exhibits an inverse linear relation with module temperature. In addition, a second-degree polynomial relation is observed between open-circuit voltage and module temperature. This variation is not too significant. The curves in Fig. 10.3 can be used for parameter estimation.

Efficiencies are presented in Fig. 10.4, where it is seen that energy efficiencies of the system vary between 7% and 12%, while the exergy efficiencies of the system, which account for all inputs, irreversibilities and thermal emissions, vary from 2% to 8%. Power conversion efficiencies for this system, which depend on fill factors, are observed to be higher than the values for energy and exergy efficiencies.

Values of 'fill factors' are determined for the system and observed to be similar to values of exergy efficiency.

10.1.5. Closure

PV cells allow use of solar energy by converting sunlight directly to electricity with high efficiency. PV systems can provide nearly permanent power at low operating and maintenance costs in an environmentally benign manner. The assessment of PV cells described here illustrates the differences between PV cell energy and exergy efficiencies. Exergy analysis should be used for PV cell evaluation and improvement to allow for more realistic planning.

10.2. Exergy analysis of solar ponds

Solar radiation is abundantly available on all parts of the earth and in many regards is one of the best alternatives to non-renewable energy sources. One way to collect and store solar energy is through the use of solar ponds which

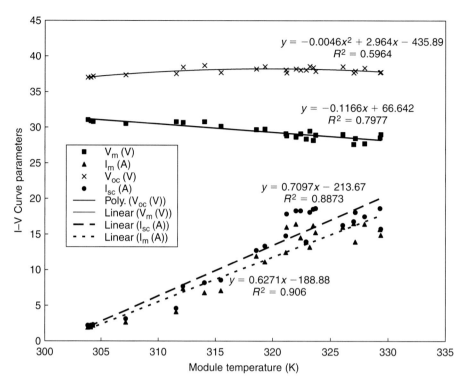

Fig. 10.3. Variation of several current–voltage (I–V) curve parameters with module temperature. Shown are data points as well as best fit curves (along with the R^2 values from the curve fitting routine).

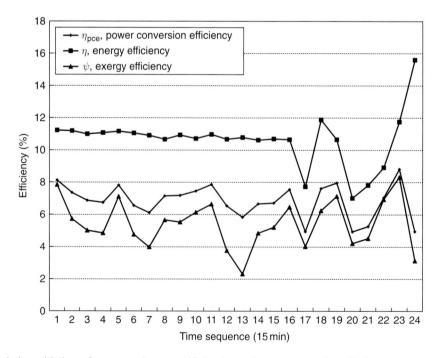

Fig. 10.4. Variation with time of energy and exergy efficiencies and power conversion efficiency.

can be used to supply thermal energy for various applications, such as process and space heating, water desalination, refrigeration, drying and power generation. Thermal energy storage has always been an important technique for energy storage. Solar ponds appear in some applications to have significant potential. The performance of a solar pond depends on its thermal energy storage capacity and its construction and maintenance costs (Dincer and Rosen, 2002; Jaefarzadeh, 2004). Performance also depends on thermophysical properties of the pond and storage fluid, and the surroundings conditions (Karakilcik, et al., 2006a, b). Solar ponds have recently received increasing attention in some applications. Numerous experimental and theoretical studies have been undertaken.

This section has two main parts. First, overall temperature distributions in a solar pond situated at Cukurova University in Adana, Turkey (35° 18′ E longitude, 36° 59′ N latitude) are measured to determine heat losses, and energy efficiencies of the zones according to the rate of incident solar radiation, absorption and transmission of the zone are examined. The data allow pond performance to be obtained experimentally for three representative months (January, May and August). Significant factors affecting performance, such as wall shading, incident solar radiation, insulation and the thicknesses of zones, are also investigated. Second, an exergy analysis of solar ponds is performed in this section and contrasted with the energy analysis. Little experimental and theoretical research has been reported on the exergetic performance of solar ponds so this section builds primarily on recent research by the authors.

10.2.1. Solar ponds

A salinity gradient solar pond is an integral device for collecting and storing solar energy. By virtue of having built-in TES, it can be used irrespective of time and season. In an ordinary pond or lake, when the sun's rays heat the water this heated water, being lighter, rises to the surface and loses its heat to the atmosphere. The net result is that the pond water remains at nearly atmospheric temperature. Solar pond technology inhibits this phenomenon by dissolving salt into the bottom layer of this pond, making it too heavy to rise to the surface, even when hot. The salt concentration increases with depth, thereby forming a salinity gradient. The sunlight which reaches the bottom of the pond is trapped there. The useful thermal energy is then withdrawn from the solar pond in the form of hot brine. The prerequisites for establishing solar ponds are: a large tract of land (it could be barren), abundant sunshine and inexpensively available salt (e.g. NaCl) or bittern.

Salt-gradient solar ponds may be economically attractive in climates with little snow and in areas where land is readily available. In addition, sensible cooling storage can be added to existing facilities by creating a small pond or lake on site. In some installations this can be done as part of property landscaping. Cooling takes place by surface evaporation and the rate of cooling can be increased with a water spray or fountain. Ponds can be used as an outside TES system or as a means of rejecting surplus heat from refrigeration or process equipment.

Being large, deep bodies of water, solar ponds are usually sized to provide community heating. Solar ponds differ in several ways from natural ponds. Solar ponds are filled with clear water to ensure maximum penetration of sunlight. The bottom is darkened to absorb more solar radiation. Salt is added to make the water more dense at the bottom and to inhibit natural convection. The cooler water on top acts as insulation and prevents evaporation. Salt water can be heated to high temperatures, even above the boiling point of fresh water.

Figure 10.5 shows a cross section of a typical salinity gradient solar pond which has three regions. The top region is called the surface zone, or upper convective zone (UCZ). The middle region is called the gradient zone, or non-convective zone (NCZ), or insulation zone (IZ). The lower region is called the heat storage zone (HSZ) or lower convective

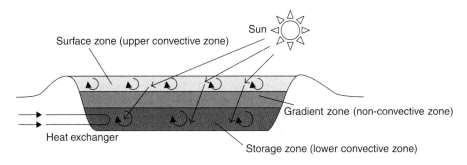

Fig. 10.5. Cross-section of a typical salinity-gradient solar pond.

zone (LCZ). The lower zone is a homogeneous, concentrated salt solution that can be either convecting or temperature stratified. Above it the non-convective gradient zone constitutes a thermally insulating layer that contains a salinity gradient. This means that the water closer to the surface is always less concentrated than the water below it. The surface zone is a homogeneous layer of low-salinity brine or fresh water. If the salinity gradient is large enough, there is no convection in the gradient zone even when heat is absorbed in the lower zone, because the hotter, saltier water at the bottom of the gradient remains denser than the colder, less salty water above it. Because water is transparent to visible light but opaque to infrared radiation, the energy in the form of sunlight that reaches the lower zone and is absorbed there can escape only via conduction. The thermal conductivity of water is moderately low and, if the gradient zone has substantial thickness, heat escapes upward from the lower zone slowly. This makes the solar pond both a thermal collector and a long-term storage device.

Further details on the three zones of solar ponds follow:

1. The UCZ is the fresh water layer at the top of the pond. This zone is fed with fresh water of a density near to the density of fresh water in the upper part to maintain the cleanliness of the pond and replenish lost water due to evaporation.
2. The NCZ or IZ lies between the LCZ and the UCZ. This zone is composed of salty water layers whose brine density gradually increases toward the LCZ. The NCZ is the key to the working of a solar pond. It allows an extensive amount of solar radiation to penetrate into the storage zone while inhibiting the propagation of long-wave solar radiation from escaping because water is opaque to infrared radiation.
3. The LCZ or HSZ is composed of salty water with the highest density. A considerable part of the solar energy is absorbed and stored in this region. The LCZ has the highest temperature, so the strongest thermal interactions occur between this zone and the adjacent insulated bottom-wall (IBW) and insulated side-walls (ISW).

Solar ponds were pioneered in Israel in the early 1960s, and are simple in principle and operation. They are long-lived and require little maintenance. Heat collection and storage are accomplished in the same unit, as in passive solar structures, and the pumps and piping used to maintain the salt gradient are relatively simple. The ponds need cleaning, like a swimming pool, to keep the water transparent to light. A major advantage of solar ponds is the independence of the system. No backup is needed because the pond's high heat capacity and enormous thermal mass can usually buffer a drop in solar supply that would force a single-dwelling unit to resort to backup heat.

10.2.2. Experimental data for a solar pond

For illustration, an experimental solar pond is considered with surface area dimensions of 2 m by 2 m and a depth of 1.5 m, as shown in Fig. 10.6. The solar pond was built at Cukurova University in Adana, Turkey. The salt-water solution is prepared by dissolving the NaCl reagent into fresh water. The thicknesses of the UCZ, NCZ and HSZ are 0.1 m, 0.6 m and 0.8 m, respectively. The range of salt gradient in the inner zones is such that the density is 1000–1045 kg/m^3 in the UCZ, and 1045–1170 kg/m^3 in the NCZ, 1170–1200 kg/m^3 in HSZ. Temperature variations are measured at the inner and outer zones of the pond. The bottom and the side-walls of the pond are plated with iron-sheets of 5 mm thickness, and contain glass wool of 50 mm thickness as an insulating layer. The solar pond is situated on a steel base 0.5 m above the ground and insulated with 20 mm thick wood slats positioned on the steel base. The inner and outer sides of the pond are covered with anti-corrosion paint. Figure 10.7 illustrates the inner zones of the solar pond.

Figure 10.8 illustrates solar radiation entering the pond, and the shading area by the south side-wall in the inner zones of the solar pond and the measurement points. The inner zones consist of 30 saline water layers of various densities. Each layer thickness is 5 cm. Temperature sensors in the zones measure the temperature distributions of the layers. Sixteen temperature distributions are located in some inner zone layers and in the insulated walls of the pond. The temperature distribution profiles are obtained using a data acquisition system (Karakilcik, 1998). To measure the temperature distributions of various regions, several temperature sensors are applied, at heights from the bottom of the pond of 0.05, 0.30, 0.55, 0.70, 0.80, 1.05, 1.35 and 1.50 m, and, from the bottom of the pond downward into the insulated bottom, at 15 and 45 mm, and for heights from the bottom of the side wall of 0, 0.35, 0.65, 0.75, 1.00 and 1.35 m.

The inner and wall temperatures of the pond are measured on an hourly basis throughout a day. The temperatures at the inner zones and ISW of the pond are measured by sensors with a range of −65°C to +155°C, and with a measurement accuracy of ±0.1°C for the temperature range of 0°C to 120°C. The sensors consist of 1N4148 semi-conductor devices with coaxial cables lengths between 17 and 20 m. Solar energy data are obtained using a pyranometer, and hourly and daily average air temperatures are obtained from a local meteorological station. Further information on

Fig. 10.6. Experimental solar pond (Karakilcik et al., 2006a).

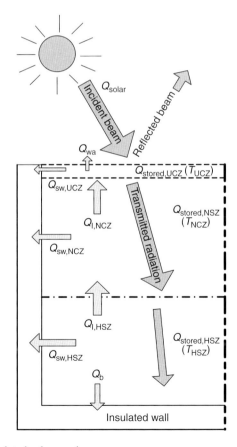

Fig. 10.7. Half-cut view of an insulated solar pond.

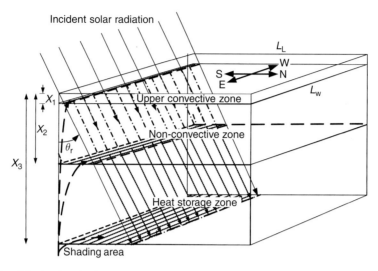

Fig. 10.8. Schematic of the insulated solar pond.

the experimental system, measurements and thermophysical properties of the utilized materials and fluids are available elsewhere (Karakilcik, 1998; Karakilcik et al., 2006a, b).

10.2.3. Energy analysis

As shown in Fig. 10.8, the UCZ, NCZ and HSZ thicknesses of the salt gradient solar pond are X_1, $X_2 - X_1$ and $X_3 - X_2$, respectively.

The working solution in the UCZ has uniform and low salinity (like seawater), while the working solution in the LCZ is stratified due to its high salinity and different density. In the NCZ, both concentration and temperature increase linearly with increasing pond depth. Part of the solar radiation incident on the solar pond is absorbed, part is reflected at the surface and the remaining part is transmitted, as illustrated in Figs. 10.9 through 10.11. In Figs. 10.9 and 10.10, most of the incident ray is transmitted through the layers and part of the transmitted ray which reaches the HSZ (Fig. 10.11) is converted to heat and stored there. The absorption by the salty water solutions changes with concentration of the solution.

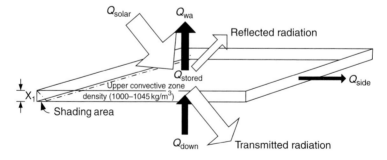

Fig. 10.9. UCZ of the solar pond.

Analysis of an experimental solar pond is generally complicated due to the differences of inner and outer conditions (e.g., pond dimensions, salty-water solutions, insulation, zone thicknesses, shading area of the layers, transmission and absorption characteristics for the layers). Here, we consider the following key parameters: zone thicknesses, temperatures in the layers, shading on the layers by the side walls, incident solar radiation absorbed by the layers, incident radiation reaching on the surface, heat losses through the ISW and thermal conductivity of the solution.

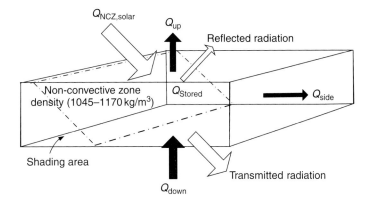

Fig. 10.10. NCZ of the solar pond.

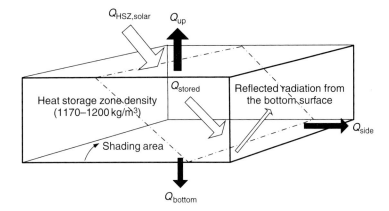

Fig. 10.11. HSZ of the solar pond.

To understand the thermal performance of a solar pond, the rates of absorption of the incident solar radiation by zone and the temperature distributions of its regions need to be determined. To realize this, the pond is treated as having three zones which are separated into 30 layer inner zones. The temperature variations of some layers depend on incident solar radiation on the horizontal surface, rates of absorption by the layers, local climate conditions, pond structure, time and insulation.

Energy efficiency for UCZ

In Fig. 10.9, energy flows for the UCZ of the pond are illustrated. Part of the incident solar radiation is reflected from the UCZ surface to air and lost. Part of the incident solar radiation is transmitted from the UCZ to the NCZ and the rest of the incident solar radiation is absorbed in the zone, heating it.

The thermal (energy) efficiency for the UCZ can generally be expressed as

$$\eta = \frac{Q_{net}}{Q_{in}} \tag{10.19}$$

Here, Q_{net} is the net heat addition to the pond and equals Q_{stored}, where

$$Q_{stored} = Q_{in} - Q_{out} = (Q_{solar} + Q_{down}) - (Q_{side} + Q_{wa}) \tag{10.20}$$

Here, Q_{stored} is the net heat stored in the UCZ, Q_{solar} is amount of the net incident solar radiation absorbed by the UCZ, Q_{down} is the total heat transmitted to the zone from the zone immediately below, Q_{side} is the total heat loss to the side walls of the pond, and Q_{wa} is the total heat lost to the surroundings from the upper layer.

Substituting Eq. (10.20) into Eq. (10.19) for the UCZ yields the following expression for the energy efficiency:

$$\eta_{UCZ} = 1 - \frac{\{Q_{side} + Q_{wa}\}}{Q_{solar} + Q_{down}}$$

and

$$\eta_{UCZ} = 1 - \frac{\{A_{01}R_{ps}[T_{ucz} - T_{side}] + U_{wa}A[T_{ucz} - T_{amb}]\}}{\left\{\beta E A_{(UCZ,I)}[1 - (1 - F)h(X_1 - \delta)] + \frac{kA}{X_1}[T_{down} - T_{ucz}]\right\}} \qquad (10.21)$$

where T_{amb} is the ambient air temperature, the value of which is taken to be that for the time of year, X_1 is the thickness of the UCZ; A_{01} is the surface area of the painted metal sheet on the side wall (and taken as $8 \times 0.05 = 0.4\,\text{m}^2$ here); δ is the thickness of the layer in the UCZ which absorbs incident long-wave solar radiation; E is the total solar radiation incident on the pond surface, A is the upper surface area of the pond; and k is the thermal conductivity of the layers in the UCZ. The term R_{ps} is the thermal resistance of the painted metal sheet surrounding the first layer and can be written as $R_{ps} = \frac{k_p k_s}{S_p k_s + S_s k_p}$.

Here k_p and k_s are thermal conductivities of the paint and iron-sheet, and S_p and S_s are the corresponding thicknesses. Also, β is the fraction of the incident solar radiation that enters the pond, and is expressed as follows (Hawlader, 1980):

$$\beta = 1 - 0.6\left[\frac{\sin(\theta_i - \theta_r)}{\sin(\theta_i + \theta_r)}\right]^2 - 0.4\left[\frac{\tan(\theta_i - \theta_r)}{\tan(\theta_i + \theta_r)}\right]^2$$

with θ_i and θ_r as the angles of incident and reflected solar radiation.

The ratio of the solar energy reaching the bottom of layer I to the total solar radiation incident on to the surface of the pond is given by Bryant and Colbeck (1977) as

$$h_I = 0.727 - 0.056 \ln\left[\frac{(X_1 - \delta)}{\cos\theta_r}\right] \qquad (10.22)$$

Here, A_{UCZ} is the net upper surface area of the UCZ (i.e., the effective area that receives incident solar radiation) and is defined as

$$A_{UCZ} = L_W[L_L - (\delta + (I - 1)\Delta x)\tan\theta_r] \qquad (10.23)$$

where θ_r is the angle of the reflected incidence, Δx is the thickness of each layer in the UCZ and taken as 0.005 m in the calculations, and L_W and L_L are the width and length of the pond, respectively.

Energy efficiency for NCZ

In Fig. 10.10, energy flows for the NCZ of the pond are illustrated. The solar radiation incident on the surface of the NCZ, which is the part of the incident solar radiation on the surface of the pond, is transmitted from the UCZ. Little of the incident solar radiation on the NCZ is reflected from the NCZ to the UCZ. The reflected part of the incident solar radiation increases the UCZ efficiency. Part of the incident solar radiation is transmitted to the HSZ while part of the incident solar radiation is absorbed by the NCZ.

In Fig. 10.10, part of the incident solar radiation is absorbed by and transmitted into the NCZ, and part of the absorbed radiation is stored in the zone. So, the NCZ is heated and the zone's temperature increases. Thus, a temperature gradient occurs in this zone. Heating increases the NCZ efficiency, which can be calculated straightforwardly with Eq. (10.19).

Following Eq. (10.20), we can write an energy balance for the NCZ as

$$Q_{net} = Q_{NCZ,solar} + Q_{down} - Q_{up} - Q_{side} \qquad (10.24)$$

where $Q_{NCZ,solar}$ is amount of the solar radiation entering the NCZ which is transmitted from the UCZ after attenuation of incident solar radiation in the UCZ, and Q_{up} is the heat loss from the NCZ to the above zone.

We can then write the energy efficiency for the NCZ as:

$$\eta_{NCZ} = 1 - \frac{\{Q_{side} + Q_{up}\}}{Q_{NCZ,solar} + Q_{down}}$$

or

$$\eta_{\text{NCZ}} = 1 - \frac{\left\{\dfrac{kA}{\Delta X}[T_{\text{UCZ}} - T_{\text{NCZ}}] + A_{01}R_{\text{ps}}[T_{\text{NCZ}} - T_{\text{side}}]\right\}}{\left\{\beta EA_{(\text{NCZ})}[(1-F)[h(X_1 - \delta) - h(X_1 - \delta + \Delta x)]] + \dfrac{kA}{\Delta X}[T_{\text{down}} - T_{\text{NCZ}}]\right\}} \tag{10.25}$$

where F is the fraction of incident solar radiation absorbed by the pond's upper layer, and $\Delta X_{\text{NCZ}} = (X_2 - X_1)$ is the thickness of the UCZ. Also, $A_{01,\text{NCZ}}$ is the surface area of the painted metal sheet on the side walls surrounding of NCZ (taken as $8 \times 0.60 = 4.8\,\text{m}^2$).

We define A_{NCZ} as the net upper surface area of the NCZ that receives the incident solar radiation as

$$A_{\text{NCZ}} = L_{\text{W}}[L_{\text{L}} - (X_1 + (I - 1)\Delta x)\tan\theta_{\text{r}}] \tag{10.26}$$

Here, I varies from 2 to 14.

Energy efficiency for HSZ

Part of the solar radiation incident on the solar pond is transmitted through the UCZ and NCZ, after attenuation, to the HSZ. In Fig. 10.11, part of the transmitted solar radiation from the NCZ to the HSZ is reflected from the bottom and the majority of the solar radiation is absorbed in the HSZ. So, the HSZ temperature is increased and a temperature gradient develops in the zone.

An energy balance for the HSZ of the solar pond can be written as

$$Q_{\text{net}} = Q_{\text{HSZ,solar}} - Q_{\text{bottom}} - Q_{\text{up}} - Q_{\text{side}} \tag{10.27}$$

where Q_{bottom} is the total heat loss to the bottom wall from the HSZ.

The energy efficiency for the HSZ of the solar pond then becomes

$$\eta_{\text{HSZ}} = 1 - \frac{(Q_{\text{bottom}} + Q_{\text{up}} + Q_{\text{side}})}{Q_{\text{HSZ,solar}}}$$

or

$$\eta_{\text{HSZ}} = 1 - \frac{\left\{AR_{\text{ps}}[T_{\text{down}} - T_{\text{HSZ}}] + \dfrac{Ak}{\Delta X_{\text{HSZ}}}[T_{\text{HSZ}} - T_{\text{up}}] + A_{01}R_{\text{ps}}[T_{\text{HSZ}} - T_{\text{side}}]\right\}}{\{\beta EA_{(\text{HCZ},I)}[(1-F)(h(X_3 - \delta))]\}} \tag{10.28}$$

where $\Delta X_{\text{HSZ}} = (X_3 - X_2)$ is the thickness of the HSZ of the pond. Also, $A_{01,\text{HSZ}}$ is the surface area of the painted metal sheet on the side walls surrounding the HSZ (taken as $8 \times 0.80 = 6.4\,\text{m}^2$). Note that the net surface area of the HSZ is equal to the net surface area at the bottom of the NCZ, i.e., $A_{\text{HSZ},I} = A_{\text{NCZ},I}$; and I varies from 15 to 30.

Results of energy analysis

Energy flows in the inner zones of the pond are illustrated in Figs. 10.9 through 10.11. The performance of the solar pond depends on not only the thermal energy flows (e.g., heat losses and heat gains in the zones), but also the incident solar radiation flows (accounting for reflection, transmission and absorption). Also, shading decreases the performance of the zones.

In Fig. 10.9, it is seen that part of the incident solar radiation is reflected on the surface, some is absorbed by the layer and part (often most) is transmitted through the UCZ to the NCZ. The average sunny area of the UCZ is determined to be $3.93\,\text{m}^2$, and the average shading area $0.07\,\text{m}^2$. The net average solar radiation incident on the sunny area of the UCZ is calculated for January, May and August as 439.42, 2076.88 and 2042.00 MJ, respectively.

The greatest part of the incident solar radiation in Fig. 10.10 is transmitted to the NCZ from the UCZ. Part of the incident solar radiation is absorbed by the NCZ layers. The incident solar radiation transmitted from the NCZ to the HSZ is significant and little incident solar radiation is reflected from the NCZ to the UCZ. The average sunny area for the NCZ is found to be $3.13\,\text{m}^2$, and the average shading area $0.87\,\text{m}^2$. The net average solar radiation on the sunny area of the NCZ is calculated for January, May and August as 351.54, 1661.50 and 1634.05 MJ, respectively.

A significant part of the incident radiation in Fig. 10.11 reaches the HSZ from the NCZ. This transmitted solar radiation from the NCZ is absorbed in the HSZ, while little of the incident solar radiation is reflected from the HSZ to the upper zones. The average sunny area for the HSZ is found to be 2.63 m^2, and the average shading area 1.37 m^2. The net average solar radiation incident on the sunny area of the HSZ is calculated for January, May and August as 193.34, 913.83 and 898.73 MJ, respectively.

The stability of the salt density distribution in a solar pond is significant (Fig. 10.12). The primary reason for differences during different months is likely the higher temperature in summer. This change is mainly attributable to the thermophysical property of the salty water, heat losses from the pond to the air, and the absorption and reflection of incident solar radiation on the surface. The reason for the fluctuations in the saline density in the upper convective and NCZ is the increase in saline density of these zones due to the evaporation of water at the upper region. These changes can be reduced by continuously adding fresh water to the top of the pond. When not using one of the salt gradient protection systems for cleaning purposes in a month, significant changes occurred in the non-convective and upper convective regions. The averaged experimental density variations of salty water vs. height from the pond bottom for 12 months (see Fig. 10.12) show little differences between the density distributions in January, April and July, due to the temperature changes and evaporation of salty water from the pond. As expected, increasing temperature decreases the density more in the summer months.

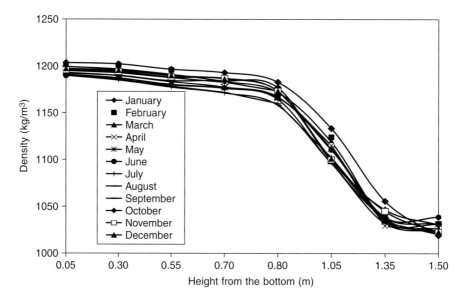

Fig. 10.12. Variation with height of salt density in the inner zones of the solar pond.

Heat losses by heat transfer from the pond during a day are determined by calculating the temperature differences for daily profiles of related months. To determine the heat losses from the inside of the solar pond, experimental temperature distribution profiles for the inner zones are obtained (see Fig. 10.13). Experimental temperature distributions are shown in Fig. 10.14 for different heights in the pond. The zone temperatures are measured throughout the months and averaged to find the monthly average temperatures at the respective points. It is clear that the zone temperatures vary with month of year, depending on the environment temperature and incoming solar radiation. The temperatures of the zones generally increase with incident solar energy per unit area of surface. Heat losses occur for each zone, with the largest in the storage zone, affecting its performance directly and significantly. To improve performance and increase efficiency, losses need to be reduced. The temperature distributions in Fig. 10.13 indicate that the temperature of the UCZ is a maximum of 35.0°C in August, a minimum of 10.4°C in January and 27.9°C in May. Similarly, the temperature of the NCZ is observed to be a maximum of 44.8°C in August, a minimum of 13.9°C in January and 37.9°C in May, while the temperature of the HSZ is observed to be a maximum of 55.2°C in August, a minimum of 16.9°C in January and 41.1°C in May. The net energy stored in the zones is calculated using property data in Table 10.1.

The energy stored in the UCZ is seen in Fig. 10.15 for January, May and August to be 3.99, 59.49 and 92.90 MJ, respectively. Similarly, the energy stored in the NCZ is seen in Fig. 10.16 for January, May and August to be 311.16,

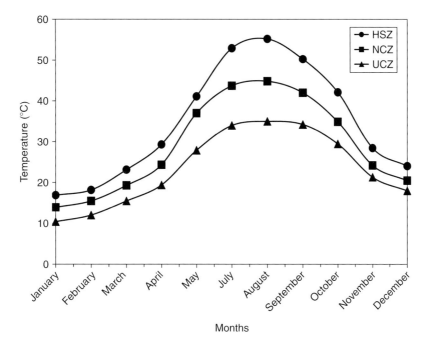

Fig. 10.13. Monthly average temperatures for the inner zones of the pond.

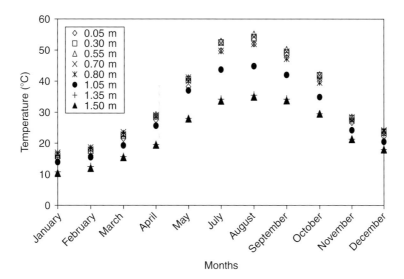

Fig. 10.14. Experimental zone temperature distributions in the inner zones of the solar pond.

Table 10.1. Thermophysical properties of water and other materials.

	Water	Saline water	Painted wall	Insulation	Air
Density (kg/m^3)	998	1185	7849	200	1.16
Thermal conductivity (J/m^1 K^1 h^1)	2160	–	21,200	143	94.68
Specific heat (J/kg^1 K^1)	4182	–	460	670	1007

Source: (Karakilcik, 1998; Dincer and Rosen, 2002).

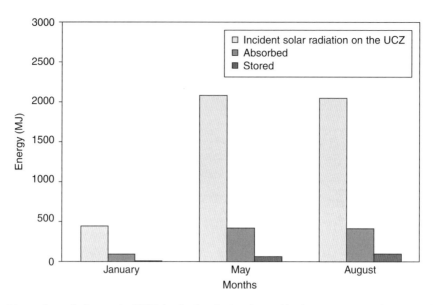

Fig. 10.15. Incident solar radiation on the UCZ that is absorbed and stored in the upper convective zone of the pond.

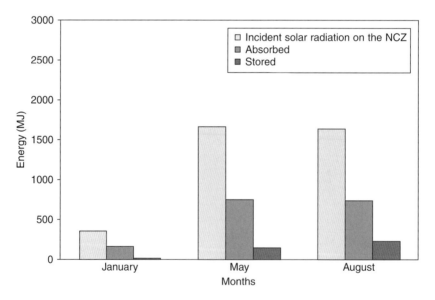

Fig. 10.16. Incident solar radiation on the UCZ that is absorbed and stored in the NCZ of the pond.

143.03 and 225.43 MJ, respectively, while the energy stored in the HSZ is seen in Fig. 10.17 for January, May and August to be 18.70, 160.31 and 252.65 MJ, respectively.

The UCZ efficiencies are seen in Fig. 10.18 to be 0.90%, 2.86% and 4.54% for January, May and August, respectively. This zone has little effect on the performance of the pond in January, and more impact in May and August. The efficiency of the UCZ is low because of the shading area rather than heat losses. The NCZ efficiencies are seen to be 3.17%, 8.60% and 13.79% for January, May and August, respectively. Shading decreases the performance of the NCZ. Shading area also has an important effect on the performance of the HSZ, for which the zone efficiencies are seen to be 9.67%, 17.54% and 28.11% for January, May and August, respectively.

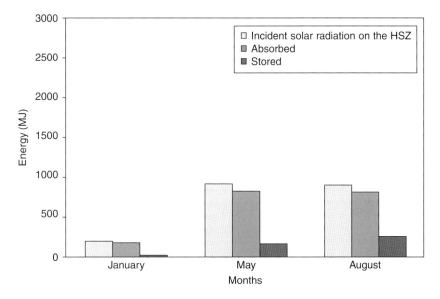

Fig. 10.17. Incident solar radiation on the UCZ that is absorbed and stored in the HSZ of the pond.

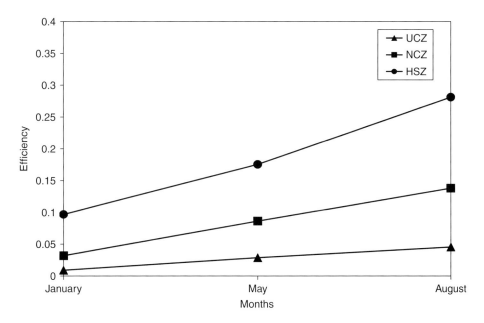

Fig. 10.18. Efficiencies of the inner zones of the pond for different months.

    A significant amount of incident solar radiation is absorbed by the HSZ in August and little of the incident solar radiation is reflected from the bottom wall of the pond. Decreasing shading area from the top to the bottom of the pond allows less solar radiation to pass through and decreases the thermal potential of the pond and hence its performance. The performance of the thermal energy storage depends on the total radiation reaching the pond's zones. The performance of the heat storage zone can be usefully determined in part using energy efficiencies. But in a solar pond, the stored energy is very low compared to incident solar radiation on the surface of the zones, so the efficiencies are also very low. The efficiencies are low in part due to the low thermal conductivity of the pond filled with salty water. The efficiencies are dependent on the temperatures of the salty water and ambient air. The temperature differences of the zones between

January, May and August alter the inner zone temperatures, the diffusion of salt molecules up from the bottom and heat losses. This analysis illustrates the effect on pond efficiency of shading by the side wall and absorption, transmission and the thicknesses of the zones.

The experimental energy efficiency profiles for the UCZ, NCZ and HSZ of the pond, for different months, are given in Fig. 10.18. The maximum energy efficiencies of the inner zones are seen to occur in August, and the minimum efficiencies in January. Although the greatest amount of solar radiation is incident on the UCZ, the lowest efficiencies are found for this zone. This is because of the zone's small thickness and its large heat losses to air from its upper surface.

The temperature distribution profiles for the inner zones usually differ, causing the zone efficiencies to differ also. Despite the decrease in solar radiation intensity when it reaches the surface of the NCZ, that zone incurs lower heat losses and thus has a higher efficiency than the UCZ. The temperature distributions thus have an important effect on the performance of the pond.

The energy efficiency of the pond is negatively affected by the energy losses due to heat transfer from the UCZ to air. A low fraction of the incident solar radiation is stored in the pond and the UCZ efficiency is negligible especially compared to that of the NCZ. The NCZ efficiency consequently has a greater effect on the performance of the pond. Most of the energy is stored in the HSZ.

The inner regions of the pond thus store more energy in August than in January due to the considerable temperature differences between the zones. Heat storage, heat losses, shading areas and solar radiation absorption should be carefully considered when determining the thermal performance of solar ponds as their effects can be significant.

10.2.4. Exergy analysis

Exergy analysis permits many of the shortcomings of energy analysis of solar pond systems to be overcome, and thus appears to have great potential as a tool for design, analysis, evaluation and performance improvement. Figure 10.19 shows the energy and exergy flows for each of the zones in the pond. An exergy analysis of each zone is presented here.

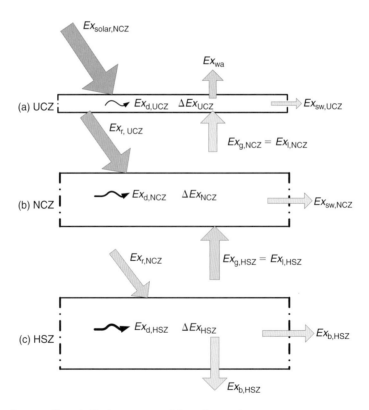

Fig. 10.19. Energy and exergy flows in the inner zones of the solar pond.

Exergy analysis for UCZ

Exergy flows in the UCZ are illustrated in Fig. 10.19a. We can write an exergy balance for the UCZ as

$$Ex_{\text{solar}} + Ex_{\text{g,NCZ}} = Ex_{\text{r,UCZ}} + Ex_{\text{d,UCZ}} + Ex_{\text{a}} + Ex_{\text{sw,UCZ}} \tag{10.29}$$

where Ex_{solar} is the exergy of the solar radiation reaching the UCZ surface, $Ex_{\text{g,NCZ}}$ is the exergy gained from the NCZ, $Ex_{\text{r,UCZ}}$ is the recovered exergy of the UCZ for the NCZ, $Ex_{\text{d,UCZ}}$ is the exergy destruction in the UCZ, $Ex_{\text{a,UCZ}}$ is the exergy loss from the UCZ to the ambient air and $Ex_{\text{sw,UCZ}}$ is the exergy loss through the side walls. Here $Ex_{\text{r,UCZ}}$ can be written according to Eq. (10.29) as

$$Ex_{\text{r,UCZ}} = Ex_{\text{ti}} - Ex_{\text{tl}} = (Ex_{\text{solar}} + Ex_{\text{g,NCZ}}) - (Ex_{\text{d,UCZ}} + Ex_{\text{a}} + Ex_{\text{sw,UCZ}}) \tag{10.30}$$

where Ex_{tl} is the total exergy losses, including exergy destruction, and Ex_{ti} is the total exergy input to the UCZ. The exergy of the solar radiation can be expressed, by modifying the expression of Petala (2003), as follows:

$$Ex_{\text{solar}} = E_{\text{net}} \left[1 - \frac{4T_0}{3T} + \frac{1}{3} \left(\frac{T_0}{T} \right)^4 \right] A_{\text{UCZ}} \tag{10.31}$$

The exergy gained from the NCZ can be expressed as

$$Ex_{\text{g,NCZ}} = m_{\text{NCZ}} C_{\text{p,NCZ}} \left[(T_{\text{m,NCZ}} - T_{\text{UCZ}}) - T_0 \left(\ln \frac{T_{\text{m,NCZ}}}{T_{\text{UCZ}}} \right) \right] \tag{10.32}$$

where E_{net} is the net incident solar radiation reaching the UCZ surface; A_{UCZ} is the net surface area of the UCZ and T is the sun's surface temperature, taken to be 6000 K (Petela, 2003); $m_{\text{NCZ}} = \rho_{\text{NCZ}} V_{\text{NCZ}}$ is the mass of salty water in the NCZ; ρ_{NCZ} is the averaged density (as seen in Table 10.2) and V_{NCZ} is the volume of the salty water in the NCZ ($V_{\text{NCZ}} = 2.4 \, \text{m}^3$).

The exergy destruction in the UCZ can be written as

$$Ex_{\text{d,UCZ}} = T_0 \Delta S_{\text{net}} \tag{10.33}$$

where ΔS_{net} is the net entropy change of the UCZ, which is $\Delta S_{\text{net}} = \Delta S_{\text{sys}} + \Delta S_{\text{surr}}$. After substituting each of the entropy change terms, Eq. (10.33) becomes

$$Ex_{\text{d,UCZ}} = T_0 \left[m_{\text{UCZ}} C_{\text{p,UCZ}} \ln \frac{T_{\text{UCZ}}}{T_0} - \left(\frac{Q_{\text{wa}}}{T_{\text{UCZ}}} + \frac{Q_{\text{sw,UCZ}}}{T_0} \right) + \left(\frac{Q_{\text{g,NCZ}}}{T_{\text{NCZ}}} + \frac{Q_{\text{sw,UCZ}}}{T_0} \right) \right] \tag{10.34}$$

In addition, we can write the exergy losses to the ambient air and through the side walls as follows:

$$Ex_{\text{a,UCZ}} = m_{\text{UCZ}} C_{\text{p,UCZ}} \left[(T_{\text{UCZ}} - T_{\text{a}}) - T_0 \left(\ln \frac{T_{\text{UCZ}}}{T_{\text{a}}} \right) \right] \tag{10.35}$$

and

$$Ex_{\text{sw,UCZ}} = m_{\text{UCZ}} C_{\text{p,sw}} \left[(T_{\text{UCZ}} - T_{\text{sw,UCZ}}) - T_0 \left(\ln \frac{T_{\text{UCZ}}}{T_{\text{sw,UCZ}}} \right) \right] \tag{10.36}$$

where $m_{\text{UCZ}} = \rho_{\text{UCZ}} V_{\text{UCZ}}$ is the mass of salty water in the UCZ; ρ_{UCZ} is the averaged density and V_{UCZ} is the volume of the salty water in the UCZ ($V_{\text{UCZ}} = 0.4 \, \text{m}^3$); $C_{\text{p,UCZ}}$ and $C_{\text{p,sw}}$ are the respective specific heats of the UCZ and insulating material; T_{a} and T_0 are the ambient temperature and the reference environment temperature, respectively and T_{UCZ}, $T_{\text{sw,UCZ}}$ and $T_{\text{m,NCZ}}$ denote the average temperatures of the UCZ, the side wall and the NCZ, respectively.

We can now define the exergy efficiency for the UCZ as the ratio of the exergy recovered from the UCZ to the total exergy input to the UCZ:

$$\psi_{\text{UCZ}} = \frac{Ex_{\text{r, UCZ}}}{Ex_{\text{ti}}} = 1 - \frac{Ex_{\text{d, UCZ}} + Ex_{\text{a}} + Ex_{\text{sw, UCZ}}}{Ex_{\text{solar}} + Ex_{\text{g, NCZ}}} \tag{10.37}$$

Table 10.2. Average monthly reference-environment temperatures and exergy contents of each zone.

	January	February	March	April	May	July	August	September	October	November	December
Reference temperature (°C)	10.0	11.0	14.2	17.6	22.0	28.0	28.0	26.0	21.0	16.0	11.0
Exergy input (UCZ) (MJ)	417.40	644.32	1160.85	1700.20	1976.24	2167.89	1982.47	1740.41	1299.94	782.72	506.14
Exergy recovered (MJ)	329.42	510.50	920.75	1347.54	1552.53	1681.57	1524.70	1344.78	1004.95	614.02	393.03
Exergy input (NCZ) (MJ)	335.05	516.70	930.67	1363.33	1588.13	1747.54	1601.34	1404.25	1048.74	629.23	407.89
Exergy recovered (MJ)	187.77	290.90	524.82	768.09	884.94	958.49	869.08	766.52	572.82	349.99	224.03
Exergy input (HCZ) (MJ)	187.77	290.98	524.82	768.09	884.94	958.50	869.08	766.52	572.82	349.99	224.03
Exergy stored (MJ)	17.12	27.19	53.15	89.27	140.79	204.40	218.00	181.39	133.28	57.03	27.92

Exergy analysis for NCZ

Fig. 10.19b shows the exergy flows in the NCZ. An exergy balance can be written as

$$Ex_{r,UCZ} + Ex_{g,HSZ} = Ex_{r,NCZ} + Ex_{d,NCZ} + Ex_{l,NCZ} + Ex_{sw,NCZ} \tag{10.38}$$

where $Ex_{r,UCZ}$ is the exergy recovered from the UCZ; $Ex_{g,HSZ}$ is the exergy gained from the HSZ, $Ex_{r,NCZ}$ is the recovered exergy of the NCZ for the HSZ, $Ex_{d,NCZ}$ is the exergy destruction in the NCZ, $Ex_{l,NCZ}$ is the exergy loss from the NCZ to the UCZ (which is equivalent to $Ex_{g,NCZ}$) and $Ex_{sw,NCZ}$ is the exergy loss through the side walls.

Here $Ex_{r,NCZ}$ can be expressed using Eq. (10.38) as

$$Ex_{r,NCZ} = Ex_{ti,NCZ} - Ex_{tl,NCZ} = (Ex_{r,UCZ} + Ex_{g,HSZ}) - (Ex_{d,NCZ} + Ex_{l,NCZ} + Ex_{sw,NCZ}) \tag{10.39}$$

where

$$Ex_{g,HSZ} = m_{HSZ}C_{p,HSZ}\left[(T_{HSZ} - T_{NCZ}) - T_0\left(\ln\frac{T_{HSZ}}{T_{NCZ}}\right)\right] \tag{10.40}$$

Here, $m_{HSZ} = \rho_{HSZ}V_{HSZ}$ is the mass of salty water in the HSZ; ρ_{HSZ} is the average density and V_{HSZ} is the volume of salty water in the HSZ ($V_{HSZ} = 3.2\,\mathrm{m}^3$).

The exergy destruction in the NCZ can then be written as

$$Ex_{d,NCZ} = T_0(\Delta S_{net,NCZ}) \tag{10.41}$$

where $\Delta S_{net,NCZ}$ is the net entropy change of the NCZ, which is $\Delta S_{net,NCZ} = \Delta S_{sys} + \Delta S_{surr}$.

The exergy losses, including the exergy destruction in the NCZ, can be derived as follows:

$$Ex_{d,NCZ} = T_0\left[m_{NCZ}C_{p,NCZ}\ln\frac{T_{m,NCZ}}{T_0} - \left(\frac{Q_{g,NCZ}}{T_{m,NCZ}} + \frac{Q_{sw,NCZ}}{T_0}\right) + \left(\frac{Q_{g,HSZ}}{T_{m,NCZ}} + \frac{Q_{sw,NCZ}}{T_0}\right)\right] \tag{10.42}$$

$$Ex_{l,NCZ} = m_{NCZ}C_{p,NCZ}\left[(T_{m,NCZ} - T_{UCZ}) - T_0\left(\ln\frac{T_{m,NCZ}}{T_{UCZ}}\right)\right] \tag{10.43}$$

$$Ex_{sw,NCZ} = m_{NCZ}C_{p,sw}\left[(T_{m,NCZ} - T_{sw,NCZ}) - T_0\left(\ln\frac{T_{m,NCZ}}{T_{sw,NCZ}}\right)\right] \tag{10.44}$$

where $C_{p,NCZ}$ is the specific heat of the NCZ and T_{HSZ} is the temperature of the HSZ.

We can now define the exergy efficiency for the NCZ as the ratio of the exergy recovered from the NCZ to the total exergy input to the NCZ:

$$\psi_{NCZ} = \frac{Ex_{r,NCZ}}{Ex_{ti}} = 1 - \frac{Ex_{d,NCZ} + Ex_{l,NCZ} + Ex_{sw,NCZ}}{Ex_{r,UCZ} + Ex_{g,HSZ}} \tag{10.45}$$

Exergy analysis HSZ

The exergy flows in the HSZ are shown in Fig. 10.19c and a zone exergy balance can be written as

$$Ex_{r,NCZ} - (Ex_{d,HSZ} + Ex_{l,HSZ} + Ex_{sw,HSZ} + Ex_{b,HSZ}) = \Delta Ex_{st} \tag{10.46}$$

where $Ex_{r,NCZ}$ is the recovered exergy from the NCZ for the HSZ, $Ex_{d,HSZ}$ is the exergy destruction in the HSZ, $Ex_{l,HSZ}$ is the exergy loss from the HSZ to the NCZ, $Ex_{sw,HSZ}$ is the exergy loss through the side walls. $Ex_{b,HSZ}$ is the exergy loss through the bottom wall and ΔEx_{st} is the exergy stored in the HSZ.

Here $Ex_{d,HSZ}$ is the exergy destruction in the HSZ which can be written as

$$Ex_{d,HSZ} = T_0(\Delta S_{net,HSZ}) \tag{10.47}$$

where $\Delta S_{net,HSZ}$ is the net entropy change of the HSZ and expressible as $\Delta S_{net,HSZ} = \Delta S_{sys} + \Delta S_{surr}$.

The exergy losses, including exergy destruction within the NCZ, can be written as follows:

$$Ex_{d,HSZ} = T_0 \left[m_{HSZ} C_{p,HSZ} \ln \frac{T_{HSZ}}{T_0} - \left(\frac{Q_{g,HSZ}}{T_{HSZ}} + \frac{Q_{sw,HSZ}}{T_0} \right) + \left(\frac{Q_b}{T_0} \right) \right] \quad (10.48)$$

$$Ex_{l,HSZ} = m_{HSZ} C_{p,HSZ} \left[(T_{HSZ} - T_{m,NCZ}) - T_0 \left(\ln \frac{T_{HSZ}}{T_{m,NCZ}} \right) \right] \quad (10.49)$$

where $C_{p,HSZ}$ is the specific heat of the salty water in the HSZ. For the side wall,

$$Ex_{sw,HSZ} = m_{HSZ} C_{p,sw} \left[(T_{HSZ} - T_{sw,HSZ}) - T_0 \left(\ln \frac{T_{HSZ}}{T_{sw,HSZ}} \right) \right] \quad (10.50)$$

Note that $Ex_{b,HSZ} = Ex_{sw,HSZ}$ due to the fact that both the side wall and the bottom layer have the same insulating materials and are surrounded by ambient air.

The exergy efficiency for the HSZ is expressible as the ratio of the exergy stored in the HSZ to the total exergy input to the HSZ which is essentially the exergy recovered from the NCZ:

$$\psi_{HSZ} = \frac{\Delta Ex_{st}}{Ex_{r,NCZ}} = 1 - \frac{\{Ex_{d,HSZ} + Ex_{l,HSZ} + Ex_{sw,HSZ} + Ex_{b,HSZ}\}}{Ex_{r,NCZ}} \quad (10.51)$$

Results of exergy analysis

Energy and exergy efficiencies are compared for the UCZ, NCZ and HSZ in the solar pond, illustrating how exergy is important for determining true magnitudes of the losses in each zone.

Figure 10.20 shows both averaged energy and exergy content variations of the pond three zones vs. month of year. The exergy content distributions in the zones are the calculated monthly average temperatures as listed in Table 10.2. The exergy contents are less than the corresponding energy contents. Although energy is conserved, some exergy is destroyed in each zone in addition to the exergy losses to the surrounding air. As seen in Fig. 10.20, the lowest-exergy contents occur in January and the highest in July. The temperature of the surroundings plays a key role since the energy and exergy losses are rejected to the ambient air. The distribution of the energy and exergy contents by month follows the solar irradiation profile closely.

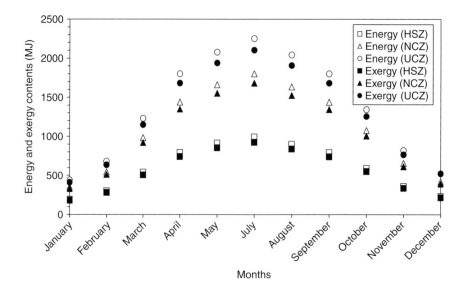

Fig. 10.20. Energy and exergy content distributions of the solar pond zones.

Figure 10.21 shows the variations of exergy input, exergy recovered and exergy destruction and losses for the UCZ over the year, except for June when measurements were not taken due to maintenance on the data acquisition system. The exergy inputs are equal to the sum of the exergy recovered and the exergy destruction and losses. For simplicity, no exergy accumulation is assumed to occur in this zone (calculations show it is less than 1%). The exergy input is highest in July when incoming solar irradiation is greatest, and the other exergy terms appear to be proportional to the input. The exergy recovered in this zone is transferred to the NCZ. The maximum and minimum exergy recovered are 1681.57 MJ in July and 392.42 MJ in January, respectively. The distribution by month is somewhat similar to the distribution in Fig. 10.20.

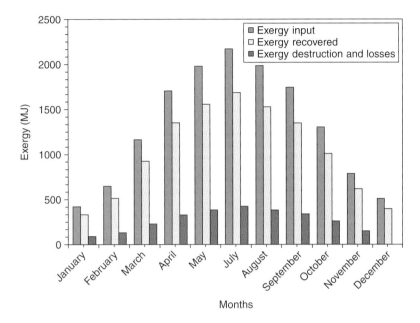

Fig. 10.21. Exergy distributions in the UCZ of the solar pond.

Figure 10.22 shows the variations of exergy input, exergy recovered and exergy destruction and losses for the NCZ over the year. Again, the exergy inputs are equal to the sum of the exergy recovered and exergy destruction and losses. No exergy accumulation is assumed. Also, the exergy is highest in July when solar irradiation is greatest and the other exergy terms are proportional to exergy input. The exergy recovered in this zone is transferred to the HSZ. The maximum and minimum exergy recovered are 958.48 MJ in July and 187.77 MJ in January, respectively. The exergy input to and recovered from this zone are listed in Table 10.2.

Figure 10.23 exhibits the distributions of exergy input, exergy stored and exergy destruction and losses for the HSZ over the year. In this zone, exergy is stored instead of recovered. This storage capability allows solar ponds to undertake daily and/or seasonal storage. The exergy input is equal to the sum of the exergy recovered and the exergy destruction and losses. The exergy stored is much smaller than the exergy input and exergy destruction and losses in the HSZ, and reaches a maximum in July of 743.10 MJ and a minimum in January of 169.68 MJ. The exergy values for each month are listed in Table 10.2.

Figure 10.24 compares the energy and exergy efficiencies for the zones over the year. As seen in the figure, the differences between energy and exergy efficiencies are small during the cooler months, and largest from May to October. As expected, the HSZ efficiencies are higher than the corresponding UCZ and NCZ efficiencies. Consequently, the inner zones of the pond store more exergy in July than in January due to the considerable temperature differences between the zones. The exergy destruction and losses significantly affect the performance of the pond and should be minimized to increase system efficiency.

10.2.5. Closure

Energy and exergy analyses have been carried out for an insulated salt gradient solar pond and its UCZ, NCZ and HSZ. Pond performance is affected strongly by the temperature of the LCZ and the temperature profile with pond

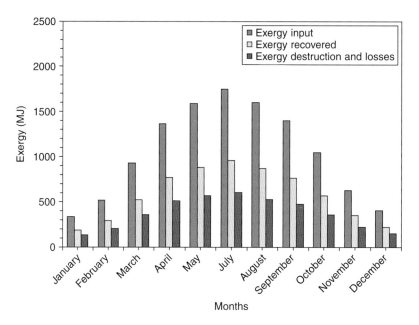

Fig. 10.22. Exergy distributions in the NCZ of the solar pond.

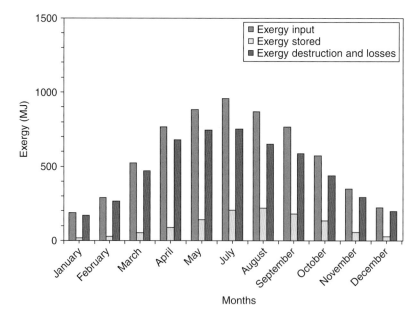

Fig. 10.23. Exergy distributions in the HSZ of the solar pond.

depth. The sunny area and the temperature of the LCZ are sensitive to wall shading. Due to the presence of insulation, heat losses from the sides and bottom of the pond are negligibly small. To increase the efficiency for the storage zone of the pond, heat losses from upper zone, bottom and side walls, reflection, and shading areas in the NCZ and HSZ should be decreased. The temperature of each layer of the inner zones depends on the incident radiation, zone thicknesses, shading areas of the zones and overall heat losses. So, to increase pond performance, the zone thicknesses should be modified to achieve higher efficiency and stability of the pond. Through careful design parameter modifications, pond performance can be maintained even if the incoming solar radiation reaching the zones is increased.

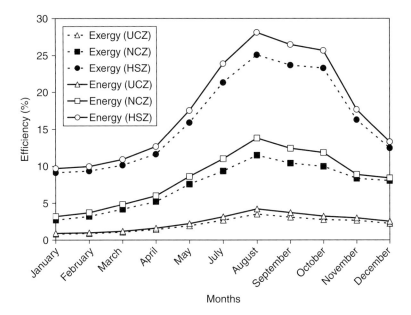

Fig. 10.24. Variation of energy and exergy efficiencies of the solar pond zones.

Exergy efficiencies are lower than the energy efficiencies for each zone of the pond due to the small magnitudes of exergy destructions in the zones and losses to the surroundings. It is important to determine the true magnitudes of these destructions and losses and minimize these for performance improvement of the pond.

Experimental data are used to determine the efficiencies for each layer of the zones for a real insulated solar pond. Several parameters for the UCZ and NCZ having influences on the thermal performance are discussed. It is shown that the introduction of the UCZ and NCZ provides many conveniences in calculating the storage efficiency in the heat storage zone, and in determining the relations with heat loads and a best operating state. Therefore, the energy and exergy efficiencies of the inner zones of a solar pond are important parameters in practical applications.

10.3. Exergy analysis of wind energy systems

Wind power is a form of renewable energy in that it is replenished daily by the sun. Warm air rises as portions of the earth are heated by the sun, and other air rushes in to fill the low-pressure areas, creating wind power. The characteristics of wind affect the design of systems to exploit its power. Wind is slowed dramatically by friction as it flows over the ground and vegetation, often causing it not to be very windy at ground level. Wind can be accelerated by major land forms, leading some regions to be very windy while other areas remain relatively calm. When wind power is converted to electricity, it can be transported over long distances and thus can serve the needs of urban centers where large populations live.

Wind energy is among the world's most significant and rapidly developing renewable energy sources. Recent techno-logical developments, concerns over fossil fuel demands and the corresponding environmental effects and the continuous increase in the consumption of conventional energy resources have reduced relative wind energy costs to economically acceptable levels in many locations. Wind energy farms, which have been installed and operated in some instances for more than 25 years, consequently, are being considered as an alternative energy source in many jurisdictions.

In practice wind power is converted to electricity by a wind turbine. In typical, modern, large-scale wind turbines, the kinetic energy of wind (the energy of moving air molecules) is converted to rotational motion by a rotor, on which is mounted a device to 'capture' the wind. This device is often a three-bladed assembly at the front of the wind turbine, but can also come in other geometries and types. The rotor turns a shaft which transfers the motion into the nacelle (the large housing at the top of a wind turbine tower). Inside the nacelle, the slowly rotating shaft enters a gearbox that greatly increases the rotational shaft speed. The output shaft rotating at a high-speed is connected to a generator that converts the rotational motion to electricity at a medium voltage (a few hundred volts). The electricity flows along heavy electric cables inside the tower to a transformer, which increases the voltage of the electric power to a level more suitable for distribution (a few thousand volts). Transformation is carried out because higher voltage electricity flows with less resistance through electric lines, generating less heat and fewer power losses. The distribution-voltage power flows through underground cables or other lines to a collection point where the power may be combined with that from other turbines. In many cases, the electricity is distributed for use to nearby farms, residences and towns. Otherwise, the distribution-voltage power is sent to a substation where its voltage is increased dramatically to transmission-voltage levels (a few hundred thousand volts) and transported through transmission lines many kilometers to distant cities and factories.

Most new and renewable energy sources, such as wind, solar, hydraulic and wave energy, are related to meteorological variables. If the meteorological characteristics of these renewable energy sources are not well known and understood, there can be important gaps in knowledge related to energy investments.

This section presents a thermodynamic analysis of wind energy using energy and exergy. The analysis provides a physical basis for understanding, refining and predicting the variations in wind energy calculations. A wind energy efficiency definition based on exergy analysis is provided.

This section contains several parts. First, wind energy and its components are discussed. Second, exergy analysis is applied to wind, and the exergy is formulated of wind energy and its components. Third, energy and exergy efficiencies are compared and shown to depend on the area considered. Last, a spatio-temporal mapping approach to wind exergy analysis is provided.

10.3.1. Wind energy systems

As a meteorological variable, wind energy refers to the energy content of wind. In electricity generation wind plays the same role as water does for hydraulic generation. Wind variables are important in such applications. Wind velocity deviation and changeability depend on time and location. Understanding such characteristics is the subject of wind velocity modeling. Determining the atmospheric boundary layer and modeling is a special consideration in wind power research. Much research has been carried out on these subjects. For instance, Petersen et al. (1998) considered wind power meteorology and sought relationships between meteorology and wind power. During the preparation of the Denmark Wind Atlas detailed research was performed on wind energy as a meteorological energy source (Petersen et al., 1981).

Meteorological variables such as temperature, pressure and moisture play important roles in the occurrence of wind. Generally, in wind engineering, moisture changeability is negligible and air is assumed to be dry. Wind as a meteorological variable can be described as a motion of air masses on a large scale with potential and kinetic energies. Pressure forces lead to kinetic energy (Freris, 1981; 1990). In wind engineering applications horizontal winds are important because they cover great areas.

The dynamic behavior of the atmosphere generates spatio-temporal variations in such parameters as pressure, temperature, density and moisture. These parameters can be described by expressions based on continuity principles, the first law of thermodynamics, Newton's law and the state law of gases. Mass, energy and momentum conservation equations for air in three dimensions yield balance equations for the atmosphere. Wind occurs due to different cooling and heating phenomena within the lower atmosphere and over the earth's surface. Meteorological systems move from one place to another by generating different wind velocities.

With the growing significance of environmental problems, clean energy generation has become increasingly important. Wind energy is clean, but it usually does not persist continually for long periods of time at a given location. Fossil fuels often must supplement wind energy systems. Many scientific studies have addressed this challenge with wind energy (e.g., Justus, 1978; Cherry, 1980; Troen and Petersen, 1989; Sahin, 2002).

During the last decade, wind energy applications have developed and been extended to industrial use in some European countries including Germany, Denmark and Spain. Successes in wind energy generation have encouraged other countries to consider wind energy as a component of their electricity generation systems. The clean, renewable and in some instances economic features of wind energy have drawn attention from political and business circles and individuals. Development in wind turbine technology has also led to increased usage. Wind turbine rotor efficiency increased from 35% to 40% during the early 1980s, and to 48% by the mid-1990s. Moreover, the technical availability of such systems has increased to 98% (Salle et al., 1990; Gipe, 1995; Karnøe and Jørgensen, 1995; Neij, 1999). Today, total operational wind power capacity worldwide has reached approximately 46,000 MW.

Koroneos et al. (2003) applied exergy analysis to renewable energy sources including wind power. This perhaps represents the first paper in the literature about wind turbine exergy analysis. But in this paper only the electricity generation of wind turbines is taken into account and the exergy efficiency of wind turbines for wind speeds above 9 m/s is treated as zero. Koroneos et al. only considered the exergy of the wind turbine, depending on electricity generation with no entropy generation analysis. In an extended version of this study, Jia et al. (2004) carried out an exergy analysis of wind energy and considered wind power for air compression systems operating over specified pressure differences, and estimated the system exergy efficiency. As mentioned before, Jia et al. wanted to estimate exergy components and to show pressure differences, and realized this situation by considering two different systems, a wind turbine and an air compressor, as a united system.

Dincer and Rosen (2005) investigated thermodynamic aspects of renewables for sustainable development. They explain relations between exergy and sustainable development. Wind speed thermodynamic characteristics are given by Goff et al. (1999), with the intent of using the cooling capacity of wind as a renewable energy source (i.e., using the wind chill effect for a heat pump system).

Although turbine technology for wind energy is advancing rapidly, there is a need to assess accurately the behavior of wind scientifically. Some of the thermodynamic characteristics of wind energy are not yet clearly understood. The capacity factor of a wind turbine sometimes is described as the efficiency of a wind energy turbine. But there are difficulties associated with this definition. The efficiency of a wind turbine can be considered as the ratio of the electricity generated to the wind potential within the area swept by the wind turbine. In this definition only the kinetic energy component of wind is considered. Other components and properties of wind, such as temperature differences and pressure effects, are neglected.

10.3.2. Energy and exergy analyses of wind energy aspects

People sense whether air is warm or cool based not only on air temperature, but also on wind speed and humidity. During cold weather, faster wind makes the air feel colder because it removes heat from our bodies faster. Wind chill is a measure of this effect, and is the hypothetical air temperature in calm conditions (air speed $V = 0$) that would cause the same heat flux from the skin as occurs for the actual air speed and temperature. The heat transfer for an air flow over a surface is slightly modified in some versions of the wind chill expression (Stull, 2000).

The present wind chill expression is based on the approaches of Osczevski (2000) and Zecher (1999), and was presented at the Joint Action Group for Temperature Indices (JAG/TI) meeting held in Toronto (2001). The JAG/TI expression makes use of advances in science, technology and computer modeling to provide a more accurate, understandable and useful formula for calculating the dangers from winter winds and freezing temperatures. In addition, clinical trials have been conducted and the results have been used to verify and improve the accuracy of the expression, which is given as

$$T_{\text{windch}} = 35.74 + 0.6215T_{\text{air}} - 35.75(V^{0.16}) + 0.4274T_{\text{air}}(V^{0.16}) \tag{10.52}$$

where the wind chill temperature T_{windch} is in °F and wind speed V is in mph.

Another wind speed factor is wind pressure. When the wind approaches an obstacle, the air flows around it. However, one of the streamlines that hits the obstacle decelerates from the upstream velocity of v_s to a final velocity of zero (or to some lower velocity). The pressure (dynamic pressure) at this stagnation point is higher than the free stream pressure (static pressure) well away from the obstacle. The dynamic pressure can be calculated from Bernoulli's equation. For flow at constant altitude, the only two terms that change in Bernoulli's equation are kinetic energy and pressure.

As explained earlier, for evaluating entropy generation we need system inlet and outlet temperature and pressure differences. Here our approach is to use the windchill effect to be able to determine the changes in heat capacities of wind. The Bernoulli equation is employed for calculating entropy generation.

Energy analysis

Wind energy E is the kinetic energy of a flow of air of mass m at a speed V. The mass m is difficult to measure and can be expressed in terms of volume V through its density $\rho = m/V$. The volume can be expressed as $V = AL$ where A is the cross-sectional area perpendicular to the flow and L is the horizontal distance. Physically, $L = Vt$ and wind energy can be expressed as

$$E = \frac{1}{2}\rho AtV^3 \tag{10.53}$$

Betz (1946) applied simple momentum theory to the windmill established by Froude (1889) for a ship propeller. In that work, the retardation of wind passing through a windmill occurs in two stages: before and after its passage through the windmill rotor. Provided that a mass m is air passing through the rotor per unit time, the rate of momentum change is $m(V_1 - V_2)$ which is equal to the resulting thrust. Here, V_1 and V_2 represent upwind and downwind speeds at a considerable distance from the rotor. The power absorbed P can be expressed as

$$P = m(V_1 - V_2)\overline{V} \tag{10.54}$$

On the other hand, the rate of kinetic energy change in wind can be expressed as

$$E_k = \frac{1}{2}m(V_1^2 - V_2^2) \tag{10.55}$$

The expressions in Eqs. (10.54) and (10.55) should be equal, so the retardation of the wind, $V_1 - \overline{V}$, before the rotor is equal to the retardation, $\overline{V} - V_2$, behind it, assuming that the direction of wind velocity through the rotor is axial and that the velocity is uniform over the area A. Finally, the power extracted by the rotor is

$$P = \rho A \overline{V}(V_1 - V_2)\overline{V} \tag{10.56}$$

Furthermore,

$$P = \rho A \overline{V}^2 (V_1 - V_2) = \rho A \left(\frac{V_1 + V_2}{2}\right)^2 (V_1 - V_2) \tag{10.57}$$

and

$$P = \rho \frac{AV_1^3}{4}[(1 + \alpha)(1 - \alpha^2)] \quad \text{where} \quad \alpha = \frac{V_2}{V_1} \tag{10.58}$$

Differentiation shows that the power P is a maximum when $\alpha = \frac{1}{3}$, i.e., when the final wind velocity V_2 is equal to one-third of the upwind velocity V_1. Hence, the maximum power that can be extracted is $\rho A V_1^3 \frac{8}{27}$, as compared with $\frac{\rho A V_1^3}{2}$ in the wind originally, i.e., an ideal windmill could extract 16/27 (or 0.593) of the power in the wind (Golding, 1955).

Exergy analysis

As pointed out earlier, energy and exergy balances for a flow of matter through a system can be expressed as

$$\sum_{in}(h + ke + pe)_{in}m_{in} - \sum_{ex}(h + ke + pe)_{ex}m_{ex} + \sum_{r}Q_r - W = 0 \tag{10.59}$$

$$\sum_{in}ex_{in}m_{in} - \sum_{ex}ex_{ex}m_{ex} + \sum_{r}Ex^Q - Ex^W - I = 0 \tag{10.60}$$

where m_{in} and m_{ex} denote mass input across port 'in' and mass exiting across port 'ex', respectively; Q_r denotes the amount of heat transfer into the system across region r on the system boundary; Ex^Q is the exergy transfer associated with Q_r; W is the work (including shaft work, electricity, etc.) transferred out of the system; Ex^W is the exergy transfer associated with W; I is the system exergy consumption; and h, ke, pe, and ex denote specific values of enthalpy, kinetic energy, potential energy and exergy, respectively. For a wind energy system, the kinetic energy and pressure terms are of particular significance.

For a flow of matter at temperature T, pressure P, chemical composition μ_j of species j, mass m, specific enthalpy h, specific entropy s, and mass fraction x_j of species j, the specific exergy can be expressed as:

$$ex = [ke + pe + (h - h_0) - T_0(s - s_0)] + \left[\sum_{j}(\mu_{j0} - \mu_{j00})x_j\right] \tag{10.61}$$

where T_0, P_0 and μ_{j00} are intensive properties of the reference environment. The physical component (first term in square brackets on the right side of the above equation) is the maximum available work from a flow as it is brought to the environmental state. The chemical component (second term in square brackets) is the maximum available work extracted from the flow as it is brought from the environmental state to the dead state. For a wind turbine, kinetic energy is dominant and there is no potential energy change or chemical component. The exergy associated with work is

$$Ex^W = W \tag{10.62}$$

The exergy of wind energy can be estimated with the work exergy expression, because there are no heat and chemical components.

Energy and exergy efficiencies

The energy (η) and exergy (ψ) efficiencies for the principal types of processes considered in this section are based on the ratio of product to total input. Here, exergy efficiencies can often be written as a function of the corresponding energy efficiencies. The efficiencies for electricity generation in a wind energy system involve two important steps:

1. *Electricity generation from shaft work*: The efficiencies for electricity generation from the shaft work produced in a wind energy system are both equal to the ratio of the electrical energy generated to the shaft work input.
2. *Shaft work production from the kinetic energy of wind*: The efficiencies for shaft work production from the kinetic energy of a wind-driven system are both equal to the ratio of the shaft work produced to the change in kinetic energy Δke in a stream of matter m_s.

The input and output variables for the system are described in Fig. 10.25. Output wind speed is estimated using the continuity equation. The total electricity generated is related to the decrease in wind potential. Subtracting the generated power from the total potential gives the wind turbine back-side wind potential (Fig. 10.25):

$$V_2 = \sqrt[3]{\frac{2(E_{\text{potential}} - E_{\text{generated}})}{\rho A t}} \qquad (10.63)$$

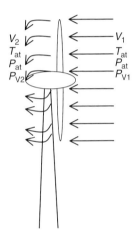

Fig. 10.25. Wind turbine and representative wind energy input and output variables.

In addition, the total kinetic energy difference gives the generated electricity which can be written as

$$\Delta KE = E_{\text{generated}} \qquad (10.64)$$

Air mass flow with time depends on density and wind speed, and can be shown as

$$\dot{m} = \rho A V \qquad (10.65)$$

The exergy of a matter flow is defined as the maximum work that can be acquired when the air flows from state (T_2, P_2) to the ambient state (T_1, P_1). The enthalpy change ΔH from state 1 and state 2 can be expressed as

$$\Delta H = \dot{m} C_{\text{p}}(T_2 - T_1) \qquad (10.66)$$

where $\dot{m}$ is mass flow rate of air, which depends on time, T_1 is the wind chill temperature at the input to the wind turbine; and T_2 is the wind chill temperature at the exit of the wind turbine. The total entropy of the system and entropy difference can be written as

$$\Delta S = \Delta S_{\text{system}} + \Delta S_{\text{surround}} \qquad (10.67)$$

$$\Delta S = \dot{m} T_{at} \left(C_p \ln\left(\frac{T_2}{T_1}\right) - R \ln\left(\frac{P_2}{P_1}\right) - \frac{Q_{loss}}{T_{at}} \right)$$ (10.68)

where

$$P_i = P_{at} \pm \frac{\rho}{2} V^2$$ (10.69)

and

$$Q_{loss} = \dot{m} C_p (T_{at} - T_{average})$$ (10.70)

Here, ΔS is the specific entropy change, T_{at} is the atmospheric temperature, P_2 is the pressure at the exit of the wind turbine for a wind speed V_2 and P_1 is the pressure at the inlet of the wind turbine for a wind speed V_1, Q_{loss} represents heat losses from the wind turbine and $T_{average}$ is the mean value of input and output wind chill temperatures. Thus, the total exergy for wind energy can be expressed using the above equations as

$$Ex = E_{generated} + \dot{m} C_p (T_2 - T_1) + \dot{m} T_{at} \left(C_p \ln\left(\frac{T_2}{T_1}\right) - R \ln\left(\frac{P_2}{P_1}\right) - \frac{Q_{loss}}{T_{at}} \right)$$ (10.71)

The first term on the right side of this equation is the generated electricity. The second and third parts are enthalpy and entropy contributions, respectively.

10.3.3. Case study

The wind energy resource and several wind energy technologies are assessed from an exergy perspective.

System considered

In order to evaluate and assess wind energy potential, a database is considered of hourly wind speed and direction measurements taken between May 2001 and May 2002 at seven stations in the northern part of Istanbul (40.97°E longitude, 29.08°N latitude). For this research, values from only one station are considered. This area comes under the influence of the mild Mediterranean climate during summer, and consequently experiences dry and hot spells for about 4 to 5 months, with comparatively little rainfall. During winter, this region comes under the influence of high-pressure systems from Siberia and the Balkan Peninsula and low-pressure systems from Iceland. Hence, northeasterly or westerly winds influence the study area, which also has high rainfall in addition to snow every year in winter. Air masses originating over the Black Sea also reach the study area (Sahin, 2002).

Results and discussion

In this section, measured generated power data from a group in Denmark are used to obtain a power curve. Pedersen et al. (1992) recommend wind turbine power curve measurements be used to determine the wind turbine required in relation to technical requirements and for approval and certification of wind turbines in Denmark. Here, output electrical power data for a 100 kW wind turbine with a rotor diameter at 18 m and hub height 30 m are given. The data power curve of this wind turbine is shown in Fig. 10.26a. The power curve exhibits two main types of behavior, depending on wind speed. At low wind speeds, power increases with wind speed until the rated power wind speed is reached. A second degree polynomial curve fit can be obtained using a least squares minimization technique. A curve is fitted between the cut-in and rated power wind speeds and its coefficient of determination (R^2) is estimated as 0.99. At high wind speeds (above 16 m/s), the power generation levels off and then tends to decrease from the rated power with increasing wind speed. The cut-out wind speed of this turbine is 20.3 m/s. In the rated wind speed region, a third degree polynomial curve is fitted and its R^2 value is calculated as 0.78. The fitted curves for electrical power generation, based on measured data, are illustrated in Fig. 10.26b.

   The exergy analysis of wind energy shows that there are significant differences between energy and exergy analysis results. According to one classical wind energy efficiency analysis technique, which examines capacity factor, the resultant wind energy efficiency is overestimated. The capacity factor normally refers to the percentage of nominal power that the wind turbine generates. The given test turbine capacity factor is also compared with modeled desired area calculations. It is seen that, as for the power curves in Fig. 10.26, there is a close relation between capacity factors. The differences between exergy and energy efficiencies are shown in Fig. 10.27. Below the cut-in wind speed (3.8 m/s)

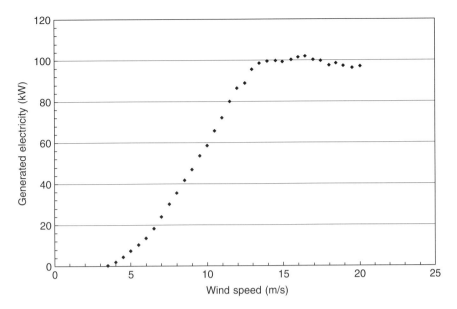

Fig. 10.26a. Test wind turbine power curve, showing electricity generated as a function of wind speed.

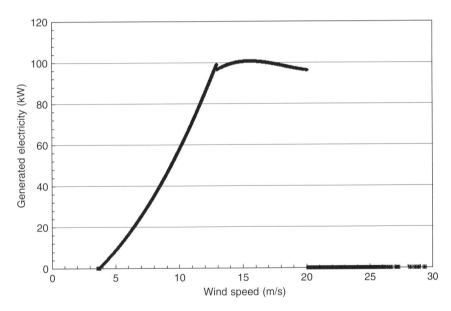

Fig. 10.26b. Test wind turbine power curve, showing regression curves for electricity generated as a function of wind speed.

and over the cut-out wind speed (20.3 m/s) electricity generation is zero, so energy and exergy efficiencies also are zero in those ranges. Since wind speed exhibits high variability during the day, with greater fluctuations than all other meteorological parameters, the fluctuations in energy and exergy efficiency values are high.

All exergy efficiencies are calculated for a selected point, and given in Fig. 10.28 as 24 hour moving average values. In this figure, the moving average values are used to show that the daily changes depend on seasonal variability and to see the periodicity of the exergy efficiencies. Moving average is a statistical method for smoothing highly fluctuating variables. In this study, 24 hour moving average is considered to illustrate daily variability. The data show that during spring and summer an approximately constant variability is observed, but in winter the fluctuations increase. In other

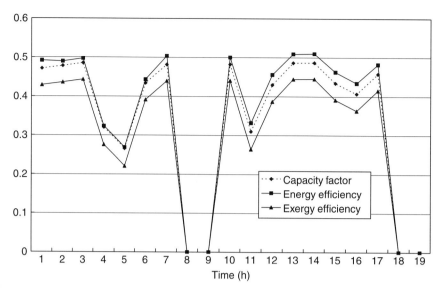

Fig. 10.27. Variation of capacity factor and energy and exergy efficiencies, using a sample set of wind data during the day.

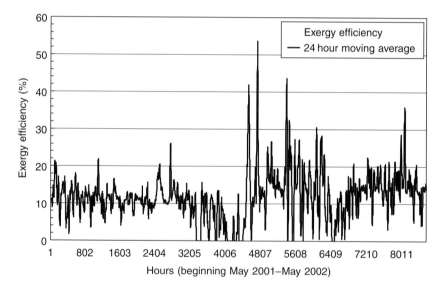

Fig. 10.28. Mean daily exergy efficiencies.

words, during high wind speeds and cold weather the efficiencies are more variable. In addition, exergy efficiencies of wind energy are low in autumn. After autumn, high wind speeds occur since the region comes under the influence of high pressure from Siberia and the Balkan Peninsula and low pressure from Iceland.

Figures 10.27 and 10.28 show the variations and large fluctuations of the efficiencies. These figures are more useful for meteorological interpretation than engineering application. For power generation application, electricity generation, which depends on the power curves, is estimated. Then, the enthalpy and entropy parts of Eq. (10.79) are calculated, and the energy and exergy efficiencies are evaluated. For each efficiency calculation, 8637 data values are employed. Then, regression analysis is applied to wind speeds between the cut-in and cut-out levels, and energy and exergy efficiencies are calculated (Fig. 10.29). The lowest efficiencies are observed at the cut-in and cut-out wind speeds. As seen in Fig. 10.30, energy and exergy efficiencies exhibit important differences at every wind speed. We suggest that exergy efficiencies

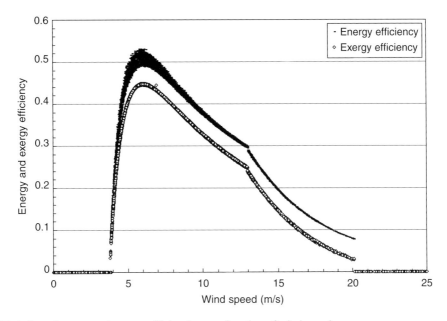

Fig. 10.29. Variation of energy and exergy efficiencies as a function of wind speed.

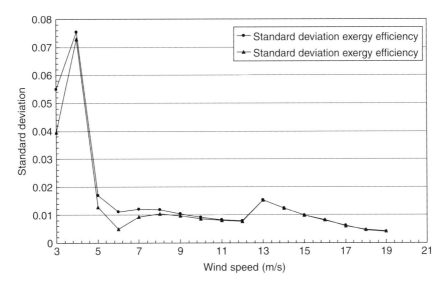

Fig. 10.30. Standard deviation of exergy and energy efficiencies with wind speed.

be used in assessments instead of energy efficiencies. Such an approach yields more realistic results and provides more information about wind energy systems.

In Fig. 10.29, it is seen that the exergy efficiency curve is smoother than the energy efficiency curve. In other words, deviations for energy efficiencies are higher than for exergy efficiencies. To illustrate these variations, mean standard deviations of these efficiencies are calculated for each wind speed interval (Fig. 10.30). It is observed that at lower wind speeds standard deviations for energy efficiencies are higher than those for exergy efficiencies. Above wind speeds of 9 m/s, the same standard deviations are observed.

In Fig. 10.31, mean energy and exergy efficiencies are presented as a function of wind speed. This figure emphasizes the differences between the efficiencies, and shows the over-estimation provided by energy efficiencies. The relative differences between energy and exergy efficiencies, where exergy efficiency is taken as the base value, are given in the same figure. There, it is seen that the relative difference is lowest at a wind speed of about 7 m/s, and increases at lower

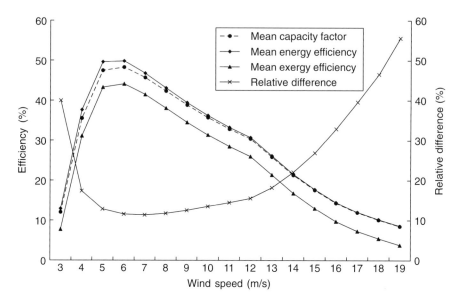

Fig. 10.31. Mean exergy and energy efficiencies, and percent differences between these values, as a function of wind speed. Mean capacity factor is also shown.

and higher wind speeds. These relative differences imply that exergy methods should be applied to wind energy systems for better understanding.

10.3.4. Spatio-temporal wind exergy maps

Most variations in atmospheric characteristics and properties depend on location. Hence, spatial modeling of wind is an important subject in wind engineering studies. Generally, spatial and temporal variations are studied separately. Spatial modeling of wind is achieved by mapping and using objective analysis methods, as reported in the meteorology and wind engineering literature. Various methods exist for data interpolation from measurement stations to any desired point (Cressman, 1959; Barnes, 1964; Schlatter, 1988).

Other estimation methods for wind properties at any desired point, where spatial correlation structure determines the weights applicable to each observation, are the optimal interpolation method of Gandin (1963); the cumulative semivariogram method of Sen (1989) and the approaches of Sen and Sahin (1998). In addition, Sahin (2002) has suggested a spatio-temporal approach based on trigonometric point cumulative semivariogram.

Geostatistics, originally proposed by Krige (1951) and developed by Matheron (1963), is now widely applied in earth sciences as a special branch of applied statistics (Davis, 1986). One of the most common mapping techniques in wind power meteorology is the European wind atlas methodology, which is based on the calculation methods of roughness change class effects and speed-up models for flow passes. It is equally important to construct a model for the effect of sheltering obstacles on the terrain, such as houses and shelter belts, through the so-called shelter model. Topography and wind climatology are essential in distinguishing landscapes. Surface wind speed time-series distribution functions are calculated by fitting the Weibull distribution with the scale, c, and the shape, k, parameters plotted at five heights, four roughness classes and eight direction sectors. The roughness change model is initially expanded to multiple roughness changes, and subsequently developed into a more general model capable of handling roughness areas extracted directly from topographical maps (Troen and Peterson, 1989).

This section describes a spatio-temporal map approach to wind exergy analysis, based on the data from an irregular set of stations scattered over an area. Other exergy analyses of wind energy generating systems do not provide exergy maps showing spatial and temporal parameters. Energy and exergy efficiency models for wind generating systems are used to produce exergy monthly maps based on Krige's method (Krige, 1951). With these maps for a specific system, exergy efficiencies at any location in the considered area can be estimated using interpolation. A case study is presented that applies these models to 21 climatic stations in Ontario, Canada to show how exergy efficiencies change and how these maps compare with energy efficiency maps.

Table 10.3. Topographical characteristics of selected meteorological stations in Ontario.

Station	Latitude (°N)	Longitude (°W)	Altitude (m)
Atikokan	48.45	91.37	395
Big Trout Lake	53.50	89.52	220
Dryden Airport	49.50	92.45	413
Kapuskasing	49.25	82.28	227
Kenora	49.47	94.22	407
Kingston	44.13	76.36	93
London	43.02	81.09	278
Moosonee	51.16	80.39	10
North Bay	46.21	79.26	358
Ottawa	45.19	75.40	116
Red Lake	51.04	93.48	375
Simcoe	46.29	84.30	187
Sault Ste Marie	42.51	80.16	241
Sioux Lookout	50.07	91.54	398
Sudbury	46.37	80.48	348
Thunder Bay	48.22	89.19	199
Timmins	48.34	81.22	295
Toronto Pearson Airport	43.40	79.38	173
Trenton	44.07	77.32	85
Wiarton	44.45	81.06	222
Windsor	42.16	82.58	190

Source: Ontario Weather Data (2004).

Generating wind energy maps

Energy and exergy efficiencies are estimated using measured generated power data from a Denmark group, as noted earlier. It is seen that capacity factors of this wind turbine system are very high even without considering enthalpy values. The capacity factor is approximately 45% for wind speeds of 8–11 m/s. Maps of estimated efficiencies for 21 stations in Ontario are subsequently developed. In this illustration, 30-year average wind speed, temperature and pressure data that were taken from Ontario Weather Data (2004) are used for these stations (Table 10.3). Wind speed values are interpolated from 10 to 30 m. The 100 kW wind turbine with a 30 m hub height is especially selected to minimize wind speed interpolation errors. This region is a lake area so interactions between water and land surfaces are very high. As a result of these topographical properties, continuous high wind speeds occur. Another important feature of this region is low temperatures with high wind speeds, leading to high wind chill temperatures.

Seasonal wind energy maps

January, April, July and October geostatistical spatio-temporal maps are developed and discussed here. These maps are intended to show the differences between energy and exergy efficiencies of a specific wind turbine system at the

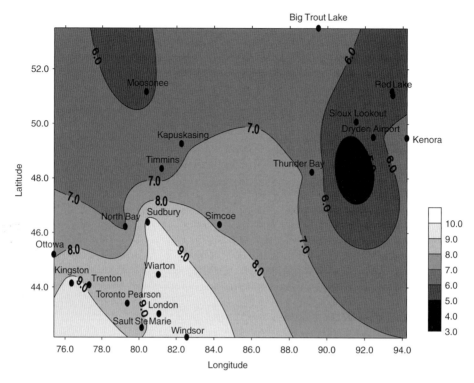

Fig. 10.32a. Map of wind speed (in units of m/s) at a height of 30 m for January for Ontario.

same conditions through exergy analysis. This analysis also gives more information and describes how efficiently wind energy is used, how much losses occur, and the locations of these losses and inefficiencies. Each month is taken to be representative of one season. The 21 stations considered are scattered throughout the map in Fig. 10.32a, where the scale of the map is given at right side. The bottom right of this map shows Lake Ontario, where climatological data are not measured, so this area is not discussed. Low wind speeds are observed in the east and north parts of Ontario in January. The monthly minimum average value observed in Atikokan in this month is below the typical wind-turbine cut-in wind speed and as a result there is no electricity generation. The monthly maximum average wind speed observed in southwestern Ontario is 9–10 m/s (Fig. 10.32a).

The estimated energy efficiencies and a corresponding map is developed for January. At low wind speeds, efficiencies are high, but this does not mean that at these values the wind turbine is more efficient than rated for that wind speed. Rather, it means that the generated electricity is low and also the potential of wind energy is low at these wind speeds. As a result, the ratio between generated electricity and potential energy is high (Fig. 10.32b). The same observations apply for exergy and, in addition, the contours for exergy efficiency are seen to be lower than those for energy efficiency for all regions. The average exergy efficiency value is 40%. This exergy map allows interpolation to be used to estimate parameter values in regions for which there are no measured data. Hence, this kind of map can be used for practical engineering applications (Fig. 10.32c).

For meaningful comparisons of energy and exergy efficiencies, the wind speed maps should be considered together. Here, differences between energy and exergy efficiencies are multiplied by 100 and divided by the highest value. Relative differences between energy and exergy efficiencies are shown in Fig. 10.32d. Large relative differences in energy efficiency values are observed, especially at low wind speeds. Contrary to this, the relative differences between energy and exergy efficiencies at high wind speeds are smaller. But these values are higher than 10% at all stations. These differences are large and should not be neglected in energy planning and management (Fig. 10.32d).

Wind speed values are clustered in three main groups. The lowest wind speed is higher than the wind speed cut-in. The highest wind speed is 10 m/s. In April, electricity can be generated at all stations. Like for the January map, the highest wind speed values in April are observed in southwestern parts of Ontario (Fig. 10.33a). In the energy efficiency map for April, the efficiencies successively increase from south to north. Because of the low wind speeds, energy efficiencies in the northern parts of this region are approximately 50%. There are also three clusters in this efficiency map (Fig. 10.33b).

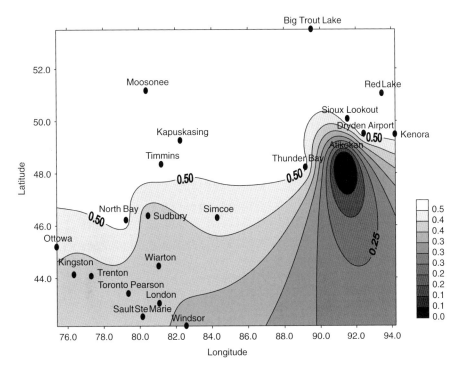

Fig. 10.32b. Energy efficiency map for Ontario for January.

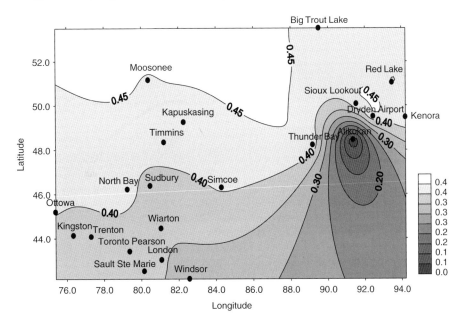

Fig. 10.32c. Exergy efficiency map for Ontario for January.

Using alternate exergy efficiency definitions, efficiencies of wind energy are decreased and two main clusters are seen in April (Fig. 10.33c). In April, the energy and exergy efficiency contours tend to align parallel to lines of constant latitude. In contrast, the relative differences between the two efficiencies are approximately aligned parallel to lines of constant longitude. The relative differences vary between 14% and 22%. In Atikokan, where the lowest wind speeds are observed, the relative difference between the two efficiencies is 22% (Fig. 10.33d).

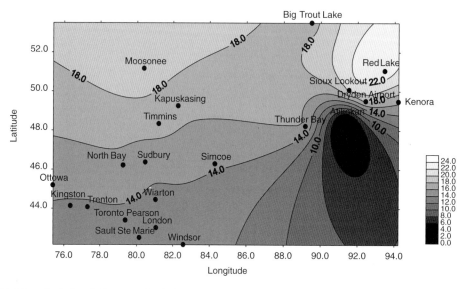

Fig. 10.32d. Map of relative differences (in %) between energy and exergy efficiencies for Ontario for January.

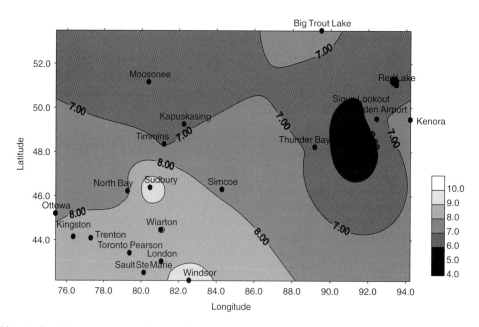

Fig. 10.33a. April wind speed map for 30 m for Ontario.

Wind speeds for July exhibit different clusters as a result of topographical effects in summer. The high heating during this month creates unstable surface conditions. The average wind speed at one station is lower than the cut-in value and as a result the energy and exergy efficiencies are zero. The highest wind speed for this month is the lowest of values for the maximums of the other months (Fig. 10.34a). The spatial distributions for energy efficiencies exhibit three clusters and the general contour values are 40–50% (Fig. 10.34b). There is an area of high energy efficiency in northwest Ontario but exergy efficiencies are lowest in this area. The dominant efficiency in July is seen to be approximately 40%, except for the eastern regions of Ontario (Fig. 10.34c). In July energy and exergy efficiencies are similar and the relative differences between these efficiencies are relatively low (Fig. 10.34d). In July, wind chill is not appreciable.

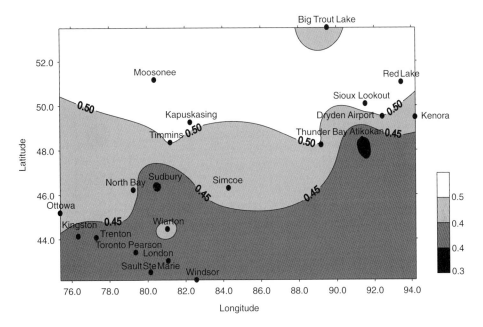

Fig. 10.33b. April energy efficiency map for Ontario.

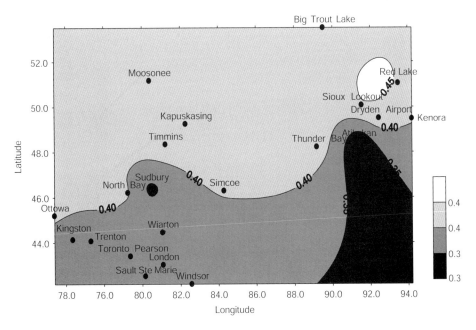

Fig. 10.33c. April exergy efficiency map for Ontario.

For October, three wind speed clusters are observed and wind power systems generate electricity in all stations (Fig. 10.35a). Energy efficiencies are grouped into two main clusters. Topographical conditions cause some localized effects at these stations in October (Fig. 10.35b). Exergy efficiencies are lower than energy efficiencies during this month. It is seen that one of the highest energy efficiency areas, which is observed in western Ontario, is less significant based on exergy (Fig. 10.35c). Without summer topographical heating, the relative differences between these efficiencies are low during October in most parts of Ontario. But wind chill becomes more appreciable during this month (Fig. 10.35d).

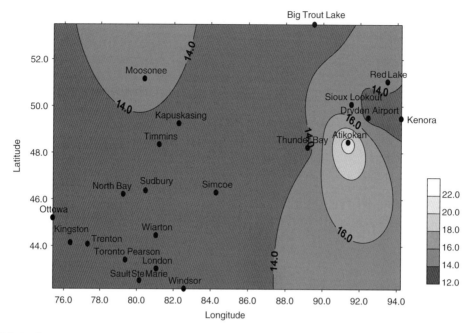

Fig. 10.33d. April energy–exergy relative errors (%) map for Ontario.

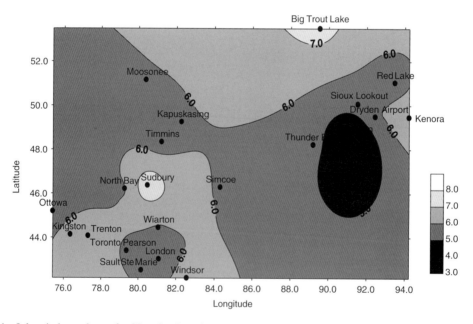

Fig. 10.34a. July wind speed map for 30 m for Ontario.

General comments

The spatio-temporal exergy maps presented here describe energetic and exergetic aspects of wind energy. Seasonal exergy and energy efficiencies are presented in the form of geostatistical maps. The application of exergy analysis for each system, and the ensuing point-by-point map analysis, adds perspective to wind power sources. Thus, exergy maps provide meaningful and useful information regarding efficiency, losses and performance for wind turbines. In addition, the approach reduces the complexity of analyses and facilitates practical analyses and applications.

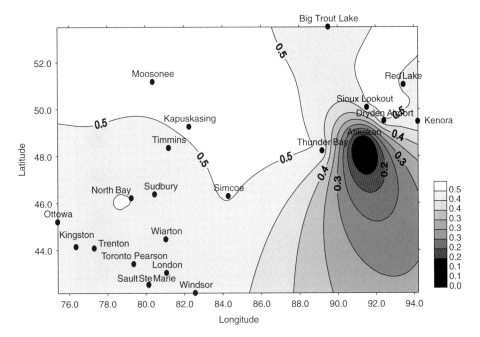

Fig. 10.34b. July energy efficiency map for Ontario.

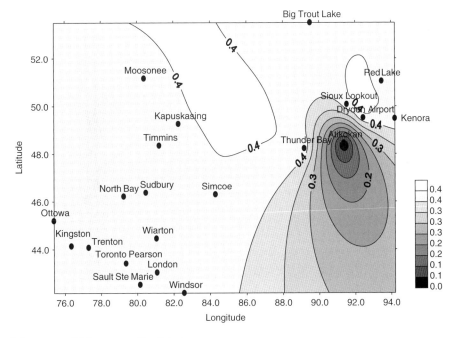

Fig. 10.34c. July exergy efficiency map for Ontario.

Some important observations can be drawn. First, the relative differences between energy and exergy efficiencies are highest in winter and lowest in summer. Second, exergy efficiencies are lower than energy efficiencies for each station for every month considered. More generally, the exergy approach provides useful results for wind energy systems, and the tools for approximating wind energy efficiencies presented here are widely applicable. Such tools can help increase the application of wind systems and optimize designs, and identify appropriate applications and optimal system arrangements.

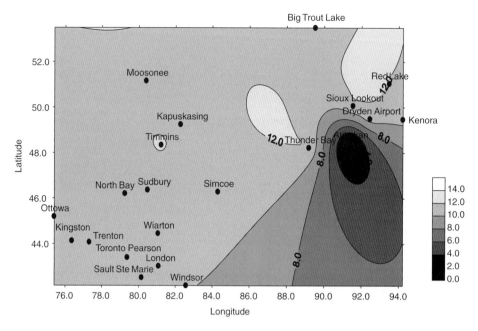

Fig. 10.34d. July energy-exergy relative errors (%) map for Ontario.

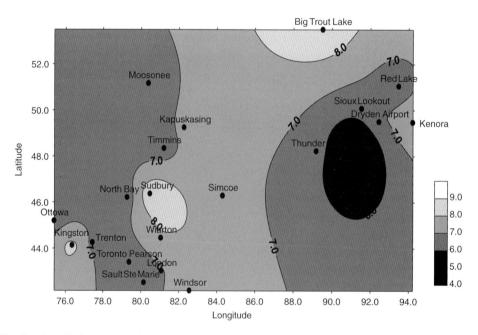

Fig. 10.35a. October wind speed map for 30 m for Ontario.

10.3.5. Closure

Exergy formulations for wind energy are developed and described that are more realistic than energy formulations. Differences are illustrated between energy and exergy efficiencies as a function of wind speed, and can be significant. Exergy analysis should be used for wind energy evaluations and assessments, so as to allow for more realistic modeling, evaluation and planning for wind energy systems. Spatio-temporal wind exergy maps provide a useful tool for assessing wind energy systems.

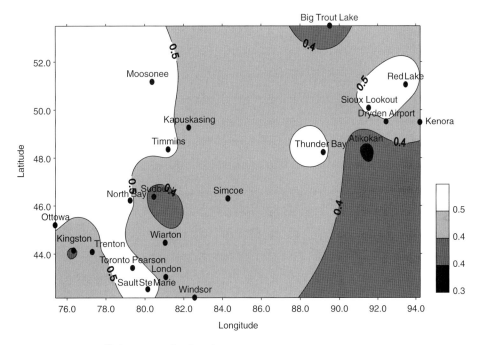

Fig. 10.35b. October energy efficiency map for Ontario.

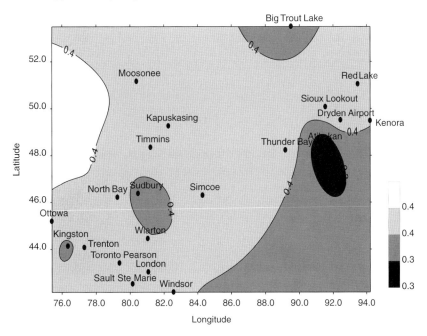

Fig. 10.35c. October exergy efficiency map for Ontario.

10.4. Exergy analysis of geothermal energy systems

The word 'geothermal' derives from the Greek words geo (earth) and therme (heat), and means earth heat. Geothermal energy is the thermal energy within the Earth's crust, i.e., the warm rock and fluid (steam or water containing large amounts of dissolved solids) that fills the pores and fractures within the rock, and flows within sand and gravel. Calculations show that the earth, originating from a completely molten state, would have cooled and become completely solid many

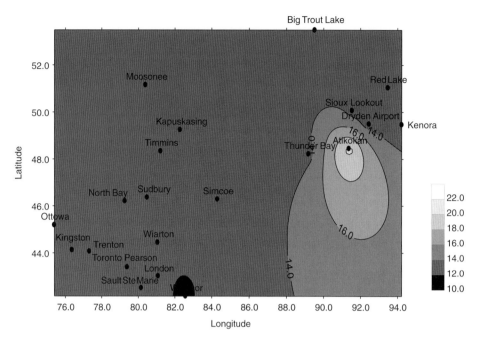

Fig. 10.35d. October energy–exergy relative errors (%) map for Ontario.

thousands of years ago without an energy input in addition to that of the sun. It is believed that the ultimate source of geothermal energy is radioactive decay within the Earth. The origin of this heat is linked with the internal structure of the planet and the physical processes occurring within it.

Geothermal energy is clean and sustainable. Geothermal energy resources are located over a wide range of depths, from shallow ground to hot water and hot rock found several kilometers beneath the Earth's surface, and down even deeper to the extremely high temperatures of molten rock called magma. Geothermal energy is to some extent renewable since a geothermal resource usually has a life of 30–50 years. The life may be prolonged by reinjection processes, which can compensate for at least part of the fluid extracted during geothermal energy use.

Geothermal energy has been used commercially for over 80 years and for four decades on the scale of hundreds of megawatts for electricity generation and direct use. The utilization of geothermal energy has increased rapidly during the last three decades. In 2000, geothermal resources had been identified in over 80 countries and utilized in 58 countries (Fridleifsson, 2001).

Most of the world's geothermal power plants were built in the 1970s and 1980s following the 1973 oil crisis. The urgency to generate electricity from alternative energy sources and the fact that geothermal energy was essentially free led to non-optimal plant designs for using geothermal resources (Kanoglu, 2002a). That era had important consequences for energy and environmental polices. Since then energy policy has been a key tool for sustainable development, given the significant role of energy in economic growth and environmental effects.

There are three general types of geothermal fields: hot water, wet steam and dry steam. Hot water fields contain reservoirs of water with temperatures between 60°C and 100°C, and are most suitable for space heating and agricultural applications. For hot water fields to be commercially viable, they must contain a large amount of water with a temperature of at least 60°C and lie within 2000 m of the surface. Wet steam fields contain water under pressure and are at 100°C. These are the most common commercially exploitable fields. When the water is brought to the surface, some of it flashes into steam, and the steam may drive turbines that produce electrical power. Dry steam fields are geologically similar to wet steam fields, except that superheated steam is extracted from the ground or an aquifer. Dry steam fields are relatively uncommon. Because superheated water explosively transforms to steam when exposed to the atmosphere, it is safer and generally more economic to use geothermal energy to generate electricity, which is more easily transported. Because of the relatively low temperature of the steam/water, geothermal energy is usually converted to electricity with an energy efficiency of 10–15%, as opposed to the 20–40% values typical of coal- or oil-fired electricity generation.

To be commercially viable, geothermal electrical generation plants must be located near a large source of easily accessible geothermal energy. A further complication in the practical utilization of geothermal energy derives from the corrosive properties of most groundwater and geothermal steam. Prior to 1950, metallurgy was not advanced enough to enable the manufacture of steam turbine blades sufficiently resistant to corrosion for geothermal uses. Geothermal energy sources for space heating and agriculture have been used extensively in Iceland, and to some degree Japan, New Zealand and the former Soviet Union. Other applications include paper manufacturing and water desalination.

Although geothermal energy is generally considered as a non-polluting energy source, water from geothermal fields often contains some amounts of hydrogen sulfide and dissolved metals, making its disposal difficult. Consequently, careful fluid treatment is required, depending on the geothermal characteristics of the location.

Global installed geothermal electrical capacity in the year 2000 was 7974 MW, and overall geothermal electrical generation was 49.3 billion kWh that year. Geothermal energy use for space heating has grown since 1995 by 12%. About 75% of global thermal use of energy production for geothermal sources is for district heating and the reminder for individual space heating (Barbier, 2002). Although the majority of district heating systems are in Europe, particularly in France and Iceland, the U.S. has the highest rate of geothermal energy use for individual home heating systems (e.g., in Klamath Falls, Oregon and Reno, Nevada). Such other countries as China, Japan and Turkey are also using geothermal district heating.

Although such systems are normally assessed with energy, a more perceptive basis of comparison is needed if the true usefulness of a geothermal energy system is to be assessed and a rational basis for the optimization of its economic value established. Energy efficiency ignores energy quality (i.e., exergy) of the working fluid and so cannot provide a measure of ideal performance. Exergy efficiencies provide comprehensive and useful efficiency measures for practical geothermal district heating systems and facilitate rational comparisons of different systems and operating conditions.

In this section, two case studies are provided: (1) energy and exergy analyses of a geothermal district heating system and (2) exergy analysis of a dual-level binary geothermal power plant.

10.4.1. Case study 1: energy and exergy analyses of a geothermal district heating system

In this case study, adapted from Oktay et al. (2007), energy and exergy analyses are carried out of the Bigadic geothermal district heating system (GDHS), including the determination of energy and exergy efficiencies.

The Bigadic geothermal field is located 38 km south of the city of Balikesir which is in the west of Turkey. The Bigadic geothermal field covers a total area of about 1 km^2. The reservoir temperature is at 110°C. As of the end of 2006, there are two wells, HK-2 and HK-3, having depths of 429 and 307 m, respectively. The well head temperature is 98°C. There are five pumps in the geothermal field, three for pumping fluids from the wells and two for pumping fluid to the mechanical room. Wells 1 and 2 are basically artesian wells through which water is forced upward under pressure. Pumps 1 and 2 were in use on the days data were taken, but Pump 3, which is designed to pump automatically when the mass flow rate requirements achieve 100 kg/s, was not in use because the mass flow rate requirements were low. The mass flow rates were 53 and 63.8 kg/s according to actual data for November and December 2006, respectively. Pump 3 generally is not used because the elevation difference between the geothermal source and the mechanical room of 200 m usually provides enough pressure to convey the fluid.

Description of the GDHS

The GDHS consists of three main parts (Fig. 10.36). In the first part, the geothermal fluid is pumped into the 'mud and gas separator unit' to separate harmful particles, and then flows to the first heat exchanger. An 18 km pipeline connects the geothermal source and the mechanical room. Over this distance, the geothermal fluid temperature decreases by 3°C to 4°C.

In the second part of the system, the geothermal fluid is cooled to approximately 44°C in the first and second heat exchangers, which are located in the mechanical room. After heat transfer occurs in the first and second heat exchangers, the geothermal fluid is sent to the first and second center pipelines. A second fluid (clean water) enters the first and second heat exchangers with a temperature of 47°C and is heated to 68°C (based on data from December 6, 2006).

In the third part, clean hot water is pumped into the heat exchangers, which are located under each building. The system is designed to have one or two heat exchangers for each building, one for space heating, and the other for hot water requirements. Presently, each building has one or two heat exchangers. Ten percent of the total residences in Bigadic have an extra heat exchanger for hot water requirements. There are three pipelines for conveying the hot water along different paths. The second and third center pipelines are in use, while the first is not yet in operation.

The second and third center pipelines are designed to transport 15,253 kW of heat to the 2200 individual residences. The indoor and outdoor design temperatures equal 20°C and −6°C, respectively. The second and third center pipelines

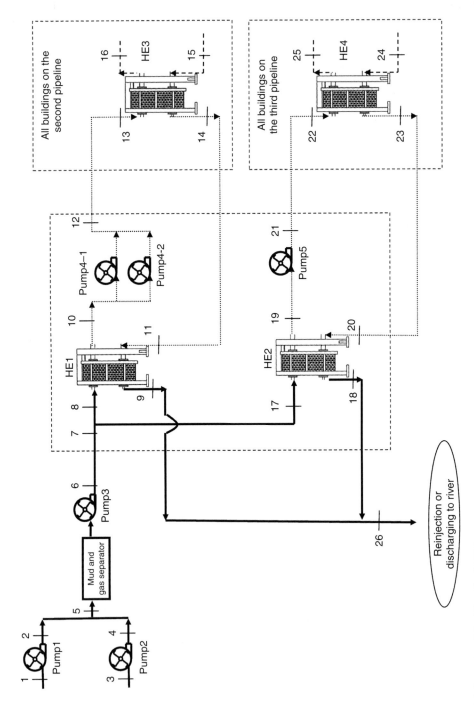

Fig. 10.36. Schematic of the Bigadic GDHS.

supply the heat requirements of the dwellings, one post office, one dormitory, eight colleges, one state hospital, two police stations and ten government buildings.

In the study, the heat exchangers for all residences are modeled as one heat exchanger. All heat from the heat exchangers on the second center pipeline is collected in one heat exchanger, called the 'Third heat exchanger'. Similarly, all heat from the heat exchangers on the third center pipeline is collected in one heat exchanger (the 'Fourth heat exchanger'). The heat is then transferred to the highest elevation and furthest points. The state hospital has both the highest elevation and the longest pipeline distance from the mechanical room and thus is a key point for calculations. Plate-type heat exchangers are used in the system. Inlet and outlet heat exchanger liquid temperatures are investigated for the state hospital.

Analysis

Mass, energy and exergy balances are written for the system and its components, following the treatments of earlier researchers (e.g., Kanoglu, 2002a; Ozgener et al., 2004; Oktay et al., 2007). The system is considered to undergo steady-state and steady-flow processes.

A mass balance for the overall geothermal system can be written as

$$\dot{m}_{T,in} = \dot{m}_{T,out} \tag{10.72}$$

where $\dot{m}_T$ denotes the total mass flow rate.

The energy and exergy of the geothermal water are calculated as

$$\dot{E}_{T,in} = \dot{m}_{tw}h_{tw} \cong \dot{m}_{tw,2}h_2 + \dot{m}_{tw,3}h_3 \tag{10.73}$$

$$\dot{Ex}_{T,in} = \dot{m}_{tw,2}[(h_{tw,2} - h_0) - T_0(s_{tw,2} - s_0)] + \dot{m}_{tw,3}[(h_{tw,3} - h_0) - T_0(s_{tw,3} - s_0)] \tag{10.74}$$

where the subscript tw denotes geothermal water and the subscripts 2 and 3 denote the working wells. Similar expressions can be written for the outlet flows of the geothermal water.

The exergy destructions in the heat exchanger, pump, pipeline and overall system are evaluated as follows:

$$\dot{Ex}_{d,he} = \dot{Ex}_{in} - \dot{Ex}_{out} \quad \text{for heat exchangers} \tag{10.75}$$

$$\dot{Ex}_{d,pu} = \dot{W}_{pu} - (\dot{Ex}_{out} - \dot{Ex}_{in}) \quad \text{for pumps} \tag{10.76}$$

$$\dot{Ex}_{d,pi} = \dot{Ex}_{in} - \dot{Ex}_{out} - \dot{Ex}^Q \quad \text{for pipes/pipelines} \tag{10.77}$$

$$\dot{Ex}_{T,d} = \dot{Ex}_{T,d,he} + \dot{Ex}_{T,d,pu} + \dot{Ex}_{T,d,pi} \tag{10.78}$$

The energy efficiency of the system is determined as

$$\eta_{sys} = \frac{\dot{E}_{T,out}}{\dot{E}_{T,in}} \tag{10.79}$$

where $\dot{E}_{T,out}$ is the total product energy output (useful heat) and $\dot{E}_{T,in}$ is the total energy input.

The exergy efficiency of a heat exchanger is defined as the ratio of the exergy output (i.e., increase in the exergy rate of the cold stream) to the exergy input (i.e., decrease in the exergy rate of the hot stream) as follows:

$$\psi_{he} = \frac{\dot{m}_{cold}(ex_{cold,out} - ex_{cold,in})}{\dot{m}_{hot}(ex_{hot,in} - ex_{hot,out})} \tag{10.80}$$

where ex is the specific exergy, expressible as

$$ex = (h - h_0) - T_0(s - s_0) \tag{10.81}$$

The exergy efficiency of the system is determined as

$$\psi_{sys} = \frac{\dot{Ex}_{T,out}}{\dot{Ex}_{T,in}} = 1 - \frac{\dot{Ex}_{d,sys} + \dot{Ex}_{nd}}{\dot{Ex}_{T,in}} \quad (10.82)$$

where the subscript nd denotes natural direct discharge.

In addition, we examine the seasonal average total residential heat demand and how it is satisfied. In the 'summer' or warmer season (i.e., when there is no need to heat dwellings), which on average has 165 days, only sanitary hot water is supplied to the residences. The total sanitary hot water load over the summer season is given by

$$\dot{E}_{smr} = N_{dw}N_{per}S\Delta T_w \, c_f \quad (10.83)$$

where N_{dw} is the average number of dwellings, N_{per} is the average number of people in each dwelling (assumed to be 4), S is the average daily usage of sanitary hot water (taken to be 50 L/person-day or 50 kg/person-day), and ΔT_w is the temperature difference between the sanitary hot water (60°C) and the tap water from the city distribution network (10°C). Thus,

$$\dot{E}_{smr} = (2200 \times 4 \times 50) \, \text{kg/day}(50°C)(4.18 \, \text{kJ/kg°C}) = 1064 \, \text{kW}$$

The total 'winter' heat demand (sanitary hot water plus space heating) can be expressed as

$$\dot{E}_{design.} = \dot{E}_{dw}N_{dw} \quad (10.84)$$

where $\dot{E}_{dw}$ is the heat load for an average (or equivalent) dwelling. Assuming there are 2200 residences, each with a maximum load of 6.9 kW, the overall winter heat load is 15.25 MW.

Equation 10.84 can also be written as

$$\dot{E}_{design} = \dot{m}c_f \Delta T_{design}N_{dw} \quad (10.85)$$

where $\Delta T_{design} = (T_{indoor} - T_{outdoor})_{design}$ is the difference between the indoor and outdoor temperatures (i.e., $20°C - (-6°C) = 26°C$).

We account for the variation of outdoor temperature through $\Delta T_{average} = (T_{indoor} - T_{outdoor})_{average}$ using average outdoor temperatures while the indoor temperature is kept constant. We now introduce the temperature ratio

$$T_R = \frac{\Delta T_{average}}{\Delta T_{design}} \quad (10.86)$$

in order to determine the average heat loads, as shown below:

$$\dot{E}_{average} = T_R \dot{E}_{design} \quad (10.87)$$

The mass flow rate can be determined from above equation as

$$\dot{m} = \frac{\dot{E}_{average}}{c_f \Delta T} \quad (10.88)$$

Table 10.4 lists the heat demand breakdown for the each month according to the average outdoor temperatures.

Results and discussion

Effect of salts and other components in the geothermal fluid on thermodynamic properties are neglected in this study. The thermodynamic properties of the geothermal fluid are taken to be those of water, properties of which are available from thermodynamic tables and software. Kanoglu (2002a) also employs this assumption in an exergy analysis of geothermal power plants.

Table 10.4. Bigadic GDHS monthly energy requirements.

Month	Average outdoor temperature (°C)	Temperature ratio, T_R	Total average energy demand (kW) (from Eqs. 10.83 and 10.87)
Winter months (from Eq. 10.87)			
October	15.1	0.188	2874.60
November	9.7	0.396	6042.53
December	6.7	0.512	7802.50
January	4.7	0.588	8975.80
February	5.4	0.562	8565.15
March	8.2	0.454	6922.52
April	13.4	0.254	3871.92
Summer months (from Eq. 10.83)			
May	17.7	–	1064.00
June	22.4	–	1064.00
July	24.5	–	1064.00
August	23.6	–	1064.00
September	19.9	–	1064.00

A parametric study of the Bigadic GDHS is presented here using data recorded in November and December 2006. Energy and exergy efficiencies and exergy destructions are determined. For each state of the geothermal fluid and hot water, the temperature, the pressure, the mass flow rate and energy and exergy rates are calculated using Engineering Equation Solver (EES). In Table 10.5 sample results are given based on data for December 2006. State 0 represents the dead state for both the geothermal fluid and hot water. The dead state conditions are taken to be 11°C and 101.3 kPa for the day considered.

An energy flow diagram for the system is illustrated in Fig. 10.37a. The thermal natural direct discharge accounts for 45.62% of the total energy input, while pump and pipeline losses account for 24.15% of the total energy input.

A detailed exergy flow diagram is given in Fig. 10.37b, and shows that 51% (corresponding to about 1468 kW) of the total exergy entering the system is lost, while the remaining 49% is utilized. The highest exergy loss (accounting for 37% of the total exergy entering) occurs from the system pipes. The second largest exergy loss is with the thermal natural direct discharge and amounts to 34% (or about 498.3 kW) of the total exergy input. This is followed by the total exergy destructions associated with the heat exchangers and pumps, which account respectively for 405.5 and 25.81 kW, or 26.3% and 1.7%, of the total exergy input to the system.

The respective energy and exergy efficiencies are found to be 30% and 36% in November and 40% and 49% in December. The reference-environment temperatures are 15.6°C in November and 11°C in December. Some may intuitively feel that having exergy efficiencies greater than energy efficiencies is not correct. In geothermal systems, however, this is common, in part due to the fact that there is a reinjection process which allows recovery of some heat, making the process/system more exergetically efficient. With Fig. 10.37b, this situation can be explained by noting that, although the input energy for the system exceeds the input exergy, the energy losses in the system are greater than the exergy losses. For November and December respectively, the percentage of energy losses are calculated as 70% and 60%, and the percentages of exergy destruction as 64% and 51%.

The energy demand remains constant during the summer months because only hot water for sanitary utilities is used. The energy demand varies during the winter months depending on outlet temperature. In Fig. 10.38a, the energy and exergy demand rates, based on data in Table 10.5, are illustrated and seen to depend on the monthly average outlet (reference) temperature.

Table 10.5. Properties of system fluids and energy and exergy rates at various locations in the Bigadic geothermal district heating system.

State no.	Fluid type	Temperature, T (°C)	Pressure, P (kPa)	Specific enthalpy, h (kJ/kg)	Specific entropy, s (kJ/kg K)	Mass flow rate, $\dot{m}$ (kg/s)	Energy rate, $\dot{E}$ (kW)	Specific exergy, ex (kJ/kg)	Exergy rate, $\dot{Ex}$ (kW)
0	TW	11	101.32	46.29	0.166	–	–	–	–
1	TW	97	101.32	406.4	1.273	35	14224.00	45.66	1598.28
2	TW	97.05	404	406.8	1.274	35	14238.00	45.78	1602.34
3	TW	96	101.32	402.2	1.261	28.8	11583.36	44.87	1292.35
4	TW	96.05	404	402.6	1.262	28.8	11594.88	44.99	1295.69
5	TW	96.64	390	405.1	1.268	63.8	25845.38	45.79	2921.10
6	TW	94.5	380	396.1	1.244	63.8	25271.18	43.60	2781.76
7	TW	90	505	377.2	1.192	63.8	24065.36	39.47	2518.13
8	TW	90	505	377.2	1.192	27.15	10240.98	39.47	1071.59
9	TW	47	450	197.2	0.665	27.15	5353.98	9.25	251.16
10	Water	68	152	284.7	0.930	55.55	15815.09	21.29	1182.77
11	Water	47	203	196.9	0.665	55.55	10937.80	8.92	495.64
12	Water	68.06	600	285.3	0.931	55.55	15848.42	21.78	1209.79
13	Water	67.1	253	281	0.919	55.55	15609.55	20.74	1152.35
14	Water	48	152	201.1	0.678	55.55	11171.11	9.43	523.86
15	Water	50	203	209.5	0.704	106.2	22248.90	10.47	1112.42
16	Water	60	182	251.2	0.831	106.2	26677.44	15.99	1698.48
17	TW	90	505	377.2	1.192	36.65	13824.38	39.47	1446.55
18	TW	47	450	197.2	0.665	36.65	7227.38	9.25	339.04
19	Water	68	152	284.7	0.930	75	21352.50	21.29	1596.90
20	Water	47	203	196.9	0.665	75	14767.50	8.92	669.18
21	Water	68.06	600	285.3	0.931	75	21397.50	21.79	1633.38
22	Water	67.1	253	281	0.919	75	21075.00	20.74	1555.83
23	Water	48	152	201.1	0.678	75	15082.50	9.43	707.28
24	Water	50	203	209.5	0.704	143.4	30042.30	10.47	1502.09
25	Water	60	182	251.2	0.831	143.4	36022.08	15.99	2293.42
26	TW	44	400	184.4	0.625	63.8	11764.72	7.58	483.83

Notes: State numbers are shown in Fig. 10.36. State zero represent the reference state. TW denotes thermal water.

Energy and exergy demands are dependent on the reference-environment temperature (e.g., the surroundings temperature). Figure 10.38a is based on average values, when the energy and exergy demands change with outlet temperature. Using this figure, a curve fitting is performed to predict the demands for varied outlet temperatures and the following correlations are obtained:

$$\dot{E} = -1.8543T^4 + 67.761T^3 - 739.89T^2 + 1787.9T + 7679.9$$

$$\dot{Ex} = -12.857T^2 - 85.102T + 1609.8$$

where T_0 is the surrounding temperature (in K), which is taken to be the reference-environment temperature. The correlations are plotted with temperatures in Celsius for convenience in practical applications.

We now link the exergy efficiency and average air temperature through the following correlation obtained by curve fitting (Fig. 10.38b):

$$\psi_c = -0.0256T^2 + 0.4038T + 50.372$$

The two cases in Fig. 10.38b are for two actual days in November and December, respectively. Note that the Bigadic GDHS does not have a reinjection section yet, so the geothermal water flows from the mechanical room to the river at 45°C. The exergy of the geothermal water entering the river is 498 kW. The exergy efficiency of the system can be increased by the addition of heat pumps and through recovery and use of the geothermal water that is flowing into the river.

The system is designed to supply the heat loads required for residences at a constant temperature with variable mass flow rates. Figure 10.39 shows both experimental (actual) and calculated exergy destructions within the system components (i.e., pumps, heat exchangers, pipelines and discharge lines), for November and December. It can be observed that: (i) both actual and calculated irreversibilities are reasonably in agreement, except for heat exchangers 1 and 2, and pipelines and (ii) the highest exergy destructions (irreversibilities) occur in pipelines and discharge lines. In those devices, there is a large room for improvement.

Furthermore we can summarize some important facts regarding the environmental benefits of the Bigadic GDHS:

- The maximum heating demand for 2200 dwellings is 15.25 MW and the energy savings achieved with the GDHS in this case amounts to 3876.24 tons of oil equivalent (TOE) per year.
- Emissions of SO_2 and CO_2 are reduced drastically. If such other fuels as coal, natural gas, fuel oil and electricity are used, the respective annual emissions of CO_2 would be 29,996, 10,236, 11,206 and 136,313 tons. Similarly, the respective annual emissions of SO_2 would be 355.49, 53.65 and 3097 tons if coal, fuel oil and electricity are used.

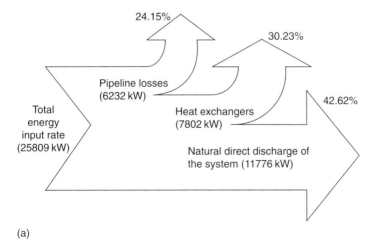

(a)

Fig. 10.37. (a) Energy flow diagram and (b) comprehensive exergy flow diagram of the Bigadic GDHS system. (HE: heat exchanger, BHK: name of well).

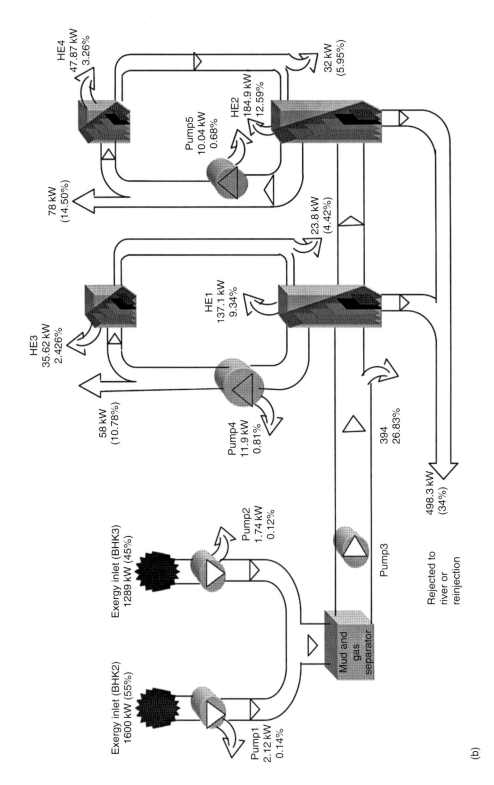

(b)

Fig. 10.37. (Continued)

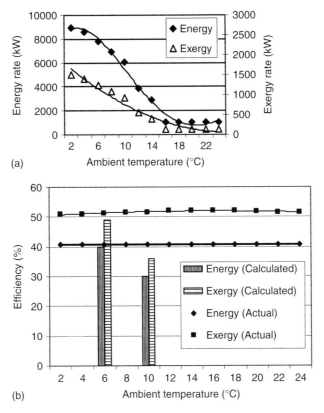

Fig. 10.38. (a) Profiles of energy and exergy rates as a function of ambient temperature, and correlation data. (b) Variation of energy and exergy efficiencies with ambient temperature. (Note: mass flow rates are controlled).

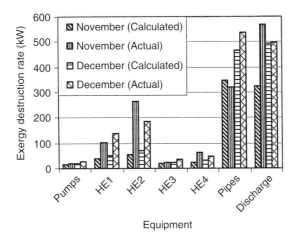

Fig. 10.39. Comparison of the exergy destructions for various components of the system.

Table 10.6 summarizes these results. Note that the emissions values in Table 10.6 are slightly different from those in the previous paragraph. This is because the values in the previous paragraph are the differences between the values listed in table for a given energy form and the geothermal energy values in the table. Using electricity clearly causes the greatest environmental problems and emissions, due to the fact that for each kilowatt of electricity generated the power plants emit about 1 kg of CO_2 and 7 g of SO_2.

Table 10.6. Summary of fuel used for heating and fuel characteristics, and corresponding emissions if the fuel is used.

Heating fuel or energy source	Fuel or energy requirement	Fuel components (%)			Emissions (ton)	
		C	S	Ash	CO_2	SO_2
Domestic coal	12,678,643 kg	67	1.5	9.8	31,090.57	380.36
Natural gas	4,177,738.6 m³	74.1	–	–	11,330.28	–
Fuel oil	3,926,182.7 kg	85.6	1	0.1	12,300.57	78.52
Electricity for resistance heating	36,422,285 kWh	Assuming a 65% efficient power plant			137,407.27	3121.91
Electricity for a ground-source heat pump	213,331.62 kWh	Assuming a heat pump with COP = 3.8			804.82	18.29
Geothermal energy	290,131 kWh/yr of electricity for pumps				1094.55	24.87

Considering system performance, only 15,253 kW (or 47.92%) of the total capacity of 31,830 kW of the current Bigadic GDHS is used. If the system worked at full capacity, the number of equivalent dwelling served would increase from 2200 to 4593. Then, 8092.54 TOE/year would be saved and, for coal, fuel oil and electric resistance heating, respectively, annual CO_2 emissions would be reduced by 62,631.64, 23,398.13 and 284,621.54 tons, while annual SO_2 emissions would be reduced by 742.26, 112.02 and 6466.54 tons.

Closing comments

The comprehensive case study presented in this section of the GDHS in Balikesir, Turkey leads to the following concluding remarks:

- Using actual thermal data from the Technical Department of the GDHS, the exergy destructions in each component and the overall energy and exergy efficiencies of the system, are evaluated for two reference temperatures (15.6°C for November, case 1, and 11°C for December, case 2).
- Energy and exergy flow diagrams clearly illustrate how much loss occurs as well as inputs and outputs. Average energy and exergy efficiencies are found to be 30% and 36% for case 1, and 40% and 49% for case 2, respectively. The key reason why the exergy efficiencies are higher is that heat recovery is used through the reinjection processes which make use of waste heat.
- The parametric study conducted shows how energy and exergy flows vary with the reference-environment temperature and that the increase in system exergy efficiency is due to the increase in the exergy input potential.

Geothermal district heating appear to be a potential environmentally benign option that can contribute to a country by providing more economic and efficient heating of residences and decreased emission rates.

10.4.2. Case study 2: exergy analysis of a dual-level binary geothermal power plant

An exergy analysis of a stillwater binary design geothermal power plant located in Northern Nevada in the US is performed using plant data taken from (Kanoglu, 2002a). The plant has a unique heat exchange design between the geothermal fluid and the working fluid as explained in the next section. A straightforward procedure for exergy analysis for binary geothermal power plants is described and used to assess plant performance by pinpointing sites of primary exergy destruction and identifying possible improvements.

Plant operation

The geothermal power plant analyzed is a binary design plant with a net electrical generation of 12.4 MW from seven identical paired units. Full power production started in April 1989. The plant operates in a closed loop with no environmental discharge and complete reinjection of the geothermal fluid. The modular power plant operates on a predominantly liquid resource at 163°C. Dry air condensers are utilized to condense the working fluid, so no fresh water is consumed. The geothermal field includes four production wells and three reinjection wells. The working (binary) fluid, isopentane, undergoes a closed cycle based on the Rankine cycle.

The plant is designed to operate with seven paired units of Level I and II energy converters. A plant schematic is given in Fig. 10.40 where only one representative unit is shown. The heat source for the plant is the flow of geothermal water (brine) entering the plant at 163°C with a total mass flow rate of 338.94 kg/s. The geothermal fluid, which remains a liquid throughout the plant, is fed equally to the seven vaporizers of Level I. Therefore, each unit receives 48.42 kg/s of geothermal fluid. The brine exits the Level I vaporizers at approximately 131°C and is fed directly to the paired Level II vaporizers where it is cooled to 100°C. The brine is then divided equally and flows in parallel to the Level I and II preheaters. These preheaters extract heat from the brine, lowering its temperature to 68°C and 65°C, respectively. The brine exiting the preheaters is directed to the reinjection wells where it is reinjected back to the ground.

In Level I, 19.89 kg/s of working fluid circulates through the cycle. The working fluid enters the preheater at 32°C and leaves at about 98°C. It then enters the vaporizer where it is evaporated at 133°C and superheated to 136°C. The working fluid then passes through the turbine and exhausts at about 85°C to an air-cooled condenser where it condenses at a temperature of 31°C. Approximately 530 kg/s of air at an ambient temperature of 13°C is required to absorb the heat

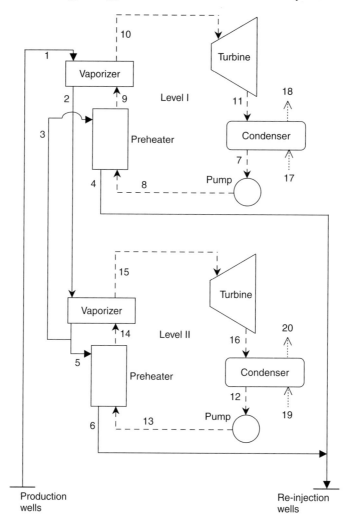

Fig. 10.40. Schematic of the binary geothermal power plant (1 of 7 units).

given up by the working fluid, raising the air temperature to 29°C. The working fluid is pumped to the preheater pressure to complete the Rankine cycle. The Level I isopentane cycle is shown on a *T–s* diagram in Fig. 10.41.

In Level II, 21.92 kg/s of working fluid cycles through the loop. The working fluid enters the preheater at 27°C and exits at 94°C. It then enters the vaporizer where it is evaporated at 98°C and slightly superheated to 99°C. The working fluid passes through the turbine, and then exhausts to the condenser at about 65°C where it condenses at a temperature of 27°C. Approximately 666 kg/s of air enters the condenser at 13°C and leaves at 26°C. The Level II isopentane cycle is shown on a *T–s* diagram in Fig. 10.42.

The saturated vapor line of isopentane is seen in Figs. 10.41 and 10.42 to have a positive slope, ensuring a superheated vapor state at the turbine outlet. Thus, no moisture is involved in the turbine operation. This is one reason isopentane is suitable as a working fluid in binary geothermal power plants. Isopentane has other advantageous thermophysical properties such as a relatively low boiling temperature that matches well with the brine in the heat exchange system and a relatively high heat capacity. Isopentane is also safe to handle, non-corrosive and non-poisonous.

The heat exchange process between the geothermal brine and isopentane is shown on a *T–s* diagram in Figs. 10.43 and 10.44 for Levels I and II, respectively. An energy balance can be written for part of the heat exchange taking place in the vaporizer of Level I using state points shown in Fig. 10.43 as

$$\dot{m}_1(h_{pp} - h_2) = \dot{m}_9(h_f - h_9) \qquad (10.89)$$

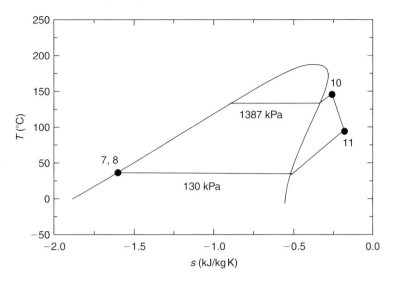

Fig. 10.41. Temperature–entropy (T–s) diagram of Level I isopentane cycle.

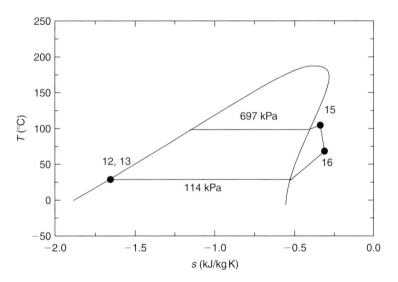

Fig. 10.42. Temperature–entropy (T–s) diagram of Level II isopentane cycle.

where h_f is the saturated liquid enthalpy of isopentane at the saturation (i.e. vaporization) temperature, 133.1°C, and the h_{pp} is the enthalpy of brine at the pinch-point temperature of the brine. Solving this equation for h_{pp}, we determine the corresponding brine pinch-point temperature T_{pp} to be 140.5°C. The pinch-point temperature difference ΔT_{pp} is the difference between the brine pinch-point temperature and the vaporization temperature of isopentane. Here, $\Delta T_{pp} = 7.4$°C. A similar energy balance for the vaporizer of Level II can be written using state points shown in Fig. 10.44 as

$$\dot{m}_2(h_{pp} - h_3) = \dot{m}_{14}(h_f - h_{14}) \tag{10.90}$$

Here, the brine pinch-point temperature is 101.3°C, the vaporization temperature in Level II is 98.4°C and the pinch-point temperature difference ΔT_{pp} is 2.9°C.

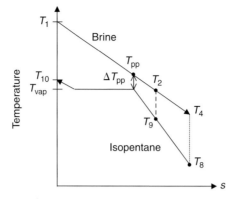

Fig. 10.43. Heat exchange process between the geothermal brine and the isopentane working fluid in Level I.

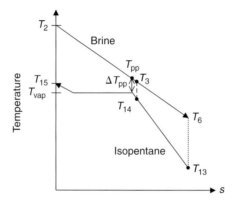

Fig. 10.44. Heat exchange process between the geothermal brine and the isopentane working fluid in Level II.

The turbine power outputs are 1271 kW in Level I and 965 kW in Level II, while the power requirements for the circulation pumps of Level I and II are 52 and 25 kWe, respectively. The net power outputs from Level I and II Rankine cycles are thus 1219 and 940 kW, respectively, giving a net power output of the combined Level I and II cycles of 2159 kW. It is estimated by plant management that approximately 200 and 190 kW power are consumed by parasitic uses in the units in Level I and II, respectively. These parasitic uses correspond to 18.1% of the net power generated in the cycle, and include fan power for the condenser and auxiliaries. There are six fans in Level I and nine in Level II. Subtracting the parasitic power from the net power generated in the cycle yields a net power output from one unit 1769 kW. Since the plant has seven identical units, the total net power output for this plant is 12,383 kW. The various power terms discussed in this section are listed in Table 10.7.

Exergy analysis

Neglecting kinetic and potential energy changes, the specific flow exergy of the geothermal fluid at any state (plant location) can be calculated from

$$ex = h - h_0 - T_0(s - s_0) \tag{10.91}$$

where h and s are the specific enthalpy and entropy of the geothermal fluid at the specified state, and h_0 and s_0 are the corresponding properties at the restricted dead state. For a mass flow rate $\dot{m}$, the exergy flow rate of the geothermal fluid

Table 10.7. Exergy rates and other properties at various plant locations for a representative unit.

State no.	Fluid	Phase	Temperature, T (°C)	Pressure, P (bar abs)	Specific enthalpy, h (kJ/kg)	Specific entropy, s (kJ/kg °C)	Mass flow rate, $\dot{m}$ (kg/s)	Specific exergy, ex (kJ/kg)	Exergy rate, $\dot{Ex}$ (kW)
0	Brine	Dead state	12.8	0.84	53.79	0.192	–	0	–
0′	Isopentane	Dead state	12.8	0.84	−377.30	−1.783	–	0	–
1	Brine	Liquid	162.8	–	687.84	1.971	48.42	125.43	6073
2	Brine	Liquid	130.7	–	549.40	1.642	48.42	81.01	3923
3	Brine	Liquid	99.9	–	418.64	1.306	24.21	46.42	1124
4	Brine	Liquid	67.8	–	283.80	0.928	24.21	19.59	474
5	Brine	Liquid	99.9	–	418.64	1.306	24.21	46.42	1124
6	Brine	Liquid	64.5	–	269.98	0.887	24.21	17.41	422
7	Isopentane	Liquid	31.0	1.30	−336.35	−1.645	19.89	1.33	27
8	Isopentane	Liquid	31.7	13.87	−333.73	−1.643	19.89	3.43	68
9	Isopentane	Liquid	97.6	13.87	−169.69	−1.157	19.89	28.55	568
10	Isopentane	Superheated vapor	136.0	13.87	167.50	−0.316	19.89	125.13	2489
11	Isopentane	Superheated vapor	85.2	1.30	103.50	−0.241	19.89	39.75	791
12	Isopentane	Liquid	26.9	1.14	−345.72	−1.676	21.92	0.80	18
13	Isopentane	Liquid	27.2	6.97	−344.56	−1.675	21.92	1.77	39
14	Isopentane	Liquid	93.7	6.97	−180.35	−1.183	21.92	25.15	551
15	Isopentane	Superheated vapor	98.7	6.97	108.48	−0.405	21.92	91.64	2009
16	Isopentane	Superheated vapor	64.6	1.14	64.50	−0.338	21.92	28.60	627
17	Air	Gas	12.8	0.84	286.29	5.703	529.87	0	0
18	Air	Gas	29.2	0.84	302.80	5.759	529.87	0.46	242
19	Air	Gas	12.8	0.84	286.29	5.703	666.53	0	0
20	Air	Gas	26.2	0.84	299.78	5.749	666.53	0.31	167

can be written as

$$\dot{Ex} = \dot{m}(ex) \tag{10.92}$$

Data for the geothermal fluid, the working fluid and air, including temperature, pressure, mass flow rate, specific exergy and exergy rate, are given In Table 10.7 following the state numbers specified in Fig. 10.40. States 0 and 0′ refer to the restricted dead states for the geothermal and working fluids, respectively. They correspond to an environment temperature of 12.8°C and an atmospheric pressure of 84 kPa, which are the values when the plant data were obtained. Properties of water are used for the geothermal fluid, so the effects of salts and non-condensable gases that might present in the geothermal brine are neglected. This simplification does not introduce significant errors in calculations since the fractions of salts and non-condensable gases are estimated by the plant managers to be small. Properties for the working fluid, isopentane, are obtained from thermodynamic property evaluation software (Friend, 1992).

The preheaters, vaporizers and condensers in the plant are essentially heat exchangers designed to perform different tasks. The exergy efficiency of a heat exchanger may be measured as the exergy increase of the cold stream divided by the exergy decrease of the hot stream (Wark, 1995). Applying this definition to the Level I vaporizer, we obtain

$$\psi_{vap\,I} = \frac{\dot{Ex}_{10} - \dot{Ex}_9}{\dot{Ex}_1 - \dot{Ex}_2} \tag{10.93}$$

where the exergy rates are given in Table 10.7. The difference between the numerator and denominator in Eq. (10.93) is the exergy destruction rate in the heat exchanger. That is,

$$\dot{I}_{vap\,I} = (\dot{Ex}_1 - \dot{Ex}_2) - (\dot{Ex}_{10} - \dot{Ex}_9) \tag{10.94}$$

Because of the complicated nature of the entire heat exchange system, the exergy efficiency and exergy destruction for the Level I vaporizer–preheater system are considered:

$$\psi_{vap-pre\,I} = \frac{\dot{Ex}_{10} - \dot{Ex}_8}{(\dot{Ex}_1 - \dot{Ex}_2) + (\dot{Ex}_3 - \dot{Ex}_4)} \tag{10.95}$$

$$\dot{I}_{vap-pre\,I} = (\dot{Ex}_1 + \dot{Ex}_3 + \dot{Ex}_8) - (\dot{Ex}_2 + \dot{Ex}_4 + \dot{Ex}_{10}) \tag{10.96}$$

The exergy efficiency of the condenser is calculated similarly. However, the exergy destruction in the condenser is approximated as the exergy decrease in exergy of isopentane across the condenser. That is, the exergy increase of the air, which is small, is neglected.

The exergy efficiency of a turbine measures of how efficiently the flow exergy of the fluid passing through it is converted to work, and can be expressed for the Level I turbine as

$$\psi_{turb\,I} = \frac{\dot{W}_{turb\,I}}{\dot{Ex}_{10} - \dot{Ex}_{11}} \tag{10.97}$$

The difference between the numerator and denominator in Eq. (10.97) is the exergy destruction rate in the turbine:

$$\dot{I}_{turb\,I} = (\dot{Ex}_{10} - \dot{Ex}_{11}) - \dot{W}_{turb\,I} \tag{10.98}$$

Analogously, the exergy efficiency and exergy destruction rate for the Level I pump can be written as

$$\psi_{pump\,I} = \frac{\dot{Ex}_8 - \dot{Ex}_7}{\dot{W}_{pump\,I}} \tag{10.99}$$

$$\dot{I}_{pump\,I} = \dot{W}_{pump\,I} - (\dot{Ex}_8 - \dot{Ex}_7) \tag{10.100}$$

Data on heat and pressure losses in pipes and valves are not available and are therefore neglected, but their effects are minor.

The exergy efficiency of the Level I isopentane cycle can be determined as

$$\psi_{\text{level I}} = \frac{\dot{W}_{\text{net I}}}{(\dot{E}x_1 - \dot{E}x_2) + (\dot{E}x_3 - \dot{E}x_4)} \tag{10.101}$$

where the denominator represents the decrease in brine exergy flow rate across the Level I vaporizer–preheater (i.e., exergy input rate to Level I). The net power of Level I is the difference between the turbine power output and the pump power input. The total exergy loss rate for the Level I cycle is approximately the exergy destruction rate, expressible as:

$$\dot{I}_{\text{level I}} = \dot{I}_{\text{pump I}} + \dot{I}_{\text{vap I}} + \dot{I}_{\text{pre I}} + \dot{I}_{\text{turb I}} + \dot{I}_{\text{cond I}} \tag{10.102}$$

The exergy efficiency of the binary geothermal power plant based on the total brine exergy flow rate decreases across the vaporizer–preheater systems of the Level I and II cycles (i.e., total exergy input rates to the Level I and II cycles) can be expressed as

$$\psi_{\text{plant,a}} = \frac{\dot{W}_{\text{net plant}}}{[(\dot{E}x_1 - \dot{E}x_2) + (\dot{E}x_3 - \dot{E}x_4)] + [(\dot{E}x_2 - \dot{E}x_3 - \dot{E}x_5) + (\dot{E}x_5 - \dot{E}x_6)]} \tag{10.103}$$

where the numerator represents the net power output from the plant, obtained by subtracting the total parasitic power, 390 kW, from the total net power output from the Level I and II cycles, 2159 kW.

The exergy efficiency of the plant can alternatively be calculated based on the brine exergy input rate to the plant (i.e. exergy rate of the brine at the Level I vaporizer inlet). That is,

$$\psi_{\text{plant,b}} = \frac{\dot{W}_{\text{net plant}}}{\dot{E}x_1} \tag{10.104}$$

When using Eq. (10.104), the exergy input rate to the plant is sometimes taken as the exergy of the geothermal fluid in the reservoir. Those who prefer this approach argue that realistic and meaningful comparisons between geothermal power plants require that the methods of harvesting the geothermal fluid be considered. Others argue that taking the reservoir as the input is not proper for geothermal power plants since conventional power plants are evaluated on the basis of the exergy of the fuel burned at the plant site (DiPippo and Marcille, 1984).

The total exergy destruction rate in the plant can be written as the difference between the brine exergy flow rate at the vaporizer inlet and the net power outputs from the Level I and Level II cycles:

$$\dot{I}_{\text{plant}} = \dot{E}x_1 - (\dot{W}_{\text{net I}} + \dot{W}_{\text{net II}}) \tag{10.105}$$

This expression accounts for the exergy losses in plant components and the exergy of the brine exiting the Level I and Level II preheaters. Here, the used brine is considered lost since it is reinjected back into the ground without attempting to use it. Some argue that the exergy of used brine is 'recovered' by returning it to the reservoir and so should not be counted as part of the exergy loss.

The exergy efficiencies and exergy destruction rates of the major plant components and the overall plant are listed in Table 10.8 for one representative unit. To pinpoint the main sites of exergy destruction and better compare the losses, an exergy flow diagram is given in Fig. 10.45.

Energy analysis

For comparison, selected energy data are provided in Table 10.8, including heat-transfer rates for vaporizers, preheaters and condensers and work rates for turbines, pumps, the Level I and II cycles, and the overall plant. Also, isentropic

Table 10.8. Selected exergy and energy data for a representative unit of the plant.[a]

Component	Exergy destruction rate (kW)	Exergy efficiency (%)	Heat transfer or work rate (kW)	Isentropic or energy efficiency (%)[b]
Vaporizer I	229.5	89.3	6703	–
Preheater I	149.9	76.9	3264	–
Vaporizer II	217.6	87.0	6331	–
Preheater II	189.5	73.0	3599	–
Preheater–vaporizer I	379.4	86.5	9967	–
Preheater–vaporizer II	407.2	82.9	9930	–
Condenser I	764	31.6	8748	–
Condenser II	610	27.4	8992	–
Turbine I	427.2	74.9	1271	70.8
Turbine II	416.9	69.8	965	66.6
Pump I	10.3	80.2	52	80.0
Pump II	4.4	82.9	25	80.0
Level I cycle	1339	43.5	1219	12.2
Level II cycle	1271	39.5	940	9.5
Level I–II cycle	2610	41.7	2159	10.9
Overall plant[c]	2610	34.2	1769	8.9
Overall plant[d]	2610	29.1	1769	5.8

[a] I and II denote that the component belongs to Level I or II, respectively.
[b] Values for turbines and pumps are isentropic efficiencies and for the Level I and II cycles and the overall plant are energy efficiencies.
[c] Based on the exergy (or energy) input to isopentane cycles.
[d] Based on the exergy (or energy) input to the plant.

efficiencies of the turbines and pumps and energy efficiencies of the Level I and II cycles and the overall plant are given. The energy efficiency of the Level I cycle is calculated as the ratio of the net power output from the Level I cycle to the heat input rate to the Level I cycle (i.e., the total heat-transfer rate in the Level I vaporizer–preheater). The energy efficiency of the plant based on the energy input rate to the plant is expressed here as

$$\eta_{plant,a} = \frac{\dot{W}_{net\ plant}}{\dot{m}_1(h_1 - h_2) + \dot{m}_3(h_3 - h_4) + \dot{m}_2(h_2 - h_5) + \dot{m}_5(h_5 - h_6)} \qquad (10.106)$$

where the terms in the denominator represent the heat-transfer rates in vaporizer I, preheater I, vaporizer II and preheater II, respectively. An alternative plant energy efficiency is

$$\eta_{plant,b} = \frac{\dot{W}_{net\ plant}}{\dot{m}_1(h_1 - h_0)} \qquad (10.107)$$

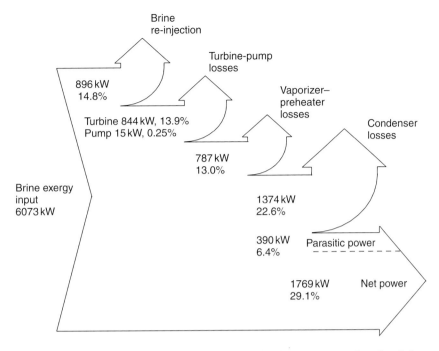

Fig. 10.45. Exergy flow diagram for the binary geothermal power plant. Percentages are based on brine exergy input.

where h_0 is the dead state specific enthalpy of the brine specified in Table 10.7. Here, the denominator represents the energy rate of the brine at the Level I vaporizer inlet.

An energy flow diagram is given in Fig. 10.46 to provide a comparison to the exergy flow diagram.

Discussion

The exergy flow diagram given in Fig. 10.45 shows that 64.5% of the exergy entering the plant is lost. The remaining 35.5% is converted to power, 18.1% of which is used for parasitic loads in the plant. The exergy efficiency of the plant is 34.2% based on the exergy input to the isopentane Rankine cycles (i.e., the exergy decreases in the brine in the vaporizer and preheater) and 29.1% based on the exergy input to the plant (i.e., the brine exergy at the Level I vaporizer inlet) (Table 10.8).

Bodvarsson and Eggers (1972) report exergy efficiencies of single- and double-flash cycles to be 38.7% and 49.0%, respectively, based on a 250°C resource water temperature and a 40°C sink temperature. Both values are significantly greater than the exergy efficiency calculated for the binary plant analyzed here. This is expected since additional exergy destruction occurs during heat exchange between the geothermal and working fluids in binary plants. DiPippo and Marcille (1984) calculate the exergy efficiency of an actual binary power plant using a 140°C resource and a 10°C sink to be 20% and 33.5% based on the exergy input to the plant and to the Rankine cycle, respectively. Kanoglu and Cengel (1999) report exergy efficiencies of 22.6% and 34.8% based on the exergy input to the plant and to the Rankine cycle, respectively, for a binary geothermal power plant with a 158°C resource and 3°C sink.

Because they use low-temperature resources, geothermal power plants generally have low energy efficiencies. Here, the plant energy efficiency is 5.8% based on the energy input to the plant and 8.9% based on energy input to the isopentane Rankine cycles. This means that more than 90% of the energy of the brine is discarded as waste.

The results suggest that geothermal resources are best used for direct heating applications instead of power generation when economically feasible. For power generation systems where used brine is reinjected back to the ground at a relatively high temperature, cogeneration in conjunction with district heating may be advantageous. The energy flow diagram in Fig. 10.46 shows that 35.2% of the brine energy is reinjected, 57.8% is rejected in the condenser, and the remainder

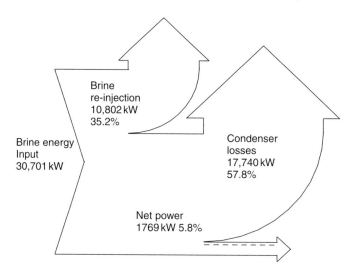

Fig. 10.46. Energy flow diagram for the binary geothermal power plant. Percentages are based on brine energy input.

is converted to power. These data provide little information on how the performance can be improved, highlighting the value of exergy analysis.

The primary exergy losses in the plant are associated with vaporizer–preheater losses, turbine-pump losses, brine reinjection and condenser losses, which represent 13.0%, 13.9%, 14.8% and 22.6% of the brine exergy input, respectively (Fig. 10.45). The exergy efficiencies of the Level I and II vaporizer–preheaters are 87% and 83%, respectively. These values are high, indicating efficient heat exchange operations. In binary geothermal power plants, heat exchangers are important components and their individual performances affect considerably overall plant performance. The exergy efficiency of the vaporizer is significantly greater than that of the preheater, mainly because the average temperature difference between the brine and the working fluid is smaller in the vaporizer than in the preheater.

The exergy efficiencies of the turbines in Levels I and II are 75% and 70%, respectively. These efficiencies indicate that the performance of the turbines can be somewhat improved. This observation is confirmed by the relatively low turbine isentropic efficiencies (in the range of 65–70%) listed in Table 10.8. That a reasonable margin for improvement exists can be seen by considering a recently built binary geothermal power plant, which has a turbine with an exergy efficiency of over 80% (Kanoglu et al., 1998). The pumps seem to be performing efficiently.

The exergy efficiencies of the condensers are in the range of 30%, making them the least efficient components in the plant. This is primarily due to the high average temperature difference between the isopentane and the cooling air. The brine is reinjected back to the ground at about 65°C. In at least one binary plant using a resource at about 160°C, the brine is reinjected at temperatures above 90°C (Kanoglu et al., 1998). Compared to this, the percent exergy loss associated with the brine reinjection is low in this plant. It is noted that condenser efficiencies are often difficult to define and interpret since the objective of a condenser is to reject heat rather than create a product.

For binary geothermal power plants using air as the cooling medium, the condenser temperature varies with the ambient air temperature, which fluctuates throughout the year and even through the day. As a result, power output decreases by up to 50% from winter to summer (Kanoglu and Cengel, 1999). Consequently, the exergy destruction rates and percentages vary temporally as well as spatially, this effect being most noticeable in the condenser.

10.5. Closing remarks

Exergy analysis is usually used to determine exergy efficiencies and identify and quantify exergy destructions so that directions for improved efficiency can be determined. These aims have been illustrated by the exergy analyses in this chapter of renewable energy systems including solar PV systems, solar ponds, wind turbines and geothermal district heating systems and power plants.

Problems

10.1 How are the energy and exergy efficiencies of solar photovoltaic systems defined?

10.2 Identify the sources of exergy loss in solar photovoltaic systems and propose methods for reducing or minimizing them.

10.3 Explain why solar photovoltaic systems are costly even though they use solar energy, which is free.

10.4 Why are the exergy efficiencies lower than the energy efficiencies for solar photovoltaic systems? Explain.

10.5 Obtain a published article on exergy analysis of solar photovoltaic systems. Using the operating data provided in the article, perform a detailed exergy analysis of the system and compare your results to those in the original article. Also, investigate the effect of varying important operating parameters on the system exergetic performance.

10.6 What is the difference between an ordinary pond or lake and a solar pond? Are solar ponds thermal energy storage systems?

10.7 How are the energy and exergy efficiencies of solar ponds defined?

10.8 Identify the operating parameters that have the greatest effects on the exergy performance of a solar pond.

10.9 Obtain a published article on exergy analysis of solar ponds. Using the operating data provided in the article, perform a detailed exergy analysis of the system and compare your results to those in the original article. Also, investigate the effect of varying important operating parameters on the system exergetic performance.

10.10 Why are exergy efficiencies lower than energy efficiencies for solar ponds?

10.11 Investigate the development of wind turbines in the last three decades. Compare the costs and efficiencies of wind turbines that existed 20 years ago to those currently being installed.

10.12 Do you agree with the statement 'the energy and exergy efficiencies of wind turbines are identical'? Explain.

10.13 What is the value of exergy analysis in assessing and designing wind turbines? What additional information can be obtained using an exergy analysis compared to an energy analysis? How can you use exergy results to improve the efficiency of wind turbines?

10.14 What is the effect of wind-chill temperature on the power generation and efficiency of a wind turbine?

10.15 How are the energy and exergy efficiencies of wind turbines defined? What is the difference between them? Which one do you expect to be greater?

10.16 What is the difference between energy-based and exergy-based spatio-temporal wind maps?

10.17 Which use of a geothermal resource at 150°C is better from an energetic and exergetic point of view: (a) district heating or (b) power generation? Explain. What would your answer be if the geothermal resource is at 90°C?

10.18 How can a geothermal resource at 150°C be used for a cooling application? How can you express the exergy efficiency of such a cooling system?

10.19 How are the energy and exergy efficiencies of geothermal district heating systems defined? Which definition is more valuable to you if you are a customer of geothermal district heat? Which definition is more valuable to you if you are an engineer trying to improve the performance of the district system?

10.20 Identify the main causes of exergy destruction in a geothermal district heating system and propose methods for reducing or minimizing them.

10.21 Geothermal resources can be classified based on the resource temperature or the resource exergy. Which classification is more suitable if geothermal energy is to be used for (a) district heating, (b) cooling and (c) power generation? Explain.

10.22 How do you explain the difference between the energy and exergy efficiencies of the geothermal district heating system considered in this chapter?

10.23 What thermodynamic cycles are used for geothermal power generation? Discuss the suitability of each cycle based on the characteristics of the geothermal resource.

10.24 Define the energy and exergy efficiencies for various geothermal power cycles. How can you express the energy and exergy of a reservoir?

10.25 Compare the energy and exergy efficiencies of a geothermal power plant. Which one is greater?

10.26 Do you support using geothermal resources below 150°C for power generation? If such an application is to occur, what is the most appropriate cycle? Explain.

10.27 Identify the main causes of exergy destruction in geothermal power plants and propose methods for reducing or minimizing them.

10.28 In a geothermal power plant using a 160°C resource, geothermal water is reinjected into the ground at about 90°C. What is the ratio of the exergy of the brine reinjected to the exergy of brine in the reservoir? How can you utilize this brine further before its reinjection?

10.29 What is the effect of ambient air temperature on the exergetic performance of a binary geothermal power plant?

10.30 Is there an optimum heat exchanger pressure that maximizes the power production in a binary geothermal power plant? Conduct an analysis to determine the optimum pressure, if one exists, for the power plant considered in this chapter.

10.31 Obtain a published article on exergy analysis of a geothermal power plant. Using the operating data provided in the article, perform a detailed exergy analysis of the plant and compare your results to those in the original article. Also, investigate the effect of varying important operating parameters on the system exergetic performance.

Chapter 11

EXERGY ANALYSIS OF STEAM POWER PLANTS

11.1. Introduction

Steam power plants are widely utilized throughout the world for electricity generation, and coal is often used to fuel these plants. Although the world's existing coal reserves are sufficient for about two centuries, the technology largely used today to produce electricity from coal causes significant negative environmental impacts. To utilize coal more effectively, efficiently and cleanly in electricity generation processes, efforts are often expended to improve the efficiency and performance of existing plants through modifications and retrofits, and to develop advanced coal utilization technologies.

Today, many electrical generating utilities are striving to improve the efficiency (or heat rate) at their existing thermal electric generating stations, many of which are over 25 years old. Often, a heat rate improvement of only a few percent appears desirable as it is thought that the costs and complexity of such measures may be more manageable than more expensive options.

To assist in improving the efficiencies of coal-to-electricity technologies, their thermodynamic performances are usually investigated. In general, energy technologies are normally examined using energy analysis. A better understanding is attained when a more complete thermodynamic view is taken, which uses the second law of thermodynamics in conjunction with energy analysis, via exergy methods.

Of the analysis techniques available, exergy analysis is perhaps the most important because it is a useful, convenient and straightforward method for assessing and improving thermal generating stations. The insights gained with exergy analysis into plant performance are informative (e.g., efficiencies are determined which measure the approach to ideality, and the causes and locations of losses in efficiency and electricity generation potential are accurately pinpointed). Exergy-analysis results can aid efforts to improve the efficiency, and possibly the economic and environmental performance, of thermal generating stations. Improvement, design and optimization efforts are likely to be more rational and comprehensive if exergy factors are considered. One reason is that exergy methods can prioritize the parts of a plant in terms of greatest margin for improvement – by focusing on plant components responsible for the largest exergy losses. For example, the authors previously showed that efficiency-improvement efforts for coal-fired electrical generation should focus on the steam generator (where large losses occur from combustion and heat transfer across large temperature differences), the turbines, the electrical generator and the transformer. In addition, however, other components should be considered where economically beneficial improvements can be identified even if they are small.

In most countries, numerous steam power plants driven by fossil fuels like oil, coal and natural gas or by other energy resources like uranium are in service today. During the past decade, many power generation companies have paid attention to process improvement in steam power plants by taking measures to improve the plant efficiencies and to minimize the environmental impact (e.g., by reducing the emissions of major air pollutants such as CO_2, SO_2 and NO_x). Exergy analysis is a useful tool in such efforts.

In this chapter, energy and exergy analyses are utilized to examine and better understand the performance of steam power plants, and to identify and evaluate possible process modifications to improve the plant efficiencies. Some alternative process configurations are then proposed. Exergy is useful for providing a detailed breakdown of the losses, in terms of waste exergy emissions and irreversibilities, for the overall plants and their components. Some illustrative examples are presented to demonstrate the importance of exergy in performance improvement of steam power plants.

11.2. Analysis

The Rankine cycle (Fig. 11.1) is used in a variety of power plants. A simple Rankine cycle consists of four main components (steam generator, turbine, condenser and pump). Additional components are usually added to enhance cycle performance and to improve efficiency.

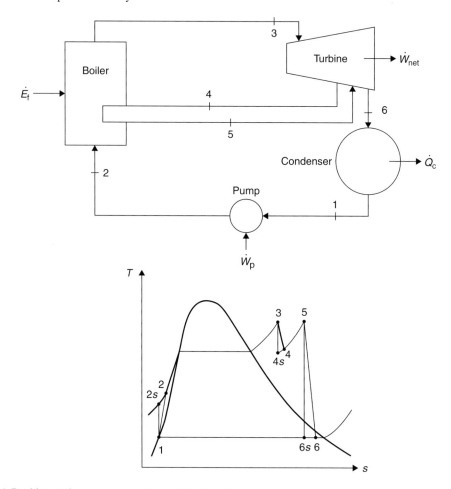

Fig. 11.1. A Rankine cycle steam power plant and its T–s diagram.

The Rankine cycle used in actual steam power plants is generally more complex, and is considered later in this chapter.

11.2.1. Balances

For each component, balances for mass, energy and exergy can be applied to find energy terms such as work output and heat addition, exergy flows and irreversibilities, and energy and exergy efficiencies. Several balances, based on energy and exergy, for the system components are given below, along with energy and exergy efficiency expressions for the overall plant.

For a steady-state process, respective balances for mass, energy and exergy can be written as follows:

$$\sum_i \dot{m}_i = \sum_e \dot{m}_e \tag{11.1}$$

$$\sum_i \dot{E}_i + \dot{Q} = \sum_e \dot{E}_e + \dot{W} \tag{11.2}$$

$$\sum_i \dot{E}x_i + \sum_j \left[1 - \frac{T_o}{T_j}\right]\dot{Q}_j = \sum_e \dot{E}x_e + \dot{W} + \dot{I} \qquad (11.3)$$

Neglecting potential and kinetic energy, Eq. (11.2) can be written as

$$\sum_i \dot{m}_i h_i + \dot{Q} = \sum_e \dot{m}_e h_e + \dot{W} \qquad (11.4)$$

and the exergy flow rate of a flowing stream of matter as

$$\dot{E}x = \dot{m}(ex) = \dot{m}(ex^{\mathrm{tm}} + ex^{\mathrm{ch}}) \qquad (11.5)$$

where the specific thermomechanical exergy ex^{tm} is expressed as

$$ex^{\mathrm{tm}} = h - h_o - T_o(s - s_o) \qquad (11.6)$$

The irreversibility rates for the plant components are assessed using a rearranged form of the exergy balance in Eq. (11.3):

$$\dot{I} = \sum_i \dot{E}x_i - \sum_e \dot{E}x_e - \dot{W} + \sum_j \left[1 - \frac{T_o}{T_j}\right]\dot{Q}_j \qquad (11.7)$$

Note that for all components except the turbines and pumps, $\dot{W}$ is equal to zero.

11.2.2. Overall efficiencies

The overall plant efficiency can be expressed as

$$\eta_{\mathrm{plant}} = \frac{\dot{W}_{\mathrm{net}}}{\dot{E}_{\mathrm{f}}} \qquad (11.8)$$

where the net power output $\dot{W}_{\mathrm{net}}$ can be written as

$$\dot{W}_{\mathrm{net}} = \dot{W}_{\mathrm{tur}}\,\eta_{\mathrm{mech}}\,\eta_{\mathrm{trans}}\,\eta_{\mathrm{gen}} - \dot{W}_{\mathrm{pump}} - \dot{W}_{\mathrm{ser}} \qquad (11.9)$$

Here, $\dot{W}_{\mathrm{tur}}$ denotes the gross power generated by the turbines, $\dot{W}_{\mathrm{pump}}$ the power used by the hot well pump and $\dot{W}_{\mathrm{ser}}$ the power for station services. Also η_{mech}, η_{trans} and η_{gen} denote the mechanical, transformer and generator efficiencies, respectively.

The plant exergy efficiency is expressible as

$$\psi_{\mathrm{plant}} = \frac{\dot{W}_{\mathrm{net}}}{\dot{E}x_{\mathrm{f}}} \qquad (11.10)$$

11.2.3. Material energy and exergy values

The energy and exergy flow rates for materials such as solids, liquids and gases within the plant can be evaluated as follows:

(i) *Solid flows*: The chemical energy for the fuel can be written as

$$\dot{E}_{\mathrm{f}} = \dot{n}_{\mathrm{f}}\overline{\mathrm{HHV}} \qquad (11.11)$$

Using Eqs. (11.5) and (11.6) and noting that the thermomechanical exergy of coal is zero at its assumed input conditions of T_o and P_o, the fuel exergy can be written as

$$\dot{E}x_{\mathrm{f}} = \dot{n}_{\mathrm{f}}\overline{ex}^{\mathrm{ch}} \qquad (11.12)$$

(ii) *Liquid and vapor flows*: All liquid flows in the plant contain H_2O. The energy flow rate of an H_2O flow can be written as

$$\dot{E} = \dot{m}(h - h_o) \qquad (11.13)$$

Using Eqs. (11.5) and (11.6) and noting that the chemical exergy of H_2O is zero, the exergy flow rate of an H_2O flow can be written as

$$\dot{Ex} = \dot{m}[h - h_o - T_o(s - s_o)] \tag{11.14}$$

(iii) *Gas flows*: The energy flow rate of a gas flow can be written as the sum of the energy flow rates for its constituents:

$$\dot{E} = \sum_i \dot{n}_i[\bar{h} - \bar{h}_o]_i \tag{11.15}$$

The exergy flow rate of a gas flow can be written with Eqs. (11.5) and (11.6) as

$$\dot{Ex}_e = \sum_i \dot{n}_i[\bar{h} - \bar{h}_o - T_o(\bar{s} - \bar{s}_o) + \overline{ex}^{ch}]_i \tag{11.16}$$

For example, modeling the coal used in a coal-fired power plant as carbon (C) and assuming complete combustion with excess air, the combustion reaction can be expressed as follows:

$$C + (1 + \lambda)(O_2 + 3.76\ N_2) \rightarrow CO_2 + \lambda O_2 + 3.76(1 + \lambda)\ N_2 \tag{11.17}$$

where λ denotes the fraction of excess combustion air. The air–fuel (AF) ratio can be written as

$$AF = \frac{\dot{m}_a}{\dot{m}_f} = \frac{\dot{n}_a M_a}{\dot{n}_f M_f} = \frac{(1 + \lambda)4.76 M_a}{M_f} \tag{11.18}$$

The energy and exergy flow rate of combustion air can be written in terms of the mole flow rate of fuel $\dot{n}_f$ using Eqs. (11.15) through (11.17) and noting that the chemical exergy of air is zero, as

$$\dot{E}_a = \dot{n}_f(1 + \lambda)[(\bar{h} - \bar{h}_o)_{O_2} + 3.76(\bar{h} - \bar{h}_o)_{N_2}] \tag{11.19}$$

and

$$\dot{Ex}_a = \dot{n}_f(1 + \lambda)\{[\bar{h} - \bar{h}_o - T_o(\bar{s} - \bar{s}_o)]_{O_2} + 3.76[\bar{h} - \bar{h}_o - T_o(\bar{s} - \bar{s}_o)]_{N_2}\} \tag{11.20}$$

It is useful to determine the hypothetical temperature of combustion gas in the steam generator without any heat transfer (i.e., the adiabatic combustion temperature) to facilitate the evaluation of its energy and exergy flows and the breakdown of the steam-generator irreversibility into portions related to combustion and heat transfer (see Table 11.1). The energy and exergy flow rates of the products of combustion can be written using Eqs. (11.15)–(11.17) as

$$\dot{E}_p = \dot{n}_f[(\bar{h} - \bar{h}_o)_{CO_2} + \lambda(\bar{h} - \bar{h}_o)_{O_2} + 3.76(1 + \lambda)(\bar{h} - \bar{h}_o)_{N_2}] \tag{11.21}$$

and

$$\dot{Ex}_p = \dot{n}_f\{[\bar{h} - \bar{h}_o - T_o(\bar{s} - \bar{s}_o) + \bar{\varepsilon}^{ch}]_{CO_2} + \lambda[\bar{h} - \bar{h}_o - T_o(\bar{s} - \bar{s}_o) + \bar{\varepsilon}^{ch}]_{O_2}$$
$$+ 3.76(1 + \lambda)[\bar{h} - \bar{h}_o - T_o(\bar{s} - \bar{s}_o) + \bar{\varepsilon}^{ch}]_{N_2}\} \tag{11.22}$$

The adiabatic combustion temperature is determined using the energy balance in Eq. (11.2) with $\dot{Q} = 0$ and $\dot{W} = 0$:

$$\dot{E}_f + \dot{E}_a = \dot{E}_p \tag{11.23}$$

Substituting Eqs. (11.11), (11.19) and (11.21) into (11.23) and simplifying yields

$$\overline{HHV} + (1 + \lambda)[(\bar{h} - \bar{h}_o)_{O_2} + 3.76(\bar{h} - \bar{h}_o)_{N_2}]_{T_a} = [(\bar{h} - \bar{h}_o)_{CO_2} + \lambda(\bar{h} - \bar{h}_o)_{O_2} + 3.76(1 + \lambda)(\bar{h} - \bar{h}_o)_{N_2}]_{T_p} \tag{11.24}$$

Here T_p can be evaluated using an iterative solution technique. Note that the flow rates of energy $\dot{E}_g$ and exergy $\dot{Ex}_g$ for the stack gas can then be evaluated using Eqs. (11.21) and (11.22) because the composition of stack gas is same as that of the product gas.

Table 11.1. Main flow data for the sample coal-fired steam power plant in Section 11.4.

Section	Mass flow rate (kg/s)	Temperature (°C)	Pressure (MPa)	Vapor fraction*
Steam generator				
Feedwater in	423.19	253.28	18.62	0
Main steam out	423.19	537.78	16.31	1
Reheat steam in	376.75	343.39	4.29	1
Reheat steam out	376.75	537.78	3.99	1
Steam turbine				
Inlet flows, by turbine section:				
High-pressure turbine	421.18	537.78	15.42	1
Intermediate-pressure turbine	376.65	537.78	3.99	1
Low-pressure turbines	314.30	251.56	0.46	1
Extraction steam flows, by destination:				
Feedwater heater no. 1 (closed)	17.08	69.89	0.03	1
Feedwater heater no. 2 (closed)	11.25	89.67	0.07	1
Feedwater heater no. 3 (closed)	17.40	166.00	0.19	1
Feedwater heater no. 4 (closed)	17.18	251.56	0.44	1
Feedwater heater no. 5 (open)	17.28	329.56	0.87	1
Feedwater heater no. 6 (closed)	18.13	419.56	1.72	1
Feedwater heater no. 7 (closed)	43.07	343.39	4.22	1
Condenser				
Condensate out	281.62	31.96	0.01	0
Cooling water in	5899.55	8.00	0.10	0
Cooling water out	5899.55	33.78	0.10	0

*A vapor fraction of 0 denotes a saturated or subcooled liquid, and of 1 denotes a dry saturated or superheated vapor.

An energy balance for the steam generator can be written as

$$\dot{E}_f - \dot{E}_g = \dot{m}_s(h_s - h_{feed}) + \dot{m}_{re}(h_{re,e} - h_{re,i}) \tag{11.25}$$

In an analysis, the flow rate of fuel (mole or mass) can be determined for a fixed steam generator output by substituting Eqs. (11.11) and (11.21) into Eq. (11.25). Then the energy and exergy flow rates of the fuel can be evaluated with Eqs. (11.11) and (11.12), of the gaseous streams (i.e., preheated combustion air, combustion gas and stack gas) with Eqs. (11.19)–(11.22), and of the H_2O streams with Eqs. (11.11) and (11.14). The irreversibilities for plant components can be assessed using Eq. (11.7), and the plant energy and exergy efficiencies evaluated using Eqs. (11.8) and (11.10), respectively.

11.3. Spreadsheet calculation approaches

The complexity of power generating units make it difficult to determine properties and quantities simply and accurately. The effort required for thermodynamic calculation during design and optimization has grown tremendously. To be competitive and to reduce planning and design time as well as errors, companies apply computer-aided methods. Such methods also facilitate optimization, which also can be time consuming.

The ability to quickly evaluate the impact of changes in system parameters is crucial to safe and efficient operation. Spreadsheet calculation schemes provide inexpensive methods for energy and exergy calculations in steam power plants and allow for visualization of operation.

Using thermodynamic water tables in Sonntag et al. (2006), the enthalpy and entropy values for each state point in a plant are calculated, and these data are entered in an EXCEL worksheet, which can be used to calculate the energy and exergy efficiencies of the cycle. A sample calculation procedure is summarized below:

- Given:

$$P_1 = 12\,\text{MPa}, \qquad\qquad T_1 = 500°\text{C}$$
$$P_2 = P_3 = 2.4\,\text{MPa}, \qquad T_3 = 500°\text{C}$$
$$P_5 = P_6 = P_{\text{sat}} = 7\,\text{kPa}, \qquad T_5 = T_{\text{sat}} = 39°\text{C}$$
$$P_4 = P_7 = P_8 = 0.15\,\text{MPa}$$

Flue gas inlet conditions: $P = 102\,\text{kPa}, T = 1500°\text{C}$
Flue gas outlet conditions: $P = 101\,\text{kPa}, T = 400°\text{C}$
- Find: w, q, η, ψ, X
- Assumptions and data determination:

$$\dot{m}_\text{w} = 1\,\text{kg/s}$$
$$T_0 = 25°\text{C} = 298.15\,\text{K}$$

Initial state of the flue gas is the reference state.
State properties are:

$$\begin{array}{ll}
h_1 = 3348\,\text{kJ/kg} & s_1 = 6.487\,\text{kJ/kg K} \\
h_2 = 2957\,\text{kJ/kg} & s_2 = 6.569\,\text{kJ/kg K} \\
h_3 = 3463\,\text{kJ/kg} & s_3 = 7.343\,\text{kJ/kg K} \\
h_4 = 2813\,\text{kJ/kg} & s_4 = 7.513\,\text{kJ/kg K} \\
h_5 = 2399\,\text{kJ/kg} & s_5 = 7.720\,\text{kJ/kg K} \\
h_6 = 163.4\,\text{kJ/kg} & s_6 = 0.5591\,\text{kJ/kg K} \\
h_7 = 163.5\,\text{kJ/kg} & s_7 = 0.5592\,\text{kJ/kg K} \\
h_8 = 462.8\,\text{kJ/kg} & s_8 = 1.4225\,\text{kJ/kg K} \\
h_9 = 474.3\,\text{kJ/kg} & s_9 = 1.4254\,\text{kJ/kg K}
\end{array}$$

- Calculation:

$$X = (h_8 - h_7)/(h_4 - h_7) = (462.8 - 163.5)/(2813 - 163.5) = 0.1130$$

$$w = w_\text{HPT} + w_\text{LPT}$$
$$= (h_1 - h_2) + [h_3 - Xh_4 - (1 - X)h_5]$$
$$= (3348 - 2957) + [3463 - 0.113 \times 2813 - (1 - 0.113) \times 2399] = 1408\,\text{kJ/kg}$$

$$q = (h_1 - h_9) + (h_3 - h_2) = (3348 - 474.3) + (3463 - 2957) = 3380\,\text{kJ/kg}$$
$$\eta = w/q = 1408/3380 = 0.4166$$

$$\dot{m}_\text{gas} = q/[C_\text{P}(T_{\text{g,o}} - T_{\text{g,i}})] = 3380/[1.1058(1500 - 400)] = 2.779\,\text{kg/s}$$
$$ex_\text{g,i} = h_\text{g,i} - T_0 \times s_\text{g,i} = 0.0 - 298.15 \times 0.0 = 0\,\text{kJ/kg}$$

$$ex_\text{g,o} = h_\text{g,o} - T_0 s_\text{g,o} = C_\text{P}[(T_{\text{g,o}} - T_{\text{g,i}}) - T_0 \times (\ln(T_{\text{g,o}}/T_{\text{g,i}}) - (k - 1)/k \times \ln(P_{\text{g,o}}/P_{\text{g,i}}))]$$
$$= 1.1058[400 - 1500 - 298.15 \times (\ln(637.15/1773.15) - (0.35/1.35) \times \ln(101/102))]$$
$$= -897.9\,\text{kJ/kg}$$

$$\dot{Ex}_\text{d} = \dot{m}_\text{gas}(ex_\text{g,i} - ex_\text{g,o}) = 2.779(0 - (-897.9)) = 2495\,\text{kW}$$
$$\psi = \dot{m}_\text{w}w/\dot{Ex}_\text{d} = 1408/2495 = 0.5643$$

11.4. Example: analysis of a coal steam power plant

The main stream data for a typical modern coal-fired steam power plant are summarized in Table 11.1. The plant is examined using energy and exergy analyses in order to (i) improve understanding of its behavior and performance and (ii) provide a base-case reference for efficiency improvement investigations.

Several assumptions and simplifications are used in the energy and exergy analyses:

- Fuel and 40% excess combustion air are supplied at environmental temperature T_0 and pressure P_0.
- The combustion air is preheated to 267°C using regenerative air heating.
- The stack-gas exit temperature is 149°C.
- All components have adiabatic boundaries.
- The turbines have isentropic and mechanical efficiencies of 91% and 99.63%, respectively.
- The generator and transformer efficiencies are both 98.32%.
- The reference-environment model used has the following property values: temperature $T_0 = 8$°C, pressure $P_0 = 101.315$ kPa (1 atm) and a chemical composition as specified elsewhere (Rosen and Dincer, 2003a,b).

Steam properties are obtained using software based on the NBS/NRC Steam Tables. Air properties are obtained based on data from the thermodynamic tables.

A schematic of the coal-fired steam power plant is shown in Fig. 11.1, and the component irreversibility rates are given in Table 11.2. Two key points are observed.

First, with the data in Table 11.1 and Eqs. (11.8) through (11.10), the plant energy efficiency η_{plant} and exergy efficiency ψ_{plant} can be evaluated as

$$\eta_{\text{plant}} = \frac{516.79 \times 0.996 \times 0.98 \times 0.98 - 0.67 - 15.06}{1238.68} = 38.9\%$$

$$\psi_{\text{plant}} = \frac{516.79 \times 0.996 \times 0.98 \times 0.98 - 0.67 - 15.06}{1292.24} = 37.3\%$$

The small difference in efficiencies is due to the specific chemical exergy of coal being slightly greater than its specific energy as measured by its high healthy value (HHV). Energy analysis indicates that the largest waste losses for the coal-fired steam power plant are attributable to the condenser, with condenser waste-heat rejection accounting for more than 50% of energy input with coal. However, exergy analysis indicates that the waste-heat emissions from the condenser, although great in quantity, are low in quality (i.e., have little exergy) because their temperatures are near to that of the environment. Therefore, condenser improvements can typically yield only slight increases in plant exergy efficiency.

Second, it can be seen from Table 11.2 that most (approximately 84%) of the total exergy destruction occurs in the steam generator, while the remainder occurs in other devices (approximately 10% in the turbine generators, 4% in the condenser and 2% in the feedwater heaters and pumps). The steam generator is thus the most inefficient plant device, with combustion in that device accounting for nearly 44% of the plant exergy consumption and heat transfer nearly 40%. As a result, significant potential exists for improving plant efficiency by reducing steam-generator irreversibilities, by modifying the combustion and/or heat-transfer processes occurring within it.

11.5. Example: impact on power plant efficiencies of varying boiler temperature and pressure

Thermodynamic parameters are taken from the Ghazlan Power Plant in Saudi Arabia (Al-Bagawi, 1994). The boiler pressure is 12.5 MPa and temperature is 510°C. The condenser pressure is 50.8 mmHg and temperature is 38.4°C. The regenerator pressure is 132 kPa. Habib et al. (1995) indicate that for maximum efficiency of a single-reheat cycle, the reheat pressure should be approximately 19% of the boiler pressure. Balances for the plant components are given in Table 11.3. For further information on the analysis details, see Dincer and Al-Muslim (2001).

In the simulation we change the inlet temperature of the high-pressure turbine (or the outlet temperature of the boiler) between 400°C and 600°C in steps of 10°C. For each temperature, the pressure is changed from 10 to 15 MPa in steps of 1 MPa. The pressure at the inlet of the low-pressure (LP) turbine is taken to be 20% of that of the high-pressure turbine. The temperature is assumed to be equal to that of the high-pressure turbine. These two conditions have been cited as important for maximizing the efficiency of the reheat cycle. The practical value of the temperature of the condenser ranges between 35°C and 40°C. For a temperature of 39°C, the water is saturated so the saturation pressure is 7 kPa.

Table 11.2. Breakdown of exergy consumption rates for devices for the sample power plant in Section 11.4.

Device	Irreversibility rate	
	(MW)	(% of total)
Steam generator		
Combustion	305.77	43.61
Heat transfer	282.81	40.34
Total	588.58	83.95
Power production devices		
High-pressure turbine	6.05	0.86
Intermediate-pressure turbine	10.04	1.43
Low-pressure turbine	36.06	5.14
Mechanical shafts	2.07	0.30
Generator	8.75	1.25
Transformer	8.75	1.25
Total	71.72	10.23
Condenser	27.85	3.97
Total	27.85	3.97
Heaters		
Closed feedwater heaters		
Low-pressure heaters		
Heater no. 0	0.1	0.01
Heater no. 1	2.01	0.29
Heater no. 2	0.75	0.11
Heater no. 3	1.55	0.22
Heater no. 4	1.40	0.20
High-pressure heaters		
Heater no. 6	2.36	0.34
Heater no. 7	2.64	0.38
Open feedwater heater		
Heater no. 5	1.51	0.22
Total	12.32	1.76
Pumps		
Hot well pump	0.32	0.05
Boiler feed pump	0.35	0.05
Total	0.67	0.10
Plant total	701.14	100.00

Table 11.3. Balances and other equations for system components in the Ghazlan Power Plant.

Component	Energy analysis	Exergy analysis
High-pressure turbine	$w_{HPT} = h_1 - h_2$	$ex_d = (h_1 - h_2) - T_0(s_1 - s_2)$
Low-pressure turbine	$w_{LPT} = h_3 - X \times h_4 - (1 - X) \times h_5$	$ex_d = (h_3 - X \times h_4 - (1 - X) \times h_5) - T_0[(s_3 - X \times s_4 - (1 - X) \times s_5)]$
Regenerator	$X = (h_8 - h_7)/(h_4 - h_7)$	$ex_d = -T_0[X \times s_4 + (1 - X) \times s_7 - s_8]$
Condenser	$m_{cw} = (1 - X)(h_5 - h_6)/(h_{co} - h_{ci})$	$ex_d = -T_0[(1 - X)(s_5 - s_6) + m_{cw}(s_{ci} - s_{co})]$
Boiler	$m_g = [(h_1 - h_9) + (h_3 - h_2)]/C_p(T_{g2} - T_{g1})$	$ex_d = -T_0[(s_9 - s_1) + (s_2 - s_3) + m_g(s_{g1} - s_{g2})]$
Pump 1	$w_{P1} = (1 - X)(h_6 - h_7)$	$ex_d = (1 - X)[(h_6 - h_7) - T_0(s_6 - s_7)]$
Pump 2	$w_{P2} = (h_8 - h_9)$	$ex_d = (h_8 - h_9) - T_0(s_8 - s_9)$

The pressure at the regenerator is set to 0.15 MPa. The water coming from the regenerator is saturated. The isentropic efficiencies of the turbines and pumps are both assumed to be 0.9.

In Fig. 11.2a, the energy efficiency is plotted against boiler temperature for values between 400°C and 600°C. The six energy efficiency profiles shown represent six different boiler pressures (10, 11, 12, 13, 14 and 15 MPa). All energy efficiency profiles increase almost linearly with the boiler temperature and the energy efficiencies vary from 0.38 to 0.45. Figure 11.2b shows the corresponding exergy efficiency distributions for several boiler pressures against the boiler temperature. The trend of the exergy efficiency profiles appears similar to that for the energy efficiency profiles. However, the exergy efficiency values vary between 0.525 and 0.6. The exergy efficiencies are higher than the energy efficiencies at different pressures, due to the fact that high-quality (or high-temperature) energy sources such as fossil fuels are used for high-temperature applications (e.g., 400–590°C). Therefore, in practical processes, both quantity and quality are accounted for well by exergy, but not by energy. The results obtained here agree well with other reports (Habib et al., 1995, 1999).

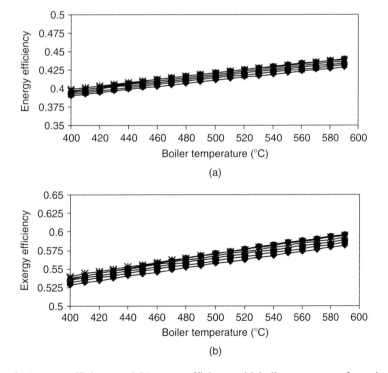

(a)

(b)

Fig. 11.2. Variation of (a) energy efficiency and (b) exergy efficiency with boiler temperature for various boiler pressures (♦10 MPa, ■11 MPa, ▲12 MPa, ×13 MPa, ∗14 MPa, ●15 MPa) for the Ghazlan Power Plant.

Energy and exergy efficiency are plotted against boiler pressure in Figs. 11.3a and 11.3b for three different temperatures (400°C, 500°C and 590°C). Although both energy and exergy efficiencies increase slightly with increasing pressure, this increase is not appreciable. The increase must be weighed against the added cost of equipment to increase the pressure. The maximum energy efficiency for the three curves occurs at a boiler pressure of 14 MPa. Therefore, this pressure can be considered as a thermodynamic optimum for such a cycle under the design conditions. Exergy efficiency profiles in Fig. 11.3b follow the same trend as energy efficiency curves in Fig. 11.3a. Moreover, the pressure of 14 MPa appears to be the optimum pressure from an exergy perspective. Because of the quality aspect, the exergy efficiencies are higher than energy efficiencies.

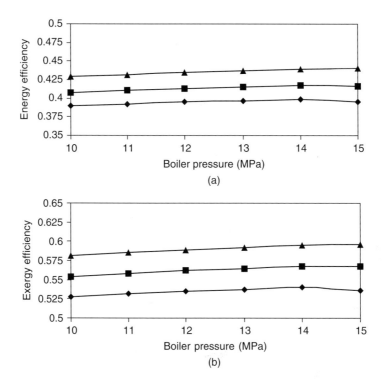

Fig. 11.3. Variation of (a) energy efficiency and (b) exergy efficiency with boiler pressure for various boiler temperatures (♦400°C, ■500°C, ▲590°C) for the Ghazlan Power Plant.

A linear relationship is observed between the heat input to the boiler and the boiler temperature at pressures ranging from 10 to 15 MPa. It is evident that increasing heat input increases boiler temperature. Although the difference between the profiles is larger at lower temperatures like 400°C, they approach each other as temperature increases and tend to match near 600°C. This means that if the system works at high temperatures (>600°C) the pressure effect becomes less dominant (Dincer and Al-Muslim, 2001).

In Figs. 11.4a and 11.4b, we compare the present energy and exergy efficiency distributions vs. the boiler temperature and pressure values with the actual data presented in Al-Bagawi (1994). The energy efficiencies are in good agreement, with differences less than 10%. For exergy efficiency, however, much greater differences appear due to the fact that the system studied by Al-Bagawi (1994) is for a three-stage Rankine cycle steam power plant with feedwater heating to increase efficiency. This modification increases the exergy efficiency significantly while the change in energy efficiency is small.

11.6. Case study: energy and exergy analyses of coal-fired and nuclear steam power plants

In this section, thermodynamic analyses and comparisons of coal-fired and nuclear electrical generating stations are performed using energy and exergy analyses. The section strives to improve understanding of the thermodynamic

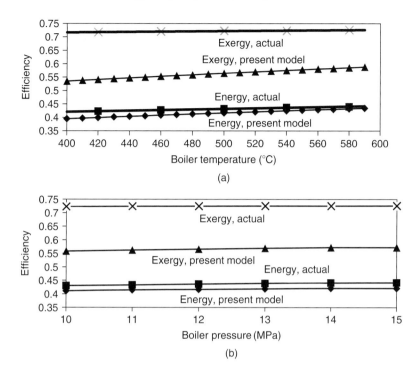

Fig. 11.4. Comparison of the energy and exergy efficiencies for the Ghazlan Power Plant with (a) boiler temperature and (b) boiler pressure, for the actual data presented by Al-Bagawi (1994) and the values calculated with the present model discussed in the text.

performance of steam power plants, and to identify areas where the potential for performance improvement is high, and trends which may aid in the design of future stations.

The coal-fired Nanticoke Generating Station (NGS) and the Pickering Nuclear Generating Station (PNGS) are selected as the representative stations on which the comparisons are based (Ontario Hydro, 1996). Both stations are located in Ontario, Canada, and are operated by the provincial electrical utility, Ontario Power Generation (formerly Ontario Hydro). Reasons these stations are selected include the following:

- The individual units in each station have similar net electrical outputs (approximately 500 MW).
- A substantial base of operating data has been obtained for them over several years (NGS has been operating since 1981, and PNGS since 1971).
- They are relatively representative of present technology.
- They operate in similar physical environments.

11.6.1. Process descriptions

Detailed flow diagrams for single units of NGS and PNGS are shown in Fig. 11.5. The symbols identifying the streams are described in Tables 11.4a and 11.4b, and the main process data in Table 11.5. Process descriptions reported previously (Ontario Hydro, 1996) for each are summarized below, in terms of the four main sections identified in the caption of Fig. 11.5. Process data are also obtained by computer simulation.

Steam generation

Heat is produced and used to generate and reheat steam. In NGS, eight pulverized-coal-fired natural circulation steam generators each produce 453.6 kg/s steam at 16.89 MPa and 538°C, and 411.3 kg/s of reheat steam at 4.00 MPa and 538°C. Air is supplied to the furnace by two 1080 kW 600 rpm motor-driven forced draft fans. Regenerative air preheaters are used. The flue gas passes through an electrostatic precipitator rated at 99.5% collection efficiency, and exits the plant through two multi-flued, 198 m high chimneys.

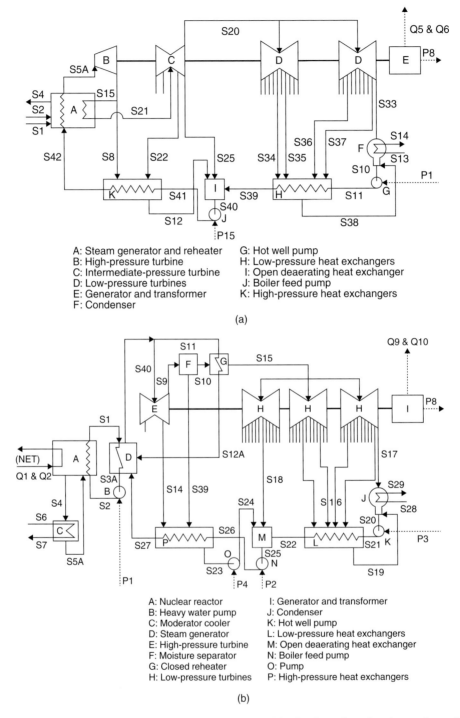

A: Steam generator and reheater G: Hot well pump
B: High-pressure turbine H: Low-pressure heat exchangers
C: Intermediate-pressure turbine I: Open deaerating heat exchanger
D: Low-pressure turbines J: Boiler feed pump
E: Generator and transformer K: High-pressure heat exchangers
F: Condenser

(a)

A: Nuclear reactor I: Generator and transformer
B: Heavy water pump J: Condenser
C: Moderator cooler K: Hot well pump
D: Steam generator L: Low-pressure heat exchangers
E: High-pressure turbine M: Open deaerating heat exchanger
F: Moisture separator N: Boiler feed pump
G: Closed reheater O: Pump
H: Low-pressure turbines P: High-pressure heat exchangers

(b)

Fig. 11.5. Process diagrams for single units of: (a) NGS and (b) PNGS. The flow of uranium into and out of the nuclear reactor, and the net heat delivered, are indicated. Lines exiting turbines represent flows of extraction steam. Stream S16 in Fig. 11.5b represents the mixed contents of the four indicated flows of extraction steam. The diagrams are each divided into four main sections: steam generation (device A for NGS and devices A–D for PNGS); power production (devices B–E for NGS, and E–I for PNGS); condensation (device F for NGS, and J for PNGS) and preheating (devices G–K for NGS, and K–P for PNGS).

Table 11.4a. Flow data for a unit in the coal-fired NGS.[a]

Flow[b]	Mass flow rate[c] (kg/s)	Temperature (°C)	Pressure (MPa)	Vapor fraction[d]	Energy flow rate[e] (MW)	Exergy flow rate[e] (MW)
S1	41.74	15.00	0.101	solid	1367.58	1426.73
S2	668.41	15.00	0.101	1.0	0.00	0.00
S3	710.15	1673.59	0.101	1.0	1368.00	982.85
S4	710.15	119.44	0.101	1.0	74.39	62.27
S5A	453.59	538.00	16.2	1.0	1585.28	718.74
S8	42.84	323.36	3.65	1.0	135.44	51.81
S10	367.85	35.63	0.0045	0.0	36.52	1.20
S11	367.85	35.73	1.00	0.0	37.09	1.70
S12	58.82	188.33	1.21	0.0	50.28	11.11
S13	18636.00	15.00	0.101	0.0	0.00	0.00
S14	18636.00	23.30	0.101	0.0	745.95	10.54
S15	410.75	323.36	3.65	1.0	1298.59	496.81
S20	367.85	360.50	1.03	1.0	1211.05	411.16
S21	410.75	538.00	4.00	1.0	1494.16	616.42
S22	15.98	423.23	1.72	1.0	54.54	20.02
S25	26.92	360.50	1.03	1.0	88.64	30.09
S33	309.62	35.63	0.0045	0.93	774.70	54.07
S34	10.47	253.22	0.379	1.0	32.31	9.24
S35	23.88	209.93	0.241	1.0	71.73	18.82
S36	12.72	108.32	0.0689	1.0	35.77	7.12
S37	11.16	60.47	0.0345	1.0	30.40	5.03
S38	58.23	55.56	0.0133	0.0	11.37	0.73
S39	367.85	124.86	1.00	0.0	195.94	30.41
S40	453.59	165.86	1.00	0.0	334.86	66.52
S41	453.59	169.28	16.2	0.0	347.05	77.57
S42	453.59	228.24	16.2	0.0	486.75	131.93
Q5					5.34	0.00
Q6					5.29	0.00
P1					0.57	0.57

(Continued)

Table 11.4a. (Continued)

Flow[b]	Mass flow rate[c] (kg/s)	Temperature (°C)	Pressure (MPa)	Vapor fraction[d]	Energy flow rate[e] (MW)	Exergy flow rate[e] (MW)
P8					523.68	523.68
P15					12.19	12.19

[a] Based on data obtained for Nanticoke Generating Station by computer simulation (Rosen, 2001) using given data (Ontario Hydro, 1973, 1983; Scarrow and Wright, 1975; Bailey, 1981; Merrick, 1984).
[b] Flow numbers correspond to those in Fig. 11.5a, except for S3, which represents the hot product gases for adiabatic combustion. Letter prefixes indicate material flows (S), heat flows (Q) and electricity flows (P).
[c] Material flow compositions, by volume, are: 100% C for S1; 79% N_2, 21% O_2 for S2; 79% N_2, 6% O_2, 15% CO_2 for S3 and S4 and 100% H_2O for other material flows.
[d] Vapor fraction indicates fraction of a vapor–liquid flow that is vapor (not applicable to S1 since it is solid). Vapor fraction is listed as 0.0 for liquids and 1.0 for superheated vapors.
[e] Energy and exergy values are evaluated using a reference-environment model (similar to the model used by Gaggioli and Petit (1977) having a temperature of 15°C, a pressure of 1 atm, and a composition of atmospheric air saturated with H_2O at 15°C and 1 atm and the following condensed phases: water, limestone and gypsum.

Table 11.4b. Flow data for a unit in PNGS.

Stream	Mass flow rate[a] (kg/s)	Temperature (°C)	Pressure (N/m²)	Vapor fraction[b]	Energy flow rate (MW)	Exergy flow rate (MW)
S1	7724.00	291.93	8.82×10^6	0.0	9548.21	2984.23
S2	7724.00	249.38	9.60×10^6	0.0	7875.44	2201.64
S3A	7724.00	249.00	8.32×10^6	0.0	7861.16	2188.64
S4	1000.00	64.52	1.01×10^5	0.0	207.02	15.99
S5A	1000.00	43.00	1.01×10^5	0.0	117.02	5.34
S6	1956.83	15.00	1.01×10^5	0.0	0.00	0.00
S7	1956.83	26.00	1.01×10^5	0.0	90.00	1.67
S9	698.00	151.83	5.00×10^5	0.88	1705.50	500.40
S10	603.00	160.00	5.00×10^5	1.0	1629.83	476.54
S11	61.00	254.00	4.25×10^6	1.0	166.88	64.62
S12A	61.00	254.00	4.25×10^6	0.0	63.57	17.78
S14	55.00	176.66	9.28×10^6	0.90	138.70	44.60
S15	603.00	237.97	4.50×10^5	1.0	1733.17	508.35
S16	83.00	60.81	2.07×10^4	0.95	204.00	28.10
S17	498.00	23.32	2.86×10^3	0.90	1125.10	44.40
S18	22.00	186.05	2.55×10^5	1.0	61.06	16.03
S19	83.00	60.81	2.07×10^4	0.0	15.89	1.13
S20	581.00	23.32	2.86×10^3	0.0	20.15	0.17

(Continued)

Table 11.4b. (Continued)

Stream	Mass flow rate[a] (kg/s)	Temperature (°C)	Pressure (N/m^2)	Vapor fraction[b]	Energy flow rate (MW)	Exergy flow rate (MW)
S21	581.00	23.40	1.48×10^6	0.0	211.55	1.13
S22	581.00	100.20	1.40×10^6	0.0	207.88	26.50
S23	150.00	134.00	3.04×10^5	0.0	75.04	12.29
S24	150.00	134.17	1.48×10^6	0.0	75.27	12.50
S25	753.00	123.69	1.40×10^6	0.0	344.21	53.16
S26	753.00	124.20	5.40×10^6	0.0	347.93	56.53
S27	753.00	163.94	5.35×10^6	0.0	476.02	96.07
S28	24073.00	15.00	1.01×10^5	0.0	0.00	0.00
S29	24073.00	26.00	1.01×10^5	0.0	1107.20	20.61
S39	95.00	160.00	6.18×10^5	0.03	75.70	23.70
S40	753.00	254.00	4.25×10^6	1.0	2060.02	797.70
Q1					1673.00	1673.00
Q2					90.00	90.00
Q9					5.56	0.00
Q10					5.50	0.00
P1					14.28	14.28
P2					3.73	3.73
P3					1.00	1.00
P4					0.23	0.23
P8					544.78	544.78

[a] All streams are modeled as 100% H_2O. Streams S1, S2, S3A, S4 and S5A are actually reactor-grade D_2O.
[b] Vapor fraction is listed as 0.0 for liquids and 1.0 for superheated vapors.

In each unit of PNGS, natural uranium is fissioned in the presence of a moderator to produce heat, which is transferred from the reactor to the boiler in the primary heat transport loop (PHTL). The flow rate of pressurized heavy water (D_2O) in the PHTL is 7724 kg/s. The D_2O is heated from 249°C and 9.54 MPa to 293°C and 8.82 MPa in the nuclear reactor. Light water steam (815 kg/s at 4.2 MPa and 251°C) is produced in the boiler and transported through the secondary heat transport loop. Spent fuel is removed from the reactor, and heat generated in the moderator is rejected.

Power production

The steam produced in the steam generation section is passed through a series of turbine generators which are attached to a transformer. Extraction steam from several points on the turbines preheats feedwater in several low- and high-pressure heat exchangers and one spray-type open deaerating heat exchanger. The LP turbines exhaust to the condenser at 5 kPa.

Each unit of NGS has a 3600 rpm, tandem-compound, impulse-reaction turbine generator containing one single-flow high-pressure cylinder, one double-flow intermediate-pressure cylinder and two double-flow LP cylinders. Steam exhausted from the high-pressure cylinder is reheated in the combustor.

Each unit of PNGS has an 1800 rpm, tandem-compound, impulse-reaction turbine generator containing one double-flow high-pressure cylinder, and three double-flow LP cylinders. Steam exhausted from the high-pressure cylinder passes through a moisture separator and a closed reheater (which uses steam from the boiler as the heat source).

Table 11.5. Main process data for single units in NGS and PNGS.

Section	NGS	PNGS
Steam generation section		
Furnace		
Coal consumption rate at full load (kg/s)	47.9	–
Flue gas temperature (°C)	120	–
Nuclear reactor		
Heavy water mass flow rate (kg/s)	–	724
Heavy water temperature at reactor inlet (°C)	–	249
Heavy water temperature at reactor outlet (°C)	–	293
System pressure at reactor outlet header (MPa)	–	8.8
Boiler (heat-exchanger component)		
Feed water temperature (°C)	253	171
Total evaporation rate (kg/s)	454	815
Steam temperature (°C)	538	251
Steam pressure (MPa)	16.9	4.2
Reheat evaporation rate (kg/s)	411	–
Reheat steam temperature (°C)	538	–
Reheat steam pressure (MPa)	4.0	–
Power production section		
Turbine		
Condenser pressure (kPa)	5	5
Generator		
Gross power output (MW)	–	542
Net power output (MW)	505	515
Condensation section		
Cooling water flow rate (m³/s)	18.9	23.7
Cooling water temperature rise (°C)	8.3	11

Condensation

Cooling water from Lake Ontario for PNGS and Lake Erie for NGS condenses the steam exhausted from the turbines. The flow rate of cooling water is adjusted so that a specified temperature rise in the cooling water is achieved across the condenser.

Preheating

The temperature and pressure of the feedwater are increased in a series of pumps and heat exchangers (feedwater heaters). The overall efficiency of the station is increased by raising the temperature of the feedwater to an appropriate level before heat is added to the cycle in the steam generation section.

11.6.2. Approach

The analyses and comparisons of NGS and PNGS are performed using a computer code developed by enhancing a state-of-the-art process simulator, i.e., Aspen Plus, for exergy analysis. The reference-environment model used by Gaggioli and Petit (1977) and Rodriguez (1980) is used in the evaluation of energy and exergy quantities, but with a reference-environment temperature T_0 of 15°C (the approximate mean temperature of the lake cooling water). The reference-environment pressure P_0 is taken to be 1 atm, and the chemical composition is taken to consist of air saturated with water vapor, and the following condensed phases at 15°C and 1 atm: water (H_2O), gypsum ($CaSO_4 \cdot 2H_2O$) and limestone ($CaCO_3$). In addition to properties in Aspen Plus data banks, which include steam properties based on the ASME steam tables, base enthalpy and chemical exergy values reported elsewhere (Rosen and Dincer, 2003a,b) are used. The base enthalpy of a component (at T_0 and P_0) is evaluated from the enthalpies of the stable components of the environment (at T_0 and P_0). The base enthalpy of a fuel is equal to the enthalpy change in forming the fuel from the components of the environment (the same environment used in exergy calculations). A compound which exists as a stable component of the reference environment is defined to have an enthalpy of zero at T_0 and P_0.

For simplicity, the net heat produced by the uranium fuel is considered the main energy input to PNGS, and D_2O is modeled as H_2O, coal as pure graphite (C) and air as 79% nitrogen and 21% oxygen by volume. Also, it is assumed that

- The turbines have isentropic and mechanical efficiencies of 80% and 95%, respectively.
- The generators and transformers are each 99% efficient, and heat losses from their external surfaces occur at 15°C (i.e., T_0).
- The input to and output from the nuclear reactor of uranium is a steady-state process.
- All heat rejected by the moderator cooler is produced in the moderator. Ontario Power Generation (1985) actually observes for each PNGS unit that, of the 90 MW rejected by the moderator cooler, 82 MW is produced in the moderator, 2.6 MW is transferred from the fuel channel to the moderator and 6.1 MW is produced in other reactor components (1.1 MW in the shield, 0.1 MW in the dump tank, 2.4 MW in the calandria and 2.5 MW in the calandria tubes) and then transferred to the moderator.

It is further assumed that the temperature at which heat can be produced by fissioning uranium can theoretically be so high that the quantities of energy and exergy of the heat can be considered equal. This assumption has a major effect on the exergy efficiencies discussed subsequently. If as an alternative fission heat is taken to be available at the temperature at which it is actually produced (i.e., at the thermal neutron flux-weighted average temperature of about 880°C), the exergy of the heat is about 75% of the energy.

11.6.3. Analysis

Energy and exergy efficiencies are evaluated as ratios of products to inputs. For the overall stations, the energy efficiency η is evaluated as

$$\eta = \frac{\text{Net energy output with electricity}}{\text{Energy input}} \tag{11.26}$$

and the exergy efficiency ψ as

$$\psi = \frac{\text{Net exergy output with electricity}}{\text{Exergy input}} \tag{11.27}$$

For most of the other plant components and sections, similar expressions are applied to evaluate efficiencies.

Efficiencies are not readily defined for the condensers, as the purpose of such devices is to reject waste heat rather than generate a product. However, the merit of the condensers with respect to the overall plant can be assessed for comparative purposes by evaluating the 'net station condenser heat (energy) rejection rate' R_{energy}, where

$$R_{energy} = \frac{\text{Heat rejected by condenser}}{\text{Net electrical energy produced}} \tag{11.28}$$

and comparing it to an analogous quantity, the 'net station condenser exergy rejection rate' R_{exergy}, where

$$R_{exergy} = \frac{\text{Exergy rejected by condenser}}{\text{Net electrical exergy produced}} \tag{11.29}$$

Table 11.6. Breakdown by section and device of exergy consumption rates (in MW) in single units of NGS and PNGS.

Section/device	NGS	PNGS
Steam generation section		
Reactor	659.0	969.7
D_2O–H_2O heat exchanger	–	47.4
D_2O pump	–	1.1
Moderator cooler	–	9.0
	Total = 659.0	Total = 1027.2
Power production section		
High-pressure turbine(s)	26.4	36.9
Intermediate-pressure turbine(s)	22.3	–
Low-pressure turbine(s)	59.2	79.7
Generator	5.3	5.5
Transformer	5.3	5.5
Moisture separator	–	0.2
Closed steam reheater	–	15.0
	Total = 118.5	Total = 142.8
Condensation section		
Condenser	43.1	24.7
	Total = 43.1	Total = 24.7
Preheat section		
Low-pressure heat exchangers	10.7	1.6
Deaerating heat exchanger	5.1	1.8
High-pressure heat exchangers	6.4	16.4
Hot well pumps	0.1	0.04
Heater condensate pumps		0.03
Boiler feed pumps	1.1	0.43
	Total = 23.4	Total = 20.8
	General Total = 844.0	General Total = 1215.5

11.6.4. Results

Simulation and analysis data (including energy and exergy values) are summarized along with specified data in Table 11.4 for the streams identified in Fig. 11.5. Exergy consumption values for the devices are listed, according to process-diagram sections, in Table 11.6. Figures 11.6 and 11.7 illustrate the net energy and exergy flows and exergy consumptions for the four main process-diagram sections described in the caption of Fig. 11.5. The data are summarized in overall energy and exergy balances in Fig. 11.8.

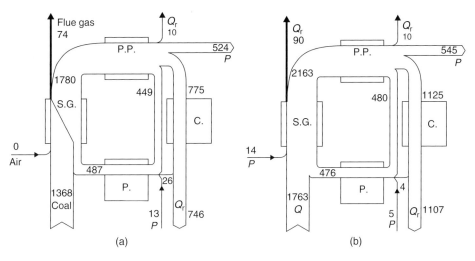

Fig. 11.6. Simplified process diagrams for single units of (a) NGS and (b) PNGS, indicating net energy flow rates (MW) for streams. Stream widths are proportional to energy flow rates. Sections of stations shown are steam generation (S.G.), power production (P.P.), condensation (C.) and preheating (P.). Streams shown are electrical power (P), heat input (Q) and heat rejected (Q_r).

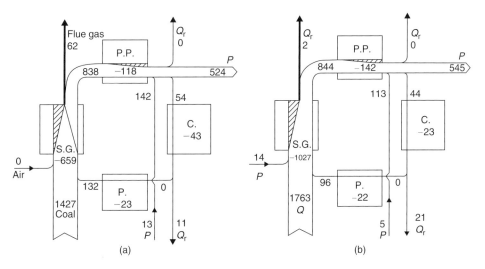

Fig. 11.7. Simplified process diagrams for single units of (a) NGS and (b) PNGS, indicating net exergy flow rates for streams and consumption rates (negative values) for devices. Stream widths are proportional to exergy flow rates, and shaded regions to exergy consumption rates. All values are in MW. Other details are as in Fig. 11.6.

Regarding result validity, it is observed that:

- Simulated stream property values are within 10% of the values measured at the stations (for properties for which data are recorded).
- Energy and exergy values and efficiencies for the overall processes and for process subsections are in broad agreement with the literature for similar processes.
- Exergy-analysis results for PNGS and NGS are relatively insensitive to the composition of the reference environment.

On the last point, it is noted that the exergies of only the coal and stack gas depend on the choice of the chemical composition of the environment, and that, for most electrical generating stations, the results of energy and exergy analyses are not significantly affected by reasonable and realistic variations in the choice of reference-environment properties.

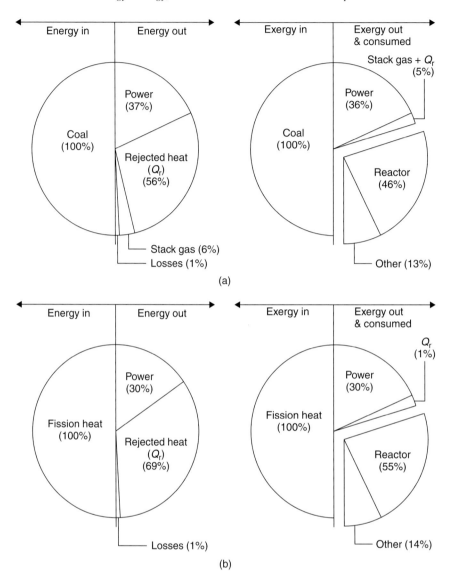

Fig. 11.8. Overall energy and exergy balances for single units of (a) NGS and (b) PNGS. The left and right halves of the energy balances represent respectively energy inputs and energy outputs. The left and right halves of the exergy balances represent respectively exergy inputs and exergy outputs and consumptions (exploded section of balance). Cooling water and air inputs are not shown because they contain zero energy and exergy. The PNGS reactor represents only the fission reactor, not all devices in the steam generation section.

11.6.5. Discussion

Overall process efficiencies

Overall energy (η) and exergy efficiency (ψ) values are evaluated for the overall processes using Eqs. (11.8) and (11.10), respectively.

For NGS, where coal is the only input source of energy or exergy,

$$\eta = \frac{(524 - 13)\ \text{MW}}{1368\ \text{MW}}(100\%) = 37\%$$

and

$$\psi = \frac{(524 - 13) \text{ MW}}{1427 \text{ MW}}(100\%) = 36\%$$

The small difference in the efficiencies is due to the fact that the specific chemical exergy of coal is slightly greater than its specific base enthalpy.

For PNGS, where fission heat is treated as the only input source of energy and exergy,

$$\eta = \frac{(545 - 19) \text{ MW}}{1763 \text{ MW}}(100\%) = 30\%$$

and

$$\psi = \frac{(545 - 19) \text{ MW}}{1763 \text{ MW}}(100\%) = 30\%$$

Ontario Power Generation (1985) reports $\eta = 29.5\%$ for PNGS. Although for each station the energy and exergy efficiencies are similar, these efficiencies differ markedly for many station sections.

Efficiencies and losses in steam generators

Exergy consumptions in the steam generation sections of the stations are substantial, accounting for 659 MW (or 72%) of the 916 MW total exergy losses for NGS, and 1027 MW (or 83%) of the 1237 MW total exergy losses for PNGS.

Of the 659 MW of exergy consumed in this section for NGS, 444 MW is due to combustion and 215 MW to heat transfer.

Of the 1027 MW of exergy consumed in this section for PNGS, 47 MW is consumed in the boiler, 9 MW in the moderator cooler, 1 MW in the heavy-water pump and 970 MW in the reactor. The exergy consumptions in the reactor can be broken down further by hypothetically breaking down into steps the processes occurring within it (Fig. 11.9): heating of the moderator, heating of the fuel pellets (to their maximum temperature of approximately 2000°C), transferring the heat within the fuel pellets to the surface of the pellets (at approximately 400°C), transferring the heat from the surface of the fuel pellets to the cladding surface (at 304°C), and transferring the heat from the cladding surface to the primary coolant and then to the preheated boiler feedwater to produce steam.

The energy and exergy efficiencies for the steam generation section, considering the increase in energy or exergy of the water as the product, for NGS are

$$\eta = \frac{[(1585 - 847) + (1494 - 1299)] \text{ MW}}{1368 \text{ MW}}(100\%) = 95\%$$

and

$$\psi = \frac{[(719 - 132) + (616 - 497)] \text{ MW}}{1427 \text{ MW}}(100\%) = 49\%$$

and for PNGS are

$$\eta = \frac{(2267 - 64 - 476) \text{ MW}}{(1763 + 14) \text{ MW}}(100\%) = 95\%$$

$$\psi = \frac{(862 - 18 - 96) \text{ MW}}{(1763 + 14) \text{ MW}}(100\%) = 42\%$$

The steam generation sections of NGS and PNGS appear significantly more efficient on an energy basis than on an exergy basis. Physically, this discrepancy implies that although 95% of the input energy is transferred to the preheated water, the energy is degraded as it is transferred. Exergy analysis highlights this degradation.

Two further points regarding PNGS are noted:

1. The step in which heat is generated by fissioning uranium (also shown for completeness in Fig. 11.9) is, by previous assumption, outside the boundary of the nuclear reactor considered here. The energy and exergy efficiencies calculated for PNGS could be significantly different if this step were considered. In this case, the energy and exergy of the fresh and spent nuclear fuel would be required. The question of what are the exergies of nuclear fuels is not completely resolved. Researchers usually only deal with the heat delivered by nuclear fuels, and most

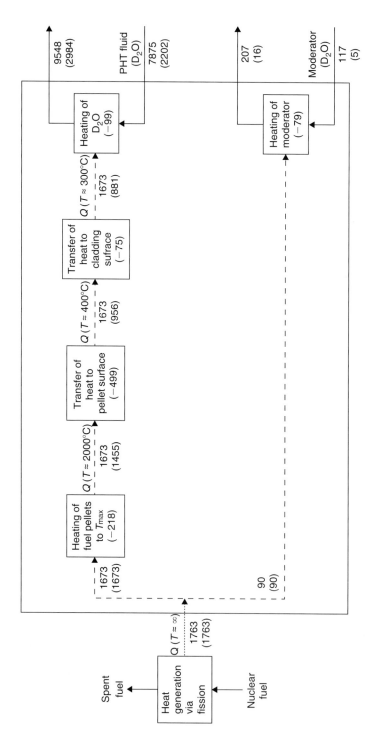

Fig. 11.9. Breakdown of the energy and exergy losses in the nuclear reactor of one PNGS unit. Material streams are represented by solid lines, and heat flows by broken lines. The heavy solid line encloses the part of the nuclear reactor considered in the present analysis. Exergy (in parentheses) and energy flow rates are indicated for streams, and exergy consumption rates (negative values in parentheses) for devices. Flows of heat Q at points in the reactor at different values of temperature T are shown. All values are in MW. PHT denotes primary heat transport.

argue that the exergy of nuclear-derived heat is equal or nearly equal to the exergy because the heat can potentially be produced at very high temperatures.

2. Since D_2O is modeled as H_2O, the chemical exergy of D_2O is neglected. Neglecting the chemical exergy of D_2O does not significantly affect the exergy-analysis results here because, since the D_2O is contained in the closed PHTL of the steam generation section and used only as a heat-transfer medium, it is only the physical exergy of the D_2O stream that is of interest.

Losses in steam condensers

In the condensers,

- A large quantity of energy enters (775 MW for each NGS unit and 1125 MW for each PNGS unit), of which close to 100% is rejected.
- A small quantity of exergy enters (54 MW for each NGS unit and 44 MW for each PNGS unit), of which 25–50% is rejected and 50–75% is internally consumed.

Thus, energy-analysis results lead to the erroneous conclusion that almost all losses in electricity generation potential for NGS and PNGS are associated with the heat rejected by the condensers, while exergy analyses demonstrate quantitatively and directly that the condensers are responsible for little of these losses (Fig. 11.8). This discrepancy arises because heat is rejected by the condensers at a temperature very near that of the environment.

The characteristics of condensers can be seen more clearly by considering the 'net station condenser heat (energy) rejection rate' R_{energy}, and the 'net station condenser exergy rejection rate' R_{exergy}. Following the expressions for these quantities in Eqs. (11.28) and (11.29), respectively, it can be shown for each coal-fired unit that

$$R_{\mathrm{energy}} = \frac{746\,\mathrm{MW}}{(524 - 13)\,\mathrm{MW}} = 1.46$$

and

$$R_{\mathrm{exergy}} = \frac{11\,\mathrm{MW}}{(524 - 13)\,\mathrm{MW}} = 0.0215$$

and for each nuclear unit that

$$R_{\mathrm{energy}} = \frac{1107\,\mathrm{MW}}{(545 - 19)\,\mathrm{MW}} = 2.10$$

and

$$R_{\mathrm{exergy}} = \frac{21\,\mathrm{MW}}{(545 - 19)\,\mathrm{MW}} = 0.0399$$

The R values indicate that the exergy rejected by the condensers is less than 4% of the net exergy produced, while the energy rejected is approximately 150–200% of the net energy produced.

Miscellaneous losses

In the power production and preheating sections of the NGS and PNGS units, energy losses were found to be very small (less than 10 MW total), and exergy losses were found to be moderately small (100–150 MW in the power production section and 20–25 MW in the preheating section). The exergy losses are almost completely associated with internal consumptions.

Environmental impact and sustainability

The environmental and sustainability insights gained through the use of exergy analysis are described for NGS in Section 3.4.

Comparison and summary

The thermodynamic behavior of the coal-fired and nuclear generating stations considered are similar in all areas, except the steam generation sections. Those differences significantly affect the efficiencies, losses and performance in these sections. Some of the differences are as follows:

- The temperatures associated with heat generation are constrained to lower values for PNGS than NGS, leading to lower efficiencies for the nuclear station.

- The heat-generation and heat-transfer mechanisms vary greatly between the stations. For PNGS, heat is generated in the calandria and transported to the steam generator, while heat is generated in the steam generator without the need for a transport step in NGS. Also, most heat transfer in the steam generator of the coal-fired station occurs from a gas to a solid, whereas the more efficient liquid-to-solid heat transfer predominates in the nuclear station.
- Thermal neutrons are absorbed and moderated for necessary operating and control processes in PNGS. This process, which is somewhat analogous to wasting fuel, is not present in the coal-fired station.
- The potential of the used form of the fuel (spent uranium for PNGS and combustion gases for NGS) is much greater for the nuclear station, where the spent fuel is highly radioactive and releases significant quantities of thermal energy for many years. These factors impact on the exergy- and energy-related performances of the stations, and should be taken into account in efforts to improve efficiencies.

In comparing the thermodynamic characteristics of coal-fired and nuclear electrical generating stations, several illuminating insights into the performance of such stations have been acquired. First, although energy and exergy efficiencies are the same for PNGS and similar for NGS, energy analyses do not systematically identify the location and cause of process inefficiencies, and exergy analyses do. That is, energy losses are associated with emissions (mainly heat rejected by condensers), and exergy losses primarily with consumptions (mainly in the reactors) and little with cooling water and stack gases. Second, since devices with the largest thermodynamic losses have the largest margins for efficiency improvement, efforts to increase the efficiencies of coal-fired and nuclear electrical generating stations should focus on the combustion and nuclear reactors, respectively. For instance, technologies capable of producing electricity without combustion (e.g., fuel cells) or utilizing heat at high temperatures could increase efficiencies significantly. This comment is, of course, overly simplistic, as such decisions require consideration of other technical and economic factors, in addition to efficiency. Third, the use of heat rejected by condensers only increases the exergy efficiencies by a few percent. Cogeneration systems, which produce heat at useful temperatures at the expense of reduced electrical output, can have greater efficiencies than conventional electrical generating stations, but the merit of cogeneration systems must be determined using exergy analyses because energy analyses tend to overstate performance.

11.7. Improving steam power plant efficiency

Electrical generating utilities usually strive to improve the efficiency (or heat rate) at their existing thermal electric generating stations, many of which are over 25 years old and mature. Often, a heat rate improvement of only a few percent appears to be desired as the costs and complexity of such measures may be more manageable than more expensive options (Rosen and Dincer, 2003b,c). At other times major efficiency improvements are sought. Exergy methods are helpful in all cases.

Modifications to increase the efficiency of thermal electrical power generation often strive to carry out one or more of the following:

- Increase the average temperature at which heat is transferred to the working fluid in the steam generator.
- Decrease the average temperature at which heat is rejected from the working fluid in the condenser.
- Reduce component inefficiencies and losses.
- Improve the integration of process flows so that would-be wastes are utilized where possible.

Many methods and technologies have been reported in the literature for improving power-plant efficiency or reducing irreversibilities.

In this section, some of the more significant thermodynamic methods to improve the efficiency of electricity generating steam power plants (particularly coal based) are surveyed and examined. We focus on minor practical improvements, which can be undertaken with limited effort and cost.

Also possible are improvements of a longer term and broader nature. These include replacing combustors with gasifiers and fuel cells, developing technologies capable of utilizing heat at increased temperatures, raising significant metallurgical temperature limits and integrating processes so that the wastes from one process become the feeds to another.

11.7.1. Exergy-related techniques

To assist in improving the efficiencies of coal-to-electricity technologies, energy analysis is normally used. A better understanding is attained when a more complete thermodynamic view is taken via exergy methods. Through the better understanding developed with exergy analysis, the efficiencies of devices and processes can usually be improved, often cost-effectively. Consequently, exergy analysis is particularly useful for (i) designing better new facilities and (ii) retrofitting or modifying existing facilities to improve them. The latter use is the focus of the present section. Some examples of how exergy methods can provide meaningful insights that can assist in increasing efficiency and achieving optimal designs for thermal power plants follow:

- Exergy-based costing and thermoeconomic analysis has been investigated and applied.
- A means of evaluating thermal power plants using an alternative second law-based efficiency measure based on a modified coefficient of performance has been proposed.
- An alternative thermodynamic approach to designing steam power plants has been suggested, and the use of 'pinch technology' in steam power plant design has been studied.

Exergy can play an important role in developing strategies and in providing guidelines for more effective use of energy in existing steam power plants. Exergy analysis also helps in improving plant efficiency by determining the origin of exergy losses, and hence providing a clearer picture. Exergy helps identify components where high inefficiencies occur, and where improvements are merited. The thermodynamic cycle can often be optimized by minimizing the irreversibilities. At full load of the cycle, the steam temperature and pressure of the boiler should be at their upper limits. However, at off-design loads, the temperature and pressure should be decreased. An exergy evaluation of a supercritical steam turbine showed that high exergy losses occur in the heat recovery steam generator and in the steam turbine. Another exergy analysis led to a great reduction in power cost. Originally, in a power plant facility, 77 MW of fuel and 2.2 MW of power were needed. After improvement by exergy analysis the fuel consumption was cut down to a value between 45 and 60 MW (Gaggioli et al., 1991).

11.7.2. Computer-aided design, analysis and optimization

Many efforts to improve coal-fired steam power plants by providing computer-aided tools for simulation, analysis and optimization have been reported. Some of these efforts have focused on processes using Rankine cycles as part or all of a power plant, and have not integrated exergy concepts. Other works have directed the computer tool at addressing exergy considerations or at ensuring a focus on exergy is a central thrust. Such computer tools can aid in developing and evaluating potential improvement measures.

11.7.3. Maintenance and control

Numerous measures related to maintenance and control are possible to reduce losses. These include: (i) reducing leaks of steam, gas, air and other substances in lines, valves, joints and devices; (ii) utilizing improved and more automated controllers, both to ensure design specifications are adhered to and to detect parameter variations which may indicate future problems; (iii) improved maintenance to ensure minor breaks are repaired and actual operating parameters match design specifications, e.g., minor modifications to the reheat steam temperature in a thermal power plant can be implemented to improve overall efficiency and (iv) periodic overhauls of devices.

11.7.4. Steam generator improvements

Steam generator efficiency has a significant effect on overall plant efficiency. The main energy processes within the steam generator are combustion and heat transfer.

Combustion-related improvements

In the combustion portion of the steam generator, fuel energy is converted to heat at appropriate conditions while minimizing losses from combustion imperfections. Efficient combustion requires temperatures sufficiently high to ignite the constituents, mixing to provide good oxygen-fuel contact and sufficient time to complete the process. For practical applications, more than the theoretical amount of air is needed for near complete combustion. However, excess air increases the quantity of stack gases and the associated waste emission. Also, irreversibility occurs during combustion even when it is complete and without excess air because the chemical reaction which the fuel and air undergo is irreversible due to the uncontrolled electron exchange between the reacting components.

One means of improving combustion efficiency involves the coal feeding system. For example, the aerodynamic separation of coal particles in coal stoker-fired boilers creates a more porous coal layer that is distributed onto a traveling fire grate, and that can result in a 2–4% increase in combustion efficiency by reducing incomplete combustion and the losses from coal powder penetration through the grate. One recently designed coal grate has a bottom section with a comb shape that is adjustable, so that the depth of the furrows in the coal layer formed on the grate can be easily changed. This device can result in a 2–4% increase in combustion efficiency through changes in the distribution of primary and secondary air, restrictions in the usage of excess combustion air and reductions in the losses from incomplete combustion (Rosen and Dincer, 2003b,c).

Combustion losses may also be reduced by improving combustion efficiency, usually through minor adjustments to the combustion chamber and related devices. Larger reductions in combustion losses are likely attainable if metallurgical temperature limits for materials used in the processes are improved so that high temperatures are permitted in the combustion chamber and related devices.

Heat transfer-related improvements

Several methods are used in industry for improving heat transfer inside a steam generator. Soot blowers can be used to keep heat exchanger surfaces in steam generators clean and thus to achieve increased efficiencies through improved heat transfer effectiveness. Specifically, coal flow dampers and soot blowers can reduce slagging so as to improve heat absorption in the lower furnace and decrease superheater attemperation spray flows. Fluidized bed combustion (FBC) also leads to heat transfer improvement. In FBC, crushed coal particles are injected in a bed of fine inert particles (e.g., sand, ash, limestone) fluidized by combustion air. Water tubes submerged in this fluidized combustion zone achieve high rates of heat transfer. FBC also allows for relatively low combustion temperatures (820–870°C) that are well below the ash fusion temperature, so firesides slagging is minimized. The higher efficiencies for FBC compared with normal combustion are in large part due to the high volumetric heat release rates and high heat transfer coefficients in the combustion zone. However, as the flame temperature tends to increase with load, the exergy-efficiency difference between FBC and normal combustion decreases as load increases (Rosen and Dincer, 2003b,c).

Increasing the average temperature at which heat is added to the power cycle by increasing the steam generator operating pressure normally raises cycle energy efficiency. However, this measure can increase moisture content and thus turbine blade erosion in the last turbine stages.

Increasing the average temperature at which heat is added to the power cycle by superheating causes the network output, the heat input and the energy efficiency to increase, while decreasing steam moisture content at the turbine exit. However, superheat temperatures are constrained by metallurgical temperature limits and other factors.

Other steam generator improvements

Some other steam generator-related efficiency-improvement measures are worth considering: (i) the amount of excess air supplied to the combustor can be reduced to decrease the exergy losses with stack gases and (ii) heat can be recovered from the spent combustion gases, before they exit the flue, to reduce fuel consumption.

Other research efforts related to the steam generator focus on understanding the exergy losses in the combustion process, burner retrofits, improving the steam generator and the heat recovery devices within it, and improving steam traps.

11.7.5. Condenser improvements

Lowering condenser operating pressure lowers the temperature of the steam within the condenser and thus the temperature at which heat is rejected from the plant. However, there is a lower limit on the condenser pressure corresponding to the temperature of the cooling medium needed for effective heat transfer. Also, lowering the condenser pressure can cause such problems as air leakage into the condenser, as well as decreased turbine efficiency and increased turbine blade erosion due to increased moisture content in the steam in the last turbine stages.

11.7.6. Reheating improvements

Reheating can affect overall plant energy efficiency and the average temperature at which heat is added to the steam generator. However, exergy destruction in steam generator, which usually accounts for more than half of the exergy destruction in the power plant, is only affected in a limited way by reheat pressures. Heat rate can be improved through reheating, depending on the respective final feed temperature for both single- or double-reheat cycles. Increasing reheat temperature reduces leakage and blading losses, and increases cycle available energy.

Some report that the most economical gain occurs when the peak reheat temperature is no more than 30 K above the saturated temperature corresponding to the steam pressure of the high-pressure turbine exhaust, but that the influence of reheat pressure on cycle performance is not great. The maximum improvement in energy efficiency and irreversible losses for single-reheat cycles appears to be attained when reheat pressures are about 19% of steam generator pressure. When a second-reheat stage is incorporated with a pressure ratio of approximately five between the first- and second-reheat stages, the energy efficiency and irreversible losses can be further improved. The optimum value of this pressure ratio is affected by reheat temperature, number of reheat stages and feedwater configuration. Improvements in plant exergy efficiency with increased reheat pressure are usually due to improvements in the turbine-cycle unit. When the optimum pressure is set for either the first or second stage reheat, a wide range of pressures are possible for the other reheat stage that cause only minimal deviations from the maximum energy efficiency and the minimum total irreversibility rate. However, there is an upper limit on the first reheat pressure due to boiler feedwater temperature and air heater constraints, and a lower limit on the second reheat pressure due to the possibility of superheated exhaust at part load increasing the LP turbine inlet temperature and leading to temper embrittlement in the LP rotor steel (Silvestri et al., 1992).

11.7.7. Regenerative feedwater heating improvements

Regenerative feedwater heating using bleed steam from the turbines increases energy efficiency by increasing the average temperature at which heat is added to the cycle. The additional work which the bleed steam could have produced by expanding further in the turbines is foregone.

Heat is transferred from the steam to the feedwater either by mixing the fluids in an open heater or without mixing in a closed heater. Open feedwater heaters are simple and inexpensive and have good heat-transfer characteristics. Closed feedwater heaters are more complex due to their internal piping, and have less effective heat transfer than open heaters since the two streams do not come into direct contact. The irreversible losses associated with heat transfer and other system processes are greater for closed heat exchangers since the extracted steam for each heater is throttled to the corresponding pressure for each LP heater (for details, see Rosen and Dincer (2003b,c)).

Incorporating heat regeneration in a steam power plant increases the energy efficiency mainly by reducing the irreversible losses in the steam generator. As the number of feedwater heaters increases, the percentage improvement provided by each additional heater decreases. It is reported (Rosen and Dincer, 2003b,c) that the energy-related optimum distribution for n feedwater heaters occurs when the total feedwater enthalpy rise experienced by feedwater is divided evenly among the heaters, and that the optimum ratio of ultimate feedwater enthalpy rise to total enthalpy rise is approximately given by

$$(h_B - h_n)/(h_B - h_0) = (n + 1)^{-1} \qquad (11.30)$$

where h denotes the specific enthalpy of feedwater, and the subscripts B, n and 0 respectively denote the saturated liquid state corresponding to the boiler pressure, the optimal feedwater condition and the saturated liquid state corresponding to the condenser pressure.

The optimum number of feedwater heaters is usually determined from economic considerations, as the use of an additional feedwater heater is normally justifiable when the fuel savings it generates are sufficient relative to its cost.

Incorporating reheating in a regenerative cycle can further improve its energy efficiency and the total irreversible losses. These improvements increase at a decreasing rate as the number of the feedwater heaters increases. Although reheating and regenerative feedwater heating each lead to efficiency improvements, the improvements are less when these measures are applied simultaneously than the sum of the improvements from applying the measures separately. One reason for this observation is that reheating results in higher extraction steam temperatures and subsequently larger temperature differences between the feedwater and the extraction steam, causing regenerative feedwater heating to be less beneficial (Silvestri et al., 1992).

11.7.8. Improving other plant components

Measures to improve other plant components using exergy analysis have been examined. For example, the use of exergy analysis to improve compression and expansion processes has been reported. Also, the optimization of heat exchangers using exergy-based criteria has been carried out in general, and specifically for feedwater heaters (Rosen and Dincer, 2003b,c).

Some other practical measures that can reduce the more significant losses in coal-fired electricity generation include the following. Heat-transfer losses can be reduced by using smaller heat-exchange temperature differences and more efficient heat exchangers. This measure also involves ensuring that heat flows of appropriate temperatures are used to heat cooler flows. Further, expansion-process losses can be reduced by improving the efficiencies of expanders. This measure may include reducing condenser pressures. Also, losses can be reduced by ensuring that no unconstrained throttling (which is highly irreversible) occurs. The use of sliding pressure operations to avoid throttling is one practical example that has significant potential to reduce exergy losses.

11.8. Closing remarks

Exergy analysis is shown in this chapter to be able to help understand the performance of steam power plants, and identify and design possible efficiency improvements. In addition, exergy methods are useful in assessing which improvements are worthwhile, and should be used along with other pertinent information to guide efficiency improvement efforts for steam power plants. Of course, measures to improve efficiency should be weighed against other factors and implemented only where appropriate.

Problems

11.1 How are the energy and exergy efficiencies of steam power plants defined? Which one is typically greater than the other? Explain.

11.2 Explain the difference between the energy and exergy efficiencies for (a) a geothermal power plant and (b) a conventional steam power plant.

11.3 Identify the sources of exergy loss in steam power plants and propose methods for reducing or minimizing them.

11.4 Which component in a steam power plant has the greatest exergy destruction? What are the causes and how can the exergy destruction be reduced or minimized?

11.5 Define the exergy efficiency of a steam turbine. Is the definition of isentropic efficiency the same as that for the exergy efficiency? Explain. Which efficiency is typically greater?

11.6 An engineer performs an exergy analysis of a steam power plant and determines the exergy efficiencies of the turbine and pump to be 90% and 65%, respectively. The engineer claims based on these results that improving the performance of the pump will increase the net power output more than improving the performance of the turbine. Do you agree with the engineer? Explain.

11.7 How can you compare the exergy efficiencies of steam power plants using as fuel (a) coal, (b) natural gas, (c) oil and (d) uranium?

11.8 What are the effects on the exergy efficiency of a steam power plant of (a) boiler pressure, (b) boiler temperature and (c) condenser pressure? Explain using the exergy destruction concept.

11.9 Compare the exergy efficiencies of steam power plants and combined cycle plants.

11.10 Identify and describe several methods for increasing the exergy efficiency of steam power plants.

11.11 Obtain a published article on exergy analysis of steam power plants. Using the operating data provided in the article, perform a detailed exergy analysis of the plant and compare your results to those in the original article. Also, investigate the effect of varying important operating parameters on the system exergetic performance.

11.12 Obtain actual operating data from a steam power plant in your area and perform a detailed exergy analysis. Discuss the results and provide recommendations based on the exergy results for improving the efficiency.

Chapter 12

EXERGY ANALYSIS OF COGENERATION AND DISTRICT ENERGY SYSTEMS

12.1. Introduction

Cogeneration is a technique for producing heat and electricity in one process that can save considerable amounts of energy. Cogeneration is often associated with the combustion of fossil fuels, but can also be carried out using some renewable energy sources and by burning wastes. The trend recently has been to use cleaner fuels for cogeneration such as natural gas. Note that the strong long-term prospects for cogeneration in global energy markets are related to its ability to provide a multitude of operational, environmental and financial benefits.

Cogeneration often reduces energy use cost-effectively and improves security of energy supply. In addition, (i) since cogeneration installations are usually located close to consumers, electrical grid losses are reduced when cogeneration is applied; (ii) cogeneration increases competition among producers and provides opportunities to create new enterprises and (iii) cogeneration is often well suited for use in isolated or remote areas.

Cogeneration is a very attractive option for facilities with high electric demands and buildings that consume large amounts of hot water and electricity every month. The higher the electric rates, the greater are the savings with cogeneration and the faster the savings pay for the initial capital investment.

The product thermal energy from cogeneration can be used for domestic hot water heating, space heating, pool and spa heating, laundry processes and absorption cooling. The more the product heat from cogeneration can be used year round in existing systems, the more financially attractive is cogeneration. Facilities that use large amounts of thermal energy each month include:

- assisted living facilities, nursing homes and senior housing,
- apartments and condominiums,
- colleges, universities and other educational institutions,
- hospitals,
- hotels,
- athletic clubs,
- industrial and waste treatment facilities,
- laundries.

Cogeneration helps overcome the main drawback of conventional electrical and thermal systems: the significant heat losses which detract greatly from efficiency. Heat losses are reduced and efficiency is increased when cogeneration is used to supply heat to various applications and facilities. The overall energy efficiency of a cogeneration system is the percent of the fuel converted to both electricity and useful thermal energy. Typical cogeneration systems have overall efficiencies ranging 65–90%.

District energy systems can utilize many energy resources, ranging from fossil fuels to renewable energy to waste heat, and are sometimes called 'community energy systems.' This is because, by linking a community's energy users together, district energy systems improve efficiency and provide opportunities to connect generators of waste energy (e.g., electric power plants or industrial facilities) with consumers who can use the waste energy. The heat in a district energy system can be used for heating or can be converted to cooling using absorption chillers or steam turbine-driven chillers.

District energy systems include both district heating and district cooling systems and distribute steam, hot water and chilled water from a central plant to individual buildings through a network of pipes. District energy systems provide space heating, air conditioning, domestic hot water and/or industrial process energy, and often are linked with electricity generation via cogeneration. With district energy, boilers and chillers in individual buildings are not required. District energy is often an attractive, efficient and environmentally benign way to reduce energy consumption. The Intergovernmental Panel on Climate Change (IPCC) identified cogeneration and district energy as a key measure for greenhouse gas reduction, and the European Commission has been developing cogeneration and district energy systems for the European Union.

District energy systems can provide other environmental and economic benefits, including:

- reduced local/regional air pollution,
- increased opportunities to use ozone-friendly cooling technologies,
- infrastructure upgrades and development that provide new jobs,
- enhanced opportunities for electric peak reduction through chilled water or ice storage,
- increased fuel flexibility,
- improved energy security.

Energy and exergy analyses of cogeneration-based district energy systems are described in this chapter. Relative to conventional systems, such integrated systems can be complex in that they often carry out the provision of electrical, heating and cooling services simultaneously. Consequently, they are sometimes referred to as trigeneration systems. This chapter describes the benefits of applying exergy analysis to such systems, and explains key exergy-based measures of the performance of such systems. A specific case is considered to illustrate the topics covered including the determination of system and component efficiencies and their improvement.

The chapter reveals insights that can aid in the design of such systems and related optimization activities, and in the selection of the proper type of systems for different applications and situations. This knowledge can help energy utilities improve existing plants where appropriate, and develop better designs.

In the example, a cogeneration-based district energy model is used to examine a range of scenarios for different systems. A design for cogeneration-based district heating (using a district heating network and heat exchangers) and district cooling (using central, electrically driven centrifugal chillers and a district cooling network), proposed by the utility Edmonton Power (1991) and MacRae (1992), is evaluated with energy and exergy analyses. Then, the design is modified by replacing the electric centrifugal chillers with heat-driven absorption chillers, and the evaluation is repeated.

Another key point raised in this chapter relates to some of the difficulties associated with the types of analysis tools often used for cogeneration/district energy systems. In general, energy technologies are normally examined thermodynamically using energy analysis, although a better understanding is attained when a more complete thermodynamic view is taken. Exergy analysis provides an additional thermodynamic perspective and, in conjunction with energy analysis, permits the performance of more complete thermodynamic analyses.

Applications of exergy analysis to cogeneration have increased in recent years (Rosen et al., 2001; Dincer et al., 2003; Rosen, 2003a,b; Rosen and Dincer, 2003d; Rosen et al., 2005), and have yielded useful results and provided meaningful insights into performance that assist in achieving optimal designs.

12.2. Cogeneration

Cogeneration, or combined heat and power (CHP), is the simultaneous production of electricity and usable heat. In conventional power plants, a large amount of heat is produced but not used. By designing systems that can use the heat, the efficiency of energy production can be increased from current levels that range from 35% to 55%, to over 80% (DOE, 2003). New technologies are making cogeneration cost-effective at smaller scales, meaning that electricity and heat can be produced for neighborhoods or even individual sites. Micro-cogeneration systems produce heat and power at site scale – for individual buildings or building complexes.

Cogeneration is a proven technology that has been around for over 100 years. Early in the 20th century, before there was an extensive network of power lines, many industries had cogeneration plants. In the U.S. the first commercial cogeneration plant was designed and built by Thomas Edison in 1882 in New York (DOE, 2003). Primary fuels commonly used in cogeneration include natural gas, oil, diesel fuel, propane, coal, wood, wood-waste and biomass. These 'primary' fuels are used to make electricity, a 'secondary' energy form. This is why electricity, when compared on a kWh to kWh basis, is typically 3–4 times more expensive than primary fuels such as natural gas.

The thermal cogeneration product is normally in the form of steam and/or hot water and the energy source is often a fossil fuel or uranium. Cogeneration has been used, particularly in industry, for approximately a century. A cogenerator can be a utility, an industry, a government or any other party.

Cogeneration systems are often based on thermal electrical generating stations (such as fossil fuel and nuclear plants), where the energy content of a resource (normally a fossil or nuclear fuel) is converted to heat (in the form of steam or hot gases) which is then converted to mechanical energy (in the form of a rotating shaft), which in turn is converted to electricity. A portion (normally 20–45%) of the heat is converted to electricity, and the remainder is rejected to the environment as waste.

Cogeneration systems are similar to thermal electricity-generation systems, except that a percentage of the generated heat is delivered as a product (normally as steam or hot water), and the quantities of electricity and waste heat produced are reduced. Overall cogeneration efficiencies (based on both the electrical and thermal energy products) of over 80% are achievable. Other advantages generally reported from cogenerating thermal and electrical energy rather than generating the same products in separate processes include: reduced energy consumption, reduced environmental emissions (due to reduced energy consumption and the use of modern technologies in large, central installations) and more economic operation. Most thermal systems for large-scale electricity generation are based on steam and/or gas turbine cycles, and can be modified relatively straightforwardly for cogeneration. Two main categories of heat demands can normally be satisfied through cogeneration: (i) residential, commercial and institutional processes, which require large quantities of heat at relatively low temperatures (e.g., for air and water heating) and (ii) industrial processes, which require heat at a wide range of temperatures (e.g., for drying, heating, boiling in, for instance, chemical processing, manufacturing, metal processing, mining and agriculture).

The use of a central heat supply to meet residential, commercial and institutional heat demands is often referred to as district heating. As well as satisfying heat demands, cogenerated heat can also drive chillers; this application (Rosen and Dimitriu, 1993) could be particularly beneficial in Ontario where the peak electrical demand is often associated with the summer cooling load.

Many general descriptions and studies of cogeneration systems have been reported, and the basic technology is well understood and proven. Numerous examples exist of large cogeneration systems: (i) a steam turbine plant in Switzerland generates 465 MW of thermal power and 135 MW of electrical power, with an overall efficiency of 75% (Horlock, 1987), (ii) a nuclear power plant in Michigan left incomplete due to lack of funding was eventually completed as a gas-fired combined cycle cogeneration plant having 12 heat recovery steam generators and gas turbines and two steam turbines, producing 1400 MW of electrical power and 285,000 kg/h of steam (Collins, 1992) and (iii) approximately 10 plants are used to generate 240 MW of electrical power and to supply 90% of the 1500 MW thermal demand for the city of Malmo, Sweden (population 250,000) (Malmo Energi AB, 1988). In the last example, fuel drives two of the plants (an extraction steam turbine plant generating 110 MW of electrical power and 240 MW of thermal power, and a back pressure steam turbine plant generating 130 MW of electrical power and 300 MW of thermal power), while the remaining plants operate on waste heat from neighboring industries (e.g., smelting, carbon-black production, sewage treatment and refuse incineration).

The size and type of a cogeneration system are normally selected to match as optimally as possible the thermal and electrical demands. Many matching schemes can be used. Systems can be designed to satisfy the electrical or thermal-base loads, or to follow the electrical or thermal loads. Storage systems for electricity (e.g., batteries) or heat (e.g., hot water or steam tanks) are often used to overcome periods when demands and supplies for either electricity or heat are not coincident. Cogeneration systems are sometimes used to supply only the peak portions of the electrical or thermal demands.

The thermal product of a cogeneration system often offsets the need for heating plants, where energy in the form of a fossil fuel or electricity is converted to heat (in the form of hot gases or another heated medium), often with an energy conversion efficiency of over 80%.

12.3. District energy

District energy (or district heating and cooling) is the utilization of one or more community energy sources to provide energy services to multiple users. This approach can replace individual, building-based furnaces, air-conditioning units, boilers and chillers. With a district energy system, thermal energy, via, e.g., hot water, steam or chilled water, is distributed by underground pipelines from the source of the energy to several buildings. Energy is then extracted at the buildings and return pipes bring the water back to the energy source to be heated or cooled again. A district energy system is capable of providing heating and/or cooling, and many are linked to electricity generation facilities and thus can also provide power.

District energy systems offer a community several advantages: (i) They free the individual building owner from the need to own and maintain a heating plant and to procure and store fuel on site. (ii) They have flexible operating systems, which can run on a variety of energy sources and can be designed to meet any consumer requirements, regardless of size. (iii) They can impact positively a community's economic situation, as well as its environmental surroundings. (iv) They provide a possible means of meeting an increase in demand for energy while, at the same time, eliminating the need to build additional power plants. (v) They can reduce local pollution levels, and incorporate limited numbers of emissions stacks which can be situated so as to disperse emissions away from municipal areas. (vi) They can use high-efficiency chillers to produce and share ice and near-freezing water efficiently and effectively using off-peak, less expensive electricity, while eliminating or reducing capital expenditures and the problems associated with the operation and maintenance of a heating and cooling system (CETC, 2003).

Due to these and other advantages, district energy has made a considerable contribution to energy conservation and environmental protection in various countries, e.g., the U.S., Canada and Denmark. In Denmark, for example, imported oil provided 92% of Danish energy supplies at the time of the first oil crisis in 1973, when the degree of energy self-sufficiency was only about 2%. At the time, district heating supplied 30% of the Danish heating market, with one-third of this heating provided via CHP. Since 1973, Denmark has achieved significant success with its energy programs (Pierce, 2002):

- Dependence on imported oil has decreased from 42% of total energy consumption to only 2%, while use of indigenous energy sources has risen from 2% of total energy consumption to 65%, respectively.
- District heating market penetration has increased from 30% to 50%, while the percentage of district heating produced by cogeneration has risen from 33% to 64%. Only 16% of district heating is now generated from fossil fuels in heat-only plants, while renewable energy now provides 20% of district heating (10% refuse incineration, 9% biomass and 1% industrial waste heat). Heat for larger urban areas is mainly provided by coal-fired CHP plants, while many smaller communities employ biomass-fired cogeneration.

District energy systems reduce greenhouse gas emissions in two ways: by replacing less efficient equipment in individual buildings with a more efficient central power plant; and by producing electricity for the central grid that can displace, for example, coal-fired and other electricity sources that involve higher greenhouse gas emissions per kilowatt-hour.

Storage of chilled water or ice is an integral part of many district cooling systems. Storage allows cooling capacity to be generated at night for use during the hottest part of the day, thereby helping manage the demand for electricity and reducing the need to build power plants (Spurr, 2003).

Figure 12.1 breaks down by sector the use of the products of district energy systems in the U.S. Most district energy output is seen to be utilized in institutional systems, which serve groups of buildings owned by one entity, such as colleges, universities, hospitals and military facilities. However, significant growth in district energy, particularly district cooling, is currently occurring in utility systems serving downtown areas.

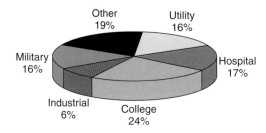

Fig. 12.1. Use by sector of district energy output in the US (Spurr, 2003).

District energy systems can use a diversity of energy resources, ranging from fossil fuels to renewable energy to waste heat. These systems are sometimes called 'community energy systems' because they link a community's energy producers and users. Such district energy systems normally increase efficiency and provide opportunities to connect generators of waste energy (e.g., electric power plants or industrial facilities) with consumers who can use that energy. The heat recovered through district energy can be used for heating or can be converted to cooling using absorption chillers or steam turbine-driven chillers.

12.4. Integrated systems for cogeneration and district energy

District energy systems are multi-building heating and cooling systems. Some district energy systems incorporate cogeneration systems. Heat is distributed by circulating hot water (or low-pressure steam) through underground piping. Some systems also include district cooling. The source of energy for district heating systems is usually a steam boiler, typically fired by natural gas, although other sources are possible. Hybrid systems, using a combination of natural gas, wood waste, municipal solid waste and waste heat from industrial sources are possible, and often more economical. District energy technologies are in common use. Because they often use waste heat, either from an existing boiler that is currently venting excess heat, or from electricity generation facilities, they are more efficient, cleaner and often more cost-effective than conventional supply systems.

12.5. Simplified illustrations of the benefits of cogeneration

Three simple illustrations are considered which demonstrate several benefits of cogeneration. In each illustration, specified demands for thermal and electrical energy are given. Then two methods are assessed for satisfying the demands, one using cogeneration, and the other based on separate processes for heat and electricity generation. Device efficiencies are specified. The illustrations highlight the reduction in energy consumption and environmental emissions, and the increase in energy efficiency, when cogeneration is substituted for separate electrical and heat generation processes. In the analyses, minor losses such as those associated with distribution are neglected.

12.5.1. Energy impacts

Illustration 1: Fuel cogeneration vs. fuel electrical generation and fuel heating This illustration considers the substitution of fuel-driven cogeneration for fuel-driven electrical generation and fuel heating (see Fig. 12.2a). A demand for 20 units of electricity and 72 units of product heat is considered. The cogeneration unit is assumed to operate like the coal cogeneration option described in Table 12.1, which has electrical and thermal efficiencies of 20% and 72%, respectively, and an overall efficiency of 92%. The separate electricity generation process is assumed to have the same efficiency as the coal-fired Nanticoke Generating Station without cogeneration (37%), and the efficiency of the separate heating process is assumed to be 85%, which is typical of a fuel-fired boiler (MacRae, 1992).

It can be shown that the cogeneration system consumes 100 units of fuel energy and loses 8 units of energy, while the two separate processes together consume 139 units of fuel energy (54 for electricity generation and 85 for heat production) and lose 47 units of energy (34 from electricity generation and 13 from heat production). Thus, cogeneration substitution decreases fuel energy consumption here by $[(139 - 100)/139] \times 100\% = 28\%$. Also, the cogeneration system efficiency (92%) exceeds that for the electricity generation system (37%), the heating system (85%), and the combined system containing the separate electricity generation and heating processes ($[92/139] \times 100\% = 66\%$).

Illustration 2: Nuclear cogeneration vs. nuclear electrical generation and fuel heating This illustration considers the substitution of nuclear cogeneration for nuclear electrical generation and fuel heating (see Fig. 12.2b). A demand for 11 units of electricity and 81 units of product heat is considered. The cogeneration unit is assumed to operate like the nuclear cogeneration option described in Table 12.1, which has electrical and thermal efficiencies of 11% and 81%, respectively, and an overall efficiency of 92%. The separate electricity generation process is assumed to have the same efficiency as Pickering Nuclear Generating Station without cogeneration (30%), and the efficiency of the separate heating process is assumed to be 85%.

It can be shown that the cogeneration unit consumes 100 units of nuclear energy and loses 8 units of energy, while the two separate processes together consume 132 units of energy (37 units of nuclear energy for electricity generation and 95 units of fuel energy for heat production) and lose 40 units of energy (26 from electricity generation and 14 for heat production). Thus, cogeneration substitution decreases fuel energy consumption here by $[(132 - 100)/132] \times 100\% = 24\%$, and eliminates fossil fuel consumption. Also, the cogeneration system efficiency (92%) exceeds that for the electricity generation system (30%), the heating system (85%) and the combined system containing the separate electricity generation and heating processes ($[92/132] \times 100\% = 70\%$).

Illustration 3: Fuel cogeneration vs. fuel electrical generation and electrical heating This illustration considers the substitution of fuel-driven cogeneration for fuel-driven electrical generation and electrical heating (see Fig. 12.2c). This illustration is identical to Illustration 1 (Fig. 12.2a), except that the separate fuel heating process is replaced by heating (at 95% efficiency) with electricity produced by the separate electrical plant.

As in Illustration 1, the cogeneration unit consumes 100 units of fuel energy and loses 8 units of energy. However, the two separate processes together consume 259 units of fuel energy (all during electricity generation, which subsequently supplies 76 units of electricity to the heating process) and lose 167 units of energy (163 from electricity

262   *Exergy: Energy, Environment and Sustainable Development*

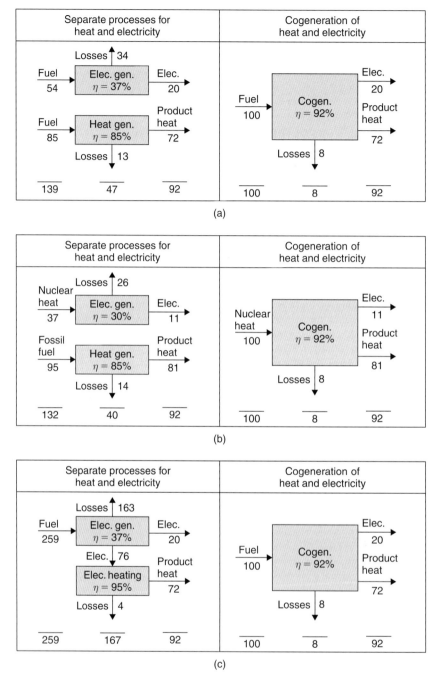

Fig. 12.2. Three illustrations of the decreased energy input required when cogeneration replaces separate heat and electricity production processes, for the same products. Relative energy units are shown for flows. Totals of input, lost and product energy values, respectively, are shown at the bottom of each diagram. (a) Illustration 1: Fuel cogeneration vs. fuel electrical generation and fuel heating; (b) Illustration 2: Nuclear cogeneration vs. nuclear electrical generation and fuel heating; (c) Illustration 3: Fuel cogeneration vs. fuel electrical generation and electrical heating.

Table 12.1. Data for sample coal and nuclear cogeneration options.

Cogeneration option	Product energy (MW)		Steam data			Efficiencies (%)		
	Electricity	Heat	Flow rate (kg/s)	Temperature (°C)	Pressure (MPa)	Electricity	Heat	Total
Coal	267	984	367	361	1.03	20	72	92
Nuclear	199	1423	603	238	0.45	11	81	92

generation and 4 from heat production). Thus, cogeneration substitution decreases fuel energy consumption here by $[(259 - 100)/259] \times 100\% = 61\%$. Also, the cogeneration system efficiency (92%) exceeds that of both the electricity generation system (37%) and the combined system containing the separate electricity generation and heating processes ($[92/259] \times 100\% = 36\%$).

12.5.2. Energy and exergy efficiencies

For the three examples, energy and exergy efficiencies for the separate processes for electricity and heat generation and for the cogeneration process are listed in Table 12.2. In evaluating the efficiencies, it is assumed that the reference-environment temperature T_0 is 15°C (288 K), the thermal product is delivered for all cases at an effective temperature T of 150°C (423 K), the energy and exergy of the fuel are identical, and the energy and exergy of 'nuclear heat' are identical. The exergetic temperature factor τ in this example is constant for all cases at

$$\tau = 1 - \frac{T_0}{T} = 1 - \frac{288\,\text{K}}{423\,\text{K}} = 0.3191$$

Table 12.2. Efficiencies for the three illustrations.

Illustration	Energy efficiency (%)		Exergy efficiency (%)		Ratio of efficiencies	
	η_{sep}	η_{cogen}	ψ_{sep}	ψ_{cogen}	η_{cogen}/η_{sep}	ψ_{cogen}/ψ_{sep}
1. Fuel electricity generation and fuel heating	66.2	92.0	30.9	43.0	1.39	1.39
2. Nuclear electricity generation and fuel heating	70.0	92.0	27.9	36.9	1.32	1.32
3. Fuel electricity generation and electrical heating	35.5	92.0	16.6	43.0	2.59	2.59

For case A, the energy and exergy efficiencies respectively for the separate processes taken as a whole are

$$\eta = \frac{20 + 72}{54 + 85}(100\%) = 66.2\%$$

$$\psi = \frac{20 + 72(0.3191)}{54 + 85}(100\%) = 30.9\%$$

and for the cogeneration process are

$$\eta = \frac{20 + 72}{100}(100\%) = 92.0\%$$

$$\psi = \frac{20 + 72(0.3191)}{100}(100\%) = 43.0\%$$

Similarly, for case B the energy and exergy efficiencies respectively for the separate processes are

$$\eta = \frac{11 + 81}{37 + 95}(100\%) = 70.0\%$$

$$\psi = \frac{11 + 81(0.3191)}{37 + 95}(100\%) = 27.9\%$$

and for the cogeneration process are

$$\eta = \frac{11 + 81}{100}(100\%) = 92.0\%$$

$$\psi = \frac{11 + 81(0.3191)}{100}(100\%) = 36.9\%$$

For case C the energy and exergy efficiencies respectively for the separate processes are

$$\eta = \frac{20 + 72}{259}(100\%) = 35.5\%$$

$$\psi = \frac{20 + 72(0.3191)}{259}(100\%) = 16.6\%$$

and for the cogeneration process are

$$\eta = \frac{20 + 72}{100}(100\%) = 92.0\%$$

$$\psi = \frac{20 + 72(0.3191)}{100}(100\%) = 43.0\%$$

The ratio of energy efficiencies for the cogeneration and separate processes are also shown in Table 12.2, along with the corresponding ratio of exergy efficiencies. It is seen that cogeneration increases both the energy and exergy efficiencies by 39% for case A, by 32% for case B and by 159% for case C.

Two primary points are illustrated in Table 12.2:

1. Cogeneration increases significantly the energy and exergy efficiency compared to separate processes for the same electrical and heating services.
2. The exergy efficiency is markedly lower than the corresponding energy efficiency for all cogeneration and non-cogeneration cases considered. This is because the thermal product, which is significantly larger than the electrical product, is delivered at a relatively low temperature (150°C) compared to the temperatures potentially achievable.

The latter point indicates that, although cogeneration improves efficiencies greatly compared to separate processes for each product, there remains a great margin for further improvement.

12.5.3. Impact of cogeneration on environmental emissions

Four key points related to environmental emissions are demonstrated by the illustrations. First, the substitution of cogeneration for separate electrical and heat generation processes for all illustrations considered leads to significant reductions in fuel energy consumption (24–61%), which approximately lead to proportional reductions in emissions. Second, the elimination of fossil fuel consumption in Illustration 2 eliminates fossil fuel emissions. Third, additional emission reductions may occur (although these are not evaluated here) when central cogeneration replaces many small heat producers. Controllable emissions are reduced at large central stations relative to small plants since better emission-control technologies as well as stricter emission limits and limit-verification mechanisms often exist at central stations. Fourth, energy losses, which relate to such environmental impacts as thermal pollution, are reduced significantly with cogeneration (by 83% for Illustration 1, 80% for Illustration 2 and 95% for Illustration 3).

12.5.4. Further discussion

Several other comments regarding the illustrations and results follow:

- An energy consumption decrease similar to that for Illustration 3 (61%) would occur if nuclear cogeneration were substituted for nuclear electrical generation and electrical heating.
- Sufficient markets must exist for cogenerated heat before cogeneration can be implemented. Potential markets, and the impacts of the various degrees of implementation possible for cogeneration, have been examined (e.g., Rosen et al., 1993).
- More detailed and comprehensive assessments are needed regarding how cogeneration can be integrated into regions before the complete effects of utility-based cogeneration can be fully understood. Other investigations directed toward obtaining a more complete understanding of these effects have been performed (e.g., Rosen and Le, 1994).
- The numerical values used in this chapter are approximate, and only intended for illustration. However, despite the approximate nature of these values and the other assumptions and simplifications introduced, the general findings remain valid when reasonably realistic alternative values are used.

The key points demonstrated by the illustrations can be summarized as follows:

- The better efficiencies and emission-reduction technologies for central cogeneration lead to reduced energy utilization and environmental emissions.
- Fuel consumptions and emissions are eliminated when nuclear-cogenerated heat is substituted for fuel heat.
- Large decreases in energy utilization and emissions occur when cogenerated heat offsets electrical heat.

12.6. Case study for cogeneration-based district energy

A major cogeneration-based district heating and cooling project in downtown Edmonton, Alberta, Canada is considered. The system (Edmonton Power, 1991; MacRae, 1992) has: (i) an initial supply capacity of 230 MW (thermal) for heating and 100 MW (thermal) for cooling; (ii) the capacity to displace about 15 MW of electrical power used for electric chillers through district cooling and (iii) the potential to increase the efficiency of the Rossdale power plant that would cogenerate to provide the steam for district heating and cooling from about 30% to 70%, respectively. The design includes the potential to expand the supply capacity for heating to about 400 MW (thermal). The design incorporates central chillers and a district cooling network. Screw chillers were to be used originally, and absorption chillers in the future.

Central chillers are often favored because: (i) the seasonal efficiency of the chillers can increase due to the ability to operate at peak efficiency more often in a central large plant and (ii) lower chiller condenser temperatures (e.g., 20°C) can be used if cooling water from the environment is available to the central plant, relative to the condenser temperatures of approximately 35°C needed for air-cooled building chillers. These two effects can lead to central large chillers having almost double the efficiencies of distributed small chillers.

There are two main stages in this analysis. First, the design for cogeneration-based district heating and cooling (Edmonton Power, 1991; MacRae, 1992) is evaluated thermodynamically. Then, the design is modified by replacing the electric centrifugal chillers with heat-driven absorption chillers (first single- and then double-effect types) and re-evaluated.

12.6.1. System description

Original system

The cogeneration-based district energy system considered here (Fig. 12.3) includes a cogeneration plant for heat and electricity, and a central electric chiller that produces a chilled fluid. Hot water is produced, to satisfy all heating requirements of the users, at a temperature and pressure of 120°C and 2 bar, respectively. The heat is distributed to the users via heat exchangers, district heating grids and user heat-exchanger substations. A portion of the cogenerated electricity is used to drive a central centrifugal chiller and the remaining electricity is used for other purposes (e.g., export, driving other electrical devices, etc.). The central chiller produces cold water at 7°C, which is distributed to users via district cooling grids.

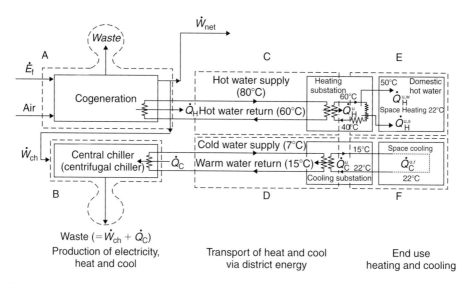

Fig. 12.3. Simplified diagram of the cogeneration-based district energy system of Edmonton Power. The system, which uses electric chillers, is divided into six subsections within three categories. On the left are production processes, including cogeneration of electricity and heat (A) and chilling (B). In the middle are district-energy transport processes, including district heating (C) and district cooling (D). On the right are end-user processes, including user heating (E) and user cooling (F).

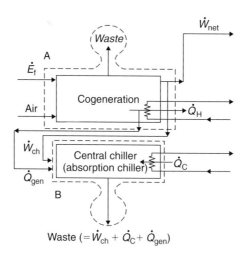

Fig. 12.4. Modified version of production processes (units A and B) for the simplified diagram in Fig. 12.3. In the modified system, the electric chillers are replaced with absorption chillers (single- or double-effect), driven mainly by heat from the cogeneration plant. The rest of the system in Fig. 12.3 (units C to F) remains unchanged in the modified system. The temperature of the heating medium supplied to the absorption chillers is higher for the double-effect chiller relative to the single-effect machine.

Modified system

For the cogeneration-based district energy system using absorption chillers, the design is modified by replacing the electric chiller with single-effect absorption chillers (see Fig. 12.4). Hot water is produced at 120°C and 2 bar to satisfy all heating requirements of the users and to drive the central absorption chiller. A small portion of the cogenerated electricity is used to drive the absorption-solution and refrigeration pumps and the remaining electricity is used for purposes other than space cooling.

This cogeneration-based district energy system is then further modified by replacing the electric centrifugal chillers with double-effect absorption chillers (see Fig. 12.4). The system is similar to the cogeneration-based district energy system using single-effect absorption chillers, except that higher-quality heat (170°C and 8 bar) is produced to drive the double-effect absorption chillers.

12.6.2. Approach and data

The plant is divided into subsections for analysis purposes. Efficiencies of the individual subsystems and the overall system are examined. Also, several selected combinations of the subsystems are evaluated to pinpoint better the locations and causes of inefficiencies.

For the analysis, the year is divided into two seasonal periods (see Table 12.3). Period 1 (October to April) has an environmental temperature of 0°C and is considered to be a winter period with only a heating demand. Period 2 (May to September) has an environmental temperature of 30°C and is considered to be a summer period with a cooling demand and a small heating demand for hot water heating. The small variations in plant efficiency that occur with changes in environmental temperature are neglected here.

Edmonton power has annual free cooling of 33 GWh/yr; the cooling requirement of the chilling plant is 169 GWh/yr. The coefficient of performance (*COP*) of the centrifugal chiller in the design is 4.5 (Edmonton Power, 1991). Thus, the annual electricity supply rate to the chiller is $\dot{W}_{ch} = 169/4.5 = 38$ GWh/yr. For the chilling operation, including free cooling and electrical cooling, $COP = (169 + 33)/38 = 202/38 = 5.32$. The net electricity output ($\dot{W}_{net}$) of the combined cogeneration/chiller portion of the system is $433 - 38 = 395$ GWh/yr, where the electrical generation rate of the cogeneration plant is 433 GWh/yr.

The overall energy efficiency of the proposed cogeneration plant is 85%, the electrical efficiency (i.e., the efficiency of producing electricity via cogeneration) is 25% and the heat production efficiency is 60%. Also, the total heating requirement of the buildings in the design region is $\dot{Q}_H = 1040$ GWh/yr for space and hot water heating, and the cooling requirement is $\dot{Q}_C = 202$ GWh/yr for space cooling (DOE, 2000). The total fuel energy input rate can be evaluated for the cogeneration plant using electric chillers as $\dot{E}_f = 1040/0.6 = 1733$ GWh/yr. Since 33 GWh/yr of this cooling is provided through free cooling, the cooling requirement of the chilling plant is 169 GWh/yr (Edmonton Power, 1991). The *COP* of the single-effect absorption chiller used here is taken to be 0.67, a typical value (Colen, 1990). Therefore, the annual heat required to drive the single-effect absorption machine is $\dot{Q}_{gen} = 169/0.67 = 252$ GWh/yr. The total fuel energy input rate to the cogeneration plant can thus be evaluated as $\dot{E}_f = (1040 + 252)/0.6 = 2153$ GWh/yr.

As mentioned above, steam is required at higher temperatures and pressures to drive the double-effect absorption chillers, and more electricity is curtailed as higher quality heat or more heat is produced. It is assumed that the overall energy efficiency of the proposed cogeneration plant is unchanged (85%) in Period 2. Only the electrical and heat efficiencies are changed due to more heat being produced in this period, when the absorption chiller is in operation. Thus, in Periods 1 and 2, respectively, the electrical efficiency (i.e., the efficiency of producing electricity via cogeneration) is 25% and 21% (Rosen and Le, 1998), and the heat production efficiency is 60% and 64% (Rosen and Le, 1995). The *COP* of the double-effect absorption chiller used here is taken to be 1.2, a typical value (Colen, 1990). Therefore, the annual heat required to drive the double-effect absorption machine is $\dot{Q}_{gen} = 169/1.2 = 141$ GWh/yr. The total fuel energy input rate to the cogeneration plant can be evaluated as the sum of the fuel energy input rate to the plant in the two periods. Thus $\dot{E}_f = 1942$ GWh/yr.

The *COP* for the chilling operation, including free cooling, using single-effect absorption cooling is $COP = 202/252 = 0.80$, and using double-effect absorption cooling is $COP = 202/141 = 1.43$. It is noted for the absorption chiller cases that, since the work required to drive the solution and refrigeration pumps is very small relative to the heat input (often less than 0.1%), this work is neglected here.

For simplicity, economics and part-load operation are not considered here, so the results and findings are thus correspondingly limited. Also, several simplifying energy-related assumptions are used to make the section concise and direct, while still permitting the differences between the energy and exergy results to be highlighted.

12.6.3. Preliminary analysis

In exergy analysis, the temperatures at different points in the system (see Figs. 12.3 and 12.4) are important. It is assumed that the average supply and return temperatures respectively are 80°C and 60°C for district heating, and 7°C and 15°C for district cooling (Edmonton Power, 1991). It is also assumed that the supply and return temperatures, respectively, are 60°C and 40°C for the user heating substation, and 15°C and 22°C for the user cooling substation. Furthermore, it is assumed that the user room temperature is constant throughout the year at 22°C. Note that the equivalent heat-transfer temperature T_{equiv} between the supply (subscript 1) and return (subscript 2) temperatures can be written as

$$T_{equiv} = \frac{h_1 - h_2}{s_1 - s_2} \qquad (12.1)$$

where h and s denote specific enthalpy and specific entropy, respectively. For district heating, the equivalent temperature is 70°C for the supply system and 50°C for the user substation, while for district cooling the equivalent temperature is 11°C for the supply system and 19°C for the user substation (Rosen and Le, 1998).

In Figs. 12.3 and 12.4, the system boundaries are for simplicity assumed to be located sufficiently far from the sources of losses that the temperature associated with such losses is equal to the temperature of the environment. The thermal exergy losses are then reduced to zero, but accounted for in the system irreversibilities.

Table 12.3 shows that 89.46% and 10.54% of the total annual heat loads occur in Periods 1 and 2, respectively. Since there is assumed to be no space heating demand in Period 2, the 10.54% quantity is taken to be the heat needs for water heating (which is assumed constant throughout the year). Table 12.3 also presents the space cooling breakdown in Period 2. Annual energy transfer rates for the cogeneration-based district energy system are shown in Table 12.4, with details distinguished where appropriate for the three chiller options considered. The data in Table 12.4 are used to calculate exergy efficiencies for the systems for each period and for the year.

Table 12.3. Monthly heating and cooling load breakdown (in %) in the design area of Edmonton, Alberta.

	Period 1 (Winter)								Period 2 (Summer)					
	Oct.	Nov.	Dec.	Jan.	Feb.	Mar.	Apr.	Total	May	June	July	Aug.	Sep.	Total
Heating load	6.90	12.73	16.83	18.67	14.05	12.95	7.34	89.46	2.39	1.56	1.34	1.92	3.33	10.54
Cooling load	0.0	0.0	0.0	0.0	0.0	0.0	0.0	0.0	10.62	22.06	32.00	26.80	8.52	100

Source: (Edmonton Power, 1991).

12.6.4. Analysis of components

Production of electricity, heat and cool

The production portion of the system consists of the cogeneration and chilling components.

Cogeneration: The total cogeneration component is considered first, and then a major section of this component, the furnace, is examined.

Total component: The electricity production rate $\dot{W}$ can be expressed for a cogeneration-based system using electric chillers as a function of the product-heat generation rate $\dot{Q}_H$ as

$$\dot{W} = \left(\frac{\eta_{elec}^{CHP}}{\eta_{heat}^{CHP}} \right) \dot{Q}_H \qquad (12.2)$$

and for a cogeneration-based system using absorption chillers as a function of the product-heat generation rates, $\dot{Q}_H$ and $\dot{Q}_{gen}$, as

$$\dot{W} = \left(\frac{\eta_{elec}^{CHP}}{\eta_{heat}^{CHP}} \right) (\dot{Q}_H + \dot{Q}_{gen}) \qquad (12.3)$$

Here, η_{elec}^{CHP} and η_{heat}^{CHP} denote respectively the electrical and heat efficiencies of the cogeneration plant.

The total energy efficiency can be written for the cogeneration plant using electric chillers as

$$\eta_{tot}^{CHP} = \frac{\dot{W} + \dot{Q}_H}{\dot{E}_f} \qquad (12.4)$$

Table 12.4. Energy transfer rates (in GWh/yr) for the cogeneration-based district energy system in Edmonton, Alberta.

Type of energy	Period 1, $T_o = 0°C$	Period 2, $T_o = 30°C$
District heating, $\dot{Q}_H$	$0.8946 \times 1040 = 930$	$0.1054 \times 1040 = 110$
Water heating, $\dot{Q}_H^{u,w}$	(22 GWh/yr/mo.) $\times$ 7 mo. $= 154$	$0.1054 \times 1040 = 110$ (or 22 GWh/yr/mo.)
Space heating, $\dot{Q}_H^{u,s}$	$930 - 154 = 776$	0
Space cooling, $\dot{Q}_C$	0	$1.00 \times 202 = 202$
Electric chiller case		
Total electricity, $\dot{W}$	$0.8946 \times 433 = 388$	$0.1054 \times 433 = 45.6$
Input energy, $\dot{E}_f$	$0.8946 \times 1733 = 1551$	$0.1054 \times 1733 = 183$
Single-effect absorption chiller case		
Heat to drive absorption chiller, $\dot{Q}_{gen}$	0	$1.00 \times 252 = 252$
Total electricity, $\dot{W}$	$0.8946 \times 433 = 388$	$25/60(110 + 252) = 151$
Input energy, $\dot{E}_f$	$0.8946 \times 1733 = 1551$	$(110 + 252)/0.6 = 603$
Double-effect absorption chiller case		
Heat to drive absorption chiller $\dot{Q}_{gen}$	0	$1.00 \times 141 = 141$
Total electricity, $\dot{W}$	$0.8946 \times 433 = 388$	$21/46 \times (110 + 141) = 82$
Input energy, $\dot{E}_f$	$0.8946 \times 1733 = 1551$	$(110 + 141)/0.64 = 391$

and for the cogeneration plant using absorption chillers as

$$\eta_{tot}^{CHP} = \frac{\dot{W} + \dot{Q}_H + \dot{Q}_{gen}}{\dot{E}_f} \tag{12.5}$$

where $\dot{E}_f$ denotes the fuel energy input rate.

The corresponding total exergy efficiency can be expressed for the cogeneration plant using electric chillers as

$$\psi_{tot}^{CHP} = \frac{\dot{W} + \tau_{Q_H}\dot{Q}_H}{(R\dot{E})_f} \tag{12.6}$$

and for the cogeneration plant using absorption chillers as

$$\psi_{tot}^{CHP} = \frac{\dot{W} + \tau_{Q_H}\dot{Q}_H + \tau_{Q_{gen}}\dot{Q}_{gen}}{(R\dot{E})_f} \tag{12.7}$$

where τ_{Q_H} and $\tau_{Q_{gen}}$ are the exergetic temperature factors for $\dot{Q}_H$ and $\dot{Q}_{gen}$, respectively. For heat transfer at a temperature T, $\tau \equiv 1 - T_o/T$ (Kotas, 1995), where T_o denotes the environmental temperature. Here, R denotes the energy grade function, values of which for most common fossil fuels are between 0.9 and 1.0 (Rosen and Le, 1995).

Furnace portion of cogeneration component: It is worthwhile to examine the characteristics of the furnace portion of the cogeneration plant, and to assess its efficiencies. The fuel used can be coal, uranium, oil, natural gas, etc. The energy efficiency of the furnace, evaluated as the ratio of furnace-delivery heat to input fuel energy, is taken to be 90%; 10% of the input energy is taken to be lost through stack gases, material wastes and miscellaneous heat losses. The energy efficiency of the boiler portion of the cogeneration plant is assumed to be 100%, as this section is considered perfectly insulated.

An exergy balance for the furnace (Fig. 12.5) can be written as

$$(R\dot{E})_f = (\tau_Q\dot{Q})_{furn} + \dot{I}_{furn} \tag{12.8}$$

where $\dot{Q}_{\text{furn}}$ denotes the furnace-delivery heat rate and $\dot{I}_{\text{furn}}$ denotes the furnace irreversibility rate. The chemical exergy of input air is zero since air is taken free from the environment. In the above equation, the exergy of the stack gases and thermal losses are neglected by locating the system boundary far from the sources of losses as shown in Fig. 12.5.

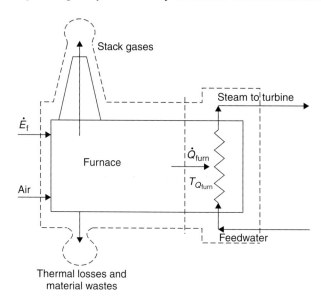

Fig. 12.5. A steam generator, simplified as a furnace (left side of vertical dashed line) and heat exchanger (right side of vertical dashed line). The vertical dashed line can be located at different places within the furnace so that the temperature associated with the furnace-delivery heat can be in part selected arbitrarily.

The energy efficiency of the furnace can be expressed as

$$\eta_{\text{furn}} = \dot{Q}_{\text{furn}}/\dot{E}_{\text{f}} \tag{12.9}$$

and the exergy efficiency as

$$\psi_{\text{furn}} = \frac{(\tau_Q \dot{Q})_{\text{furn}}}{(R\dot{E})_{\text{f}}} \tag{12.10}$$

or, using Eq. (12.8), as

$$\psi_{\text{furn}} = \frac{\eta_{\text{furn}}}{R_{\text{f}}} \tau_{Q_{\text{furn}}} = \frac{\eta_{\text{furn}}}{R_{\text{f}}} \left(1 - \frac{T_{\text{o}}}{T_{Q_{\text{furn}}}}\right) \tag{12.11}$$

The exergy efficiency is seen from the above equation to be proportional to both the energy efficiency and the exergetic temperature factor for the furnace-delivery heat, and inversely proportional to the energy grade function. The exergy efficiency, unlike the energy efficiency, takes into account the environmental temperature T_{o} and the furnace-delivery heat temperature $T_{Q_{\text{furn}}}$. As the energy grade function often has a value of unity, the deviation between the energy and exergy efficiencies of the furnace mainly depends on the ratio of temperatures $T_{\text{o}}/T_{Q_{\text{furn}}}$. The exergy and energy efficiencies can be almost equal if $T_{Q_{\text{furn}}}$ is high enough, while the energy efficiency greatly exceeds the exergy efficiency when $T_{Q_{\text{furn}}}$ is low (i.e., near T_{o}). In general, the adiabatic flame temperature of fossil fuels is much higher than the environmental temperature, so the furnace energy and exergy efficiencies can take on similar numerical values.

To illustrate, efficiencies are presented in Table 12.5 for values of the furnace-delivery heat temperature ranging from the high temperatures near the core of the furnace (2000 K) to the relatively lower temperatures of the combustion gases near to the walls of the cogeneration system steam generator (1000 K). As the furnace-delivery heat temperature increases, the energy efficiency stays fixed at 90% (as described earlier) while the exergy efficiency increases toward 90%, reaching 76% for a furnace-heat delivery temperature of 2000 K. The actual efficiencies for the furnace portion of the cogeneration system depend on where the boundary between the furnace and steam generator is chosen to be located.

Table 12.5. Efficiencies for the furnace section of the cogeneration plant, for a range of furnace-delivery heat temperatures.*

Temperature of furnace-delivery heat, $T_{Q_{\text{furn}}}$ (K)	Efficiency (%)	
	Energy (η)	Exergy (ψ)
1000	90	63
1500	90	72
2000	90	76

* Here the environmental temperature for summer conditions is used (i.e., $T_0 = 30°C$), and the energy and exergy values are taken to be equal for the 'fuel' resource.

Chilling: The exergy efficiency can be written for the chilling operation using electric chillers as

$$\psi_{\text{ch}} = \frac{-\tau_{Q_C}\dot{Q}_C}{\dot{W}_{\text{ch}}} \tag{12.12}$$

and using absorption chillers as

$$\psi_{\text{ch}} = \frac{-\tau_{Q_C}\dot{Q}_C}{\tau_{Q_{\text{gen}}}\dot{Q}_{\text{gen}}} \tag{12.13}$$

Transport and utilization of heat and cool

The transport portion of the system involves district heating and cooling, while the utilization portion includes end-use heating and cooling.

District heating and cooling: District heating utilizes hot water supply and warm water return pipes, while district cooling utilizes cold water supply and cool water return pipes. The pipes are assumed here to be perfectly insulated so that heat loss or infiltration during fluid transport can be neglected. Hence, the energy efficiencies of the district heating and cooling portions of the system are both 100%. The exergy efficiency can be evaluated for district heating as

$$\psi_{\text{DH}} = \frac{\tau_{Q_H^u}\dot{Q}_H^u}{\tau_{Q_H}\dot{Q}_H} \tag{12.14}$$

and for district cooling as

$$\psi_{\text{DC}} = \frac{-\tau_{Q_C^u}\dot{Q}_C^u}{-\tau_{Q_C}\dot{Q}_C} \tag{12.15}$$

End-use heating and cooling: Heat loss and infiltration for the end-use heating and cooling components are assumed negligible here so that their energy efficiencies are 100%. The exergy efficiency can be expressed for end-use heating as

$$\psi_{\text{UH}} = \frac{\tau_{Q_H^{u,s}}\dot{Q}_H^{u,s} + \tau_{Q_H^{u,w}}\dot{Q}_H^{u,w}}{\tau_{Q_H^u}\dot{Q}_H^u} \tag{12.16}$$

and for end-use cooling as

$$\psi_{\text{UC}} = \frac{-\tau_{Q_C^{u,r}}\dot{Q}_C^{u,r}}{-\tau_{Q_C^u}\dot{Q}_C^u} \tag{12.17}$$

The left and right terms in the numerator of Eq. (12.16) represent the thermal exergy supply rates for space and hot water heating, respectively.

12.6.5. Analysis of overall system

Since there are three different products generated (electricity, heat and cool), application of the term energy efficiency here is prone to be misleading, in part for the same reason that 'energy efficiency' is misleading for a chiller. Here, an overall system 'figure of merit' f_{sys} is used, and calculated as follows (Rosen and Le, 1995):

$$f_{sys} = \frac{\dot{W}_{net} + \dot{Q}_H^{u,s} + \dot{Q}_H^{u,w} + \dot{Q}_C^{u,r}}{\dot{E}_f} \tag{12.18}$$

The corresponding exergy-based measure of efficiency is simply an exergy efficiency, and is evaluated as

$$\psi_{sys} = \frac{\dot{W}_{net} + \tau_{Q_{vH}^{u,s}}\dot{Q}_H^{u,s} + \tau_{Q_H^{u,w}}\dot{Q}_H^{u,w} - \tau_{Q_C^{u,r}}\dot{Q}_C^{u,r}}{(R\dot{E})_f} \tag{12.19}$$

12.6.6. Effect of inefficiencies in thermal transport

The hot water supply and warm water return pipes in district heating and cooling systems can be on or below ground. For a realistic analysis, the heat loss or gain during fluid transport should be considered, even if the pipes are insulated. Hence, the energy efficiencies of the district heating and cooling portions of the system are both in reality less than 100%. The thermal interaction between heating/cooling pipe and ground can be very complex. An analysis of the process should include weather data for the ground–air boundary interaction and soil thermal and hydraulic properties for heat and moisture transport in soils.

The exergy efficiency can be evaluated by extending Eqs. (12.14) and (12.15) to account for thermal losses/gains for district heating as

$$\psi_{DH} = \frac{\tau_{Q_H^u}\dot{Q}_H^u - \tau_{Q_H^l}\dot{Q}_H^l}{\tau_{Q_H}\dot{Q}_H} \tag{12.20}$$

and for district cooling as

$$\psi_{DC} = \frac{-\tau_{Q_C^u}\dot{Q}_C^u + \tau_{Q_C^g}\dot{Q}_C^g}{-\tau_{Q_C}\dot{Q}_C} \tag{12.21}$$

The right terms in the numerator of Eqs. (12.20) and (12.21) represent the piping thermal exergy loss and gain in the ground, respectively.

The corresponding exergy-based measure of efficiency is simply an overall system exergy efficiency, and is evaluated by similarly extending Eq. (12.19) as

$$\psi_{sys} = \frac{\dot{W}_{net} + \tau_{Q_H^{u,s}}\dot{Q}_H^{u,s} + \tau_{Q_H^{u,w}}\dot{Q}_H^{u,w} - \tau_{Q_C^{u,r}}\dot{Q}_C^{u,r} - \tau_{Q_H^l}\dot{Q}_H^l + \tau_{Q_C^g}\dot{Q}_C^g}{(R\dot{E})_f} \tag{12.22}$$

12.6.7. Analyses of multi-component subsystems

Several multi-component subsystems within the overall cogeneration-based district energy system can be identified which have important physical meanings. The most significant of these subsystems are identified and discussed in the next section. The energy and exergy efficiencies for these subsystems are based on the efficiencies presented in previous sections for the overall system and its components.

12.6.8. Results

For the cogeneration-based district energy system using electric chillers, single-effect absorption chillers and double-effect absorption chillers, Tables 12.6 through 12.9 list the energy and exergy efficiencies evaluated for the main system components, for several subsystems comprised of selected combinations of the components, and for the overall system.

Efficiencies are presented in Table 12.6 for each of the six main components of the cogeneration-based district energy system (Figs. 12.3 and 12.4), for the three chiller cases considered. Also listed in Table 12.6 are efficiencies,

Table 12.6. Efficiencies for the six components and three main function-based subsystems of the cogeneration-based district energy system, considering three types of chillers.

Subsystem	Energy efficiency, η (%)			Exergy efficiency, ψ (%)		
	Centrifugal chiller	1-stage absorption chiller	2-stage absorption chiller	Centrifugal chiller	1-stage absorption chiller	2-stage absorption chiller
Production of electricity, heat and cool						
Cogeneration	85	85	85	37	37	37
Chilling	450*	67*	120*	36	23	30
Combined cogeneration and chilling	94	83	88	35	35	35
Transport of heat and cool						
District heating	100	100	100	74	74	74
District cooling	100	100	100	58	58	58
Combined district heating and cooling (i.e., district energy)	100	100	100	73	73	73
End-use heating and cooling						
End-use heating	100	100	100	54	54	54
End-use cooling	100	100	100	69	69	69
Combined end-use heating and cooling	100	100	100	55	55	55

* These are coefficient of performance (COP) values when divided by 100.

Table 12.7. Efficiencies for the overall system and the subsystems representing the heating and cooling sides of the cogeneration-based district energy system.

Subsystem	Energy efficiency, η (%)			Exergy efficiency, ψ (%)		
	Centrifugal chiller	1-stage absorption chiller	2-stage absorption chiller	Centrifugal chiller	1-stage absorption chiller	2-stage absorption chiller
Heating side (cogeneration, district heating and end-use heating)	85	85	85	30	31	31
Cooling side (chilling, district cooling and end-use cooling)	532*	80*	143*	14	9	12
Overall system	94	83	88	28	29	29

* These are coefficient of performance (COP) values when divided by 100.

broken down by function category (production, transport, use), for the three main subsystems identified in Figs. 12.3 and 12.4, i.e.,

- production of electricity, heat and cool (including cogeneration and chilling),
- transport of heat and cool (consisting of district heating and cooling, also known in combination as district energy),
- end-use heating and cooling (for space and hot water heating, and space cooling).

Table 12.8. Efficiencies for the thermal-energy distribution portion of the system (i.e., the combined transport and end-use subsystems).

Subsystem	Energy efficiency, η (%)			Exergy efficiency, ψ (%)		
	Centrifugal chiller	1-stage absorption chiller	2-stage absorption chiller	Centrifugal chiller	1-stage absorption chiller	2-stage absorption chiller
District and end-use heating	100	100	100	40	40	40
District and end-use cooling	100	100	100	41	41	41
Combined district energy* and end-use heating and cooling	100	100	100	40	40	40

* District energy is combined district heating and cooling.

Table 12.9. Efficiencies for subsystems of the cogeneration-based district energy system, selected to reflect the perspective of a production utility.[a]

Subsystem	Energy efficiency, η (%)			Exergy efficiency, ψ (%)		
	Centrifugal chiller	1-stage absorption chiller	2-stage absorption chiller	Centrifugal chiller	1-stage absorption chiller	2-stage absorption chiller
Cogeneration and district heating	85	85	85	34	35	34
Chilling and district cooling	532[b]	80[b]	143[b]	21	14	18
Combined cogeneration, chilling and district energy[c]	94	83	88	32	32	32

[a] District energy production utilities are usually responsible for producing thermal energy and transporting it to users, but not for the end-use processes.
[b] These are coefficient of performance (COP) values when divided by 100.
[c] District energy is combined district heating and cooling.

Efficiencies of the overall cogeneration-based district energy system, for the three chiller cases considered, are presented in Table 12.7. The efficiencies for the heating and cooling sides of the overall system are also presented in Table 12.7.

Efficiencies are presented in Table 12.8 for the portion of the cogeneration-based district energy system involving the distribution of thermal energy (heat or cool). This subsystem comprises the district and end-use heating and cooling components. Efficiencies for the heating and cooling portions of this thermal energy distribution subsystem are also given in Table 12.8. The principal process occurring in all of the components of this subsystem is heat transfer.

Utilities which provide electricity, heating and cooling services by operating district energy systems are normally mainly concerned with the processes involved in producing these energy forms and transporting them to users. The end-use of the energy commodities is left to the users. Hence, from the perspective of district energy utilities, it is useful to know the efficiencies of the combined production and transport subsystem. Energy and exergy efficiencies for this subsystem, for the three chiller cases considered, are given in Table 12.9. The efficiencies for the heating and cooling portions of this subsystem are also included in Table 12.9.

12.6.9. Discussion

Overall energy efficiencies (Table 12.7) are seen to vary, for the three system alternatives considered, from 83% to 94%, and exergy efficiencies from 28% to 30%. Tables 12.6 through 12.9 demonstrate that energy efficiencies do not provide meaningful and comparable results relative to exergy efficiencies when the energy products are in different forms. For example, the energy efficiency of the overall process using electric chillers is 94%, which could lead one to believe that the system is very efficient. The exergy efficiency of the overall process, however, is 28%, indicating that the process is far from ideal thermodynamically. The exergy efficiency is much lower than the energy efficiency in part because heat is being produced at a temperature (120°C) higher than the temperatures actually needed (22°C for space heating and 40°C for hot water heating). The low exergy efficiency of the chillers (see Table 12.6) is largely responsible for the low exergy efficiency for the overall process.

The exergy efficiencies of the chilling, district cooling and end-use cooling subsystems respectively are 36%, 58% and 69% (see Table 12.6). For the combination that includes all three subsystems mentioned above, the exergy efficiency takes on a relatively low value of 14% (Table 12.7). This low efficiency value can be explained by noting that the cool water supply temperature (11°C) needed for space cooling (to 22°C) is relatively near to the environmental temperatures (in summer). The exergy of the cool is small compared with the work input to drive the electric centrifugal chiller. The excess exergy input via work is destroyed due to irreversibilities.

The exergy-based efficiencies in Tables 12.6 through 12.9 are generally different than the energy-based ones because the energy efficiencies utilize energy quantities which are in different forms, while the exergy efficiencies provide more meaningful and useful results by evaluating the performance and behavior of the systems using work equivalents for all energy forms. The exergy and energy for electricity are the same while the exergy for the thermal energy forms encountered here is less than the corresponding energy.

The results for cogeneration-based district energy systems using absorption chillers (single-effect and double-effect types) and using electric chillers are, in general, similar.

Generally, the results appear to indicate that the three integrated cogeneration and district energy systems considered have similar efficiencies. It is likely, therefore, that the choice of one option over another will be strongly dependent on economics and other factors (e.g., environmental impact, space availability, noise limitations, etc.).

Finally, integrated district energy systems may involve thermal energy storage. For example, a ground-coupled heat pump system can extract low-grade heat, which may be deposited in the ground during summer using the waste heat from a central chiller and/or by natural means, for space heating in winter. This low-grade heat can also be extracted using a heat pump for domestic hot water during both winter and summer. Utilizing thermal energy storage may increase energy and exergy efficiencies of building energy systems.

12.7. Closing remarks

The efficiencies and losses presented of the many complex components and subsystems that comprise cogeneration and district energy systems highlight the important insights provided by exergy analysis. This is particularly true when these systems are integrated since different energy forms are simultaneously produced in cogeneration-based district energy systems. The ways in which energy and exergy values differ is shown in Table 12.10, where the energy grade function, defined as the ratio of exergy to energy for a substance or energy form, is shown. Exergy analysis therefore provides more meaningful efficiencies than energy analysis, and pinpoints the locations and causes of inefficiencies more accurately. The results indicate that the complex array of energy forms involved in cogeneration-based district energy systems make them difficult to assess and compare thermodynamically without exergy analysis. This difficulty is primarily attributable to the different nature and quality of the three product energy forms: electricity, heat and cool. This understanding is

Table 12.10. Values of energy grade function for various forms of energy.*

Energy form	Energy grade function (R)
Electricity	1.0
Natural gas	0.913
Steam (100°C)	0.1385
Hot water (66°C)	0.00921
Hot air (66°C)	0.00596

* For a reference-environment temperature of $T_o = 30°C$.

important for designers of such systems in development and optimization activities and in selecting the proper type of system for different applications and situations.

Problems

12.1   How are the energy and exergy efficiencies of cogeneration plants defined? Provide definitions considering a cogeneration plant with (a) power and heat outputs and (b) power and cooling outputs.

12.2   Using typical operating values, compare the exergetic performance of the following cogeneration systems involving power and heat outputs: (a) steam turbine, (b) gas turbine, (c) diesel engine and (d) geothermal.

12.3   What are the major benefits of applying exergy analysis to cogeneration systems?

12.4   Why is it difficult to assess and compare cogeneration systems thermodynamically without exergy analysis? Explain.

12.5   How do cogeneration systems allow better matching of source and application so that exergy losses are lower compared to separate processes for electricity generation and heating? Explain.

12.6   How do cogeneration plants help reduce or minimize harmful emissions? Explain.

12.7   Identify the sources of exergy loss in cogeneration plants and propose methods for reducing or minimizing them.

12.8   What is the effect of the temperature of the heat supplied to the district on the exergy efficiency of a cogeneration plant? Explain.

12.9   What is the effect of the ratio of heat and power outputs on the performance of cogeneration systems? Explain.

12.10 Identify all existing cogeneration systems in your country or state and determine the rates of power and heat production for each system.

12.11 For the case studies considered in this chapter, the exergy efficiency is markedly lower than the corresponding energy efficiency for all cogeneration and non-cogeneration cases. Explain the reasons for this large difference.

12.12 Compare the energy and exergy efficiencies of cogeneration and combined cycle plants.

12.13 Obtain a published article on exergy analysis of cogeneration plants. Using the operating data provided in the article, perform a detailed exergy analysis of the plant and compare your results to those in the original article. Also, investigate the effect of varying important operating parameters on the system exergetic performance.

12.14 Obtain actual operating data for a cogeneration plant in your area and perform a detailed exergy analysis using these data. Discuss the results and provide recommendations based on the exergy results for improving the efficiency.

12.15 What is trigeneration? Is it acceptable to use the term cogeneration for trigeneration?

12.16 How can you express the energy and exergy efficiencies of a system with power, heat and cold products? Assume that cooling is accomplished by an absorption cooling system.

12.17 How can you compare the exergy values for the power, heat and absorption cold products of a single system?

12.18 What are the heat-source temperature requirements for single-effect and double-effect absorption chillers? For a given cooling task, which system will have the higher exergy efficiency? Explain.

12.19 How do cogeneration and trigeneration systems allow better matching of energy source and application so that exergy losses are lower compared to separate processes for electricity generation, heating and cooling? Explain.

12.20 Identify sources of exergy loss in cogeneration plants and propose methods for reducing or minimizing them.

Chapter 13

EXERGY ANALYSIS OF CRYOGENIC SYSTEMS

13.1. Introduction

Cryogenics is associated with low temperatures, usually below $-150°C$ (123 K). The general scope of cryogenic engineering is the design, development and improvement of low-temperature systems and components. Applications of cryogenic engineering include liquefaction and separation of gases, high-field magnets and sophisticated electronic devices that use the superconductivity of materials at low temperatures, space simulation, food freezing, medical uses such as cryogenic surgery, and various chemical processes (ASHRAE Refrigeration, 2006).

Progress in the production of cryogenic temperatures is based to a great extent on advances in thermodynamics. A solid understanding of the thermal processes within a refrigeration cycle is not possible without a sound knowledge of the laws of thermodynamics and cyclic processes. During the early years of classical thermodynamics, between 1842 and 1852, Julius Robert Mayer, James Prescott Joule, Rudolf Clausius, William Thomson and Hermann von Helmholtz published important findings concerning the first and second laws of thermodynamics. Before that time, the main approaches involved trial and error and many experiments. Even after, necessary data were often missing and only became available much later (Foerg, 2002).

The liquefaction of gases has always been an important area of refrigeration since many important scientific and engineering processes at cryogenic temperatures depend on liquefied gases. At temperatures above the critical point, a substance exists in the gas phase only. The critical temperatures of helium, hydrogen and nitrogen are $-268°C$, $-240°C$ and $-147°C$, respectively. Therefore, none of these substances exist in liquid form at atmospheric conditions. Furthermore, low temperatures of this magnitude cannot be obtained by ordinary refrigeration techniques.

To avoid heat leaks into cryogenic storage tanks and transfer lines, high-performance materials are needed that provide high levels of thermal insulation. A good understanding of thermal insulation is important for the development of efficient and low-maintenance cryogenic systems.

In today's world, cryogenics and low-temperature refrigeration are taking on increasingly significant roles. From applications in the food industry, energy and medical technologies to transportation and the space shuttle, requirements exist for cryogenic liquids to be stored and transferred.

In this chapter, a comprehensive exergy analysis is presented of a multistage cascade refrigeration cycle used for natural gas liquefaction as a cryogenic system, based on a report by Kanoglu (2002b). The multistage cascade cryogenic system is described and an exergy analysis of the cycle components and the minimum work required for liquefaction are provided.

13.2. Energy and exergy analyses of gas liquefaction systems

Several cycles, some complex and others simple, exist for the liquefaction of gases. Here, we consider Linde–Hampson cycle shown schematically and on a $T–s$ diagram in Fig. 13.1, to illustrate energy and exergy analyses of liquefaction cycles. Makeup gas is mixed with the uncondensed portion of the gas from the previous cycle, and the mixture at state 1 is compressed by an isothermal compressor to state 2. The temperature is kept constant by rejecting compression heat to a coolant. The high-pressure gas is further cooled in a regenerative counter-flow heat exchanger by the uncondensed portion of gas from the previous cycle to state 3, and throttled to state 4, which is a saturated liquid–vapor mixture state. The liquid (state 6) is collected as the desired product, and the vapor (state 5) is routed through the heat exchanger to cool the high-pressure gas approaching the throttling valve. Finally, the gas is mixed with fresh makeup gas, and the cycle is repeated.

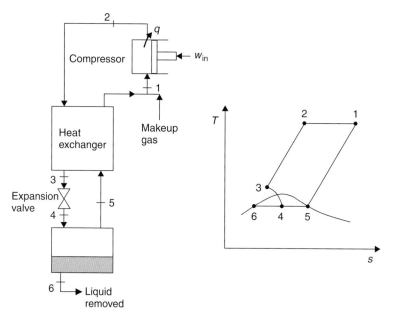

Fig. 13.1. Schematic and temperature–entropy diagram for a simple Linde–Hampson liquefaction cycle.

The refrigeration effect for this cycle may be defined as the heat removed from the makeup gas in order to turn it into a liquid at state 6. Assuming ideal operation for the heat exchanger (i.e., the gas leaving the heat exchanger and the makeup gas are at the same state as state 1, which is the compressor inlet state), the refrigeration effect per unit mass of the liquefied gas is given by

$$q_{\mathrm{L}} = h_1 - h_6 = h_1 - h_{\mathrm{f}} \quad \text{(per unit mass of liquefaction)} \tag{13.1}$$

where h_{f} is the specific enthalpy of saturated liquid that is withdrawn. From an energy balance on the cycle, the refrigeration effect per unit mass of the gas in the cycle may be expressed as

$$q_{\mathrm{L}} = h_1 - h_2 \quad \text{(per unit mass of gas in the cycle)} \tag{13.2}$$

The maximum liquefaction occurs when the difference between h_1 and h_2 (i.e., the refrigeration effect) is maximized. The ratio of Eqs. (13.2) and (13.1) is the fraction of the gas in the cycle that is liquefied, y. That is,

$$y = \frac{h_1 - h_2}{h_1 - h_{\mathrm{f}}} \tag{13.3}$$

An energy balance on the heat exchanger gives

$$h_2 - h_3 = x(h_1 - h_5)$$

where x is the quality of the mixture at state 4. Then the fraction of the gas that is liquefied may also be determined from

$$y = 1 - x \tag{13.4}$$

An energy balance on the compressor gives the work of compression per unit mass of the gas in the cycle as

$$w_{\mathrm{in}} = h_2 - h_1 - T_1(s_2 - s_1) \tag{13.5}$$

The last term in this equation is the isothermal heat rejection from the gas as it is compressed. Assuming air behaves as an ideal gas during this isothermal compression process, the compression work may also be determined from

$$w_{in} = RT_1 \ln(P_2/P_1) \tag{13.6}$$

The coefficient of performance (COP) of this ideal cycle is then given by

$$COP = \frac{q_L}{w_{in}} = \frac{h_1 - h_2}{h_2 - h_1 - T_1(s_2 - s_1)} \tag{13.7}$$

In liquefaction cycles, an efficiency parameter used is the work consumed in the cycle for the liquefaction of a unit mass of the gas. This is expressed as

$$w_{in \, per \, mass \, liquefied} = \frac{h_2 - h_1 - T_1(s_2 - s_1)}{y} \tag{13.8}$$

As the liquefaction temperature decreases the work consumption increases. Noting that different gases have different thermophysical properties and require different liquefaction temperatures, this work parameter should not be used to compare the work consumptions of the liquefaction of different gases. A reasonable use is to compare the different cycles used for the liquefaction of the same gas.

Engineers are usually interested in comparing the actual work used to obtain a unit mass of liquefied gas and the minimum work requirement to obtain the same output. Such a comparison may be performed using the second law. For instance, the minimum work input requirement (reversible work) and the actual work for a given set of processes may be related to each other by

$$w_{actual} = w_{rev} + T_0 s_{gen} = w_{rev} + ex_{dest} \tag{13.9}$$

where T_0 is the environment temperature, s_{gen} is the specific total entropy generation and ex_{dest} is the specific total exergy destruction during the processes. The reversible work for the simple Linde–Hampson cycle shown in Fig. 13.1 may be expressed by the stream exergy difference of states 1 and 6 as

$$w_{rev} = ex_6 - ex_1 = h_6 - h_1 - T_0(s_6 - s_1) \tag{13.10}$$

where state 1 has the properties of the makeup gas, which is essentially the dead state. This expression gives the minimum work requirement for the complete liquefaction of a unit mass of the gas. An exergy efficiency may be defined as the reversible work input divided by the actual work input, both per unit mass of liquefaction:

$$\psi = \frac{w_{rev}}{w_{actual}} = \frac{h_6 - h_1 - T_0(s_6 - s_1)}{(1/y)[h_2 - h_1 - T_1(s_2 - s_1)]} \tag{13.11}$$

We present a numerical example for the simple Linde–Hampson cycle shown in Fig. 13.1. It is assumed that the compressor is reversible and isothermal; the heat exchanger has an effectiveness of 100% (i.e., the gas leaving the liquid reservoir is heated in the heat exchanger to the temperature of the gas leaving the compressor); the expansion valve is isenthalpic; and there is no heat leak to the cycle. The gas is air, at 25°C and 1 atm at the compressor inlet and at 20 MPa at the compressor outlet. With these assumptions and specifications, the various properties at the different states of the cycle and the performance parameters discussed above are determined to be

$$
\begin{array}{lll}
h_1 = 298.4 \, \text{kJ/kg} & s_1 = 6.86 \, \text{kJ/kg K} & q_L = 34.9 \, \text{kJ/kg gas} \\
h_2 = 263.5 \, \text{kJ/kg} & s_2 = 5.23 \, \text{kJ/kg K} & w_{in} = 451 \, \text{kJ/kg gas} \\
h_3 = 61.9 \, \text{kJ/kg} & s_f = 2.98 \, \text{kJ/kg K} & COP = 0.0775 \\
h_4 = 61.9 \, \text{kJ/kg} & T_4 = -194.2°C & w_{in} = 5481 \, \text{kJ/kg liquid} \\
h_5 = 78.8 \, \text{kJ/kg} & x_4 = 0.9177 & w_{rev} = 733 \, \text{kJ/kg liquid} \\
h_f = -126.1 \, \text{kJ/kg} & y = 0.0823 & \psi = 0.134
\end{array}
$$

The properties of air and other substances are obtained using EES software (Klein, 2006). This analysis is repeated for different fluids and the results are listed in Table 13.1.

Table 13.1. Performance parameters for a simple Linde–Hampson cycle for various fluids.

	Nitrogen	Air	Fluorine	Argon	Oxygen	Methane
Liquefaction temperature T_4(°C)	−195.8	−194.2	−188.1	−185.8	−183.0	−161.5
Fraction liquefied y	0.0756	0.0823	0.0765	0.122	0.107	0.199
Refrigeration effect q_L (kJ/kg gas)	32.6	34.9	26.3	33.2	43.3	181
Specific work input w_{in} (kJ/kg gas)	468	451	341	322	402	773
COP	0.0697	0.0775	0.0771	0.103	0.108	0.234
Specific work input w_{in} (kJ/kg liquid)	6193	5481	4459	2650	3755	3889
Minimum specific work input w_{rev} (kJ/kg liquid)	762	733	565	472	629	1080
Exergy efficiency ψ (%)	12.3	13.4	12.7	17.8	16.8	27.8

We observe in Table 13.1 that, as the boiling temperature decreases, the fraction of the gas that is liquefied, the COP and the exergy efficiency decrease. The exergy efficiency values are low, indicating significant potential for improving performance, and thus decreasing the need for work consumption. Noting that the cycle considered in this numerical example involves a reversible isothermal compressor and a 100% effective heat exchanger, the exergy efficiency figures here are optimistic. In practice, an isothermal compression process may be approached using a multistage compressor. For a high effectiveness, the size of heat exchanger must be large, meaning a higher cost. The work consumption may be decreased by replacing the expansion valve with a turbine. Expansion in a turbine usually results in a lower temperature with respect to an expansion valve while producing work, and thus decreasing the total work consumption in the cycle. The complexity and added cost associated with using a turbine as an expansion device is only justified in large liquefaction systems (Kanoglu, 2001, 2002b). In some systems both a turbine and an expansion valve are used to avoid problems associated with liquid formation in the turbine.

The effect of liquefaction temperature on the liquefied mass fraction and COP is illustrated in Fig. 13.2 while Fig. 13.3 shows the effect of liquefaction temperature on the exergy efficiency for various gases. These figures are obtained using the cycle in Fig. 13.1. As the liquefaction temperature increases the liquefied mass fraction, the COP and the exergy efficiency increase.

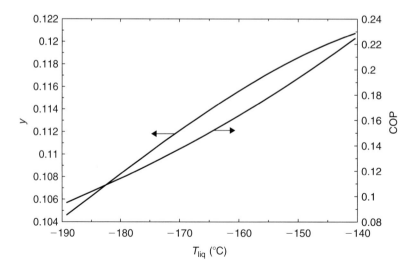

Fig. 13.2. Variation of liquefied mass fraction and COP with liquefaction temperature for oxygen.

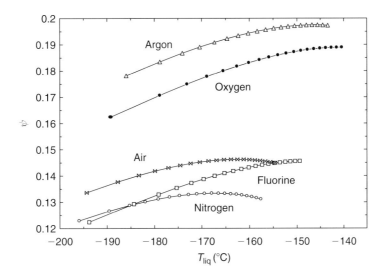

Fig. 13.3. Relation between exergy efficiency and liquefaction temperature for various gases.

13.3. Exergy analysis of a multistage cascade refrigeration cycle for natural gas liquefaction

An exergy analysis is presented of a multistage cascade refrigeration cycle used for natural gas liquefaction, based on a report by Kanoglu (2002b). Multistage cascade refrigeration is described and an exergy analysis of the cycle components and the minimum work required for liquefaction are provided.

13.3.1. Background

Natural gas is a mixture of methane (60–98%), small amounts of other hydrocarbons and various amounts of nitrogen, carbon dioxide, helium and other trace gases. Natural gas can be stored as compressed natural gas (CNG) at pressures of 16–25 MPa and around room temperature, or as liquefied natural gas (LNG) at pressures of 70–500 kPa and temperatures of around −160°C or lower. Natural gas can be transported as a gas in pipelines or as a liquid. In the latter case, the gas is liquefied using unconventional refrigeration cycles and then transported, often by marine ships in specially made insulated tanks. It is returned to a gaseous state in receiving stations for end use.

Several refrigeration cycles, working on different refrigerants, can be used for natural gas liquefaction including the mixed-refrigerant cycle, the cascade cycle and the gas expansion cycle. The first cycle used for natural gas liquefaction was a multistage cascade refrigeration cycle using three refrigerants (propane, ethane or ethylene and methane) in the individual refrigeration cycles that make up the overall cycle.

Much work is consumed to produce LNG at about −160°C from natural gas at atmospheric temperature in the gas phase (Finn et al., 1999). Reducing the work consumed in the cycle can help reduce the cost of LNG. Exergy analysis can help design, optimize and assess such systems, identifying the locations of exergy destruction and thereby highlighting directions for potential improvement. For such work consuming processes, exergy analysis helps determine the minimum work required for a desired result.

13.3.2. Description of the cycle

Figure 13.4 shows a schematic of the cascade refrigeration cycle and its components. The cycle consists of three subcycles, each using a different refrigerant. In the first cycle, propane leaves the compressor at a high temperature and pressure and enters the condenser where cooling water or air is used as a coolant. The condensed propane enters an expansion valve where its pressure is decreased to the evaporator pressure. As the propane evaporates, the heat of evaporation comes from the condensing ethane, cooling methane and cooling natural gas. Propane leaves the evaporator and enters the compressor, thus completing the cycle.

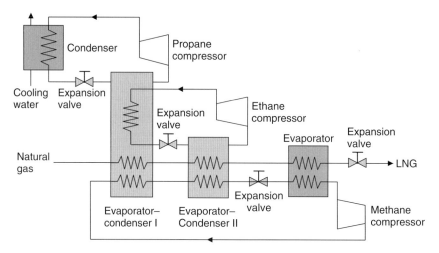

Fig. 13.4. Schematic of a cascade refrigeration cycle (showing only one stage for each refrigerant cycle for simplicity).

The condensed ethane expands in an expansion valve and evaporates as methane condenses and natural gas is further cooled and liquefied. Finally, the methane expands and evaporates as natural gas is liquefied and subcooled. As methane enters the compressor to complete the cycle, the pressure of the LNG is dropped in an expansion valve to the storage pressure. The three refrigerant cycles have multistage compression and expansion usually with three stages, and consequently three evaporation temperature levels for each refrigerant. The mass flows in each stage are usually different.

Natural gas from the pipeline undergoes a process in which acid gases are removed and pressure is increased to an average value of 40 bar before entering the cycle.

13.3.3. Exergy analysis

The exergy flow rate of a fluid in a control volume can be written, neglecting kinetic and potential energies, as

$$\dot{Ex} = \dot{m}ex = \dot{m}[h - h_0 - T_0(s - s_0)] \tag{13.12}$$

where T_0 is the reference (dead) state temperature, h and s are the specific enthalpy and entropy of the fluid at the specified state, respectively, and h_0 and s_0 are the corresponding properties at the dead state. Multiplying the specific flow exergy ex by the mass flow rate gives the exergy flow rate.

The specific exergy change between two states (e.g., inlet and outlet) can be expressed as

$$\Delta ex = h_2 - h_1 - T_0(s_2 - s_1) \tag{13.13}$$

Some exergy is lost during the process due to entropy generation. The specific irreversibility can be written as $i = T_0 \Delta s = T_0 s_{gen}$, where s_{gen} is the entropy generation. Two main causes of entropy generation in the process are friction and heat transfer across a finite temperature difference. Heat transfer is generally accompanied by exergy transfer, which is given by

$$ex_q = \int \delta_q \left(1 - \frac{T_0}{T}\right) \tag{13.14}$$

where δq is differential heat transfer and T is the source temperature where heat transfer takes place. Heat transfer is assumed to occur in surroundings at T_0. If the transferred heat is lost, Eq. (13.14) expresses a thermal exergy loss.

Exergy destruction and exergetic efficiency expressions are presented below for the cycle components in Fig. 13.4.

Evaporators and condensers

The evaporators and condensers in the system are treated as heat exchangers. There are four evaporator–condenser systems in the cycle. The first, evaporator–condenser I, is the evaporator of the propane cycle and the condenser of the ethane and methane cycles. Similarly, evaporator–condenser II is the evaporator of the ethane cycle and the condenser of the methane cycle. The third system is the evaporator of the methane cycle and the fourth system is the condenser of the propane cycle where cooling water is used as a coolant. An exergy balance for an evaporator–condenser expresses the exergy destroyed in the system as the difference of the exergies of incoming and outgoing streams. For evaporator–condenser I,

$$\dot{I} = \dot{Ex}_{in} - \dot{Ex}_{out} = \left[\sum (\dot{m}_p ex_p) + \sum (\dot{m}_e ex_e) + \sum (\dot{m}_m ex_m) + (\dot{m}_n ex_n)\right]_{in}$$
$$- \left[\sum (\dot{m}_p ex_p) + \sum (\dot{m}_e ex_e) + \sum (\dot{m}_m ex_m) + (\dot{m}_n ex_n)\right]_{out} \tag{13.15}$$

where the subscripts in, out, p, e, m and n denote inlet, outlet, propane, ethane, methane and natural gas, respectively. The summations account for the fact that there are three stages in each refrigerant cycle with different pressures, evaporation temperatures and mass flow rates.

The exergetic efficiency of a heat exchanger can be defined as the ratio of total outgoing stream exergies to total incoming stream exergies. For evaporator–condenser I,

$$\psi = \frac{\sum (\dot{m}_p ex_p)_{out} + \sum (\dot{m}_e ex_e)_{out} + \sum (\dot{m}_m ex_m)_{out} + (\dot{m}_n ex_n)_{out}}{\sum (\dot{m}_p ex_p)_{in} + \sum (\dot{m}_e ex_e)_{in} + \sum (\dot{m}_m ex_m)_{in} + (\dot{m}_n ex_n)_{in}} \tag{13.16}$$

An alternate definition for the exergy efficiency of a heat exchanger is the ratio of the increase in the exergy of the cold fluid to the decrease in the exergy of the hot fluid (Wark, 1995). For evaporator–condenser I, the only fluid with an exergy increase is propane while the exergies of ethane, methane and natural gas decrease. Therefore, the alternate efficiency can be written as

$$\psi = \frac{\sum (\dot{m}_p ex_p)_{out} - \sum (\dot{m}_p ex_p)_{in}}{\sum (\dot{m}_e ex_e)_{in} - \sum (\dot{m}_e ex_e)_{out} + \sum (\dot{m}_m ex_m)_{in} - \sum (\dot{m}_m ex_m)_{out} + (\dot{m}_n ex_n)_{in} - (\dot{m}_n ex_n)_{out}} \tag{13.17}$$

The values of efficiencies calculated using these two approaches are usually similar. In this analysis, the second approach is used.

The exergy destruction rate and exergy efficiency for evaporator–condenser II are expressible as

$$\dot{I} = \dot{E}_{in} - \dot{E}_{out} = \left[\sum (\dot{m}_e ex_e) + \sum (\dot{m}_m ex_m) + (\dot{m}_n ex_n)\right]_{in} - \left[\sum (\dot{m}_e ex_e) + \sum (\dot{m}_m ex_m) + (\dot{m}_n ex_n)\right]_{out} \tag{13.18}$$

$$\psi = \frac{\sum (\dot{m}_e ex_e)_{out} - \sum (\dot{m}_e ex_e)_{in}}{\sum (\dot{m}_m ex_m)_{in} - \sum (\dot{m}_m ex_m)_{out} + (\dot{m}_n ex_n)_{in} - (\dot{m}_n ex_n)_{out}} \tag{13.19}$$

The exergy destruction rate and exergy efficiency can be written for the evaporator of the methane cycle as

$$\dot{I} = \dot{Ex}_{in} - \dot{Ex}_{out} = \left[\sum (\dot{m}_m ex_m) + (\dot{m}_n ex_n)\right]_{in} - \left[\sum (\dot{m}_m ex_m) + (\dot{m}_n ex_n)\right]_{out} \tag{13.20}$$

$$\psi = \frac{\sum (\dot{m}_m ex_m)_{out} - \sum (\dot{m}_m ex_m)_{in}}{(\dot{m}_n ex_n)_{in} - (\dot{m}_n ex_n)_{out}} \tag{13.21}$$

and for the condenser of the propane cycle as

$$\dot{I} = \dot{Ex}_{in} - \dot{Ex}_{out} = \left[\sum (\dot{m}_p ex_p) + (\dot{m}_w ex_w)\right]_{in} - \left[\sum (\dot{m}_p ex_p) + (\dot{m}_w ex_w)\right]_{out} \tag{13.22}$$

$$\psi = \frac{(\dot{m}_w ex_w)_{out} - (\dot{m}_w ex_w)_{in}}{\sum (\dot{m}_p ex_p)_{in} - \sum (\dot{m}_p ex_p)_{out}} \tag{13.23}$$

where the subscript 'w' denotes water.

Compressors

There is one multistage compressor in the cycle for each refrigerant. The total work consumed in the cycle is the sum of work inputs to the compressors. The minimum work input for the compressor occurs when no irreversibilities occur and exergy destruction is correspondingly zero. In reality, irreversibilities occur due to friction, heat loss and other dissipative effects. The exergy destructions in the propane, ethane and methane compressors, respectively, can be expressed as

$$\dot{I}_p = \dot{Ex}_{in} - \dot{Ex}_{out} = \sum (\dot{m}_p ex_p)_{in} + \dot{W}_{p,in} - \sum (\dot{m}_p ex_p)_{out} \tag{13.24a}$$

$$\dot{I}_e = \dot{Ex}_{in} - \dot{Ex}_{out} = \sum (\dot{m}_e ex_e)_{in} + \dot{W}_{e,in} - \sum (\dot{m}_e ex_e)_{out} \tag{13.24b}$$

$$\dot{I}_m = \dot{Ex}_{in} - \dot{Ex}_{out} = \sum (\dot{m}_m ex_m)_{in} + \dot{W}_{m,in} - \sum (\dot{m}_m ex_m)_{out} \tag{13.24c}$$

where $\dot{W}_{p,in}$, $\dot{W}_{e,in}$ and $\dot{W}_{m,in}$ are the actual power inputs to the propane, ethane and methane compressors, respectively. The exergy efficiency of a compressor can be defined as the ratio of the minimum work input to the actual work input. The minimum work is simply the exergy difference between the actual inlet and exit states. Applying this definition to the propane, ethane and methane compressors, the respective exergy efficiencies become

$$\psi_p = \frac{\sum (\dot{m}_p ex_p)_{out} - \sum (\dot{m}_p ex_p)_{in}}{\dot{W}_{p,in}} \tag{13.25a}$$

$$\psi_e = \frac{\sum (\dot{m}_e ex_e)_{out} - \sum (\dot{m}_e ex_e)_{in}}{\dot{W}_{e,in}} \tag{13.25b}$$

$$\psi_m = \frac{\sum (\dot{m}_m ex_m)_{out} - \sum (\dot{m}_m ex_m)_{in}}{\dot{W}_{m,in}} \tag{13.25c}$$

Expansion valves

In addition to the expansion valves in the refrigerant cycles, an expansion valve is used to reduce the pressure of LNG to the storage pressure. Expansion valves are essentially isenthalpic devices with no work interaction and negligible heat interaction with the surroundings. Exergy balances can be used to write the exergy destruction rates for the propane, ethane, methane and LNG expansion valves:

$$\dot{I}_p = \dot{Ex}_{in} - \dot{Ex}_{out} = \sum (\dot{m}_p ex_p)_{in} - \sum (\dot{m}_p ex_p)_{out} \tag{13.26a}$$

$$\dot{I}_e = \dot{Ex}_{in} - \dot{Ex}_{out} = \sum (\dot{m}_e ex_e)_{in} - \sum (\dot{m}_e ex_e)_{out} \tag{13.26b}$$

$$\dot{I}_m = \dot{Ex}_{in} - \dot{Ex}_{out} = \sum (\dot{m}_m ex_m)_{in} - \sum (\dot{m}_m ex_m)_{out} \tag{13.26c}$$

$$\dot{I}_n = \dot{Ex}_{in} - \dot{Ex}_{out} = \sum (\dot{m}_n ex_n)_{in} - \sum (\dot{m}_n ex_n)_{out} \tag{13.26d}$$

The exergy efficiency of an expansion valve can be defined as the ratio of the total exergy output to the total exergy input. The exergy efficiencies for the expansion valves considered here are thus

$$\psi_p = \frac{\sum (\dot{m}_p ex_p)_{out}}{\sum (\dot{m}_p ex_p)_{in}} \tag{13.27a}$$

$$\psi_e = \frac{\sum (\dot{m}_e ex_e)_{out}}{\sum (\dot{m}_e ex_e)_{in}} \tag{13.27b}$$

$$\psi_m = \frac{\sum (\dot{m}_m ex_m)_{out}}{\sum (\dot{m}_m ex_m)_{in}} \tag{13.27c}$$

$$\psi_n = \frac{\sum (\dot{m}_n ex_n)_{out}}{\sum (\dot{m}_n ex_n)_{in}} \tag{13.27d}$$

Overall cycle

The exergy destruction in the overall cycle is the sum of exergy destructions in all internal devices (i.e., condensers, evaporators, compressors and expansion valves). This cycle exergy destruction can be obtained using the preceding exergy destruction expressions. The exergy efficiency of the overall cycle can be defined as

$$\psi = \frac{\dot{E}x_{out} - \dot{E}x_{in}}{\dot{W}_{actual}} = \frac{\dot{W}_{actual} - \dot{I}_{total}}{\dot{W}_{actual}} \tag{13.28}$$

where the numerator expresses the exergy difference (or the actual work input to the cycle $\dot{W}_{actual}$ less the total exergy destruction $\dot{I}$). The actual work input to the overall cycle is the sum of the work inputs to the propane, ethane and methane compressors:

$$\dot{W}_{actual} = \dot{W}_{p,in} + \dot{W}_{e,in} + \dot{W}_{m,in} \tag{13.29}$$

The exergy efficiency of the cycle can also be expressed as

$$\psi = \frac{\dot{W}_{min}}{\dot{W}_{actual}} \tag{13.30}$$

where $\dot{W}_{min}$ is the minimum work input to the cycle, which represents the minimum work for the liquefaction process.

13.3.4. Minimum work for the liquefaction process

The exergy efficiency of the natural gas liquefaction process can be defined as the ratio of the minimum work required to produce a certain amount of LNG to the actual work input. An exergy analysis on the cycle permits the minimum work input to be determined. The liquefaction process essentially involves the removal of heat from the natural gas. Therefore, the minimum work for the liquefaction process can be determined utilizing a reversible or Carnot refrigerator. The minimum work input is simply the work input required for the Carnot refrigerator for a given heat removal, and can be expressed as

$$w_{min} = \int \delta_q \left(1 - \frac{T_0}{T}\right) \tag{13.31}$$

where δq is the differential heat transfer and T is the instantaneous temperature at the boundary where the heat transfer takes place. Note that T is smaller than T_0 for a liquefaction process and to yield a positive work input, the sign of heat transfer must be negative since it is a heat output. The evaluation of Eq. (13.31) requires a knowledge of the functional relationship between the heat transfer δq and the boundary temperature T, which is usually not available.

As seen in Fig. 13.4, natural gas flows through three evaporator–condenser systems in the multistage refrigeration cycle before it is fully liquefied. Thermodynamically, this three-stage heat removal from natural gas can be accomplished using three Carnot refrigerators as seen in Fig. 13.5a. The first Carnot refrigerator receives heat from the natural gas and supplies it to the heat sink at T_0 as the natural gas is cooled from T_1 to T_2. Similarly, the second Carnot refrigerator receives heat from the natural gas and supplies it to the heat sink at T_0 as the natural gas is cooled from T_2 to T_3. Finally, the third Carnot refrigerator receives heat from the natural gas and supplies it to the heat sink at T_0 as the natural gas

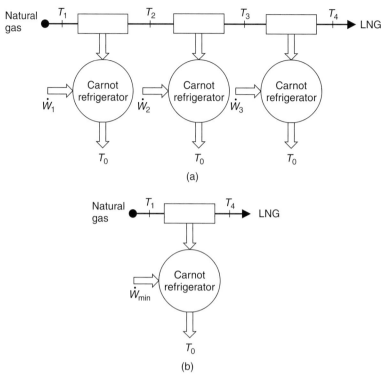

Fig. 13.5. Determination of minimum work for (a) the cycle considered and (b) an equivalent system in terms of initial and final states.

is further cooled from T_3 to T_4, where it exits as LNG. The work rates required by the first, second and third Carnot refrigerators, respectively, can be determined as

$$\dot{W}_1 = \dot{m}_n(ex_1 - ex_2) = \dot{m}_n[h_1 - h_2 - T_0(s_1 - s_2)] \tag{13.32a}$$

$$\dot{W}_2 = \dot{m}_n(ex_2 - ex_3) = \dot{m}_n[h_2 - h_3 - T_0(s_2 - s_3)] \tag{13.32b}$$

$$\dot{W}_3 = \dot{m}_n(ex_3 - ex_4) = \dot{m}_n[h_3 - h_4 - T_0(s_3 - s_4)] \tag{13.32c}$$

The total power input, which expresses the minimum power input for the liquefaction process, is the sum of the above terms:

$$\dot{W}_{min} = \dot{W}_1 + \dot{W}_2 + \dot{W}_3 = \dot{m}_n(ex_1 - ex_4) = \dot{m}_n[h_1 - h_4 - T_0(s_1 - s_4)] \tag{13.33}$$

This minimum power can alternatively be obtained using a single Carnot refrigerator that receives heat from the natural gas and supplies it to a heat sink at T_0 as the natural gas is cooled from T_1 to T_4. Such a Carnot refrigerator is equivalent to the combination of three Carnot refrigerators shown in Fig. 13.5b. The minimum work required for the liquefaction process depends only on the properties of the incoming and outgoing natural gas and the ambient temperature T_0.

A numerical value of the minimum work can be calculated using typical values of incoming and outgoing natural gas properties. When entering the cycle, the pressure of the natural gas is around 40 bar and the temperature is approximately the same as the ambient temperature (i.e., $T_1 = T_0 = 25°C$). Natural gas leaves the cycle liquefied at about 4 bar and 150°C. Since the natural gas in the cycle usually consists of more than 95% methane, it is assumed for simplicity that the thermodynamic properties of natural gas are the same as those for methane. Using these inlet and exit states, the minimum work input to produce a unit mass of LNG can be determined from Eq. (13.33) as 456.8 kJ/kg. The heat removed from the natural gas during the liquefaction process is determined from

$$\dot{Q} = \dot{m}_{\mathrm{n}}(h_1 - h_4) \tag{13.34}$$

For the inlet and exit states of natural gas described above, the heat removed from the natural gas can be determined from Eq. (13.34) to be 823.0 kJ/kg. That is, the removal of 823.0 kJ/kg of heat from the natural gas requires a minimum of 456.8 kJ/kg of work. Since the ratio of heat removed to work input is defined as the COP of a refrigerator, this corresponds to a COP of 1.8. This relatively low COP of the Carnot refrigerator used for natural gas liquefaction is expected due to high difference between the temperatures T and T_0 in Eq. (13.31). An average value of T can be obtained from the definition of the COP for a Carnot refrigerator, which is expressed as

$$\mathrm{COP}_{\mathrm{R,rev}} = \frac{1}{T_0/T - 1} \tag{13.35}$$

Using this equation with COP = 1.8 and $T_0 = 25°C$ we determine $T = -81.3°C$. This is the temperature a heat reservoir would have if a Carnot refrigerator with a COP of 1.8 operated between this reservoir and another reservoir at 25°C. Note that the same result could be obtained by writing Eq. (13.31) in the form

$$w_{\min} = q\left(1 - \frac{T_0}{T}\right) \tag{13.36}$$

where $q = 823.0$ kJ/kg, $w_{\min} = 456.8$ kJ/kg and $T_0 = 25°C$.

We now investigate how the minimum work changes with natural gas liquefaction temperature. We take the inlet pressure of natural gas to be 40 bar, the inlet temperature to be $T_1 = T_0 = 25°C$, and the exit state to be saturated liquid at the specified temperature. The properties of methane are obtained using EES software (Klein, 2006). Using the minimum work relation in Eq. (13.33) and the reversible COP relation in Eq. (13.35), the plots in Fig. 13.6 are obtained.

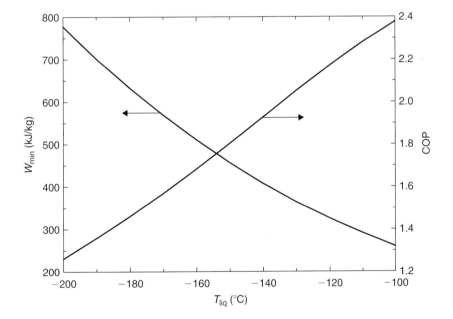

Fig. 13.6. Variation of minimum specific work and COP with natural gas liquefaction temperature.

The minimum work required to liquefy a unit mass of natural gas increases almost linearly with decreasing liquefaction temperature as shown in Fig. 13.6. Obtaining LNG at 200°C requires three times the minimum work to obtain LNG at −100°C. Similarly, obtaining LNG at −150°C requires 1.76 times the minimum work to obtain LNG at −100°C.

The COP of the Carnot refrigerator increases almost linearly with liquefaction temperature as shown in Fig. 13.6. The COP decreases almost by half when the liquefaction temperature decreases from $-100°C$ to $-200°C$. These figures show that the maximum possible liquefaction temperature should be used to minimize the work input. In other words, LNG should not be liquefied to lower temperatures than needed.

For the typical natural gas inlet and exit states specified in the previous section, the minimum work is determined to be 456.8 kJ/kg of LNG. A typical actual value of work input for a cascade cycle used for natural gas liquefaction is given by Finn et al. (1999) to be 1188 kJ/kg of LNG. Then the exergetic efficiency of a typical cascade cycle can be determined from Eq. (13.30) to be 38.5%. The actual work input required depends mainly on the feed and ambient conditions and the compressor efficiency.

13.3.5. Discussion

Recent advances have made it possible to replace the JT expansion valve of the cycle with a cryogenic hydraulic turbine. The same pressure drop as in the JT valve is achieved with the turbine while producing power. Using the same typical values as before we take the cryogenic turbine inlet state to be 40 bar and $-150°C$. Assuming isentropic expansion to a pressure of 4 bar, the work output is calculated to be 8.88 kJ/kg of LNG. When this work is subtracted from the work input, this corresponds to a decrease of 2% in the minimum work input. The use of a cryogenic turbine results in 4% extra LNG production in an actual natural gas liquefaction plant (Verkoelen, 1996). Also, for the same inlet conditions, the temperature of LNG at the cryogenic turbine exit is shown to be lower than that at the expansion valve exit (Kanoglu, 2000).

The main site of exergy destruction in the cycle is the compressors. Improvements to the exergy efficiencies of the compressors will reduce the work input for the liquefaction process. Having three-stage evaporation for each refrigerant in the cascade cycle results in a total of nine evaporation temperatures. Also, having multiple stages makes the average temperature difference between the natural gas and the refrigerants small. This results in a smaller exergy destruction in the evaporators since the greater the temperature difference the greater the exergy destruction. As the number of evaporation stages increases the exergy destruction decreases. However, adding more stages means additional equipment cost. More than three stages for each refrigerant is usually not justified.

13.4. Closing remarks

Exergy analyses of cryogenic systems, particularly gas liquefaction systems, have been described. Exergy analysis is particularly important for cryogenic applications because the exergy of a cryogenic substance or fuel becomes increasingly significant as its temperature decreases well below the environment temperature.

Problems

13.1 What is the difference between a refrigeration system and a gas liquefaction system? Which system typically involves higher exergy destruction and thus lower exergy efficiency?

13.2 Compare the following processes: (a) transportation of natural gas in pipelines and (b) liquefying natural gas and then transporting it in tanks.

13.3 Compare the following processes from an exergetic point of view: (a) transportation of natural gas as a gas in tanks and (b) liquefying natural gas and then transporting it in tanks.

13.4 How can you express the COP and exergy efficiency of a gas liquefaction process? Can the COP and exergy efficiency be greater than 1?

13.5 Provide three alternative definitions of exergy efficiency for a gas liquefaction system?

13.6 Write an expression for the minimum work for the liquefaction of a gas? What is the relationship between this expression and the exergy of the heat removed from the gas during the liquefaction process?

13.7 Which is greater in a gas liquefaction system: the heat removal per unit mass of the gas or the heat removal per unit mass of liquefaction? Can one be determined from the other?

13.8 How does the exergy efficiency of a gas liquefaction system change with liquefaction temperature?

13.9 Investigate various cycles used for gas liquefaction and compare them from an exergetic perspective.

13.10 What is the importance of liquefying hydrogen for a future hydrogen economy? Can the Linde–Hampson process described in this chapter be used for hydrogen liquefaction? If not, what modifications would be needed to allow for hydrogen liquefaction? Which cycles are currently used for hydrogen liquefaction?

13.11 Can an absorption cooling system be used for gas liquefaction? Explain.

13.12 Compare the work required to compress a gas in a gas liquefaction cycle using (a) an isothermal compressor and (b) an isentropic compressor.

13.13 Do you favor replacing the expansion valve with a turbine in a gas liquefaction system? Explain from an exergetic perspective.

13.14 Provide some guidelines in the selection of refrigerants for use in oxygen, methane and hydrogen liquefaction systems?

13.15 Investigate the use of cryogenic turbines in natural gas liquefaction systems.

13.16 Which components of the natural gas liquefaction system considered in this chapter involve greater exergy destructions? Provide methods for reducing or minimizing the exergy losses.

13.17 Obtain a published article on exergy analysis of a gas liquefaction system. Using the operating data provided in the article, perform a detailed exergy analysis of the system and compare your results to those in the original article. Also, investigate the effect of varying important operating parameters on the system exergetic performance.

13.18 Obtain actual operating data from a gas liquefaction system and perform a detailed exergy analysis. Discuss the results and provide recommendations based on the exergy results for improving the efficiency.

Chapter 14

EXERGY ANALYSIS OF CRUDE OIL DISTILLATION SYSTEMS

14.1. Introduction

Petroleum refining is the process of separating the many compounds present in crude petroleum. The principle used in refining is that the longer the carbon chain, the higher is the temperature at which the compounds will boil. In a refining system, crude petroleum is heated so that the compounds in it convert to gases. The gases pass through a distillation column and become cooler as their height increases. When a gaseous compound cools below its boiling point, it condenses. The condensed liquids are drawn off the distillation column at various heights.

Although all fractions of petroleum find uses, the greatest demand is for gasoline. Crude petroleum contains only 30–40% gasoline. Transportation demands require that over 50% of the crude oil be converted into gasoline. To meet this demand some petroleum fractions are converted to gasoline. This may be done in several ways: 'cracking,' i.e., breaking down large molecules of heavy heating oil; 'reforming,' i.e., changing the molecular structures of low-quality gasoline molecules; and 'polymerization,' i.e., forming longer molecules from smaller ones.

For example if pentane is heated to about 500°C the covalent carbon–carbon bonds begin to break during the cracking process. Many kinds of compounds including alkenes are made during this cracking process. Alkenes are formed because there is not enough hydrogen to saturate all bonding positions after the carbon–carbon bonds are broken.

Crude oil has little economic value and has no practical applications in its original state. Even after separation of gas, water, H_2S and other components, crude oil is still a mixture of thousands of hydrocarbons ranging from very light to very heavy components. A refinery complex converts crude oil to useful products, e.g., liquefied petroleum gas (LPG), kerosene, diesel fuel, gasoline, jet fuels, asphalt, etc. To accomplish this conversion, crude oil undergoes numerous chemical and physical processes in different parts of a refinery (Al-Muslim et al., 2003). Processing of crude oil into its constituents occurs in distillation columns, after the oil has passed through desalination and cleaning processes. Working at temperatures of up to 400°C, the gaseous crude oil is transferred into distillation columns and then condensed at varying temperatures and pressures as part of the refining process. Different condensates have varying boiling points and the higher the boiling point the higher the gas ascends before condensing. For optimum operation of a distillation column, accurate temperature, pressure and flow measurements are required.

The boiling points of organic compounds sometimes help indicate other physical properties. A liquid boils when its vapor pressure is equal to the atmospheric pressure. Vapor pressure is determined by the kinetic energy of molecules. Kinetic energy is related to temperature and the mass and velocity of the molecules. When the temperature reaches the boiling point, the average kinetic energy of the liquid particles is sufficient to overcome the forces of attraction that hold molecules in the liquid state. When molecules in the liquid state have sufficient kinetic energy, they may escape from the surface and turn into a gas. Molecules with the most independence in individual motion typically achieve sufficient kinetic energy (velocity) to escape at lower temperatures. The vapor pressure for such a compound is higher and it therefore will boil at a lower temperature.

The utilization of thermodynamic analysis to improve efficiency has increased in the industrial world for many reasons, including the following: easily accessible energy resources are limited and environmental policies are becoming stricter. Increases in oil prices in the 1970s and rising concerns in the 1980s and 1990s about the adverse environmental impact caused by energy systems have caused considerable effort to be dedicated to improving efficiencies of existing and future designs of energy systems, using thermodynamic and other tools.

Energy analysis is conventionally used to optimize the yield of desired products in crude oil distillation. However, for economic, resource scarcity and environmental reasons, we seek to optimize the utilization of energy resources, and exergy analysis has become an increasingly popular tool for such activities. However, most attention is still focused on using energy analysis and many studies have been undertaken on energy analyses of various thermodynamic systems and processes in the petroleum and petrochemical industries. Relatively limited work has been reported on exergy analyses of distillation processes and units.

A detailed exergy analysis of crude oil distillation systems is described in this chapter, based on earlier reports (Al-Muslim et al., 2003, 2005; Al-Muslim and Dincer, 2005). The effects of varying key system parameters (e.g., distillation column temperature and pressure) on system efficiencies at various conditions are highlighted.

14.2. Analysis approach and assumptions

The three governing equations commonly used in thermodynamic analysis of systems, conservation of mass and energy and non-conservation of entropy, are applied here to crude oil distillation. For the process, steady-state, steady-flow behavior is assumed with negligible changes in kinetic and potential energies. The ambient conditions are as follows: $T_0 = 25°C = 298.15$ K and $P_0 = 101$ kPa. For every individual component, the three balances are applied and quantities such as heat added, exergy loss rate and exergy efficiency are evaluated. After simplification, the mass, energy and exergy balances, respectively, are

$$\sum_i \dot{m}_i = \sum_e \dot{m}_e \tag{14.1}$$

$$\sum_i \dot{E}_i + \dot{Q}_{cv} = \sum_e \dot{E}_e + \dot{W}_{cv} \tag{14.2}$$

$$\sum_i \dot{Ex}_i + \sum_j (1 - T_0/T_j)\dot{Q}_{cv} = \sum_e \dot{Ex}_e + \dot{W}_{cv} + \dot{I}_{cv} \tag{14.3}$$

where the exergy losses from a control volume and total exergy are given, respectively, as

$$\dot{I}_{cv} = \dot{W}_{cv}^{rev} - \dot{W}_{cv} \tag{14.4}$$

$$Ex = U + P_0 V - T_0 S + \sum_i \mu_{0,i} N_i \tag{14.5}$$

The exergy efficiency can be expressed as

$$\psi = \sum_e \dot{Ex}_e / \sum_i \dot{Ex}_i \tag{14.6}$$

14.3. Description of crude oil distillation system analyzed

Crude oil distillation is the first step in a refinery complex. A physical process separates crude oil into different fractions depending on the difference of boiling temperatures of its constituents. In most distillation plants, crude oil is processed in two towers: the atmospheric tower where light hydrocarbons are separated and the vacuum tower where heavier hydrocarbons are separated. The products of crude oil distillation can be final products or feedstocks to other plants.

14.3.1. Overall system

The crude oil distillation plant has many components, e.g., crude oil furnace, distillation towers and a heat exchanger network (HEN). Figure 14.1 illustrates a schematic diagram of the crude oil distillation system considered here. The system consists of two crude oil distillation units (the atmospheric distillation unit (ADU) and the vacuum distillation unit (VDU)), two crude oil furnaces and a HEN. The HEN is not shown, but its effect via utilizing high-temperature product streams to preheat the crude oil is analyzed. A crude oil mass flow rate of 507 kg/s (300,000 barrels per day) is

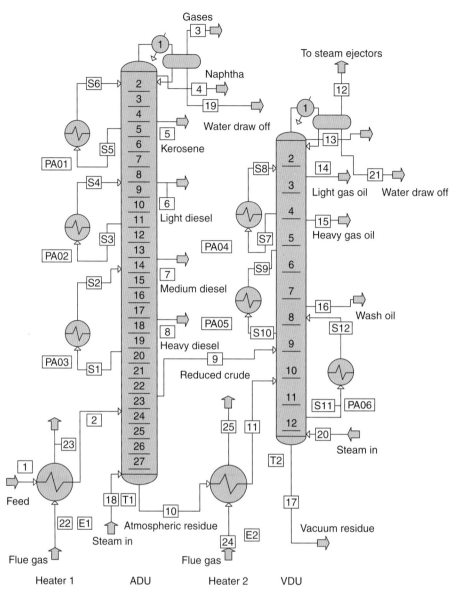

Fig. 14.1. Model of a crude oil distillation plant (Al-Muslim et al., 2003, 2005).

considered. The system is modeled to facilitate analysis, and a description of the components and parameters follows below (for details, see Al-Muslim et al., 2003).

14.3.2. System components

Crude oil furnace 1 (Heater 1)

Crude oil is heated in this furnace using the hot exhaust of fuel combustion. The crude oil starts vaporizing as heat is input. At the heater outlet, the crude oil is not totally vaporized, being typically 75% vapor. Typical outlet temperatures are 350–380°C. The charge (crude oil) can be fed to the heater directly from storage tanks at atmospheric temperature, but in most plants the crude oil passes through a series of preheaters that utilize the high temperatures of the distillation column product streams. The outlet temperature of the preheating heat exchangers can reach 280°C.

Atmospheric distillation unit

The ADU is a long column with many trays and operates at or above atmospheric pressure (typically 200 kPa). The trays have bubble caps or holes to allow vapor to pass through them and are either of the two-pass or four-pass types. Crude oil enters the ADU in a tray above the bottom in the so-called flash zone. The crude oil vaporizes increasingly as it rises. Liquid, referred to as atmospheric residue, is drawn from the bottom tray. As the vapor rises, it passes through the trays and cools as it comes in contact with the liquid.

Crude fractions settle in trays in the rectifying section and are drawn off at five liquid side cuts depending on their average boiling point. The light products, which have low boiling points, tend toward the top and the heavier products, with relatively higher boiling points tend toward the bottom. The side cuts are, from heavy to light: heavy diesel, medium diesel, light diesel and kerosene. The very light products, e.g., butane and lighter in addition to light naphtha, exit as vapor at the top of the column. The atmospheric overhead is partially condensed in heat exchangers. When cooled, the naphtha condenses and exits as a liquid while the lighter products remain as gases. Uncondensed vapor flows to the fuel gas system, which is used as fuel for the furnaces.

In our model, a 27-tray column is assumed. The crude oil is introduced in tray no. 5. The side cut trays are dedicated as follows: no. 7 for residue, no. 10 for heavy diesel, no. 14 for medium diesel, no. 19 for light diesel and no. 24 for kerosene. The very light materials exit from the overhead tray (no. 1) and the very heavy materials from the bottom tray (no. 27).

Heat is removed from the column through the overhead condenser and a number of pump-around circuits, typically three. The column overhead vapor passes through heat exchangers where it cools and condenses. A portion of the condensed liquid is returned to the top of the column at a lower temperature. The pump-around circuit draws liquid from a certain tray, cools the liquid in heat exchangers and returns the liquid to a tray above the original tray at a lower temperature. The pump-around circuit is characterized by the withdraw tray, the return tray, the mass flow rate and the return temperature. The ADU has three pump-around circuits (PA01, PA02, PA03), which are described in Table 14.1.

Table 14.1. Data for the pump-around circuits of the ADU and VDU.

Pump-around circuit	Withdraw tray	Return tray	Mass flow rate (kg/s)	Return temperature (°C)
PA01	5	3	0.1	50
PA02	10	7	0.4	125
PA03	17	13	0.3	220
PA04	3	2	0.2	75
PA05	6	4	0.08	245
PA06	9	7	0.03	200

Superheated steam is introduced from the bottom of the column at about 350°C, so as to reduce the partial pressure in the column and thus enhance vaporization and separation of the crude oil. The temperature profile is designed so that the lowest temperature is well above the dew point temperature of the steam at column pressure, to ensure it will not condense in the column. Steam is typically supplied at about 1 kg steam/100 kg of crude oil. The important parameters of the ADU, which must be considered in design, are: number of trays, crude oil entrance (flash zone) and locations and draw-off rates of side cuts.

Parameters that are frequently varied during plant operation are temperature profile and pressure profile of the column. The temperature profile can be controlled by varying the flow rate in the pump-around circuits and return temperature. Pressure profile can be controlled by varying column overhead pressure. Those profiles significantly affect column operation and the ability to produce the desired products.

Crude oil furnace 2 (Heater 2)

The purpose of this furnace is to heat the bottom residue from the ADU from 350°C to 400°C. Its operation is similar to that for Heater 1.

Vacuum distillation unit

The VDU operates at vacuum pressure to help in separating heavy hydrocarbons and allows lower temperatures for distillation than are possible in the ADU. The vacuum pressure is typically 18 kPa at the vacuum column flash zone and 10 kPa at the column top. The vacuum is created by placing a series of ejectors in service. The VDU uses the same operating principles as the ADU but with fewer trays and side cuts.

In our model, a 12-tray column is assumed. Part of the residue is charged directly from the ADU at tray no. 7 and the other part is charged from the furnace at tray no. 2. The three side cuts are, in order of heavy to light: wash oil at tray no. 4, heavy vacuum gas oil at tray no. 7 and light vacuum gas oil at tray no. 10. The vacuum residue leaves from the bottom tray (no. 12) and usually is transferred to an asphalt plant.

Superheated steam is introduced at the bottom of the column. The function of the steam is the same as in the ADU. Heat from the VDU is removed mainly by the pump-around circuits. The VDU has three pump-around circuits (PA04, PA05, PA06), which are shown in Table 14.1. Again, the temperature and pressure profiles of the column are the key parameters used to control the operation of the unit.

Heat exchanger network

The HEN has two purposes: (i) cooling product streams from their boiling point temperatures to about 60°C and (ii) preheating the crude oil from atmospheric temperatures to about 280°C. The heat exchangers are normally of the shell and tube type.

14.4. System simulation

In the model, 17 state points are identified for the distillation process itself, four state points for steaming and four state points for heating. There are three pump-around circuits for each of the ADU and VDU. Important parameters for the study are temperature, pressure and flow rates for each stream, including the pump-around circuits.

SimSci/PROII software (2000) is used to simulate the system to determine the temperature, pressure, enthalpy and entropy at the side cuts. The program is flexible and can model numerous refinery processes in detail, ranging from crude oil characterization and preheating to complex reaction and separation units.

Several simulations are carried out. The first simulation is for plant operating conditions. Then, the input parameters are varied to investigate how they affect the energy and exergy efficiencies and irreversibility rates of individual components and the overall system. Table 14.2 summarizes the parameters for the first simulation case.

14.5. Energy and exergy analyses

Energy and exergy analyses described of the main system components and the overall crude oil distillation system.

14.5.1. Crude heating furnace (E1)

Applying the energy balance in Eq. (14.2) and assuming an adiabatic process, the hot air mass flow rate $\dot{m}_9$ is evaluated as

$$\dot{m}_9 = \dot{m}_1 \frac{(h_2 - h_1)}{(h_{10} - h_9)} \tag{14.7}$$

Equaion (14.3) is used to find the exergy consumption rate, noting that no chemical exergy change occurs in the heater since the chemical composition of crude does not change:

$$\dot{I}_{E1} = \dot{m}_1(ex_1 - ex_2) + \dot{m}_9(ex_{10} - ex_9) \tag{14.8}$$

The exergy efficiency is determined with Eq. (14.6):

$$\psi_{E1} = \left(\sum_i \dot{Ex}_i - \dot{I}_{E1} \right) \Big/ \sum_i \dot{Ex}_i \tag{14.9}$$

Table 14.2. Input data for simulation of the crude oil distillation plant at normal operating conditions.

State	Stream	Phase	Temperature (°C)	Pressure (kPa)	Mass flow rate (kg/s)
1	Crude feed	Liquid	25	101	507
2	Crude heated	Mixed	350	101	507
3	Offgas	Gas	70	205	2
4	Naphtha	Liquid	70	205	5
5	Kerosene	Liquid	90	210	15
6	Light diesel	Liquid	110	215	35
7	Medium diesel	Liquid	130	220	75
8	Heavy diesel	Liquid	160	225	55
9	Reduced oil	Liquid	260	230	60
10	Atmospheric residue	Liquid	350	235	260
11	Atmospheric residue heated	Mixed	400	235	260
12	Vacuum gas	Gas	200	10	10
13	Vacuum condensate	Liquid	200	10	2
14	Light gas oil	Liquid	220	12	25
15	Heavy gas oil	Liquid	290	14	25
16	Wash oil	Liquid	310	16	50
17	Vacuum residue	Liquid	380	18	140
18	Steam	Gas	350	500	5
19	Water	Liquid	70	101	2
20	Steam	Gas	400	205	2
21	Water	Liquid	70	101	0.5
22	Flue gas in	Gas	1100	101	600
23	Flue gas out	Gas	350	101	600
24	Flue gas in	Gas	1100	101	60
25	Flue gas out	Gas	350	101	60

Source: Al-Muslim et al. (2003).

where the exergy input rate for the crude heating furnace is given by

$$\sum_i \dot{E}x_i = \dot{m}_1 ex_1 + \dot{m}_9 ex_{10} \tag{14.10}$$

14.5.2. Atmospheric distillation unit (T1)

Equation (14.2) is applied to find the heat transfer from the column:

$$\dot{Q}_{T1} = \dot{m}_3 h_3 + \dot{m}_4 h_4 + \dot{m}_5 h_5 + \dot{m}_6 h_6 + \dot{m}_7 h_7 + \dot{m}_8 h_8 - \dot{m}_2 h_2 \tag{14.11}$$

With Eqs. (14.3) and (14.5), we obtain the physical and chemical exergy consumption rates, respectively:

$$\dot{I}_{T1,ph} = \dot{m}_2 ex_{2,ph} - \dot{m}_3 ex_{3,ph} - \dot{m}_4 ex_{4,ph} - \dot{m}_5 ex_{5,ph} - \dot{m}_6 ex_{6,ph} - \dot{m}_7 ex_{7,ph} - \dot{m}_8 ex_{8,ph} + (1 - T_0/T)\dot{Q}_{T1} \quad (14.12)$$

$$\dot{I}_{T1,ch} = \dot{m}_2 ex_{2,ch} - \dot{m}_3 ex_{3,ch} - \dot{m}_4 ex_{4,ch} - \dot{m}_5 ex_{5,ch} - \dot{m}_6 ex_{6,ch} - \dot{m}_7 ex_{7,ch} - \dot{m}_8 ex_{8,ch} \quad (14.13)$$

The exergy efficiency is found with Eq. (14.6) as

$$\psi_{T1} = \left[\sum_i \dot{E}x_i + (1 - T_0/T)\dot{Q}_{T1} - \dot{I}_{E1} \right] \Big/ \sum_i \dot{E}x_i \quad (14.14)$$

where the exergy input rate for the ADU is

$$\sum_i \dot{E}x_i = \dot{m}_2 \varepsilon_2 \quad (14.15)$$

14.5.3. Overall exergy efficiency

Taking the entire system as a control volume, external heat transfer is observed to occur at T1. The overall internal exergy consumption rates are equal to the sum of the internal exergy consumption rates of the individual components. Thus,

$$\sum_i \dot{E}x_i = \dot{m}_1 ex_1 + \dot{m}_{10} ex_{10} \quad (14.16)$$

$$\dot{I}_{overall} = \dot{I}_{E1} + \dot{I}_{T1,ph} + \dot{I}_{T1,ch} \quad (14.17)$$

$$\psi_{overall} = \left[\sum_i \dot{E}x_i + (1 - T_0/T)\dot{Q}_{T1} - \dot{I}_{overall} \right] \Big/ \sum_i \dot{E}x_i \quad (14.18)$$

14.6. Results and discussion

14.6.1. Simulation results

A simulation of the system in Fig. 14.1 was carried out by Al-Muslim et al. (2003, 2005) for actual operating conditions. Input data are given in Table 14.2 and the simulation results in Table 14.3.

Detailed simulation results are reported in Al-Muslim (2002) and Al-Muslim et al. (2003), where the different streams are identified with their properties (e.g., temperature, pressure, composition). The reports also describe the operations occurring in the plant devices, and include simulation input data and output results (e.g., component data, calculation sequence, heat exchanger summary, column summary, stream molar component rates and stream summary).

14.6.2. Energy and exergy results

By applying the energy and exergy balances from earlier sections, we find energy and exergy efficiencies and irreversibility rates for individual components and the overall system. To determine the contribution of chemical exergy to the total exergy loss, the calculations for exergy are made twice: (i) with the chemical exergy term and (ii) without the chemical exergy loss inherent to the separation process. The results are shown in Table 14.4.

The energy efficiency of the ADU is 49.7% and of the VDU is 57.9%. The ADU energy efficiency is lower because the main separation occurs there. The energy efficiency of the overall system is 51.9%. The energy efficiencies of the heaters are not included as we assume adiabatic heat transfer in the process. The greatest exergy consumption occurs in the ADU, with 56% of the total exergy consumption. This is again due to the fact that the main separation takes place in the ADU. These losses are composed of physical and chemical exergy losses. The chemical exergy losses are 6.8% of the total exergy losses. The irreversibility losses in the VDU are significant, at 26% of the total exergy consumption,

Table 14.3. Simulation results for crude oil distillation plant under normal operating conditions.

Steam number	Description	Phase	Temperature (°C)	Pressure (kPa)	Mass flow rate (kg/s)	Specific entropy (kJ/kg K)	Specific enthalpy (kJ/kg)
1	Feed	Liquid	25	102	507.6221	4.440448	34.39055
2	Feed	Vapor	352.3746	102	507.6221	6.615505	1040.125
3	Offgas	Vapor	68.00522	206.6044	1.09×10^{-12}	4.216201	760.3895
4	Naphtha	Mixed	68.00522	206.8148	0.566983	2.509089	164.5537
5	Kerosene	Mixed	68.91907	208.4696	15.56693	3.039461	348.8646
6	Light diesel	Mixed	68.86612	211.7791	32.14957	2.323762	182.8285
7	Medium diesel	Mixed	91.40	215.91	75.812	4.2661	186.98
8	Heavy diesel	Liquid	157.1815	219.2254	53.9682	5.015905	323.7495
9	Reduced	Liquid	261.4419	223.3622	58.0393	5.878275	555.9205
10	ADU bottom	Liquid	342.2222	228.8781	260.1879	6.557474	750.3401
11	ADU bottom	Liquid	408.8889	228.8781	260.1879	6.858651	945.6614
12	Offgas	Vapor	203.2595	11.10056	10.21269	6.830466	1125.868
13	Condensed	Liquid	203.2595	11.10056	0.834219	5.768256	398.4854
14	Light vacuum gas oil	Liquid	219.95	11.100	22.617	5.8602	438.40
15	Heavy vacuum gas oil	Liquid	286.3149	11.92794	39.7135	6.238811	602.1979
16	Wash oil	Liquid	312.649	13.16899	46.74861	6.426736	666.1357
17	VDU bottom	Liquid	299.5011	15.23742	142.2034	6.386489	628.2944
18	Steam	Vapor	353.3333	515.0106	5.415384	7.651651	3172.316
19	Water	Water	68.00522	206.8148	0.517354	0.96201	284.4449
20	Steam	Vapor	768	413.6856	2.141962	8.831165	4075.276
21	Water	Water	n/a	n/a	0	n/a	n/a
22	Flue	Vapor	1100	102	600	7.986952	1180.016
23	Flue exhaust	Vapor	353.3333	102	600	7.100531	329.1276
24	Flue2	Vapor	1100	102	60	7.986952	1180.016
25	Flue2 exhaust	Vapor	357	102	60	7.106715	333.0128

Source: Al-Muslim et al. (2003).

but lower than those of the ADU. This is because less separation is involved. The chemical exergy losses represent 9.1% of the total VDU exergy losses. The exergy consumption is 16% of the total for Heater 1 and 2% for Heater 2. This is because most of the heating load is carried in Heater 1. In Heater 2 the flow is almost half of that in Heater 1 and the temperature rise is less. Only physical exergy losses are present here as there are no separation processes in the heater.

The components with the greatest irreversibilities have the lowest exergy efficiencies. The exergy efficiencies are 43.3% for the ADU, 50.1% for the VDU, 82.1% for Heater 1, 95.6% for Heater 2 and 23.3% for the overall system. The overall exergy efficiency is lower than the component efficiencies. When more components are added to the system, the

Table 14.4. Model results for crude oil distillation plant under normal operating conditions.

	Heater1	ADU	Heater2	VDU	Overall
$\Sigma \dot{H}_{in}$ (kW)	725467	545170	266030	287044	822177
$\Sigma \dot{H}_{out}$ (kW)	725467	270692	266030	166148	426803
$\dot{Q}_{cv}$ (kW)	0	274478	0	120896	395374
η		0.497		0.579	0.519
$\Sigma \dot{Ex}_{ph,in}$ (kW)	549311	460180	211396	233428	632531
$\Sigma \dot{Ex}_{ph,out}$ (kW)	444036	201557	201436	124561	149806
I_{ph} (kW)	105275	341129	9960	151841	608204
$\Sigma \dot{Ex}_{in}$ (kW)	589496	500365	223817	248621	672716
$\Sigma \dot{Ex}_{out}$ (kW)	484221	216749	213858	124560.5	149806
I (kW)	105275	366121	9960	167033	648389
ψ	0.821	0.433	0.956	0.501	0.223
I_{ch} (kW)	0	24993	0	15192.3	40185
$\% I_{ch}^{*}$	0	6.8	0	9.1	6.2

* Denotes the percentage contribution of I_{ch} to I.

overall exergy efficiency typically decreases further (unless high-efficiency measures are taken). This is because losses accumulate for each additional component added.

14.6.3. Impact of operating parameter variations

It is often useful to investigate the influence on system performance of varying operating conditions, particularly the temperature and pressure profiles of the distillation column. The temperature profile is controlled by the pump-around circuits. The key operating parameters are the pump-around flow rate and return temperature. The effect of varying these parameters on energy and exergy efficiencies of both distillation columns and the overall system are illustrated in this section.

The ADU has three pump-around circuits: PA01, PA02 and PA03. Figure 14.2 illustrates the effects of changing the mass flow rate of PA03, as a representative case, on the energy and exergy efficiencies of the ADU and the overall system. For the ADU, the operating condition, which is the middle point on each trend line, is the optimum condition. The energy efficiency at this point is 4–9% higher than at adjacent points, the exergy efficiency is 6–10% higher than at adjacent points and the irreversibility rate is 3–12% lower than at adjacent points. As the operating condition departs from the optimum, more inputs are required by the overhead condenser for the same product yield. For the overall system, there is no significant variation in energy and exergy efficiencies with variations in PA03 mass flow rate. However the irreversibility rate follows the same trend as for the ADU, with minimum irreversibilities being 5–10% lower than at adjacent operating condition points.

The VDU has three pump-around circuits: PA04, PA05 and PA06. Figure 14.3 illustrates the effects of varying the mass flow rate of PA04, as a representative case, on the energy and exergy efficiencies of the VDU and the overall system. For the VDU, the energy efficiency varies by less than 2%, the exergy efficiency by less than 5% and hence the irreversibility rate by less than 2%. The variations for the overall system of energy and exergy efficiencies and irreversibility rate are all less than 2%. This is because the heat duty on the VDU overhead condenser is negligible and most of the heat duty is carried by the pump-around circuits.

Figure 14.4 shows the effect of varying the PA03 return temperature on the energy efficiency, exergy efficiency and irreversibility rate of the ADU and the overall system. The operating condition is again the approximate optimum condition. The energy efficiency at this point is 4–11% higher than at adjacent points, the exergy efficiency is 5–14%

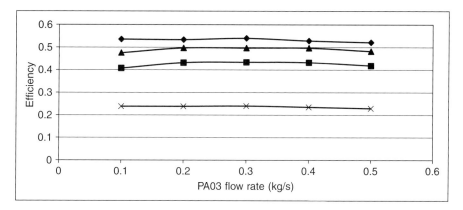

Fig. 14.2. Variation of energy and exergy efficiencies for the overall crude oil distillation system and its main components with pump-around PA03 mass flow rate (♦ overall energy efficiency, ▲ atmospheric distillation unit energy efficiency, ■ atmospheric distillation unit exergy efficiency and × overall exergy efficiency).

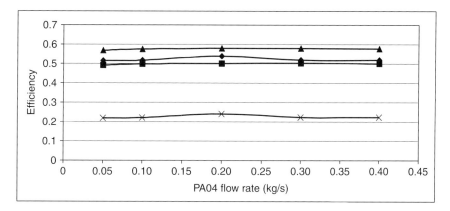

Fig. 14.3. Variation of energy and exergy efficiencies for the overall crude oil distillation system and its main components with pump-around PA04 mass flow rate (▲ vacuum distillation unit energy efficiency, ♦ overall energy efficiency, ■ vacuum distillation unit exergy efficiency and × overall exergy efficiency).

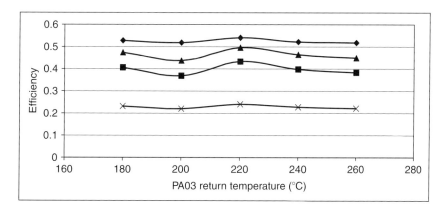

Fig. 14.4. Variation of energy and exergy efficiencies for the overall crude oil distillation system and its main components with pump-around PA03 return temperature (♦ overall energy efficiency, ▲ atmospheric distillation unit energy efficiency, ■ atmospheric distillation unit exergy efficiency and × overall exergy efficiency).

higher than at adjacent points and the irreversibility rate is 8–14% lower than at adjacent points. This result occurs because, as the operating condition departs from the optimum, additional input is required by the overhead condenser for the same product yield. For the overall system, there is no significant change in energy and exergy efficiencies with the PA03 return temperature. However, the irreversibility rate follows the same trend as the ADU, with minimum irreversibilities at the operating condition, where the irreversibility rate is 4–9% lower than at adjacent points.

Figure 14.5 illustrates the effect of varying the return temperature of PA04 on the energy and exergy efficiencies and the irreversibility rate of the VDU and the overall system. Varying the return temperatures of PA04, PA05 and PA06 has no significant effect on the energy and exergy efficiencies and irreversibility rate of the VDU and the overall system. As for the VDU, the energy and exergy efficiencies and the irreversibility rate variations are within 2%. The variations for the overall system are within 1%. This is because heat duty on the VDU overhead condenser is negligible and most of the heat duty is carried by the pump-around circuits.

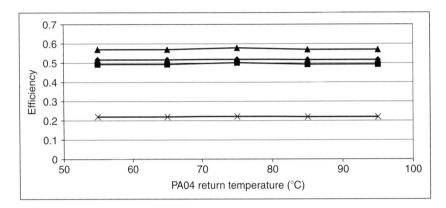

Fig. 14.5. Variation of energy and exergy efficiencies for the overall crude oil distillation system and its main components with pump-around PA04 return temperature (▲ vacuum distillation unit energy efficiency, ◆ overall energy efficiency, ■ vacuum distillation unit exergy efficiency and × overall exergy efficiency).

Another significant operating parameter is the distillation pressure. The pressure profile of the distillation column is controlled through the overhead pressure of the column. The vapor flow to the condenser can be restricted to increase the pressure. As the column overhead pressure increases, the yield of heavy hydrocarbons increases and of light hydrocarbons decreases. The effect of varying ADU and VDU overhead pressures on the energy and exergy efficiencies of both distillation columns and the overall system is illustrated in Figs. 14.6 and 14.7. The operating overhead pressure is the approximate optimum pressure. The energy efficiency at this pressure is 4–13% higher than at adjacent points, the exergy efficiency is 6–16% higher than at adjacent points and the irreversibility rate is 5–21% lower than at adjacent points. Departures from the optimum operating condition require more input to the overhead condenser for the same product yield. For the overall system, there is no significant change in energy and exergy efficiencies with column overhead pressure. However the irreversibly rate follows the same trend as for the ADU, with minimum irreversibilities occurring at the operating condition, where the irreversibility rate is 4–13% lower than at adjacent points. As shown in Fig. 14.7, the VDU overhead pressure has little effect on the energy and exergy efficiencies and the irreversibility rate of the VDU and the overall system. For the VDU, the energy and exergy efficiencies and the irreversibility rate variations are within 5%. These variations for the overall system are within 2%. This result is attributable to the fact that heat duty for the VDU overhead condenser is negligible and most of the heat duty is carried by the pump-around circuits.

14.6.4. Result limitations

In some cases, there are significant differences between the calculated results and actual data, due to the fact that it is difficult to determine specific entropy values for the different flows in the distillation process. The flows are mixtures of thousands of hydrocarbons and the reference state is defined with different parameters in terms of temperature, pressure and chemical composition. Moreover, the compositions of products of the distillation plants vary substantially in the same plant depending on properties of the crude oil used and operating conditions.

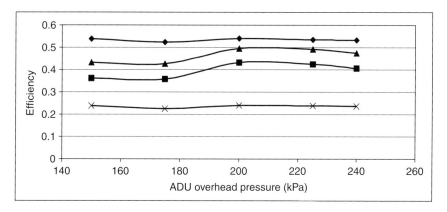

Fig. 14.6. Variation of energy and exergy efficiencies for the overall crude oil distillation system and its main components with ADU overhead pressure (♦ overall energy efficiency, ▲ vacuum distillation unit energy efficiency, ■ vacuum distillation unit exergy efficiency and × overall exergy efficiency).

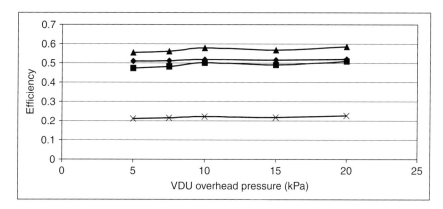

Fig. 14.7. Variation of energy and exergy efficiencies for the overall crude oil distillation system and its main components with VDU overhead pressure (▲ vacuum distillation unit energy efficiency, ♦ overall energy efficiency, ■ vacuum distillation unit exergy efficiency and × overall exergy efficiency).

This limitation is further illustrated by noting that the refinery overall efficiency found by Anaya et al. (1990) is 8.6%, while that found by Cornelissen (1997) is 5.18%. In the model presented in this chapter, the overall efficiency is 14%. This is due to the use of different calculation methods and model assumptions, relating to plant configuration, type of crude oil, product temperatures and normal operating conditions. More accurate results can be obtained with more detailed analyses. Nonetheless, the present results are illustrative both qualitatively and quantitatively.

14.7. Closing remarks

Energy and exergy analyses are conducted of crude oil distillation, focusing on the main devices: an ADU and a VDU and two heaters. Simulations determine the effect of varying operating conditions (e.g., temperature and pressure profiles of the columns) on energy and exergy efficiencies and irreversibility rates of the individual components and the overall system. The temperature and pressure profiles of the ADU at the normal operating condition are observed to provide the approximate optimum energy efficiency and exergy efficiencies and irreversibility rates. Changing the temperature profile and pressure profile of the VDU does not lead to significant changes in the energy and exergy efficiencies and irreversibility rates.

Problems

14.1 What is the importance of crude oil distillation? Explain.

14.2 How can exergy analysis be used in improving the efficiency of an oil distillation process?

14.3 How are the energy and exergy efficiencies for an oil distillation process defined? Define each efficiency and explain each term used in the definitions. What are typical values of energy and exergy efficiency for actual oil distillation systems?

14.4 Which components of an oil distillation system typically involve greater exergy destructions? Provide methods for reducing or minimizing the exergy losses in these components.

14.5 What are the important operating parameters in oil distillation systems? What are the effects of varying these parameters on the system energy and exergy efficiencies.

14.6 Are the components in an oil distillation unit with greater exergy destructions necessarily those with lower exergy efficiencies?

14.7 Can you compare the exergetic performance of small- and large-scale oil distillation units? Explain.

14.8 Identify several methods for reducing or minimizing the exergy destructions in oil distillation units.

14.9 Obtain a published article on exergy analysis of oil distillation plants. Using the operating data provided in the article, perform a detailed exergy analysis of the plant and compare your results to those in the original article. Also, investigate the effect of varying important operating parameters on the system exergetic performance.

14.10 Obtain actual operating data from an oil distillation plant and perform a detailed exergy analysis. Discuss the results and provide recommendations based on the exergy results for improving the efficiency.

Chapter 15

EXERGY ANALYSIS OF FUEL CELL SYSTEMS

15.1. Introduction

Energy is an important factor in interactions between nature and society and is considered a key resource for economic development. Environmental concerns encompass a range of pollutants and hazards to health and the environment. Environmental degradation often occurs locally, and sometimes regionally or globally. Many environmental issues are associated with the production, transformation and use of energy. Acid rain, stratospheric ozone depletion and global climate change, for instance, are all related to energy use.

Technologies using fossil fuels are major sources of air pollutants and contribute significantly to environmental concerns such as regional acidification and climate change. These concerns combined with uncertainties about fossil fuel reserves and increasing oil prices have increased interest in alternative fuels and energy conversion technologies that limit environmental impact, provide energy security, and facilitate economic growth and sustainable development.

The fuel cell is an energy conversion technology that has received considerable attention recently, particularly as a potential replacement for conventional fossil fuel-driven technologies.

The principles of power generation with a fuel cell were discovered over 160 years ago by a Welsh judge, Sir William Grove. Until recently, the use of fuel cells was confined to the laboratory and to space applications, where they provide electricity, heat and water. Space uses have occurred since the 1960s because fuel cells were considered less risky and more reliable than other options. At that time, the technology was in its infancy and too expensive for terrestrial applications. Recently, interest in fuel cells has increased sharply and progress toward commercialization has accelerated. Today, practical fuel cell systems are becoming available and are expected to attract a growing share of the markets for automotive power and generation equipment as costs decrease to competitive levels.

The fuel cell is an electrochemical energy conversion device, which converts the chemical energy of hydrogen directly and efficiently into electrical energy while emitting only waste heat and liquid water. Fuel cells are efficient and generate electricity in one step, with no moving parts. In many cases, e.g., for proton exchange membrane (PEM) fuel cells, fuel cells operate at low temperatures. This contrasts notably with the combustion process employed in traditional power plants, where a fuel is burned at high temperature to create heat, which is converted to mechanical energy and then to electricity. Since fuel cells do not combust fossil fuels, they emit none of the acid rain or smog-producing pollutants that are the by-products of burning coal or oil or natural gas.

There are many kinds of fuel cells. Two important types are the proton exchange membrane (PEM) fuel cell and the solid oxide fuel cell (SOFC). In principle, a PEM fuel cell operates like a battery but, unlike a battery, it does not run down or require recharging, and produces energy in the form of electricity and heat as long as fuel is supplied. The fuel cell converts chemical energy directly into electricity by combining oxygen from the air with hydrogen gas without combustion. If pure hydrogen is used, the only material output is water and almost no pollutants are produced. Very low levels of nitrogen oxides are emitted, but usually in the undetectable range. The hydrogen can be produced from water using renewable energy forms like solar, wind, hydro or geothermal energy. Hydrogen also can be extracted from hydrocarbons, including gasoline, natural gas, biomass, landfill gas, methanol, ethanol, methane and coal-based gas.

In this chapter, exergy analyses are presented of two systems: a PEM fuel cell power system for a light-duty fuel cell vehicle and a combined SOFC–gas turbine system. Parametric studies are also conducted to investigate the effect of operating-condition variations on the performance of the systems.

15.2. Background

In a fuel cell, electrical power is produced through the following electrochemical reaction of hydrogen and oxygen:

$$H_{2(g)} + \frac{1}{2}O_{2(g)} \Rightarrow H_2O_{(l)} + \text{Electrical power} + \text{Waste heat} \qquad (15.1)$$

15.2.1. PEM fuel cells

PEM fuel cells (membrane or solid polymer) operate at relatively low temperatures (about 90°C), have high power densities, can vary their output quickly to meet shifts in power demand, and are suited for many applications (particularly automobiles, where quick start-up is required). According to the U.S. Department of Energy, PEM fuel cells 'are the primary candidates for light-duty vehicles, for buildings and potentially for much smaller applications such as replacements for rechargeable batteries.' The PEM is a thin plastic sheet through which hydrogen ions can pass. The membrane is coated on both sides with highly dispersed metal alloy particles (mostly platinum) that are active catalysts. Hydrogen is fed to the anode side of the fuel cell where, due to the effect of the catalyst, hydrogen atoms release electrons and become hydrogen ions (protons). The electrons travel in the form of an electric current that can be utilized before it returns to the cathode side of the fuel cell where oxygen is fed. The protons diffuse through the membrane to the cathode, where the hydrogen atom is recombined and reacted with oxygen to produce water, thus completing the overall process.

PEM fuel cells have received considerable attention recently, notably as a potential replacement for the conventional internal combustion engine (ICE) in transportation applications. This development is especially important since fossil fuel-driven automobiles are major emitters of air pollution and contribute significantly to environmental problems.

PEM fuel cell powered automobiles using hydrogen offer several advantages including efficient and environmentally benign operation, quick start-up, compatibility with renewable energy sources and power densities competitive with those for ICEs. Major barriers hindering the commercialization of PEM fuel cell powered automobiles are cost and hydrogen infrastructure. The cost of a PEM fuel cell powered automobile can be reduced by improving the performance of the PEM fuel cell itself, and exergy analysis can assist in such activities.

15.2.2. Solid oxide fuel cells

SOFCs generate electricity, usually at high temperatures and for stationary applications. SOFCs utilize a solid oxide, usually doped zirconia, as the electrolyte. They operate at atmospheric or elevated pressures at a temperature of approximately 800–1000°C. At these temperatures, the electrolyte becomes sufficiently conductive to oxide ions. The temperature of exhaust gases from the cells is 500–850°C, a temperature which is attractive for cogeneration applications or for use in bottoming cycles for all-electric power plants. The SOFC conducts oxygen ions from an air electrode (cathode), where they are formed, through a solid electrolyte to a fuel electrode (anode). There, they react with CO and H_2 contained in the fuel gas to deliver electrons and produce electricity. Reformation of natural gas or other fuels containing hydrocarbons can be accomplished within the generator, thus eliminating the need for an external reformer. Individual cells are bundled into an array of series-parallel electrically connected cells forming a semi-rigid structure that comprises the basic generator building block.

Several features of SOFC technology make it attractive for utility and industrial applications. SOFCs exhibit high tolerances to fuel contaminants. The high temperature of the reaction does not require expensive catalysts and permits direct fuel processing in the fuel cells. The solid oxide electrolyte is very stable. Because no liquid phases are present in the electrolyte, many of the problems associated with electrode flooding, electrolyte migration and catalyst wetting are avoided. Cell components of the SOFC can be fabricated in a variety of self-supporting shapes and configurations that may not be feasible with fuel cells employing liquid electrolytes.

Despite the ability of SOFCs to generate electricity, their implementation in industry can be more effective in combination with traditional gas turbines cycles (Larminie and Dicks, 2003). In this chapter, a type of integrated energy system involving SOFCs and gas turbines is examined. In the coupled gas turbine cycles considered, the exhaust gases from SOFCs are utilized, making the completeness of fuel conversion in the SOFC stack less essential. This coupling increases the power of the combined unit and decreases the size and the cost of the SOFC stack, which is a significant advantage today.

A necessary step when using natural gas (mainly methane) in SOFCs is its preliminary conversion to hydrogen and carbon monoxide. State-of-the-art Ni–YSZ (yttria-stabilized zirconia) anodes permit methane conversion directly on the anode surfaces (internal reforming), so that the electrochemical and reforming processes proceed simultaneously. Hengyong and Stimming (2004) report that to perform methane conversion and avoid catalyst carbonization, the molar ratio between methane and steam (or steam with carbon dioxide) should be 1:2 or higher at the SOFC inlet.

15.3. Exergy analysis of a PEM fuel cell power system

This section describes a thermodynamic model of a PEM fuel cell power system for transportation applications. A performance analysis is performed, based on the comprehensive study by Hussain et al. (2005), considering the operation of all components in the system, and a parametric study is carried out to examine the effect of varying operating conditions (e.g., temperature, pressure and air stoichiometry) on the energy and exergy efficiencies of the system. Further, thermodynamic irreversibilities in each component of the system are determined. For further details and discussion, see Hussain et al. (2005).

15.3.1. System description

The PEM fuel cell power system for light-duty vehicles shown in Fig. 15.1, taken from Ballard (2004), is considered. The system consists of two major parts: the PEM fuel cell stack and the system module. A cooling pump is also employed. The system module includes the air compressor, the heat exchanger, humidifiers and the cooling loop.

The PEM fuel cell stack module is the heart of the power system. There, pressurized, humidified air and hydrogen are supplied from the system module, and electrical power is produced via the electrochemical reaction of hydrogen and oxygen in Eq. (15.1). Waste heat produced in the stack module is removed through the cooling loop.

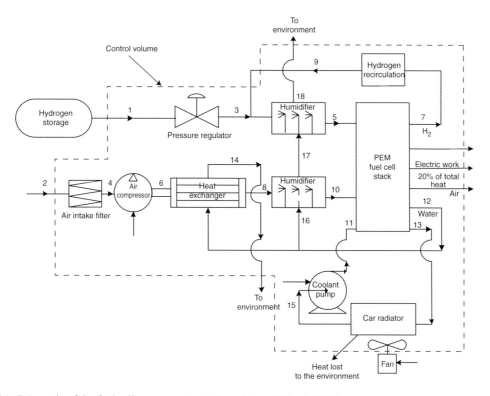

Fig. 15.1. Schematic of the fuel cell power system (adapted from Ballard, 2004).

The air compressor in the system module provides pressurized oxygen in the form of air, to the stack. The pressurized air is cooled in a heat exchanger and humidified in a humidifier before entering the stack. Similarly, compressed hydrogen stored on-board is humidified in a humidifier before entering the stack. Humidification of inlet streams is necessary to prevent dehydration of the membranes in the fuel cell stack. Not all the hydrogen supplied to the fuel cell reacts in the fuel cell stack; unreacted hydrogen leaving the stack is recirculated.

The cooling loop removes the heat produced by the exothermic reaction of hydrogen and oxygen, and consists of a radiator, a cooling pump and a radiator fan. Coolant (water/glycol) passes through the stack to remove waste heat.

15.3.2. PEM fuel cell performance model

The PEM fuel cell performance model developed by Baschuk and Li (2003) is used to model and simulate the fuel cell stack. The model predicts the voltage of a single cell for any specified operating condition. The voltage of the overall stack is obtained by multiplying the single cell potential with the number of cells in the stack. The output voltage of a fuel cell can be represented as

$$E(I) = E_r - E_{irr} \tag{15.2}$$

where E_r denotes the reversible voltage of the cell and E_{irr} the irreversible voltage loss or overpotential due to catalyst layers, electron migration in the bipolar plates and electrode backing, and proton migration in the polymer electrolyte membrane. These are described below.

Reversible cell voltage (E_r)

The reversible cell voltage is the cell potential obtained at a thermodynamic reversible condition, and can be expressed as

$$E_r = 1.229 + 0.85 \times 10^{-3}(T - 295.15) + 4.31 \times 10^{-5}T \ln\left[\left(\frac{C_{H_2}}{22.22}\right)\left(\frac{C_{O_2}}{7.033}\right)^{\frac{1}{2}}\right] \tag{15.3}$$

Irreversible cell voltage loss or overpotential (E_{irr})

The irreversible cell voltage loss or overpotential is composed of activation overpotential (η_{act}) due to catalyst layers, ohmic overpotential (η_{ohmic}) due to electron migration in the bipolar plates and electrode backing and proton migration in the polymer electrolyte membrane, and concentration overpotential (η_{con}) due to the mass transfer limitations at higher current densities. That is,

$$E_{irr} = \eta_{act} + \eta_{ohmic} + \eta_{con} \tag{15.4}$$

Activation overpotential (η_{act})

The activation overpotential is associated with the catalyst layers. It takes into account electrochemical kinetics, and electron and proton migration, and is composed of both the anode and cathode catalyst layer activation overpotentials:

$$\eta_{act} = \eta_{act}^a + \eta_{act}^c \tag{15.5}$$

where η_{act}^a and η_{act}^c are the activation overpotentials in the anode and cathode catalysts layers, respectively.

Ohmic overpotential (η_{ohmic})

The ohmic overpotential can be expressed as

$$\eta_{ohmic} = \eta_{bp}^a + \eta_{bp}^c + \eta_e^a + \eta_e^c + \eta_m \tag{15.6}$$

where η_{bp}^a and η_{bp}^c are the ohmic losses of the anode and cathode bipolar plates, respectively. The ohmic losses of the anode and cathode electrode backing layers are denoted by η_e^a and η_e^c. The overpotential due to the polymer electrolyte membrane is denoted η_m. Detailed descriptions of expressions for these overpotentials can be found elsewhere (Baschuk and Li, 2003).

Concentration overpotential (η_{con})

Concentration overpotential is associated with mass transfer limitations at higher current densities, and is composed of both the anode and cathode concentration overpotentials:

$$\eta_{con} = \eta_{con}^a + \eta_{con}^c \qquad (15.7)$$

where η_{con}^a and η_{con}^c are the anode and cathode concentration overpotentials, respectively.

Stack power ($\dot{W}_{stack}$)

The power produced by a single cell is expressible as

$$\dot{W}_{cell} = E(I) \times I \times A_{cell} \qquad (15.8)$$

where I is the current density and A_{cell} is the geometric area of the cell.

The stack power is obtained by multiplying the single cell power by the number of fuel cells in the stack n_{fc}:

$$\dot{W}_{stack} = n_{fc} \times \dot{W}_{cell} \qquad (15.9)$$

The power system under consideration here has 97 cells of 900 cm² geometric area in the stack, producing a net system power of 68 kW at $I = 1.15$ A/cm² and $E = 0.78$ V.

15.3.3. Analysis

Assumptions

The assumptions made in the analysis are as follows (as outlined by Hussain et al., 2005):

- The hydrogen storage cylinder or tank is at a constant pressure of 10 bar and temperature of 298 K.
- The isentropic efficiencies of the compressor, cooling pump and radiator fan are 70%.
- 20% of the total heat produced by the fuel cell stack is lost due to convection and radiation (Cownden et al., 2001).
- The temperatures at the inlet and outlet of the coolant circulation pump are equal.
- The environmental (restricted) state is at standard temperature and pressure conditions, i.e., 298 K and 1 atm.
- Moist atmospheric air with the composition given in Table 15.1 is used as the dead (unrestricted) state.
- Kinetic and potential energies are negligibly small.
- The system and its components are taken to be at steady-state so that time derivatives are zero.

Table 15.1. Mole fractions and chemical exergy of the components at dead state.

Component, i	N_2	O_2	H_2O	CO_2	Ar
Mole fraction, $x_{00,i}$	0.775	0.206	0.018	0.0003	0.0007
Specific chemical exergy, $ex_{ch,i}$ (J/mole)	631.51	3914.26	9953.35	20108.5	17998.14

The net power produced by the fuel cell power system is obtained by deducting the parasitic loads from the gross stack power. For the present system, the net system power is expressible as

$$\dot{W}_{net} = \dot{W}_{stack} - \dot{W}_{ac} - \dot{W}_{cp} - \dot{W}_{rf} \qquad (15.10)$$

where $\dot{W}_{stack}$ denotes the stack power, and $\dot{W}_{ac}$, $\dot{W}_{cp}$ and $\dot{W}_{rf}$ denote respectively the power input to the air compressor, the cooling pump and the radiator fan.

The governing thermodynamic equations for the system and its individual components are used to obtain the corresponding exergy balances. Exergy balances and energy and exergy efficiencies for the system and its components are presented below.

Overall system

An exergy balance for the overall system can be written as

$$\dot{N}_1 ex_1 + \dot{N}_2 ex_2 + \dot{N}_9 ex_9 - \dot{N}_{air} ex_{air} - \dot{N}_{14} ex_{14} - \dot{N}_{18} ex_{18}$$
$$-\dot{W}_{net} - (1 - T_0/T_{stack})(0.2 \times \dot{Q}_{stack}) - (1 - T_0/T_{radiator})\dot{Q}_{radiator} - \dot{I}_{system} = 0 \tag{15.11}$$

where subscripts 1, 2, 14 and 18 denote system states shown in Fig. 15.1. Also, $\dot{Q}_{stack}$ and $\dot{Q}_{radiator}$ are the rates heat is produced by the stack and rejected by the radiator to the environment, respectively, and $\dot{I}_{system}$ is the internal rate of exergy destruction.

Energy and exergy efficiencies of the system are defined as follows:

$$\eta_{system} = \frac{\dot{W}_{net}}{\dot{N}_1 h_1 + \dot{N}_9 h_9} \tag{15.12}$$

$$\psi_{system} = \frac{\dot{W}_{net}}{\dot{N}_1 ex_1 + \dot{N}_9 ex_9} \tag{15.13}$$

System components

Analyses of system components are carried out to assess their performances and contributions to the overall system.

15.3.4. Results and discussion

The analysis presented above is integrated with the fuel cell performance model of Baschuk and Li (2003), and applied to the system with the fuel cell stack operating at varying temperatures, pressures and fuel–air stoichiometric ratios. The base-case operating conditions of the system are listed in Table 15.2.

Table 15.2. Base-case operating conditions.

T (°C)	80
P (atm)	3
Fuel stoichiometry, S_{fuel}	1.1
Air stoichiometry, S_{air}	2.0

Overall system

The variations of net system and gross stack power with current density at the base-case operating conditions are shown in Fig. 15.2. At a current density of 1.15 A/cm², the net power produced by the system is approximately 68 kW. The difference between the gross stack power and the net system power is observed to increase with increasing current density (i.e., external load). This phenomenon is due to the increased parasitic loads with increased external loads.

The variations of system energy and exergy efficiencies with current density, at the base-case operating conditions, are shown in Fig. 15.3. The maximum system energy and exergy efficiencies are 42.3% and 49.6%, respectively, at a current density of 0.42 A/cm². The system energy and exergy efficiencies at a typical cell voltage of 0.78 V and current density of 1.15 A/cm² are found to be 37.7% and 44.2%, respectively. System energy and exergy efficiencies both increase with

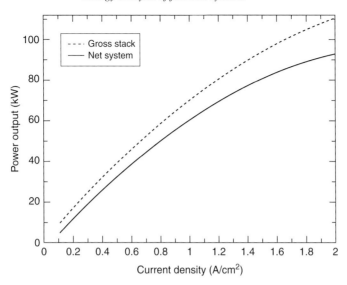

Fig. 15.2. Variation of power output with current density at base-case operating conditions.

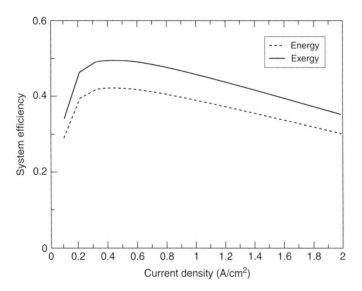

Fig. 15.3. Variation of system efficiencies with current density at base-case operating conditions.

current density at lower current densities. After reaching a maximum, these parameters decrease with current density. At lower current densities, the molar flow rate of fuel consumed by the stack and the power required by the auxiliary devices are low. Both energy and exergy efficiencies reach peak at a current density of $0.42 \, \text{A/cm}^2$. These efficiencies decrease with increasing current density beyond this value due to the increase in parasitic load and molar consumption of fuel. Exergy efficiencies are higher than the corresponding energy efficiencies due to the lower exergy values of hydrogen at states 1 and 9 in Fig. 15.1 compared to the enthalpy values.

Figures 15.4 and 15.5 show energy and exergy efficiencies of the system at different operating temperatures of the fuel cell stack. The operating pressure is fixed at 3 atm, and air and fuel stoichiometries are kept constant at 1.1 and 2.0, respectively, as proposed by Hussain et al. (2005). The energy and exergy efficiencies of the system both are seen to increase with increasing temperature because irreversible losses of the fuel cell stack decrease with increasing temperature, which in turn reduces the system irreversibility.

Figures 15.6 and 15.7 show the variations of system energy and exergy efficiencies with current density at different operating stack pressures. The operating temperature is 80°C and the air and fuel stoichiometries are 1.1 and 2.0,

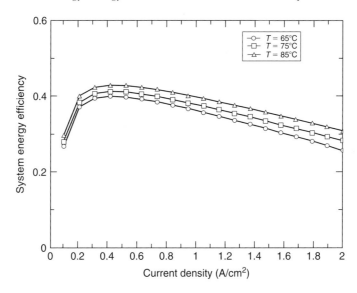

Fig. 15.4. Variation of system energy efficiency with current density at different operating temperatures of the fuel cell stack.

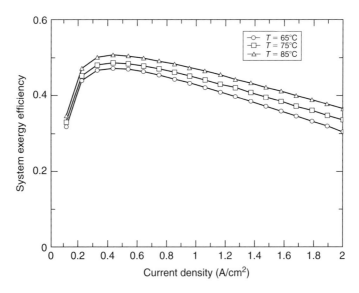

Fig. 15.5. Variation of system exergy efficiency with current density at different operating temperatures of the fuel cell stack.

respectively. With increasing pressure, the energy and exergy efficiencies of the system both increase. This is due to a significant increase in the gross stack power caused by a decrease in irreversible losses, especially those associated with anode and cathode overpotentials, with increasing pressure. When pressure increases, the concentrations of reactants at reaction sites increase, and irreversible losses in the form of anode and cathode overpotentials decrease, enhancing the performance of the fuel cell stack. Although, high-pressure operation requires pressurization of inlet streams, which increases the parasitic load in the form of power input to the compressor, the net power produced by the system nonetheless increases with increasing pressure.

The variations of system energy and exergy efficiencies with current density at different air stoichiometries are shown in Figs. 15.8 and 15.9. The operating temperature and pressure are fixed at 80°C and 3 atm, respectively, and the

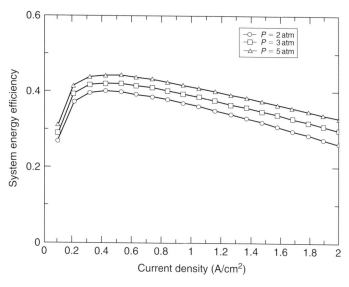

Fig. 15.6. Variation of system energy efficiency with current density at different operating pressures of the fuel cell stack.

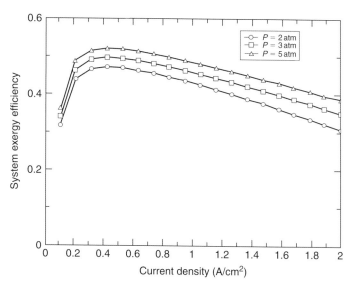

Fig. 15.7. Variation of system exergy efficiency with current density at different operating pressures of the fuel cell stack.

fuel stoichiometry at 1.1. Almost no appreciable increase is observed in energy and exergy efficiencies with increased stoichiometry of air. The molar flow rate of air increases with increasing air stoichiometry, resulting in a decrease in cathode overpotential and an increase in gross power produced by the stack. Although the gross stack power increases, the net system power increase is not significant. For instance, increasing the air stoichiometry from 3.0 to 4.0 causes almost no increase in the net power produced by the system. This phenomenon is again due to an increase in parasitic load which offsets the increase in gross stack power with air stoichiometry.

System components

Figure 15.10 shows an exergy flow diagram of the system at a particular operating condition. The value inside each component denotes the exergy consumption (or irreversibility) rate. The system energy and exergy efficiencies are 37.7% and 44.2%, respectively. The largest irreversibility rate is found in the fuel cell stack. Other major irreversibility rates occur in the fuel humidifier, where hydrogen from the storage cylinder and exhaust hydrogen not utilized in the fuel cell

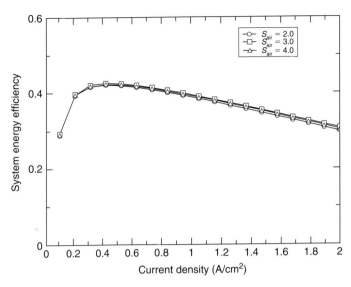

Fig. 15.8. Variation of system energy efficiency with current density at different air stoichiometries of the fuel cell stack.

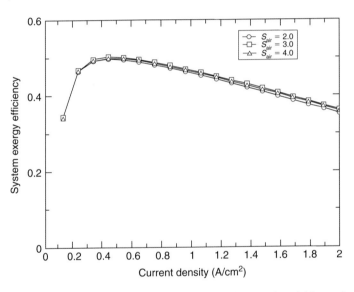

Fig. 15.9. Variation of system exergy efficiency with current density at different air stoichiometries of the fuel cell stack.

stack are mixed and humidified before being fed to the stack. The greatest potential for enhancing the performance of the system involves reducing the irreversibility rate in the fuel cell stack. In some instances, such measures can also reduce costs and thereby aid efforts to commercialize fuel cell power systems in transportation applications.

15.3.5. Closure

The energy and exergy analyses of a PEM fuel cell power system for light-duty vehicles described here help understand system performance and illustrate the effects of varying operating conditions on energy and exergy efficiencies of the system. The largest exergy consumption rate occurs in fuel cell stack, so reducing exergy consumption rates there has the potential to significantly improve system efficiencies. With increased external load (current density), the difference between the gross stack power and net system power increases as a result of increased parasitic loads. Energy and exergy efficiencies of the system increase with increased stack operating temperature. Although high-pressure operation

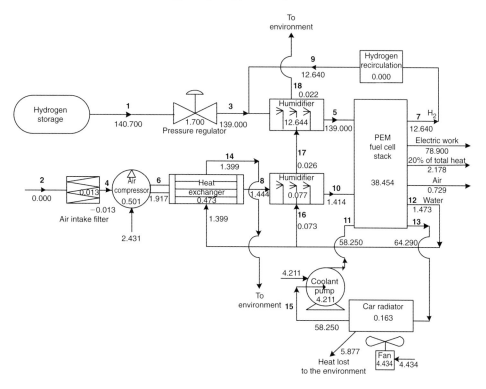

Fig. 15.10. Exergy flow diagram of the system at base-case operating conditions ($I = 0.938$ A/cm^2 and $V = 0.466$ V). The values in components denote their exergy consumption rates. All units are in kW (Hussain et al., 2005). Identifying numbers for flows are shown in bold.

increases the parasitic load in the form of power input to the compressor, the net power produced by the system increases with increasing pressure, resulting in increasing energy and exergy efficiencies. The efficiencies do not increase with increasing air stoichiometry.

15.4. Energy and exergy analyses of combined SOFC–gas turbine systems

Two combined SOFC–gas turbine systems are considered. The primary difference between the systems is that they provide a molar ratio of 1:2 between methane and steam (or steam with carbon dioxide) at the SOFC inlet in different ways. One system involves recycling the exhaust gases around the anodes of the SOFC stack and the other produces the required steam in the coupled gas turbine cycle.

   This section describes these systems and compares them in terms of exergy and energy efficiencies and electrical power generation, based on a recent study by Granovskii et al. (2006c, 2008). In the comparison, fixed SOFC stacks with equal exergy and energy efficiencies and electrical work generation capacities are considered.

15.4.1. Description of systems

Two SOFC–gas turbine systems are presented in Figs. 15.11a and 15.11b. The initial stream of natural gas, after compression in device 5, heating in device 7 and mixing with steam in device 14 for scheme (a) or with exhaust gases for scheme (b), is directed to the anodes of the SOFC stack (device 1). There, two processes occur simultaneously: conversion of methane into a mixture of carbon monoxide and hydrogen on the surface of the anodes and electrochemical oxidation of the resultant mixture with oxygen. The oxidation reaction is accompanied by electricity generation in the SOFCs. The anode exhaust gaseous flow is directed to the combustion chamber (device 2), where the remainder of the conversion products combust with air.

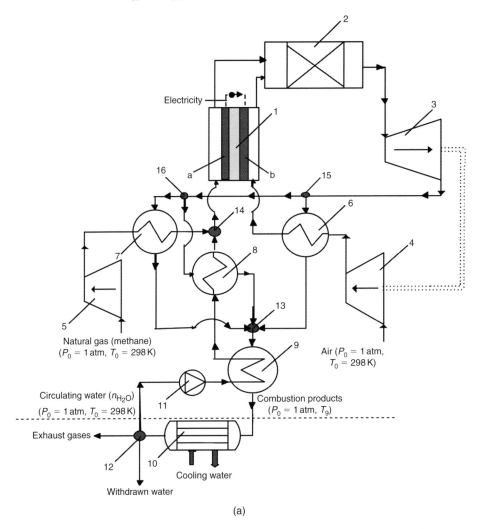

Fig. 15.11. Two combined SOFC–gas turbine systems: (a) with steam generation in the gas turbine cycle; (b) with recycling of exhaust gases around the anodes of the SOFC stack. Numbers indicate devices, as follows: 1 – SOFC stack; 2 – combustion chamber; 3 – turbine; 4, 5 – compressors; 6 – recuperator; 7 – fuel preheater; 8(a) – steam superheater; 8(b) – water evaporator and superheater; 9(a) – evaporator; 9(b) – steam turbine; 10 – condenser; 11 – pump; 12(a) – separator; 12(b), 13(b), 15(a), 16(a) – flow divider valves; 13(a), 14(a, b), 15(b) – mixers; a, b – anode and cathode of SOFC stack, respectively.

For scheme (a) the water after pumping in device 11 is evaporated in device 9, superheated in a heat exchanger (device 8), mixed with methane in device 14 and directed to the anodes of the SOFC stack (device 1).

For scheme (b) the exhaust gaseous flow, which still contains fuel, from the anodes of the SOFC stack is divided by a flow divider valve (device 13) into two flows. The first flow is directed to the combustion chamber (device 2) and the second is recycled and mixed with the input methane in device 14 and directed back to the anodes of the SOFC stack (device 1).

Air is compressed in device 4, heated in a recuperator (device 6), and directed to the cathodes of the SOFCs (device 1). In the SOFCs, oxygen is utilized, and oxygen-depleted air is heated, mixes with the remainder of the conversion products from the SOFC anodes and enters the combustion chamber (device 2). The combustion products expand in a turbine (device 3), and are divided for scheme (a) by the flow divider valves (devices 15 and 16) into three flows, and for scheme (b) into two flows (device 12). Then the flows are directed into three heat exchangers (devices 6, 7 and 8) for scheme

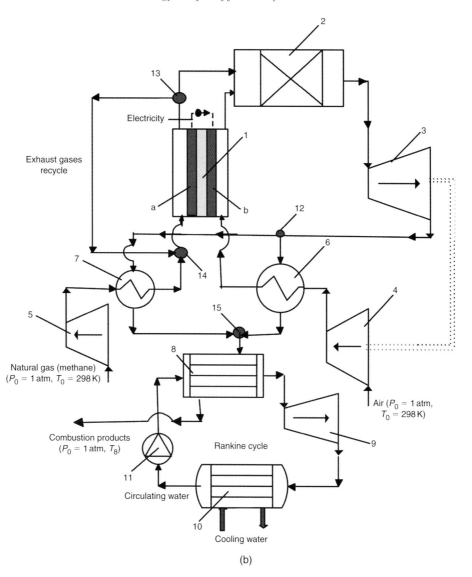

Fig. 15.11. (Continued)

(a) or into two heat exchangers (devices 6 and 7) for scheme (b), where they heat the input flows of air, methane and water (scheme (a)) and air and methane (scheme (b)). Subsequently, all flows of combustion products are mixed again in device 13 (scheme (a)) and in device 15 (scheme (b)). Then the heat of the combined flow is employed to evaporate circulation water in device 9 (scheme (a)) or to produce electricity in the bottoming Rankine cycle (scheme (b)). The bottoming Rankine cycle consists of a water evaporator and superheater (device 8), steam turbine (device 9), condenser (device 10) and water pump (device 11).

For scheme (a) after the evaporator (device 9), the combustion products are cooled in a condenser (device 10) and then divided into the three parts in a separator (device 12): recycled water, withdrawn water and exhaust gases.

15.4.2. Analysis

A schematic of the SOFC with internal reforming and recycling of part of the exhaust gases is shown in Fig. 15.12. During operation, a neutral molecule of oxygen takes four electrons from the porous cathode of the SOFC, depletes

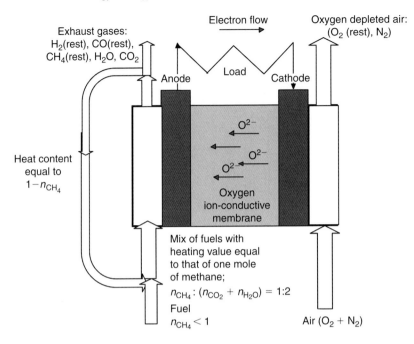

Fig. 15.12. Operation of a SOFC with recycle of exhaust gases and with internal methane reforming to hydrogen and carbon monoxide. The SOFC is illustrated for a fuel load of one mole of methane.

into negative charged ions, conducts through the electrolyte (ion-conductive membrane), returns electrons to the external circuit and oxidizes the products of methane conversion (H_2 and CO) on the anode surface. Recycling of the combustion products provides the necessary molar ratio of methane to the sum of steam and carbon dioxide at the inlets of the fuel cells, which may not be lower than 1:2. This experimentally demonstrated minimum ratio by Hengyong and Stimming (2004) is required for reforming methane into hydrogen and carbon monoxide and avoiding carbon deposition on the porous anodes surfaces.

To compare the two schemes (Figs. 15.11a and 15.11b), the efficiencies of the SOFC stack are taken to be the same, meaning that at equal energy content of the inlet flows the same power is generated. For the scheme in Fig. 15.11b, recycling some of the combustion product flow, normally containing a significant amount of combustible components, decreases the input methane flow. The input flow of methane for the scheme in Fig. 15.11b is determined by (i) the energy content of the recycled exhaust gases and (ii) the concentrations of steam and carbon dioxide in the total flow of exhaust gases. Then, in line with Fig. 15.12 for the scheme in Fig. 15.11b, the molar fraction of recycled exhaust gases α and the molar methane input flow rate $\dot{n}_{CH_4}^b$ are determined as follows:

$$\dot{n}_i^r = \alpha \dot{n}_i^{ex} \tag{15.14}$$

$$(\dot{n}_{CH_4}^b + \dot{n}_{CH_4}^r)/(\dot{n}_{CO_2}^r + \dot{n}_{H_2O}^r) = 1/2 \tag{15.15}$$

$$\dot{n}_{CH_4}^r q_{CH_4}^{LHV} + \dot{n}_{CO}^r q_{CO}^{LHV} + \dot{n}_{H_2}^r q_{H_2}^{LHV} + \dot{n}_{CH_4}^b q_{CH_4}^{LHV} = \dot{n}_{CH_4}^a q_{CH_4}^{LHV} \tag{15.16}$$

where $\dot{n}_{CH_4}^a$ is the molar methane input flow rate for the scheme in Fig. 15.11a; $\dot{n}_i^r$ and $\dot{n}_i^{ex}$ are the molar flow rates of components in the recycled and exhaust flows, respectively; and q_i^{LHV} is the lower heating value of the ith component in the flows considered. The left side of Eq. (15.16) represents the sum of lower heating values of the combustible components in the recycled and input flows. Equations (15.14) and (15.15) account for the requirement for a specific ratio of methane to steam plus carbon dioxide in the SOFC inlet flow and Eq. (15.16) is introduced to ensure the two presented schemes are compared for equally productive SOFC stacks.

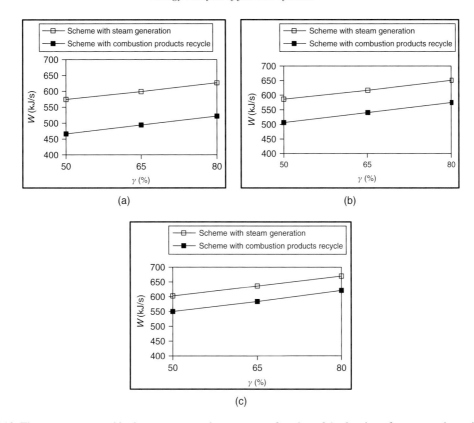

Fig. 15.13. The power generated in the two compared systems as a function of the fraction of oxygen γ that oxidizes the fuel yielding energy which is directly converted to electricity in the SOFC stack. The moles of oxygen n_{O_2} conducted through the SOFC electrolyte per heating value of one mole of methane (scheme in Fig. 15.11a) or the mix of fuels with a heating value equal to one mole of methane (scheme in Fig. 15.11b) at the SOFC inlet are considered as follows: (a) 1.2 moles of oxygen; (b) 1.4 moles of oxygen and (c) 1.6 moles of oxygen.

The lower heating value of methane is $q_{CH_4}^{LHV} = 802.6\,\text{kJ/mol}$, as defined by the oxidation reaction:

$$CH_4 + 2O_2 \rightarrow CO_2 + 2H_2O_{(g)} + \Delta H(-802.6\,\text{kJ}) \tag{15.17}$$

Analogously the lower heating values of hydrogen and carbon monoxide, respectively, are $q_{H_2}^{LHV} = 241.8\,\text{kJ/mol}$ and $q_{CO}^{LHV} = 283.0\,\text{kJ/mol}$.

The following assumptions are applied in the exergy analyses of the designs in Fig. 15.11: (i) gases are modeled as ideal, (ii) energy losses due to mechanical friction are negligible, (iii) the work of the water pump is negligible compared to the work of the turbines and compressors, (iv) thermodynamic and chemical equilibria are achieved at the outlet of the SOFC stack and (v) all combustible components are combusted completely in the combustion chamber. The fourth assumption is based on the fact that the catalytic conversion of methane to hydrogen and carbon monoxide is faster than the electrochemical processes inside the fuel cell (Dicks, 1998) and, therefore, the composition of the mixture at the anode outlet is close to the equilibrium one.

The parameters that characterize the combined power generation cycle and their values are listed in Table 15.3. Typical values of η_t and η_{cmp} are considered and the temperature of gases at the SOFC inlet, based on values often cited in the literature, is taken to be 700°C (i.e., $T_1^{in} = 973\,\text{K}$) (Haile, 2003). The outlet temperature for the turbines (device 3 in Fig. 15.11) (T_3^{out}) is taken to be 1023 K to provide a temperature difference between the input and output flows in the heat exchangers (devices 6a, 7a, 8a, 6b and 7b) of 50°C, which is generally acceptable for such heat exchange processes (Hinderink et al., 1996). Since the temperature of the gaseous mixture decreases as it expands in the turbines, the temperatures at the turbine outlets define the pressure drop (P_{max}/P_{min}) in the gas turbine cycle.

Table 15.3. Parameter values for the SOFC–gas turbine power generation systems.

Parameter	Value
Isentropic efficiency of turbines, η_t	0.93
Isentropic efficiency of compressors, η_{cmp}	0.85
Minimum pressure in the gas turbine cycle, P_{min} (atm)	1
Temperature at the turbine outlet, T_3^{out} (K)	1023
Temperature of SOFC inlet streams, T_1^{in} (K)	973
Standard temperature, T_0 (K)	298
Standard pressure, P_0 (atm)	1
Ratio of one mole of methane (scheme in Fig. 15.11a) or the combustible mixture with a heat content equal to that of one mole of methane (scheme in Fig. 15.11b) at the SOFC inlet to the moles of air	1:20
Air composition, volume percentage	21% O_2, 79% N_2

15.4.3. Thermodynamic model of the SOFC stack

The efficiency of the SOFC stack can be defined by the following two parameters (related to the heat content of one mole of methane): the total molar flow rate of oxygen $\dot{n}_{O_2}$ conducted through the electrolyte (ion-conductive membrane) and the fraction of the conducted oxygen flow γ that oxidizes the fuel yielding energy which is directly converted to electricity. The remainder is the oxygen flow which oxidizes fuel, but where the released energy provides heating of the input flows of methane and air and drives the endothermic reforming of methane to hydrogen and carbon monoxide. This heat is related to the activation, ohmic and concentration losses. If the SOFC stack is included in the combined SOFC–gas turbine system these losses do not directly corresponded to exergy losses (the loss in ability to produce electrical work). This is because this heat is employed in the gas turbine cycle and used to heat the input flows of fuel and air and drive the endothermic reactions of methane conversion to carbon monoxide and hydrogen, except the second one:

$$CH_4 + H_2O \leftrightarrow CO + 3H_2 + \Delta H(206\,kJ) \tag{15.18}$$

$$CO + H_2O \leftrightarrow CO_2 + H_2 + \Delta H(-41\,kJ) \tag{15.19}$$

For an adiabatic SOFC stack, the following energy balance can be written:

$$\Delta H + W_{el} = 0 \tag{15.20}$$

where W_{el} denotes electrical power and ΔH the rate of the enthalpy change in the SOFC stack. The calculations are made for an input methane flow rate $\dot{n}_{CH_4}^a$ of one mole per second for the scheme in Fig. 15.11a. Since the efficiencies of the SOFC stack are taken to be the same, the input flow rate of methane $\dot{n}_{CH_4}^b$ for the scheme in Fig. 15.11b is obtained by solving Eqs. (15.14) through (15.16). The electrical power W_{el} is determined as a percentage of the energy flow rate equal to the flow rate of one mole of methane with a lower heating value $q_{CH_4}^{LHV}$ in line with the reaction in Eq. (15.18) and by using the given values of $\dot{n}_{O_2}$ and γ.

The enthalpy change is a function of temperatures T_1^{in} and T_1^{out} and the compositions of all the flows at the inlet and outlet of the SOFC stack:

$$\Delta \dot{H} = f(T_1^{\text{in}}, T_1^{\text{out}}, \text{composition}^{\text{in}}, \text{composition}^{\text{out}}) \tag{15.21}$$

Here, the temperature T_1^{out} defines the composition of the methane conversion products (according to assumption (iv)), so that the solution of Eq. (15.21) gives the temperature at the SOFC outlet. Details on the thermodynamic calculations for the SOFC stack and the overall system are described elsewhere (Granovskii et al., 2006a, b). The electrical power produced is related to the operational-circuit fuel cell voltage V_s as follows:

$$V_s = \frac{\dot{W}_{\text{el}}}{\dot{n}_{O_2} N_e F} \tag{15.22}$$

where $\dot{n}_{O_2}$ is the molar flow rate of oxygen conducted through the electrolyte of the fuel cell, N_e is the number of moles of electrons transmitted into a circuit chain by one mole of oxygen (which is 4) and F is the Faraday constant (the charge of one mole of electrons).

15.4.4. Exergy balances for the overall systems

For the power generation scheme with steam generation presented in Fig. 15.11a, an exergy balance is considered for the part of the system above the dashed line, i.e., excluding the condenser and the separator (devices 10 and 12). This division implies that the thermal exergy of the combustion products transmitted to the cooling water in the condenser is not utilized in this system. For the scheme with the recycle (Fig. 15.11b), the temperature of the combustion products leaving the Rankine cycle is taken as equal to 100°C ($T_{\text{cmb}}^{\text{out}} = T_8 = 373$ K) (this condition favors this scheme) and the efficiency of their transformation into mechanical work to be 40% (Cengel and Turner, 2005).

The exergy balance in the case where only mechanical and electrical work are produced is expressible for both cases as follows:

$$\Delta \dot{Ex} = \sum \dot{Ex}_{\text{in}} - \sum \dot{Ex}_{\text{out}} = \sum_j \dot{W}_j + \sum_j \dot{Ex}_{D_j} \tag{15.23}$$

where $\sum \dot{Ex}_{\text{in}}$ is the sum of the input exergy flow rates of methane, air and water for the scheme in Fig. 15.11a and methane and air for the scheme in Fig. 15.11b, at standard conditions (T_0, P_0). That is,

$$\dot{Ex}_{\text{in}}^{\text{CH}_4} = \dot{n}_{\text{CH}_4}(h_{\text{CH}_4}(T_0) - T_0 s_{\text{CH}_4}(T_0, P_0)) \tag{15.24}$$

$$\dot{Ex}_{\text{in}}^{\text{Air}} = \sum_i \dot{n}_i^{\text{Air}}(h_i^{\text{Air}}(T_0) - T_0 s_i^{\text{Air}}(T_0, P_i)) \tag{15.25}$$

$$\dot{Ex}_{\text{in}}^{\text{H}_2\text{O}} = \dot{n}_{\text{H}_2\text{O}}(h_{\text{H}_2\text{O}}(T_0) - T_0 s_{\text{H}_2\text{O}}(T_0, P_0)) \tag{15.26}$$

Here, $\sum \dot{Ex}_{\text{out}}$ is the exergy flow rate of the combustion products directed to the condenser (Fig. 15.11a) or leaving the Rankine cycle (Fig. 15.11b):

$$\sum \dot{Ex}_{\text{out}} = \sum_i \dot{n}_i^{\text{cmb}}(h_i^{\text{cmb}}(T_{\text{cmb}}^{\text{out}}) - T_0 s_i^{\text{cmb}}(T_{\text{cmb}}^{\text{out}}, P_i)) \tag{15.27}$$

Also, $\sum_j \dot{W}_j$ is the sum of the power generated in the turbines and SOFCs, and consumed in the compressors (with a negative sign), and $\sum_j \dot{Ex}_{D_j}$ denotes the exergy destruction rate of the system which is calculated as the sum of the exergy destruction rates in each of the system devices, where

$$\dot{Ex}_{D_j} = T_0 \Delta \dot{S}_j \tag{15.28}$$

Here, $\Delta \dot{S}_j$ is the entropy generation rate in the jth device of the schemes considered, and $\dot{n}_i^{\text{Air}}$ and $\dot{n}_i^{\text{cmb}}$ denote the molar flow rates of the components constituting the flows of air (oxygen and nitrogen) and combustion products, and h_i and s_i denote their specific enthalpies and entropies, respectively.

The two schemes are considered in order to compare their electrical work generation capacities (the efficiency of mechanical work conversion into electricity is higher than 97%) and exergy and energy efficiencies. It is seen from Eq. (15.23) that the power generated $\sum_j \dot{W}_j$ increases with increasing rate of the exergy change $\Delta \dot{E}x$ and with decreasing exergy destruction rate $\sum_j \dot{E}x_{D_j}$ in the system. The rate of the exergy change $\Delta \dot{E}x$ in the system at fixed input and output temperatures and pressures increases with increasing flow rate of methane $\dot{n}_{CH_4}$ and, by extension, flow rates of air, water and combustion products (Eqs. (15.24) through (15.27)).

15.4.5. Results and discussion

The modeling results for the two systems considered are listed in Tables 15.4a and 15.4b. The input data are presented in Table 15.3 and in the first two columns in Tables 15.4a and 15.4b. The first column provides the molar flow rate of oxygen $\dot{n}_{O_2}$ conducted through the electrolyte of the SOFC stack per mole of methane (Fig. 15.11a, Table 15.4a) or per the mixture of the combustible components with an energy content equal to that of one mole of methane (Fig. 15.11b, Table 15.4b). According to the basic principles of SOFC operation, all oxygen conducted through the electrolyte is completely reacted in the anode compartments. Therefore, the total fuel energy or exergy consumed in the SOFC stack can be defined by the flow of oxygen conducted through the electrolyte $\dot{n}_{O_2}$ (column 1 in Tables 15.4a and 15.4b). The second column relates to the efficiency of fuel exergy or energy conversion to electricity in the fuel cells. The parameter γ identifies the fraction of oxygen which oxidizes the fuel yielding energy which is directly converted to electricity. The remainder is the oxygen which oxidizes fuel, but where the released energy heats the input flows of methane and air and drives the endothermic reforming of methane into hydrogen and carbon monoxide. When the SOFCs are the only source of electrical work, this heat corresponds to the activation, ohmic and concentration losses.

It can be seen in Tables 15.4a and 15.4b for equal values of $\dot{n}_{O_2}$ and γ that the electrical power $\dot{W}_{el}$ generated in the SOFC stack is the same for both schemes. This phenomenon occurs because the heat contents of the input fuel flows are equal under these conditions. However, the same capacities of electricity generation by the SOFC stacks are achieved at different input flow rates of methane, $\dot{n}_{CH_4}^a$ or $\dot{n}_{CH_4}^b$, for the overall schemes. This leads to different powers generated by the entire combined scheme, since $\dot{W} = \sum_j \dot{W}_j$. As can be seen in Fig. 15.13 the scheme with steam generation (Fig. 15.11a) yields more power $\dot{W}$.

Efficiencies

The energy and exergy efficiencies (η and ψ) of the combined systems are expressed as follows:

$$\eta^i = \frac{\sum_j \dot{W}_j^i}{\dot{n}_{CH_4}^i q_{CH_4}^{LHV}} = \frac{\dot{W}^i}{\dot{n}_{CH_4}^i q_{CH_4}^{LHV}} = \frac{w^i}{q_{CH_4}^{LHV}} \tag{15.29}$$

$$\psi^i = \frac{\sum_j \dot{W}_j^i}{\dot{n}_{CH_4}^i ex_{CH_4}^0} = \frac{\dot{W}^i}{\dot{n}_{CH_4}^i ex_{CH_4}^0} = \frac{w^i}{ex_{CH_4}^0} \tag{15.30}$$

where the index i denotes a for the scheme in Fig. 15.11a or b for the scheme in Fig. 15.11b, $\sum_j \dot{W}_j^i$ is the sum of electrical and mechanical power generated by the system, $\dot{n}_{CH_4}^i$ is the molar flow rate of methane consumed in the system (see Tables 15.4a and 15.4b), $q_{CH_4}^{LHV}$ is the lower heating value of methane (802.6 kJ/mol), $ex_{CH_4}^0$ is the standard exergy of methane (818.1 kJ/mol) and w^i is the work (electricity) produced per mole of methane consumed in the combined system. In the case when only electrical energy is generated in a system, it follows from Eqs. (15.29) and (15.30) that the energy and exergy efficiencies are very close to each other.

Work considerations

As can be seen in Fig. 15.14 the scheme with exhaust gas recycle (Fig. 15.11b) yields more work w per mole of methane consumed and has the higher energy and exergy efficiencies (see Tables 15.4a and 15.4b) but lower power $\dot{W}$.

Table 15.4a. Operating parameters of the SOFC–gas turbine cycle for the system with steam generation (Fig. 15.11a).[a]

$\dot{n}_{O_2}$ (mol/s)	γ (%)	T_1^{out} (K)	T_2^{out} (K)	P_{max} (atm)	$\dot{W}_{el}$ (kW)	$\sum_j \dot{W}_j$ (kW)	$\sum_j \dot{Ex}_{D_j}$ (kJ/s)	$\dot{n}_{CH_4}$ (mol/s)	$\dot{n}_{H_2O}$ (mol/s)	V_s (volt)	η	ψ
1.2	50	1147	1585	8.7	240.3	575.7	210.2	1	3.5	0.52	0.72	0.71
	65	1069	1523	7.0	312.4	600.4	190.5	1	3.0	0.675	0.75	0.74
	80	990	1457	5.5	384.8	625.5	170.0	1	2.5	0.83	0.78	0.77
1.4	50	1217	1217	7.7	280.5	589.5	199.3	1	3.2	0.52	0.73	0.72
	65	1127	1127	5.9	364.7	618.8	175.9	1	2.6	0.675	0.77	0.76
	80	1032	1032	4.4	448.8	648.6	151.2	1	2.0	0.83	0.81	0.80
1.6	50	1289	1289	6.8	320.7	603.3	188.4	1	2.9	0.52	0.75	0.74
	65	1188	1188	4.9	416.6	636.8	160.8	1	2.2	0.675	0.79	0.78
	80[b]	1078	1078	3.43	512.9	671.6	131.9	1	1.5	0.83	0.84	0.83

[a] $\dot{n}_{O_2}$ is the molar flow rate of oxygen conducted through the electrolyte of the SOFC stack for one mole of methane (scheme in Fig. 15.11a, Table 15.4a) or for the mixture of combustible components with an energy content equal to that of one mole of methane (scheme in Fig. 15.11b, Table 15.4b), γ is the fraction of oxygen which oxidizes the fuel yielding energy which is directly converted to electricity in the SOFC stack, T_1^{out} and T_2^{out} are temperatures at the SOFC stack and combustion chamber outlets, respectively; P_{max} is the maximum pressure in the cycle; $\dot{W}_{el}$ is the electric power generated in the SOFC stack; $\dot{Ex}_q$ is the exergy flow rate of the heat transferred into the bottoming Rankine cycle; $\sum_j \dot{W}_j$ is the total power generated in the combined system; $\sum_j \dot{Ex}_{D_j}$ is the total rate of exergy losses in the combined systems; $\dot{n}_{CH_4}$ is the basic molar flow rate of methane consumed in the combined system upon which calculations are made; $\dot{n}_{H_2O}$ is the molar flow rate of pressurized steam; V_s is the operational circuit voltage of the SOFC stack and η and ψ are the energy and exergy efficiencies of the scheme, respectively.
[b] For values in this row, a ratio of methane:steam equal to 1:2 or higher at the SOFC stack inlet is not maintained.

Table 15.4b. Operating parameters of the SOFC–gas turbine cycle for the system with recycling of combustion products (Fig. 15.11b).*

$\dot{n}_{O_2}$ (mol/s)	γ (%)	T_1^{out} (K)	T_2^{out} (K)	P_{max} (atm)	$\dot{W}_{el}$ (kW)	$\dot{Ex}_q$ (kJ/s)	$\sum_j \dot{W}_j$ (kW)	$\sum_j \dot{Ex}_{D_j}$ (kJ/s)	$\dot{n}_{CH_4}$ (mol/s)	V_s (volt)	η	ψ
1.2	50	1202	1479	5.6	240.3	55.3	466.4	162.9	0.80	0.52	0.73	0.72
	65	1117	1396	4.2	312.4	38.5	494.9	134.4	0.80	0.675	0.77	0.76
	80	1030	1311	3.1	384.8	23.9	523.8	105.5	0.80	0.83	0.82	0.81
1.4	50	1262	1481	5.7	280.5	58.6	509.6	165.4	0.86	0.52	0.74	0.73
	65	1163	1384	4.1	364.7	38.9	542.6	132.3	0.86	0.675	0.79	0.78
	80	1063	1285	2.9	448.8	21.8	576.1	98.8	0.86	0.83	0.83	0.82
1.6	50	1323	1475	5.6	320.7	60.7	548.9	167.5	0.91	0.52	0.75	0.74
	65	1210	1364	3.8	416.6	38.1	586.6	129.8	0.91	0.675	0.80	0.79
	80	1097	1252	2.5	512.9	19.3	624.7	91.7	0.91	0.83	0.86	0.85

* Parameters are described in Table 15.4a.

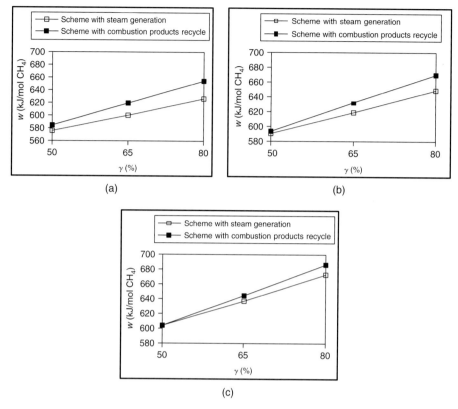

Fig. 15.14. The work w produced per mole of methane consumed in the two compared systems as a function of the fraction of oxygen γ which oxidizes the fuel yielding energy that is directly converted to electricity in the SOFC stack. The moles of oxygen $\dot{n}_{O_2}$ conducted through the SOFC electrolyte per heating value of one mole of methane (scheme in Fig. 15.11a) or the mix of fuels with a heating value equal to one mole of methane (scheme in Fig. 15.11b) at the SOFC inlet are considered as follows: (a) 1.2 moles of oxygen; (b) 1.4 moles of oxygen and (c) 1.6 moles of oxygen.

We now consider which scheme (Fig. 15.11a or 15.11b) permits a higher reduction in natural gas consumption (per unit time), $\Delta\dot{G}_{CH_4}^{i}$, in the case of its implementation instead of a contemporary combined gas turbine–steam cycle with its highest thermal efficiency of $\eta^{gt} \approx 0.55$:

$$\Delta\dot{G}_{CH_4}^{i} = \frac{\dot{W}^i}{q_{CH_4}^{LHV}}\left(\frac{1}{\eta^{gt}} - \frac{1}{\eta^i}\right) \tag{15.31}$$

Here, i denotes a (scheme in Fig. 15.11a) or b (scheme in Fig. 15.11b), $\dot{W}^i$ is the power output and η^i is the energy efficiency. The relative reduction in fuel consumption for the scheme in Fig. 15.11a relative to the scheme in Fig. 15.11b, β, can be expressed as:

$$\beta = \frac{\Delta\dot{G}_{CH_4}^{a} - \Delta\dot{G}_{CH_4}^{b}}{\Delta\dot{G}_{CH_4}^{b}} \tag{15.32}$$

Expressing the power generated in the systems through their energy efficiencies in line with Eq. (15.29), we obtain the following relationship:

$$\frac{\dot{W}^a}{\dot{W}^b} = \frac{\dot{n}_{CH_4}^{a}\eta^a q_{CH_4}^{LHV}}{\dot{n}_{CH_4}^{b}\eta^b q_{CH_4}^{LHV}} = \frac{\dot{n}_{CH_4}^{a}\eta^a}{\dot{n}_{CH_4}^{b}\eta^b} \tag{15.33}$$

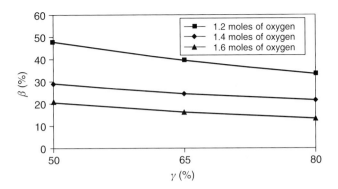

Fig. 15.15. The relative reduction in natural gas consumption β as a result of substitution of the scheme in Fig. 15.11b with the scheme in Fig. 15.11a, as a function of the fraction γ of oxygen that oxidizes the fuel yielding energy which is directly converted to electricity in the SOFC stack at different oxygen conductivity values $\dot{n}_{O_2}$.

After substitution of Eq. (15.31) into Eq. (15.32), the following expression for β is derived using Eq. (15.33):

$$\beta = \frac{\dot{n}_{CH_4}^a \eta^a}{\dot{n}_{CH_4}^b \eta^b} \frac{1/\eta^{gt} - 1/\eta^a}{1/\eta^{gt} - 1/\eta^b} - 1 \tag{15.34}$$

Figure 15.15 shows the relative reduction in fuel consumption β as a function of the fraction of oxygen γ associated with electricity generation. It can be seen from this figure that at lower values of γ and oxygen conductivity $\dot{n}_{O_2}$, the relative reduction in fuel consumption for the scheme in Fig. 15.11a compared to the scheme in Fig. 15.11b can reach about 20%. At higher values of γ and oxygen conductivity $\dot{n}_{O_2}$, the difference in β becomes less significant, remaining in the range of 3–8%.

From a practical point of view, the scheme in Fig. 15.11a (relative to that in Fig. 15.11b) allows the generation of more power with the same SOFC stack but with some decrease in the efficiency of fuel energy conversion to electricity.

15.4.6. Closure

Integrated systems involving SOFCs and gas turbines are analyzed and compared in this section. When using natural gas in SOFCs, it is necessary first to convert it to hydrogen and carbon monoxide. For effective methane conversion and to avoid catalyst carbonization, the molar ratio between methane and steam (or steam with carbon dioxide) should be 1:2 or higher at the SOFC inlet. Two approaches for providing this desirable ratio in combined SOFC–gas turbine systems are compared here. The first approach involves generation of the required steam in the coupled gas turbine cycle and the second, which is more traditional, involves recycling part of exhaust gases back to the SOFC stack. Overall energy and exergy efficiencies are compared for the system.

15.5. Closing remarks

Energy and exergy analyses of PEM fuel cells and SOFCs are described in this chapter, demonstrating the ability of exergy analysis to improve understanding of a system, highlight potential beneficial improvements and compare alternatives.

Problems

15.1   List some present day application areas of fuel cells and possible future applications.

15.2   Calculate the exergy destruction during the adiabatic combustion of methane with a stoichiometric amount of air. How can you reduce or minimize this exergy destruction?

15.3   Compare fuel cell technology to combustion devices from an exergetic point of view?

15.4   What are the current energy and exergy efficiencies of fuel cell automobiles?

15.5   What is the current status of fuel cell technology? What are the obstacles facing use of this technology today?

15.6   Describe the operating principle of a proton exchange membrane (PEM) fuel cell. What are the advantages of PEM fuel cells?

15.7   Provide expressions for energy and exergy efficiencies of a PEM fuel cell. Explain each term in the expressions. Which efficiency is typically greater and why?

15.8   Identify the operating parameters that have the greatest effects on the exergetic performance of a PEM fuel cell?

15.9   Describe the operation of a combined SOFC–gas turbine system.

15.10   Provide expressions for energy and exergy efficiencies of a combined SOFC–gas turbine system. Explain each term in the expressions.

15.11   Identify the operating parameters that have the greatest effects on the exergetic performance of a combined SOFC–gas turbine system?

15.12   Identify methods for reducing or minimizing the exergy destructions in fuel cell systems.

15.13   Obtain a published article on exergy analysis of a fuel cell system. Using the operating data provided in the article, perform a detailed exergy analysis of the system and compare your results to those in the original article. Also, investigate the effect of varying important operating parameters on the system exergetic performance.

Chapter 16

EXERGY ANALYSIS OF AIRCRAFT FLIGHT SYSTEMS

16.1. Introduction

As the airline industry evolves, those who manage it require a good knowledge of its many facets. One important aspect is aircraft performance, which plays a vital role in the economic fortunes of airlines and aircraft designers and manufacturers. Decisions require an appreciation of such performance characteristics as the relationship between fuel burn, distance and permissible payload for a given aircraft.

An engineer who designs power plants for aircraft makes decisions regarding type, configuration, size and arrangement, accounting for the expected performance of the final product. For this reason, the engineer needs to be familiar with aircraft and power plant performance characteristics and the impact of design factors on these characteristics. A sound knowledge of aircraft performance is also necessary for aircraft operators. Airlines need this information to determine how aircraft can be operated efficiently, economically, safely and with little environmental impact. Similarly, armed services need to understand aircraft performance characteristics in order to use aircraft in a manner that provides the greatest possible advantage and most effective support.

Consequently, an engineer needs to understand aircraft performance issues for design and analysis as well as evaluation of finished aircraft. Similarly, a good knowledge of performance characteristics and limitations of various classes of aircraft is needed to establish sound strategies for commercial and military applications. Exergy analysis can help provide this information, as it is a useful tool for design, analysis and performance improvement of aircraft power plants and flight systems.

An exergy analysis is presented in this chapter of an aircraft with a turbojet engine over a flight. The application of exergy to aerospace engines has been somewhat limited. Early efforts include those of Clarke and Horlock (1975) and Lewis (1976). These were followed by applications of exergy to various types of aerospace engines (turbojet, turbofan, scramjet) (e.g., Malinovskii, 1984; Kresta, 1992; Murthy, 1994; Brilliant, 1995a, b;), which followed the earlier approaches.

The extension of exergy analyses from ground-based systems to the aerospace engine differs from the traditional analysis of terrestrial systems in two main ways:

1. The aerospace engine is typically based on the open Brayton cycle, with the production of thrust generally involving the ejection of exhaust gases at high temperatures and velocities. This process has high exergy losses in the exhaust (Etele and Rosen, 2000), which are separate from the losses incurred through irreversibilities within the system. This high exhaust loss, which is particular to the aerospace engine, leads to low exergy efficiencies.
2. The operating environment varies significantly during the operation of aerospace engines during a flight, which differs notably from the situation for most ground-based processes where the environment usually remains relatively constant. Exergy analysis requires a definition of the reference environment, which is often modeled as the ambient environment since it is the actual environment in which the system operates and all exchanges of matter and energy occur. The exergy analyses cited earlier use a single typical operating environment for each engine considered, emulating the approach of ground-based systems, where sufficient analysis accuracy and realism is obtained with a single reference environment. However, ambient pressure and temperature variations over typical operating ranges for aerospace engines (sea level to 15,000 m) are often significant and can affect analysis accuracy if ignored.

The high exhaust loss noted in the first point has led to further research in this area to develop a revised second-law-based method for evaluating efficiencies based on an earlier concept described by Curran (1973). In these efforts the second law analysis approach is modified for aerospace engines, by comparing the desired output not to the overall exergy input but to the output of an idealized version of the engine under consideration. Then the engine efficiency is not 'penalized' unreasonably for the large exhaust exergy. This distinction between the applications of exergy analysis to aerospace vs. terrestrial systems has received much attention, with analyses of aerospace engines indicating a large portion of their inefficiency to be inherent in the manner in which they operate. Developing a method of assessing efficiencies and making comparisons that accounts for this operation have been the focus of recent exergy-based work.

   This chapter focuses on the second major distinction in the exergy analysis of aerospace as opposed to ground-based systems, the variation in the operating environment. Since a fixed reference environment is unrealistic for most aerospace applications, the reference environment can be modeled as the ambient operating environment by allowing it to vary as the operating environment changes. The reference-environment conditions can range from those at sea level to, in some instances, the near absolute zero temperature and vacuum conditions of space. The use of a reference environment fixed at a single operating condition has the advantages of reduced calculation complexity and convenient engine assessments for flight altitudes as high as low Earth orbit and beyond, and the disadvantage of having a reference environment different from the actual operating environment.

   The effect is assessed here of different reference-environment models on the exergy efficiencies of a turbojet engine during a flight. Continually varying and constant reference environments are considered. The chapter builds on previous work (Etele and Rosen, 1999) by focusing on the significantly varying operating environment encountered when applying exergy analysis to aerospace engines over a flight, following the work of Clarke and Horlock (1975).

16.2. Exergy analysis of a turbojet

The application of exergy to a turbojet engine, during a single operating condition and over a flight, is described in this section. This treatment follows closely the theoretical analysis of Clarke and Horlock (1975). A typical turbojet is shown in Fig. 16.1.

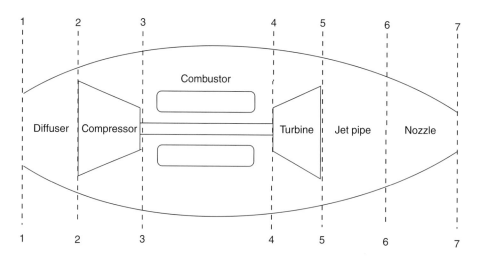

Fig. 16.1. Typical turbojet engine.

16.2.1. Exergy flows through a turbojet

The exergy entering a system either leaves the system, is accumulated in the system or is consumed due to irreversibilities. For a material flow, the specific exergy function ζ can be expressed as

$$\zeta = h^{\circ} - T_{\infty}s \tag{16.1}$$

where T_{∞} denotes the free stream temperature (which is the same as the reference-environment temperature T_{o}), s, the specific entropy and h°, the specific total or stagnation enthalpy as seen by an observer at rest in the reference environment.

The specific exergy ex can be written for a single constituent i of a flow as the difference between the specific exergy functions at the initial and reference states:

$$ex_i = \zeta_{\text{initial}} - \zeta_\infty \tag{16.2}$$

If the flow is treated as a perfect gas,

$$ex_i = c_{p_i}(T_i - T_\infty) - T_\infty \left(c_{p_i} \ln \frac{T_i}{T_\infty} - R_i \ln \frac{p_i}{p_\infty} \right) + \frac{c_i^2}{2} \tag{16.3}$$

where T_i, p_i, c_{pi}, R_i and c_i denote respectively for constituent i the temperature, pressure, specific heat at constant pressure, gas constant and absolute velocity with respect to a fixed reference environment. Also, p_∞ denotes the free stream pressure (which is the same as the reference-environment pressure p_o).

An exergy balance can be written for a general control volume in motion as

$$\left(\sum_i \dot{m}_i ex_i \right)_{\text{incoming}} \geq P_S + P_T + \sum_i q_i \frac{T_i - T_\infty}{T_i} + \left(\sum_i \dot{m}_i ex_i \right)_{\text{outgoing}} \tag{16.4}$$

where P_S and P_T denote respectively the shaft and thrust power extracted from the control volume, q_i denotes the heat-transfer rate across the control volume at a point where the temperature is T_i and ex_i and m_i denote respectively the specific exergy and mass flow rate of constituent i in the mixture. The difference between the left and right-hand sides of Eq. (16.4) is equal to the irreversibility of the system. The equality in Eq. (16.4) applies for an ideal system, and the inequality for real systems.

The thrust power across any component in a turbojet engine, where the mass flow rate is constant across the component boundaries and the flight velocity of the engine is U, can be written as

$$P_T = \dot{m} U (V_{\text{outgoing}} - V_{\text{incoming}}) \tag{16.5}$$

where U denotes the flight velocity and V, the flow velocity, relative to the control volume boundaries (where the flow is parallel to the flight direction).

The exergy input to a turbojet engine is provided by fuel and is mainly chemical exergy (although some physical exergy may exist due to the difference between the fuel storage conditions and the reference environment). The specific exergy expression in Eq. (16.3) does not account for the fuel chemical exergy as it implicitly assumes that the substances considered exist in the reference environment. To determine the specific fuel exergy it can be separated into several terms, which provide additional understanding about the total fuel exergy:

$$ex_{\text{fuel}} = ex_{\text{std}} + ex_{\text{rel}} + ex_{\text{vel}} \tag{16.6}$$

where

$$ex_{\text{std}} = \sum_i (\beta_i h_{i_{\text{std}}} - \gamma_i h_{i_{\text{std}}}) - T_\infty \sum_i (\beta_i s_{i_{\text{std}}} - \gamma_i s_{i_{\text{std}}}) \tag{16.7}$$

$$ex_{\text{rel}} = \sum_i (\beta_i \Delta h_{i_1} - \gamma_i \Delta h_{i_\infty}) - T_\infty \sum_i (\beta_i \Delta s_{i_1} - \gamma_i \Delta s_{i_\infty}) \tag{16.8}$$

$$ex_{\text{vel}} = \sum_i \beta_i \frac{c_i^2}{2} \tag{16.9}$$

Here, β_i denotes the mass of constituent i in a unit mass of fuel, γ_i denotes the mass of constituent i produced by the complete combustion of a unit mass of fuel and $h_{i\text{std}}$ and $s_{i\text{std}}$ are the specific enthalpy and entropy of constituent i.

Note that γ_i can be positive or negative as it represents the net products of combustion. For O_2, for example, $\gamma_{O_2} = -4$ when CH_4 is the fuel, as no mass units of O_2 are present in the fuel and during complete combustion four mass units of O_2 are consumed per unit mass of CH_4.

The specific exergy expression in Eq. (16.3) is useful when the chemical composition of the substance under consideration is the same in both the operating environment and the reference environment. For a turbojet, this condition holds in all stations up to the combustion chamber. Past this point a further term must be added to Eq. (16.3) to account for the chemical exergy created by the change in the chemical composition of the working fluid during the combustion process. The chemical exergy exists because the mole fractions of each constituent in the working fluid after combustion differ from those of the same constituent in the reference environment. Thus a different specific exergy function is defined for the working fluid after combustion:

$$ex = \bar{c}_p(T - T_\infty) - T_\infty(\bar{c}_p \ln \frac{T}{T_\infty} - \bar{R} \ln \frac{p}{p_\infty}) + \frac{c^2}{2} + T_\infty \left(\sum_i \lambda_i R_i \ln \frac{x_i}{x_{i\infty}} \right)$$ (16.10)

where the barred values pertain to the working fluid as a whole; c, T and p denote the absolute velocity, temperature and pressure of the working fluid, respectively; λ_i denotes the mass fraction of constituent i per unit of post-combustion working fluid; and x_i and $x_{i\infty}$ denote the mole fractions of constituent i in the working fluid and the reference environment, respectively.

16.2.2. Exergy efficiencies for a turbojet

The rational efficiency, defined as the ratio of useful or desired work obtained from the system to the total quantity of incoming exergy, is used here for assessing and comparing different systems (Clarke and Horlock, 1975; Czysz and Murthy, 1991; Murthy, 1994). For a turbojet, the useful work is the thrust and

$$\psi = \frac{P_T}{\dot{Ex}_{incoming}}$$ (16.11)

where the incoming exergy rate accounts for the incoming air and fuel as follows:

$$\dot{Ex}_{incoming} = \dot{m}_{fuel} ex_{fuel} + \dot{m}_{air} ex_{air}$$ (16.12)

Since the exergy of the reference environment is zero, the free stream exergy term is zero when the operating and reference environments are identical, and the incoming exergy is solely associated with the fuel. For an adiabatic engine where no shaft power is extracted, the thrust power produced and the total losses together are equal to or less than the incoming exergy (see Eq. (16.4)). The rational efficiency can then be written as

$$\psi = 1 - \frac{\sum \text{Exergy loss rate}}{\dot{m}_{fuel} ex_{fuel} + \dot{m}_{air} ex_{air}}$$ (16.13)

where the total exergy loss rate in the numerator of the rightmost term is expressed as the sum of the loss rates incurred in each engine component.

To assess the rational efficiency of an engine over an entire flight, Eq. (16.11) is modified. When considering an instant in time, the rational efficiency expression in Eq. (16.11) is appropriate because it includes instantaneous values of thrust power and incoming exergy flow rate. For an entire flight, however, cumulative measures of these quantities are required, and the cumulative rational efficiency ψ_{cum} can be expressed as

$$\psi_{cum} = \frac{\int_0^t P_T(t) dt}{\int_0^t \dot{Ex}_{incoming}(t) dt}$$ (16.14)

where the numerator is the integral of the instantaneous thrust power over the flight (from an initial time of 0 to final time t) and the denominator is the integral of the instantaneous incoming exergy flow rate over the flight. The instantaneous and cumulative rational efficiencies, in Eqs. (16.11) and (16.14) respectively, are equal only at the beginning of a flight.

16.2.3. Impact of environment on turbojet assessment

The analyses presented here involve both operating and reference environments. The operating environment temperature and pressure are those for the current altitude, as the actual performance of the turbojet is dependent on the incoming flow conditions. The thrust produced and the thermodynamic properties at the various engine stations are determined using operating environment values. Results of exergy analyses, however, depend on engine performance (and hence the operating environment) and the reference environment. Reference environment changes do not affect quantities such as thrust but cause efficiencies and losses calculated using thrust to vary, sometimes significantly.

16.3. Flight characteristics

To facilitate examining the effects of different reference-environment models on the accuracy of exergy analysis results over a flight, the flight characteristics are established. A cruising altitude of 15,000 m is considered over a ground distance of approximately 3500 km (approximately the distance between Toronto and Vancouver) with both the departure and destination aerodromes assumed to be at sea level. The total flight time is approximately four hours. A schematic of the turbojet with key states identified is shown in Fig. 16.1, and operating parameters for a flight are listed in Table 16.1.

The aircraft ascends at a constant rate of climb of 3000 m/min to cruising altitude over a period of 5 min. The descent portion of the flight is accomplished using a constant descent angle of 10° under cruise power conditions. The engine operating parameters in climb are different from those in cruise (see Table 16.1 for details) but because a cruising descent is used, the engine operating parameters in both cruise and descent are identical.

16.4. Cumulative rational efficiency

The cumulative rational efficiency in Eq. (16.14) is used here as a measure of merit for the overall engine during the flight. The behavior of this efficiency for variable and constant reference environments is examined.

16.4.1. Variable reference environment

The variable reference-environment curve in Fig. 16.2 shows the cumulative rational efficiency of the turbojet decreasing rapidly at the beginning of the flight and then leveling off asymptotically. At the start of the flight (distance = 0 km and altitude = sea level), the cumulative rational efficiency is identical to the instantaneous rational efficiency obtained

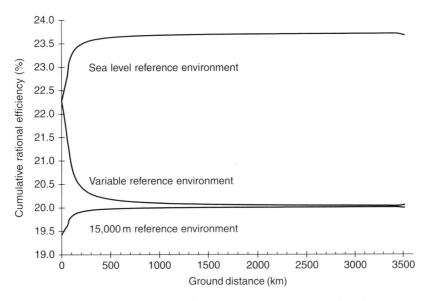

Fig. 16.2. Variation of turbojet cumulative rational efficiency over a flight range of 3500 km at a cruising altitude of 15,000 m, for various reference environments.

Table 16.1. Specified turbojet operating parameters for a complete flight.

Section	Engine component	Performance criteria: climb [cruise]	Mach number (at exit plane): climb [cruise]
∞	Free stream	–	0.80 [0.80]
∞–1	External diffusion	$p_1^o/p_\infty^o = 1.00$ [1.00]	0.70 [0.70]
1–2	Internal diffuser	$p_2^o/p_1^o = 0.95$ [0.95]	0.50 [0.40]
2–3	Compressor	$p_3^o/p_2^o = 26$ [20] $\eta_c = 0.85$ [0.90]	0.50 [0.40]
3–4	Combustor	$p_4^o/p_3^o = 0.90$ [0.95] $f = 1/55$ [1/50] $Q_R = 51{,}445$ kJ/kg	0.35 [0.30]
4–5	Turbine	$\eta_t = 0.88$ [0.92]	0.50 [0.40]
5–6	Jet pipe	$p_6^o/p_5^o = 0.98$ [0.98]	0.40 [0.30]
6–7	Nozzle	$\eta_n = 0.98$ [0.98] $p_7 = p_\infty$	–

with Eq. (16.11) at 22.27% (see Fig. 16.3). As the aircraft climbs and reaches cruise altitude, the cumulative rational efficiency decreases to a value of 20.04% at a distance of 3445 km. The maximum cumulative rational efficiency variation is therefore 2.23% over the entire flight. The instantaneous rational efficiency values also vary by approximately the same amount, decreasing from 22.27% at sea level to 20.57% at the end of the climb segment (a distance of 73 km), and dropping further to 20.02% as the engine operating parameters are modified for the cruise condition. The maximum instantaneous rational efficiency variation is 2.25%.

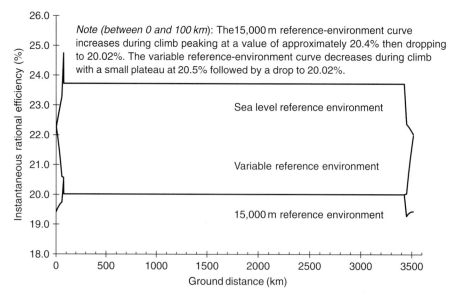

Fig. 16.3. Variation of turbojet instantaneous rational efficiency over a flight range of 3500 km at a cruising altitude of 15,000 m, for various reference environments.

That the cumulative rational efficiency at 3445 km (20.04%) is almost identical to the instantaneous rational efficiency during cruise (20.02%) is to be expected given the length of the flight. Since the majority of the aircraft operating time is at cruising conditions which are constant, any variations caused by the climbing and descending portions of the flight are overwhelmed by the much longer cruise segment. This observation is evident from the shape of the cumulative rational efficiency curve. At the beginning of the flight when the aircraft has spent no time cruising, the climb conditions dominate the behavior of the cumulative curve. Thus the rapidly decreasing instantaneous rational efficiency during the climb portion of the flight (Fig. 16.3, for ground distances from 0 to 73 km) dominates the behavior of the cumulative curve in this region (Fig. 16.2, for ground distances from 0 to 73 km). At the end of the flight the instantaneous rational efficiency increases due to the descent in the same manner that it decreased during the climb segment. However even with this rapid increase, the effect of the instantaneous rational efficiency is much less pronounced on the cumulative curve as only a very small increase in the cumulative rational efficiency is seen in Fig. 16.2 starting at 3445 km.

Generally the more time spent under cruising conditions, the more the cumulative efficiency results tend to reflect the instantaneous results during cruise (which are constant).

The stabilizing and averaging nature of cumulative results de-emphasizes the sudden variations in instantaneous efficiencies. Specifically, the cumulative efficiencies somewhat mask (i) the sharp decrease in instantaneous efficiency during the climb segment of the flight and (ii) the small instantaneous-efficiency plateau observed as the engine enters the tropopause under climb conditions (past this plateau, the engine switches operating parameters from climb to cruise settings, thus creating the discontinuous (vertical) change in the instantaneous rational efficiency). This stabilizing and averaging effect is even more noticeable during the descent portion of the flight, as previously mentioned, as only a small increase in the cumulative rational efficiency is observed in Fig. 16.2 despite the relatively large increase in the instantaneous efficiencies seen in Fig. 16.3.

Note that although the aircraft starts to descend at a flight distance of 3425 km, the instantaneous rational efficiency changes very little at this point. It is not until the troposphere is reached at a distance of 3445 km that the instantaneous efficiency starts to increase rapidly.

In Fig. 16.3, the 15,000 m and variable reference-environment curves are identical during cruise, but the 15,000 m reference-environment curve starts to decrease at the start of the descent whereas the variable reference-environment curve does not increase dramatically until a small distance further where the aircraft re-enters the troposphere. This delay in increasing instantaneous efficiency for the variable reference-environment curve is due to the fact that in the tropopause the instantaneous efficiency is nearly constant and as such no change is visible.

16.4.2. Constant reference environment

The use of a constant sea level reference environment to evaluate the cumulative rational efficiency produces errors in both numerical accuracy and predicted trends. At an operating altitude of sea level (for a distance traveled of 0 km), the variable and sea level curves in Fig. 16.2 are identical at a value of 22.27%. However, whereas the variable reference-environment curve indicates that the engine efficiency decreases as the flight progresses, the sea level reference-environment curve shows the opposite trend, with the curve reaching a maximum value of 23.71% at a ground distance of 3425 km, a variation of 1.44%. The cumulative sea level curve starts to decrease at the start of the descent due to the marked change in instantaneous rational efficiency shown in Fig. 16.3 at the start of the descent. This behavior is in contrast to that for the variable cumulative rational efficiency curve which reaches a minimum at the point the aircraft descends into the troposphere, at a ground distance of 3445 km.

The cumulative sea level reference-environment curve tends asymptotically toward the instantaneous sea level reference-environment value during cruise (23.72%) and, as shown by the value of the cumulative rational efficiency at 3425 km, this value is nearly reached. The maximum error (the maximum difference between the cumulative rational efficiencies at variable and sea level reference environments) occurs at the start of the descent portion of the flight and is equal to 3.67%. This result is different from the instantaneous results, where the maximum error occurs at the end of the climb segment (73 km) while the engine is still operating under climb conditions. In this case, the use of a sea level reference environment predicts an instantaneous rational efficiency of 24.75%, which, when compared to the value predicted for the variable reference-environment curve of 20.57%, yields a maximum error of 4.18%. However comparing the instantaneous results during cruise, the error between using a variable and sea level reference environment is 3.70%. This is the asymptotic limit for the cumulative results, i.e., the maximum value which is approached but never reached of the error between the cumulative curves as the flight distance is increased.

The reasons for the increasing cumulative rational efficiency when using a constant sea level reference environment while cruising at an altitude of 15,000 m are the same as those for the instantaneous rational efficiency results. The use of this reference environment creates the 'illusion' of negative exergy entering the engine with the air flow at all altitudes above sea level. As the flight time increases (which requires the aircraft altitude to increase to the cruising height), the quantity of this negative exergy increases, causing the cumulative rational efficiency to increase. Since the entire flight is spent at altitudes above sea level, the engine continues to 'ingest' negative exergy, resulting in a total accumulation of approximately $-2.40\,$GJ. This fictitious exergy is significant in quantity, representing approximately 15.72% of the total exergy input through the fuel of 15.27 GJ. (*Note*: the exergy input with the fuel evaluated using a variable reference environment is approximately 15.19 GJ.)

The use of a constant 15,000 m reference environment produces a cumulative rational efficiency curve with a shape similar to the constant sea level reference-environment curve, but displaced negatively on the efficiency axis. This result is to be expected, as the use of a 15,000 m reference environment at an operating altitude of sea level creates the 'illusion' of positive exergy in the incoming air flow. This added exergy decreases the rational efficiency compared to the case for a variable reference environment, yielding a value of 19.42% at sea level (for both the instantaneous and cumulative values). However, as the flight time increases during climb, the reference and operating environments approach and eventually meet at the cruising altitude, thus eliminating the fictitious positive exergy in the incoming air flow. In this case, the total accumulation of fictitious exergy is approximately 0.05 GJ compared to the cumulative exergy input through the fuel of approximately 15.19 GJ. Thus the fictitious exergy represents a much smaller percentage of the total actual exergy input, approximately 0.32%.

The largest difference between the 15,000 m and variable reference-environment cumulative rational efficiencies (Fig. 16.2) occurs at sea level and is equal to 2.85% (since this point is the beginning of the flight, it also has the largest difference in the instantaneous values). The cumulative 15,000 m reference-environment curve increases with altitude (again predicting the opposite trend from the variable reference-environment case) to a value of 20.01% at 3425 km which is very near the asymptotic value of 20.02% (the instantaneous cruise value using a 15,000 m reference environment). Thus the predicted variation in the cumulative rational efficiency over the entire flight is 0.59%.

16.5. Cumulative exergy loss

Since the cumulative exergy efficiency for the turbojet engine over the considered flight is approximately 20–24%, the cumulative exergy loss is approximately 76–80% of the exergy input.

The cumulative exergy loss is made up of two main parts:

1. More than half is associated with the emission of exhaust gases.
2. The remaining exergy loss is almost entirely due to exergy destruction within the engine.

16.6. Contribution of exhaust gas emission to cumulative exergy loss

The contribution of the exhaust gas emission to the cumulative exergy loss is examined here. The percentage of exergy contained and accumulated in the exhaust over an entire flight is expressed as a percentage of the total cumulative incoming exergy. The variation of cumulative exhaust emission exergy over the flight is illustrated in Fig. 16.4 for various reference environments.

16.6.1. Variable reference environment

The variable reference-environment curve in Fig. 16.4 shows that the cumulative exhaust exergy percentage increases at the beginning of the flight, and then levels off asymptotically to a constant value. The rapid increase in exhaust exergy percentage between distances of 0 and 73 km is due to the increasing altitude during this phase of flight, when the reference-environment pressure and temperature decrease and the exergy of the exhaust gases correspondingly increase. At sea level (a distance of 0 km) 50.3% of the cumulative exergy input is lost through the exhaust while at 3445 km this value increases to 56.4%, a variation of 6.1%. The cumulative exhaust loss curve asymptotically approaches the instantaneous exhaust loss percentage during cruise of 56.5% as the flight distance is increased. There is a small decrease beyond a distance of 3445 km in the instantaneous percentage of the input exergy contained in the exhaust because descent occurs thereby lowering the exhaust exergy (due to higher reference-environment pressure and temperature). However, due to the much greater time spent at cruising conditions, the short duration of this phase of flight has little impact on the cumulative results.

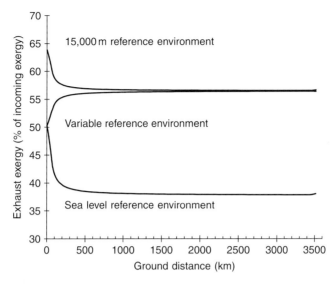

Fig. 16.4. Variation of cumulative exhaust emission exergy over a flight range of 3500 km at a cruising altitude of 15,000 m, for various reference environments.

16.6.2. Constant reference environment

With a constant sea level reference environment, the cumulative exhaust exergy percentage decreases as the flight progresses, going from a value of 50.3% at a ground distance of 0 km (sea level) to a value of 37.9% at 3425 km, a variation of 12.4%. This trend is opposite to that exhibited by the variable reference-environment curve in Fig. 16.4 where exhaust emissions contain increasing exergy as the flight progresses.

Since the decrease in exhaust exergy becomes greater as the difference between the operating and reference-environment pressures increases (due to the fact that the exhaust gases are expanded to the operating environment pressure which decreases as the altitude increases while the reference-environment pressure remains constant), the constant sea level curve in Fig. 16.4 decreases as the aircraft climbs (between 0 and 73 km). Also, although the exhaust gases are at a higher temperature than the reference-environment temperature at sea level, the thermal portion of the exhaust exergy is still decreased when compared to the variable reference-environment case. This decrease occurs because the sea level temperature is higher than the temperature at 15,000 m, thus decreasing the apparent thermal difference.

At the start of the flight the values of the exhaust exergy as a percentage of incoming exergy for both the sea level and variable reference environments are the same. However, as the flight progresses, the sea level curve diverges from the variable curve and reaches a maximum difference at 3425 km of 18.5%.

For the constant 15,000 m reference environment in Fig. 16.4, the cumulative exhaust exergy percentage decreases from 63.8% at the beginning of the flight to a value of 56.5% at a distance of 3425 km, a variation of 7.3%. As with the use of a constant sea level reference environment, this choice of reference environment leads to the cumulative exhaust exergy decreasing with increasing flight distance, a trend opposite to that observed for the variable reference-environment curve. In this case, however, the reference-environment temperature and pressure are initially lower than the operating environment values, so there is a negative pressure difference at sea level which tends to increase the value of the exhaust exergy. This effect is responsible for the apparent increased exhaust exergy percentage seen at the start of the flight for the 15,000 m reference-environment curve. As the flight distance increases, the difference between the operating and reference environments decreases (a trend opposite to that for the constant sea level case), reducing the effect of the fictitious pressure difference and hence causing the exhaust exergy percentage to decrease and approach the variable reference-environment values. Thus the maximum difference between the 15,000 m and variable reference-environment values of exhaust exergy percentage is 13.6% and occurs at the beginning of the flight, as opposed to the constant sea level curve which has a maximum difference at the end of the cruise segment of the flight.

The cumulative exhaust exergy curves in Fig. 16.4 show the advantage of using a constant reference environment with conditions equal to those at the cruising altitude, if it proves impractical to use a variable reference environment. The cumulative rational efficiency and cumulative exhaust exergy percentage for a constant 15,000 m reference environment are almost identical to those for the variable reference environment, when the cruising distances are sufficiently large.

16.7. Closing remarks

Exergy analysis is useful for understanding the efficiencies of aircraft and their engines; and can assist in improvement efforts. The understanding provided in this chapter of the effect of varying reference environment on exergy efficiencies and losses is helpful in engine design work and provides a more comprehensive assessment of performance, allowing an engine to be better tailored to the types of flights and operating conditions it will encounter.

It is advantageous to use cumulative, rather than instantaneous, rational efficiencies to evaluate engine performance over an entire flight for two main reasons:

1. The sharp changes and irregularities in the instantaneous efficiencies with flight distance (Fig. 16.3) are put into better perspective in terms of their impact on engine efficiency over an entire flight by weighting them appropriately. For example, the peak instantaneous rational efficiency of 24.75% at 73 km under climb conditions when using a constant sea level reference environment is not noticeable in the cumulative results because it occurs for such a short duration.
2. The cumulative results demonstrate more clearly the advantage of using a constant reference environment equivalent to the cruising altitude conditions.

On the latter point, it is noted that, from the instantaneous viewpoint alone, both the sea level and 15,000 m reference environments attain somewhat similar maximum errors (approximately 4% and 3%, respectively), suggesting erroneously that either choice of constant reference environment would produce similar cumulative errors. However, since the majority of the flight is conducted at a cruise altitude of 15,000 m, the cumulative efficiency errors for the sea level reference environment increase with distance traveled, while those for the 15,000 m reference environment decrease. This error reduction is clearly seen in Fig. 16.3 where the 15,000 m and variable reference-environment curves converge with distance traveled and almost intersect, whereas the sea level reference-environment curve diverges from the variable curve and asymptotically approaches a 3.70% error. The dependence of the cumulative rational efficiency errors on distance is clearly visible in Fig. 16.2. At a distance of approximately 1000 km, for example, the cumulative rational efficiency for the 15,000 m reference environment exhibits little of the error it had during the initial portion of the flight due to the reference environment, while at the same distance the sea level reference-environment curve is near to its maximum error.

Problems

16.1 Explain why automobiles normally use internal combustion engines while most aircraft use gas turbines? Identify and compare typical exergy efficiencies for each application.

16.2 How are the energy and exergy efficiencies for a turbojet engine defined? What are typical values of energy and exergy efficiency for actual engines?

16.3 What would be an appropriate reference environment for exergy analyses of aircraft and their engines? What is the effect of a variable environment on the engine exergetic performance?

16.4 Describe the operation of a turbojet engine and provide expressions for its energy and exergy efficiencies.

16.5 Discuss if the assumption of a constant reference environment leads to significant errors in exergy analyses of turbojet engines.

16.6 What are the main locations of exergy losses in an aircraft turbojet engine? Propose methods for reducing or minimizing these losses.

16.7 What are the important operating parameters in turbojet engines? What are the effects on engine energy and exergy efficiencies of varying these parameters?

16.8 Are the components in a turbojet engine with greater exergy destructions necessarily those components with lower exergy efficiencies?

16.9 How can you meaningfully compare the exergetic performance of small- and large-size aircraft engines?

16.10 Obtain a published article on exergy analysis of aircraft engines. Using the operating data provided in the article, perform a detailed exergy analysis of the engine and compare your results to those in the original article. Also, investigate the effect of varying important operating parameters on the engine exergetic performance.

16.11 Obtain actual operating data from an aircraft engine and perform a detailed exergy analysis. Discuss the results and provide recommendations based on the exergy results for improving the efficiency.

Chapter 17

EXERGOECONOMIC ANALYSIS OF THERMAL SYSTEMS

17.1. Introduction

In the analysis and design of energy systems, techniques are often used which combine scientific disciplines (mainly thermodynamics) with economic disciplines (mainly cost accounting) to achieve optimum designs. For energy conversion devices, cost accounting conventionally considers unit costs based on energy. Many researchers have recommended that costs are better distributed among outputs if cost accounting is based on the thermodynamic quantity exergy. One rationale for this statement is that exergy, but not energy, is often a consistent measure of economic value. In addition, exergy-based economic-analysis methodologies exist (e.g., exergoeconomics, thermoeconomics).

Another approach for discussing the merits of thermoeconomics identifies as important the ratio of thermodynamic loss rate to capital cost. The approach involves examining data for devices in systems, and showing that correlations exist between capital costs and specific second-law-based thermodynamic losses (i.e., total and internal exergy losses).

A brief summary is presented here of existing analysis techniques which integrate exergy and economics. The goals of most such techniques include the determination of:

- The appropriate allocation of economic resources so as to optimize the design and/or operation of a system.
- The economic feasibility and profitability of a system (by obtaining the actual costs of products, and their appropriate prices).

As pointed out earlier, cost accounting for energy conversion devices conventionally considers unit costs based on energy. Since exergy is more often than energy a consistent measure of economic value, various researchers have recommended that costs are better distributed among outputs based on exergy. In addition, many of these researchers have developed methods of performing economic analyses based on exergy, which are referred to by a variety of names (e.g., thermoeconomics, second law costing, cost accounting and exergoeconomics). These analysis techniques have the following common characteristics:

- They combine exergy and economic disciplines to achieve the objectives listed above.
- They recognize that exergy, not energy, is the commodity of value in a system, and they consequently assign costs and/or prices to exergy-related variables.

Tsatsaronis (1987) identifies four main types of analysis methodologies, depending on which of the following forms the basis of the technique:

1. Exergy-economic cost accounting.
2. Exergy-economic calculus analysis.
3. Exergy-economic similarity number.
4. Product/cost efficiency diagrams.

A number of researchers have developed methods of performing economic analyses based on exergy. These are not all reviewed here, but the following important points are noted:

- General discussions of the analysis techniques appear in several textbooks (e.g., Bejan, 1982; Szargut et al., 1988; Kotas, 1995).

- Several detailed reviews of these analysis techniques have been published (e.g., Bejan, 1982; Tsatsaronis, 1987; Kotas, 1995; Rosen and Dincer, 2003a,b). These reviews include discussions, comparisons and critiques of the different techniques.
- Many articles on specific topics in this field have been published, both separately and in volumes devoted in large part to exergy methods.

In this chapter, the relations between thermodynamic losses and capital costs are considered and examined for systems and their constituent devices. For illustration, several modern electrical generating stations are examined, and possible generalizations in the relation between thermodynamic losses and capital costs are suggested. The considered electrical generating stations operate on a range of fuels (coal, oil, uranium).

This chapter provides insights into the relations between energy and exergy losses and capital costs for electrical generating stations, in particular, and for energy systems, in general. The insights can assist in integrating thermodynamics and economics in the analysis and design of energy systems.

This chapter also highlights the merits of second-law analysis over the more conventional first-law analyses. Proponents of second-law analysis conventionally argue that its use can help improve process performance. Generally, however, the application of second-law analysis to existing 'mature' technologies does not lead to significant design modifications or performance improvements. Consequently, these arguments and demonstrations alone do not convince many non-users of second-law analysis of the merits of using second-law along with conventional first-law analysis techniques.

The approach presented here examines thermodynamic and economic data for mature devices, and shows that correlations exist between capital costs and thermodynamic losses for devices. The existence of such correlations likely implies that designers knowingly or unknowingly incorporate the recommendations of second-law analysis into process designs indirectly (Rosen, 1986).

17.2. Economic aspects of exergy

Exergy is a useful concept in economics. In macroeconomics, exergy offers a way to reduce resource depletion and environmental destruction, by such means as an exergy tax. In microeconomics, exergy has been combined beneficially with cost–benefit analysis to improve designs. By minimizing the life-cycle cost (LCC), we find the 'best' system given prevailing economic conditions and, by reducing exergy losses, we often reduce environmental effects.

Designing efficient and cost-effective systems, which also meet environmental requirements, is one of the foremost challenges that engineers face. Given the world's finite natural resources and large energy demands, it is important to understand the mechanisms which degrade energy and resources and to develop systematic approaches for improving systems while simultaneously reducing environmental impact. Exergy combined with economics (both macro- and micro-) provides a powerful tool for the systematic study and optimization of systems.

17.2.1. Exergy and economics

A number of researchers of the thermoeconomic aspects of energy systems cite Georgescu-Roegen (1971) as the father of the thermodynamics of economics and a pioneer in this field. Exergy and microeconomics form the basis of thermoeconomics (Evans and Tribus, 1962), which is also called exergoeconomics (Bejan et al., 1996) and exergonomics (Yantovskii, 1994). The concept of utility is a central concept in macroeconomics and is also closely related to exergy. An exergy tax is an example of how exergy can be introduced into macroeconomics.

Some noteworthy comments by researchers in the areas of exergy and economics, and their relations to optimization activities, are presented below.

Wall (1993) captures many of the key relations between exergy and economics when he points out that 'the concept of exergy is crucial not only to efficiency studies but also to cost accounting and economic analyses. Costs should reflect value, and since value is not generally associated with energy but with exergy, assignments of cost to energy lead to misappropriations, which are common and often gross. Using exergy content as a basis for cost accounting can help management price products and evaluate profits. Exergy can also assist in making operating and design engineering decisions, and design optimization. Exergy provides a rational basis for evaluating fuels and resources; process, device and system efficiencies; dissipations and their costs; and the value and cost of systems outputs.'

Sciubba (2001) summarizes exergy and its relation to several economic and environmental factors: 'Already fifty years ago, energy conversion systems were the target of a detailed analysis based on second law concepts. The analysis indicated that the relevant design procedures of the time neglected to recognise that the irreversibility in processes and components depend on the energy "degradation rate" and not only on the ratio between the intensities of the output and input flows, and that there is a scale of energy quality that can be quantified by an entropy analysis. In essence, the legacy of this approach, universally accepted today, is that the idea of "conversion efficiency" based solely on the first law considerations is erroneous and misleading. This method evolved throughout the years into the so-called "availability analysis," later properly renamed "exergy analysis," and it has had a very profound impact on the energy conversion system community, to a point that it is difficult today to find a design standard which does not make direct or indirect use of exergetic concepts in its search for an "optimal" configuration. The same method has been extended to "complex systems," like an industrial settlement, a complete industrial sector, and even entire nations, and most recently has been brought to the attention of energy agencies as a proposed legislative tool for energy planning and policy making.'

Extending these thoughts to economics, Sciubba (2001) goes on to state that exergy analysis 'has always been regarded as unable to determine real design optima, and therefore its use has been associated with customary monetary cost-analysis; it was only recently that a complete and theoretically sound tool, based on a combination of mixed economic and thermodynamic methods and properly named "thermoeconomics," was developed to industry standards. In this approach, efficiencies are calculated via an exergy analysis, and "non-energetic expenditures" (financial, labor and environmental remediation costs) are explicitly related to the technical and thermodynamic parameters of the process under consideration: the optimisation consists of determining the design point and the operative schedule that minimise the overall (monetary) cost, under a proper set of financial, normative, environmental and technical constraints. In spite of a long tradition of contrary opinion, exergy seems indeed to possess an intrinsic, very strong and direct correlation with economic values: one of the goals of the Extended Exergy Accounting method (EEA) is to exploit this correlation to develop a formally complete theory of value based indifferently on an exergetic or on a monetary metric (that is, a general valuing or pricing method in which kJ/kg or kJ/kW are consistently equivalent to $/kg and $/kW respectively), and it is based on the fundamental idea that, while exergetic and monetary costs may have the same morphology (they represent the amount of resources that must be "consumed" to produce a certain output), their topology (structure) may be different, leading to the possibility of different optimal design points.'

Finally, Sciubba (2001) comments how remarkable it is that 'another topic of paramount importance in the engineering field can be successfully tackled by EEA methods: this is the environmental issue, taken in its extended meaning of "impact of anthropic activities on the pre-existing environment." A critical analysis of the leading engineering approaches to the environmental issue is also reported in that work. It is stated that EEA, which proposes a different quantifier (the "extended exergy content") for the analysis of processes and plants, can be regarded as a successful combination of the methods put forth. The EEA has indeed incorporated some elements of existing methods like Life-Cycle Analysis, Cumulative Exergy Analysis, Emergy Analysis, Extended Exergy Analysis, Complex Systems, and can thus be properly considered as a synthesis of the pre-existing theories and procedures of "Engineering Cost Analysis," from which it has endeavored to extract the most successful characteristics, as long as they were suitable for a consistent and expanded formulation based on the new concept of "Extended Exergy."' Sciubba (2001) proceeds to explain the how the name of EEA developed, noting that 'the attribute "extended" refers to the additional inclusion in the exergetic balance of previously neglected terms (corresponding to the so-called non-energetic costs, to labor and to environmental remediation expenditures); the word "accounting" has been suggested as a reminder that exergy does not satisfy a balance proper, in that the unavoidable irreversibilities which characterize every real process irrevocably destroy a portion of the incoming exergy; it is also a reminder that the exergy destruction is the basis for the formulation of a theory of "cost," because it clearly relates the idea that to produce any output, some resources have to be "consumed."'

Sciubba (2001) also deals specifically with the problem of an 'optimal design' satisfying two requirements: (a) performing as specified by design data and abiding by all constraints and (b) displaying the most desirable behavior under a certain set of operative conditions. This 'optimality' is not always expressed by a well-posed (in a mathematical sense) objective function: in practice, vaguely formulated optimization criteria are often the basis of the design, which nevertheless cannot be regarded as anything else than an 'optimal' one, however fuzzy or incompletely identified this optimum may be.

17.2.2. Energy and exergy prices

The selection of energy sources for industrial and other uses is primarily governed by prices. Energy conversion systems thus place demands on the energy supply system. Sometimes, energy conversion systems are shown to be uneconomical over the long term, e.g., prices are incorrect or insufficient as a basis for planning. One example of this situation is when prices are set based on short-sighted political assessments or on insufficient knowledge of the resource in question and the consequences of its use. It is therefore important to find more sound methods for price setting.

Prices based on exergy values can be designed so as to foster resource saving and efficient technology. The prices of physical resources ought to be set more in relation to their physical value, i.e., exergy. The differences between the energy and exergy values for several common fuels, which can affect price setting, are briefly summarized below, based on a case study for Göteborg, Sweden (Wall, 1997):

- *Electricity*: Electrical energy is in theory totally convertible to work. The energy price is therefore also the exergy price. The price of electricity varies considerably with several factors, including the capacity (or maximum power output), fuel consumption and fixed and variable costs for the system.
- *Gasoline and diesel fuel*: Gasoline consists mainly of octane (C_8H_{18}) for which the exergy content is about 94% of the energy value. The exergy content of diesel fuel (42.7 MJ/kg) is about 104% of its energy content based on lower heating value. The exergy value is higher than the energy value since, among other reasons, the partial pressure of carbon dioxide is included in the exergy calculation.
- *Fuel oil*: The exergy content of fuel oil is about 97% of its energy content (based on lower heating value). This value is approximately valid for heavier oils.
- *Town gas*: Town gas consists of roughly 65% hydrogen, 20% carbon dioxide and other substances. The specific energy and exergy contents of town gas are about 92.2 and 75.5 MJ/kg, respectively. Thus, the exergy content constitutes about 82% of the energy content.
- *Coal*: The exergy content and price of coal vary for each coal type dramatically.
- *Wood products*: The fuel value of wood products varies considerably depending on water content. In the case of wood with a 50% water content, the energy and exergy values are 12.4 and 12.1 MJ/kg, respectively. These values are valid when the resulting water vapor used is condensed. It is difficult to calculate a relevant price due to the relatively small amounts of wood that are sold for space heating purposes.
- *District heating*: The exergy of a given quantity of district heat can be calculated as:

$$Ex = Q(1 - T_{od})/([T_{su} - T_{od}]\ln[T_{su}/T_{re}])$$

where subscripts 'od,' 'su' and 're' represent outdoor, supply and return, respectively. The supply and return temperatures within the district heating system are regulated with respect to the outdoor temperature. For example, the supply temperature is maintained at about 85°C at outdoor temperatures above 2°C and is subsequently raised in inverse proportion to the outdoor temperature, up to 120°C at an outdoor temperature of -20°C. The exergy content thus varies with the outdoor temperature. For Göteborg, the mean exergy content is calculated at about 17% of the total heat quantity required for the heating season. The district heating subscriber in Göteborg is charged a fixed rate plus an estimated energy rate. It can be seen that the price of exergy is thus six times more than that of energy.

The relative energy and exergy prices for various energy commodities studied by Wall (1997) are listed in Table 17.1, after converting values from SEK to U.S. $ using the exchange rate for the year of 2000. The highest energy price is seen to be that of gasoline, which is a refined fuel with special areas of use. The lowest energy price is that of paper, which is also probably the most expensive and least efficient to handle as fuel. The prices of coal and wood products are low.

The differences in energy and exergy prices in Table 17.1 are small for each of the energy sources except district heating. The district heating subscriber pays much more for exergy than other energy users. In such cases, a consumer's heating bill often can be reduced by using a heat pump. Not only is it often cost-effective for a consumer to obtain heat via a heat pump rather than subscribe to district heating, but it is also cost-effective to use heat pumps in district heating systems. In this case, the energy and exergy prices for district heating would be different than listed in Table 17.1. A conscientious energy policy could speed up development of a movement toward efficient and resource-saving technologies by basing prices on exergy, rather than energy.

17.3. Modeling and analysis

The exergoeconomic methodology is described by considering the balance equations for appropriate quantities.

17.3.1. Fundamental relationships

As pointed out earlier, a general balance for a quantity in a system may be written as

$$\text{Input} + \text{Generation} - \text{Output} - \text{Consumption} = \text{Accumulation} \qquad (17.1)$$

Table 17.1. Average relative energy and exergy prices* of some
common energy forms for Göteborg, Sweden.

Energy commodity	Energy price ($/GJ)	Exergy price ($/GJ)
Electricity	1.00	1.00
Gasoline	1.67	1.78
Diesel	0.85	0.81
Fuel oil #1	0.81	0.83
Fuel oil #3–4	0.65	0.66
Town gas	0.90	1.10
Coal	0.28	0.29
Firewood	0.29	0.29
Paper	0.21	0.22
Wood paper	0.47	0.49
District heating	0.83	4.89

* The energy price for electricity is taken as 1 $/GJ and the energy and exergy
prices for other commodities are accordingly calculated relative to the price
for electricity.
Source: Wall (1997).

Here, input and output refer respectively to quantities entering and exiting through system boundaries, generation and consumption refer respectively to quantities produced and consumed within the system, and accumulation refers to build up (either positive or negative) of the quantity within the system. Differential and integral forms of the general balance may be written. The terms in Eq. (17.1) are written as rates in the differential form:

$$\text{Input rate} + \text{Generation rate} - \text{Output rate} - \text{Consumption rate} = \text{Accumulation rate} \tag{17.2}$$

and as amounts in the integral form:

$$\text{Amount input} + \text{Amount generated} - \text{Amount output} - \text{Amount consumed} = \text{Amount accumulated} \tag{17.3}$$

The differential balance describes what is happening in a system at a given instant of time, and the integral balance describes what happens in a system between two instants of time. Differential balances are usually applied to continuous processes, and integral balances to batch processes. For steady-state processes, the accumulation rate term in the differential balance is zero.

Thermodynamic balances

Energy can be neither generated nor consumed, while exergy is consumed during a process due to irreversibilities. Consequently, Eq. (17.1) can be written for these quantities as

$$\text{Energy input} - \text{Energy output} = \text{Energy accumulation} \tag{17.4}$$

$$\text{Exergy input} - \text{Exergy output} - \text{Exergy consumption} = \text{Exergy accumulation} \tag{17.5}$$

The output terms in Eqs. (17.4) and (17.5) can be separated into product and waste components as follows:

$$\text{Energy output} = \text{Product energy output} + \text{Waste energy output} \tag{17.6}$$

$$\text{Exergy output} = \text{Product exergy output} + \text{Waste exergy output} \tag{17.7}$$

Economic balances

Cost is an increasing, non-conserved quantity. The general balance in Eq. (17.1) can be written for cost as

$$\text{Cost input} + \text{Cost generation} - \text{Cost output} = \text{Cost accumulation} \tag{17.8}$$

Cost input, output and accumulation represent respectively the cost associated with all inputs, outputs and accumulations for the system. Cost generation corresponds to the appropriate capital and other costs associated with the creation and maintenance of a system. That is,

$$\text{Cost generation} = \text{Capital cost of equipment} + \text{All other creation and maintenance costs} \tag{17.9}$$

Other costs include, for example, interest and insurance costs. The 'cost generation rate' term in a differential cost balance represents the total cost generation levelized over the operating life of the system. The 'amount of cost created' term in an integral cost balance represents the portion of the total cost generation accounted for in the time interval under consideration.

17.3.2. Definition of key terms

Two types of thermodynamic losses are considered. These are defined here, with the aid of differential forms of the thermodynamic balances already described in Eqs. (17.4)–(17.7). For simplicity, it is assumed here that no losses are associated with the accumulation terms in the energy and exergy balances in Eqs. (17.4) and (17.5).

Energy losses can be identified directly from the energy balances in Eqs. (17.4) and (17.6). For convenience, the energy loss rate for a system is denoted in the present analysis as $\dot{L}_{en}$ (loss rate based on energy). As there is only one loss term, the 'waste energy output,' in Eq. (17.6),

$$\dot{L}_{en} \equiv \text{Waste energy output rate} \tag{17.10}$$

Exergy losses can be identified from the exergy balances in Eqs. (17.5) and (17.7). There are two types of exergy losses: the 'waste exergy output' in Eq. (17.7), which represents the loss associated with exergy that is emitted from the system, and the 'exergy consumption' in Eq. (17.5), which represents the internal exergy loss due to process irreversibilities. These two exergy losses sum to the total exergy loss. Hence, the loss rate based on exergy, $\dot{L}_{ex}$, is defined as

$$\dot{L}_{ex} \equiv \text{Exergy consumption rate} + \text{Waste exergy output rate} \tag{17.11}$$

The capital cost is defined here using the cost balances in Eqs. (17.8) and (17.9) and is denoted by K. Capital cost is simply that part of the cost generation attributable to the cost of equipment:

$$K \equiv \text{Capital cost of equipment} \tag{17.12}$$

Capital cost is for simplicity the only economic term considered here. It is noted that with this approach there is no need to know either the costs associated with inputs or the cost allocations among outputs. It can be argued that it is more rational to use the entire cost generation term, as defined by Eq. (17.9), in place of capital cost. The principal reason that capital costs are used here is that the use of the cost generation term increases significantly the complexity of the analysis, since numerous other economic details (interest rates, component lifetimes, salvage values, etc.) must be fully known. There are two main justifications for this simplification:

1. Capital costs are often the most significant component of the total cost generation. Hence, the consideration of only capital costs closely approximates the results when cost generation is considered.
2. Cost generation components other than capital costs often are proportional to capital costs. Hence, the trends described are in qualitative agreement with those identified when the entire cost generation term is considered.

For a thermal system operating normally in a continuous steady-state, steady-flow process mode, the accumulation terms in Eqs. (17.1)–(17.5) and (17.8) are zero. Hence all losses are associated with the already discussed terms $\dot{L}_{en}$ and $\dot{L}_{ex}$. The energy and exergy loss rates can be obtained through the following equations:

$$\dot{L}_{en} = \sum_{\text{inputs}} \text{Energy flow rates} - \sum_{\text{products}} \text{Energy flow rates} \tag{17.13}$$

$$\dot{L}_{ex} = \sum_{inputs} \text{Exergy flow rates} - \sum_{products} \text{Exergy flow rates} \qquad (17.14)$$

where the summations are over all input streams and all product output streams. Equations (17.13) and (17.14) are obtained by rearranging Eqs. (17.4) through (17.7), (17.10) and (17.11).

17.3.3. Ratio of thermodynamic loss rate to capital cost

A parameter, R, is defined as the ratio of thermodynamic loss rate $\dot{L}$ to capital cost K as follows:

$$R \equiv \dot{L}/K \qquad (17.15)$$

The value of R generally depends on whether it is based on energy loss rate (in which case it is denoted R_{en}), or exergy loss rate (R_{ex}), as follows:

$$R_{en} \equiv \dot{L}_{en}/K \qquad (17.16)$$

and

$$R_{ex} \equiv \dot{L}_{ex}/K \qquad (17.17)$$

Note that one can consider four main thermodynamic loss rates (Rosen and Dincer, 2003a,b): energy ($\dot{L}_{en}$); exergy ($\dot{L}_{ex}$); internal exergy ($\dot{L}_{ex-i}$), i.e., exergy consumptions due to process irreversibilities within the system; and external exergy ($\dot{L}_{ex-e}$), i.e., the waste outputs of exergy across a system boundary. Note that $\dot{L}_{ex-i} + \dot{L}_{ex-e} = \dot{L}_{ex}$ for a system and, since energy is conserved, all energy losses are associated with external waste emissions. Also, capital cost (K) values for the equipment in a system are considered and, to simplify the economic portion of the work, other costs (e.g., interest and insurance costs) are not considered. Values of the parameter R based on energy loss rate, and on total, internal and external exergy loss rates are considered. In investigating sets of R values, maximum (R_{max}), minimum (R_{min}), mean (R_m), standard deviation (SD(R)) and coefficient of variation (CV(R)), which is the ratio of standard deviation to mean, are considered.

17.4. Key difference between economic and thermodynamic balances

An important difference exists between the cost balance and the energy and exergy balances. In the latter two, the values associated with all quantities are defined by scientific relationships. For the cost balance, however, only cost input and generation are defined. The distribution of costs over outputs and accumulations is not defined. Costs are allocated subjectively, depending on the type and purpose of the system and other economic considerations. For example, costs may be distributed proportionally to all outputs and accumulations of a quantity (such as mass, energy or exergy), or all non-waste outputs and accumulations of a quantity.

Consider, for example, a coal-fired electrical generating station operating in a steady-state, steady-flow mode (Fig. 17.1). Clearly, subjective decisions must be made regarding the allocation of costs among outputs (i.e., the electricity, stack gas and cooling water outputs may each be allocated part of the input and generation costs). The allocations depend on the uses for the outputs. The exhaust cooling water, for example, may be treated either as a waste, in which case it may be reasonable to allocate none of the costs to it, or as a by-product, in which case it may be reasonable to allocate part of the total costs to it. Figure 17.2 presents energy, exergy and cost balances for Nanticoke Generating Station, a typical coal-fired electrical generating station (for details, see Section 11.6). Thermodynamic data for this station are used in the example in the next section. In Fig. 17.2, the input and generation costs are allocated to the product electricity, and the exhaust cooling water and stack gas are treated as wastes.

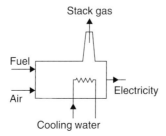

Fig. 17.1. A coal-fired electrical generating station. Minor flows such as ash and miscellaneous heat losses are not shown.

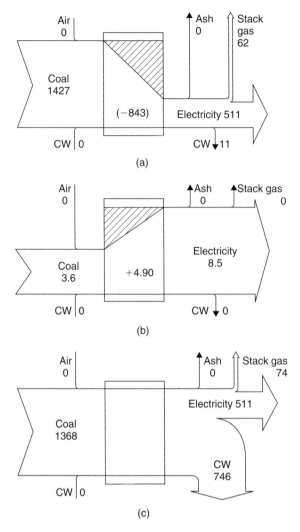

Fig. 17.2. Energy, cost and exergy balances for the coal-fired NGS. The rectangle in the center of each diagram represents the station. Widths of flow lines are proportional to the relative magnitudes of the represented quantities. CW denotes cooling water. (a) Exergy balance showing flow rates (positive values) and consumption rate (negative value in parentheses, denoted by hatched region) of exergy (in MW); (b) cost balance showing flow rates (positive values) and creation rate (positive value in parentheses, denoted by hatched region) of cost (in Canadian ¢/kWh); (c) energy balance showing flow rates of energy (in MW). Adapted from Rosen and Scott (1987) where further details are available with costs modified to 2002 Canadian cents (as explained in the text).

The costs shown in Fig. 17.2 are in 2002 Canadian cents. These costs are evaluated by modifying the original data, in 1982 Canadian cents, using the Consumer Price Index (CPI) tabulated by Statistics Canada (accessible via www.statcan.ca). The CPI data, which represent changes in prices of all goods and services purchased for consumption, indicate that $1.00 in 1982 has the same buying power as $1.82 in 2002.

17.5. Example: coal-fired electricity generation

The exergoeconomic methods described in this chapter are illustrated for the case of a coal-fired electrical generating station.

17.5.1. Plant description and data

The thermodynamic and economic data used in the analysis and the corresponding sources of the data are given, and the means by which the data are categorized and relevant quantities evaluated are described.

Thermodynamic data

Thermodynamic data for the coal-fired Nanticoke Generating Station are used. The station is described in Section 11.6 but, to assist the reader, some of the main process data for a station unit are summarized in Table 17.2. A detailed flow diagram for a single station unit is shown in Fig. 11.5a, and the corresponding symbols identifying the streams are described in Table 11.4a. For convenience of the reader, the station figure and flow data are repeated here as Fig. 17.3 and Table 17.3, respectively.

The main findings of the energy and exergy analyses of the station in Section 11.6 are as follows:

- For the overall plant, the energy efficiency, defined as the ratio of net electrical energy output to coal energy input, is 37.4%, and the corresponding exergy efficiency is 35.8%.
- In the steam generators, the energy and exergy efficiencies are evaluated, considering the increase in energy or exergy of the water as the product. The steam generators appear significantly more efficient on an energy basis

Table 17.2. Main process data for a unit (500 MW net electrical output) in a coal-fired generating station.*

Mass flow rates (kg/s)	Temperatures (°C)	Pressures (MPa)
47.9 (for coal at full load)	120 (for flue gas)	16.9 (for primary boiler steam)
454 (for primary boiler steam)	253 (for boiler feedwater)	4.0 (for reheat boiler steam)
411 (for reheat boiler steam)	538 (for primary boiler steam)	0.005 (for condenser)
18,636 (for cooling water)	538 (for reheat boiler steam)	
	8.3 (for rise in cooling water)	

* Based on data for Nanticoke Generating Station in Section 11.6.

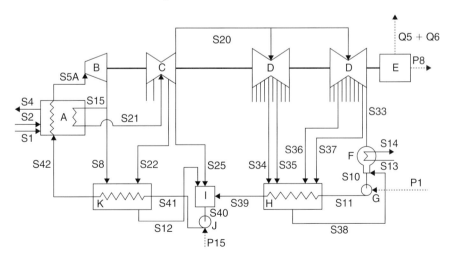

Fig. 17.3. Process flow diagram for a single unit of the coal-fired electrical generating station. Symbols identifying devices are explained in the legend and identifying flows are explained in Table 11.4a. Lines exiting turbines represent flows of extraction steam. A: steam generator and reheater, B: high-pressure turbine, C: intermediate-pressure turbine, D: low-pressure turbines, E: generator and transformer, F: condenser, G: hot well pump, H: low-pressure heat exchangers, I: open deaerating heat exchanger, J: boiler feed pump and K: high-pressure heat exchangers.

Table 17.3. Flow data for a unit (500 MWe net output) in a coal-fired generating station.[a]

Flow[b]	Mass flow rate[c] (kg/s)	Temperature (°C)	Pressure (MPa)	V.F.[d]	Energy flow rate[e] (MW)	Exergy flow rate[e] (MW)
S1	41.74	15.00	0.101	solid	1367.58	1426.73
S2	668.41	15.00	0.101	1.0	0.00	0.00
S3	710.15	1673.59	0.101	1.0	1368.00	982.85
S4	710.15	119.44	0.101	1.0	74.39	62.27
S5A	453.59	538.00	16.2	1.0	1585.28	718.74
S8	42.84	323.36	3.65	1.0	135.44	51.81
S10	367.85	35.63	0.0045	0.0	36.52	1.20
S11	367.85	35.73	1.00	0.0	37.09	1.70
S12	58.82	188.33	1.21	0.0	50.28	11.11
S13	18,636.00	15.00	0.101	0.0	0.00	0.00
S14	18,636.00	23.30	0.101	0.0	745.95	10.54
S15	410.75	323.36	3.65	1.0	1298.59	496.81
S20	367.85	360.50	1.03	1.0	1211.05	411.16
S21	410.75	538.00	4.00	1.0	1494.16	616.42
S22	15.98	423.23	1.72	1.0	54.54	20.02
S25	26.92	360.50	1.03	1.0	88.64	30.09
S33	309.62	35.63	0.0045	0.93	774.70	54.07
S34	10.47	253.22	0.379	1.0	32.31	9.24
S35	23.88	209.93	0.241	1.0	71.73	18.82
S36	12.72	108.32	0.0689	1.0	35.77	7.12
S37	11.16	60.47	0.0345	1.0	30.40	5.03
S38	58.23	55.56	0.0133	0.0	11.37	0.73
S39	367.85	124.86	1.00	0.0	195.94	30.41
S40	453.59	165.86	1.00	0.0	334.86	66.52
S41	453.59	169.28	16.2	0.0	347.05	77.57
S42	453.59	228.24	16.2	0.0	486.75	131.93
Q5					5.34	0.00
Q6					5.29	0.00
P1					0.57	0.57
P8					523.68	523.68
P15					12.19	12.19

[a] Based on data obtained for Nanticoke Generating Station by computer simulation (Rosen and Scott, 1998; Rosen, 2001) using given data (Ontario Hydro, 1973, 1983; Scarrow and Wright, 1975; Bailey, 1981; Merrick, 1984).

- (94.6%) than on an exergy basis (49.5%), indicating that although most of the input energy is transferred to the preheated water, the energy is degraded as it is transferred. Most of the exergy losses in the steam generators are associated with internal consumptions (mainly due to combustion and heat transfer).
- In the condensers, a large quantity of energy enters (about 775 MW for each unit), of which close to 100% is rejected; and a small quantity of exergy enters (about 54 MW for each unit), of which about 25% is rejected and 75% is internally consumed.
- In other plant devices, energy losses are very small (about 10 MW total), and exergy losses are moderately small (about 150 MW total) and almost completely associated with internal consumptions.

Economic data

In this example, typical economic data for similar power plants, rather than exact economic data are used. Consequently, capital cost data are used for the coal-fired Harry Allen Station in southern Nevada. This station contains three units of 500 MWe net output each, which are similar to the Nanticoke units. Capital cost data from Bechtel Power Corporation (1982) and Nevada Power Company are used, as listed by Tsatsaronis and Winhold (1985) in Table 6 of a paper describing an exergoeconomic analysis of the Harry Allen Station. The relevant data are given in Table 17.4, following the device categorization described in the next section. The costs shown in Table 17.4 (as well as Tables 17.5 and 17.6 and Figs. 17.4 and 17.5) are in 2002 U.S. dollars. These costs are evaluated by modifying the original data in 1982 U.S. dollars in Tsatsaronis and Winhold (1985) and Bechtel Power Corporation (1982), using the CPI tabulated by the U.S. Department of Labor's Bureau of Labor Statistics (accessible at www.bls.gov/cpi). The CPI data, which represents changes in prices of all goods and services purchased for consumption, indicate that $1.00 in 1982 has the same buying power as $1.86 in 2002.

17.5.2. Data categorization

In the present analysis, the generating station is subdivided into the following devices:

- Turbine generators.
- Steam generators (including the steam generator and reheater, and the air preheater and fan).
- Preheating devices (including all heat exchangers and pumps used for preheating).
- Condensers.
- Overall station (including the above devices plus all other plant devices).

Waste emissions are taken to be output cooling water for the condenser, stack gas for the steam generator and both quantities for the total plant. Miscellaneous heat losses are treated as waste emissions for all devices.

Ambiguity exists regarding what should be taken to be the material waste emissions for the turbine generators and preheating devices. Consequently, three cases are considered for the turbine generators:

1. None of the steam exhausted or extracted from the turbines is taken to be a waste emission (Case I).
2. Only the steam exhausted to the condenser from the low-pressure (LP) turbine is taken to be a waste emission (Case II).
3. All steam exhausted or extracted from the turbines is taken to be a waste emission (Case III).

Table 17.3. (Continued)

---

[b] Flow numbers correspond to those in Fig. 17.3, except for S3, which represents the hot product gases for adiabatic combustion. Letter prefixes indicate material flows (S), heat flows (Q) and electricity flows (P).

[c] Material flow compositions, by volume, are: 100% C for S1; 79% N_2, 21% O_2 for S2; 79% N_2, 6% O_2, 15% CO_2 for S3 and S4; 100% H_2O for other material flows.

[d] Vapor fraction (V.F.) indicates fraction of a vapor–liquid flow that is vapor (not applicable to S1 since it is solid). Vapor fraction is listed as 0.0 for liquids and 1.0 for superheated vapors.

[e] Energy and exergy values are evaluated using a reference-environment model (similar to the model used by Gaggioli and Petit (1977) having a temperature of 15°C, a pressure of 1 atm and a composition of atmospheric air saturated with H_2O at 15°C and 1 atm and the following condensed phases: water, limestone and gypsum.

Table 17.4. Breakdown by device of capital costs (in 10^6\$) for a unit of the coal-fired generating station.[a]

Turbine generators		63.93
Main condensers and auxiliaries		8.69
Steam generators		153.95
Steam generators and reheaters, air preheaters	148.74	
Air fan	5.21	
Preheating devices		16.02
Low-pressure pumps	1.30	
Intermediate-pressure pumps	2.42	
High-pressure pumps and driving turbine	6.38	
Preheater #1	0.56	
Preheater #2	0.56	
Preheater #3	0.47	
Preheater #4	0.61	
Preheater #5	0.52	
Preheater #6	0.54	
Preheater #7	2.66	
Other plant devices		31.95
Other plant equipment	19.08	
Unaccounted processing units[b]	12.87	
All main devices in overall plant		274.54

[a] Costs have been modified to 2002 U.S. dollars (as explained in the text), and are based on literature for the Harry Allen Station (Bechtel Power Corporation, 1982; Tsatsaronis and Winhold, 1985), particularly Table 6 of Tsatsaronis and Winhold (1985) (which presents costs for the total three units in the station).
[b] 'Unaccounted processing units' is an approximated value, which accounts for the portions of the remaining costs in Table 6 of Tsatsaronis and Winhold (1985) that should be applied to the 'other plant devices' group.

Also, two cases are considered for the preheating devices:

1. No waste emissions exist (Case I).
2. The stream flowing from the preheating devices to the condenser (S38 in Fig. 17.3) is taken to be a waste emission (Case II).

Since the station operates normally in a continuous steady-state, steady-flow process mode, the accumulation terms in Eqs. (17.1)–(17.5) and (17.8) are zero. Hence all losses are associated with the already discussed terms $\dot{L}_{en}$ and $\dot{L}_{ex}$. The energy and exergy loss rates are evaluated by considering devices or groups of devices in Fig. 17.3 and the data in Table 17.3. In Eqs. (17.13) and (17.14) the summations are over all input streams and all product output streams.

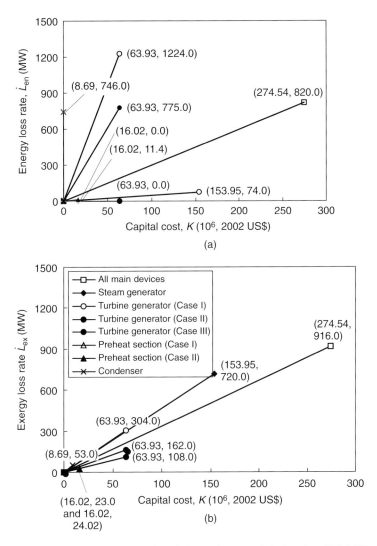

Fig. 17.4. Thermodynamic loss rate as a function of capital cost for several devices in a 500 MW unit of the coal-fired Nanticoke Electrical Generating Station. (a) Energy loss rate and (b) exergy loss rate. Costs have been modified to 2002 U.S. dollars (as explained in the text).

17.5.3. Results and discussion

Values of $\dot{L}_{en}$, $\dot{L}_{ex}$ and K for the unit considered of the coal-fired electrical generation station are presented in Table 17.5. Plots of thermodynamic loss rate are presented in Fig. 17.4 as a function of capital cost for the overall generating station and the following station devices: turbine generators, steam generators (Cases I–III), preheating devices (Cases I and II) and condensers. Energy loss rate (Fig. 17.4a) and exergy loss rate (Fig. 17.4b) are considered. Figure 17.4 is based on the data in Table 17.5.

Values of the thermodynamic-loss-rate-to-capital-cost ratios R_{en} and R_{ex} for the devices in the generating station are listed, along with values of K, $\dot{L}_{en}$ and $\dot{L}_{ex}$, in Table 17.5, and plotted in Fig. 17.5. Note that the value of R for a device is given by the slope of the line in Fig. 17.4 for the device.

For each of the eight device cases considered, Table 17.6 presents statistical data for the two thermodynamic-loss-rate-to-capital-cost ratios, R_{en} and R_{ex}. The statistical quantities considered in Table 17.6 are: minimum, maximum, mean,

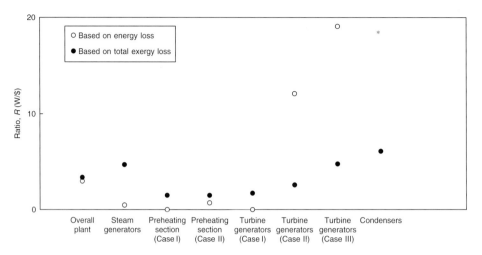

Fig. 17.5. Values of thermodynamic-loss-rate-to-capital-cost ratio, R, for several devices in a 500 MW unit of the coal-fired NGS. Costs have been modified to 2002 U.S. dollars (as explained in the text). Note that * shows $R = 85.8$ W/$ based on energy loss.

Table 17.5. Capital cost and thermodynamic loss data for several devices in a unit of the coal-fired generating station.[a]

Device[b]	K ($10^6$$)	$\dot{L}_{en}$ (MW)	R_{en} (W/$)	$\dot{L}_{ex}$ (MW)	R_{ex} (W/$)
Steam generators	153.95	74.0	0.481	720.0	4.68
Turbine generators					
• Case I	63.93	0.0	0.0	108.0	1.69
• Case II	63.93	775.0	12.1	162.0	2.53
• Case III	63.93	1224.0	19.1	304.0	4.76
Main condensers (and auxiliaries)	8.69	746.0	85.8	53.0	6.10
Preheating devices					
• Case I	16.02	0.0	0.00	23.0	1.44
• Case II	16.02	11.4	0.71	24.2	1.51
Overall station	274.54	820.0	2.99	916.0	3.34

[a] Costs have been modified to 2002 U.S. dollars (as explained in the text).
[b] Device descriptions and capital costs correspond to those detailed in Table 17.4.

standard deviation and coefficient of variation. As CV(R) is a measure of the relative variation in a set of values about the mean and is independent of scale of measurement, it permits comparisons of the variations in several sets of data.

In the present undertaking, $n = 8$, since each set of R values contains eight values, corresponding to the values for the turbine generators (Cases I, II and III), the steam generators, the preheating devices (Cases I and II), the main condensers and auxilliaries, and the overall station.

The statistical data in Table 17.6 indicate that the relative spread in thermodynamic-loss-rate-to-capital-cost ratios for different devices in the station is large when the ratio is based on energy loss (i.e., for R_{en}) and small when based on exergy loss (i.e., for R_{ex}). This observation is supported by the range of maximum to minimum values, which is much

Table 17.6. Statistical data for thermodynamic loss-rate-to-capital-cost ratio values for several devices in a unit of the coal-fired generating station.[a]

Parameter	Based on energy loss	Based on exergy loss
Minimum, R_{min} (W/\$)	0.0	1.44
Maximum, R_{max} (W/\$)	85.8	6.10
Mean, R_m (W/\$)[b]	15.1	3.26
Standard deviation, SD(R) (W/\$)[b]	27.4	1.64
Coefficient of variation, CV(R) (%)[b]	181.5	50.31

[a] Costs have been modified to 2002 U.S. dollars (as explained in the text).
[b] The mean R_m for the set of n values, $R_1, R_2, \ldots R_n$, is a measure of the center of the set. The standard deviation, SD(R), of the set is a measure of the absolute variation in the set of R values about the mean R_m. The coefficient of variation, CV(R), is the standard deviation as a percentage of the mean.

greater for R_{en} values than for R_{ex} values. That is,

$$(R_{en})_{max} - (R_{en})_{min} \gg (R_{ex})_{max} - (R_{ex})_{min} \qquad (17.18)$$

The observation is also supported by the coefficients of variation, which indicate that the relative variation is much greater in the set of R_{en} values than in the set of R_{ex} values. That is,

$$CV(R_{en}) \gg CV(R_{ex}) \qquad (17.19)$$

For the turbine generators and preheating devices, the energy loss rate $\dot{L}_{en}$ is sensitive to the definition of material waste outputs (recall that three cases of material waste outputs are considered for the turbine generators, and two cases for the preheating devices). The exergy loss rate $\dot{L}_{ex}$ is relatively less sensitive. Since the capital cost K is constant for each device for all cases considered, the corresponding thermodynamic-loss-rate-to-capital-cost ratios (R_{en} and R_{ex}) exhibit sensitivities similar to those exhibited by $\dot{L}_{en}$ and $\dot{L}_{ex}$.

The above results suggest that, for devices in modern coal-fired electrical generating stations, a systematic correlation exists for capital cost and exergy loss, but not for capital cost and energy loss.

The results further suggest for these devices that values of R_{ex} approximately conform to a particular value, denoted here by R_{ex}^*. The value of R_{ex}^* does not necessarily identify the 'ideal' value of R_{ex}. The meaning of R_{ex}^* may be better understood by considering that coal-fired electrical generating stations are and have been for some time widely used. Consequently, the design of the overall plant and individual devices can be viewed as 'successful,' and in that sense represents the appropriate optimum, as determined by market and other forces.

17.6. Case study: electricity generation from various sources

The methodology from the previous section (Rosen, 1991; Rosen and Dincer, 2002) is applied to single units of oil-fired (Lennox), coal-fired (Lakeview and Nanticoke) and nuclear (Bruce B) electrical generating stations in Ontario. Data have been obtained from Ontario Power Generation, formerly Ontario Hydro (e.g., 1969, 1970, 1973, 1979, 1983, 1985, 1990, 1991, 1996). Table 17.7 presents selected primary characteristics of the four stations. Note that the lakeview station no longer exists, and the Lennox station now uses natural gas as well as oil.

For all four stations, devices are separated according to the breakdown used in the previous section (Rosen, 1990, 1991). For the Bruce B station, however, the steam generators include the nuclear reactor, the primary heat-transport-loop pump and the moderator cooler, rather than the air preheater and fan. In addition, the deaerator and the high-pressure (HP) and LP preheaters are considered for all four stations. In both preheater instances, two cases are considered: (I) no material wastes exit and (II) the flow exiting the preheaters is treated as a waste. Additional devices considered for some stations include the condensate pumps (for Lennox, Nanticoke and Bruce B) and the boiler feed pumps (for Lennox, Lakeview and Bruce B).

Two additional device breakdowns are considered for Lennox only: (i) the gland seal condenser (GSC) and (ii) the LP preheaters with the GSC. In both instances, two cases are considered: (I) no material wastes exit and (II) flows to

Table 17.7. Selected information for four electrical generating stations.

Station	Fuel	Net electrical power output (MW)	Energy efficiency (%)	Exergy efficiency (%)
Lennox	Oil	512	37.0	34.8
Lakeview	Coal	307	35.5	33.2
Bruce B	Uranium	842	31.0	31.0
Nanticoke	Coal	500	37.4	35.8

Source: (Rosen, 1990, 1991).

the condenser are treated as wastes. Also, two additional device breakdowns, each with two cases, are considered for Bruce B only: the HP and LP preheaters (Case I – no material wastes exit and Case II – flows from the LP preheaters to the deaerator and from the HP preheaters to the condenser are treated as wastes); and the moisture separator and heat exchangers (Case I – no material wastes exit and Case II – flows from the moisture separator and heat exchanger that do not enter the LP turbine are treated as wastes).

17.6.1. Results and discussion

Values for devices of K and several $\dot{L}$ and R parameters are listed in Tables 17.8 through 17.11 for the four generating stations. Statistical data for R values are presented for the four stations individually and combined in Table 17.12 (using Case I values where multiple cases exist).

The costs shown in Tables 17.8 through 17.11 are in 2002 Canadian dollars. These costs are evaluated by modifying the original data using the CPI tabulated by Statistics Canada (accessible at www.statcan.ca). The original costs for the Lennox station (Table 17.8) are in 1973 Canadian dollars, for the Lakeview station (Table 17.9) are in 1969 Canadian dollars, for the Bruce B station (Table 17.10) are in 1976 Canadian dollars and for the Nanticoke station (Table 17.11) are in 1982 U.S. dollars. The 1982 U.S.–Canada exchange rate is used to convert the costs in Table 17.11 to Canadian dollars. The CPI and exchange-rate data indicate that $1.00 (Canadian) in 1969 has the same buying power as $5.02 (Canadian) in 2002, $1.00 (Canadian) in 1973 has the same buying power as $4.14 (Canadian) in 2002, $1.00 (Canadian) in 1976 has the same buying power as $3.14 (Canadian) in 2002 and $1.00 (U.S.) in 1982 has the same buying power as $2.27 (Canadian) in 2002.

The results of the previous work suggest a systematic correlation may exist between capital cost and exergy loss (both total and internal) for devices in modern electrical generating stations, but not between capital cost and energy loss or external exergy loss. This observation is supported by the maximum, minimum and coefficient of variation data in Table 17.12, which show the relative spread in values for R_{ex} and R_{ex-i} to be smaller than the spreads for R_{en} or R_{ex-e}.

Values for R_{ex} and R_{ex-i} are seen to be similar for most cases, as are the statistical quantities calculated for sets of R_{ex} and R_{ex-i} values (see Table 17.12).

The results suggest that the values of R_{ex} and R_{ex-i} for a group of devices in a coal-fired station may approximately conform to particular appropriate values, R_{ex}^* and R_{ex-i}^*. This statement appears to be applicable to the technologies analyzed in the present section. The similar behavior between R_{ex} and R_{ex-i}, as discussed previously, likely implies that values for R_{ex}^* and R_{ex-i}^* behave similarly, and are similar in value. The values of R_{ex}^* and R_{ex-i}^* do not necessarily identify 'ideal' values of R_{ex} and R_{ex-i}. The meaning of R_{ex}^* and R_{ex-i}^* may be better understood by considering that coal-fired, oil-fired and nuclear electrical generating stations are and have been for some time widely used. Consequently, the design of the overall plant and individual devices can be viewed as 'successful,' and in that sense represents an appropriate optimum. Thus R_{ex}^* and R_{ex-i}^* likely reflect the appropriate trade-off between exergy losses and capital costs which is practised in successful plant designs.

17.6.2. Relations for devices in a single generating station

For a given station the mean thermodynamic-loss-rate-to-capital-cost ratios based on total (or internal) exergy loss (see Table 17.12) are similar to the overall-station ratio values based on total (or internal) exergy loss (see Tables 17.8 through 17.11). That is, $R_{m,ex} \approx (R_{ex})_s$ and $R_{m,ex-i} \approx (R_{ex-i})_s$, where the mean is for all station devices considered, and the subscript 's' refers to the overall station. In addition, Tables 17.8 through 17.12 show that a corresponding similarity

Table 17.8. Device parameter values for Lennox Oil-Fired Generating Station (using 2002 Canadian dollars).*

Device	Case number	K (10^6\$)	$\dot{I}_{en}$ (MW)	R_{en} (W/\$)	$\dot{I}_{ex}$ (MW)	R_{ex} (W/\$)	$\dot{I}_{ex\text{-}i}$ (MW)	$R_{ex\text{-}i}$ (W/\$)	$\dot{I}_{ex\text{-}e}$ (MW)	$R_{ex\text{-}e}$ (W/\$)
Steam generators		60.80	207.56	3.41	819.24	13.47	769.74	12.66	49.5	0.81
Turbines	I	67.22	0	0	67.70	1.01	67.70	1.01	0	0
	II	67.22	676.95	10.07	107.66	1.60	67.70	1.01	39.96	0.59
	III	67.22	1087.95	16.18		3.56	67.70	1.01	171.65	2.55
Preheating section	I	5.37	0	0	18.05	3.36	18.05	3.36	0	0
	II	5.37	6.25	1.16	18.29	3.44	18.05	3.36	0.24	0.044
Condenser		3.13	657.37	210.25	39.53	12.64	23.57	7.54	15.96	5.11
High-pressure preheater	I	1.77	0	0	5.06	2.85	5.06	2.85	0	0
	II	1.77	39.64	22.34	13.89	7.83	5.06	2.85	8.83	4.98
Low-pressure preheater	I	1.35	0	0	5.43	4.02	5.43	4.02	0	0
	II	1.35	6.12	4.53	5.67	4.19	5.43	4.02	0.24	0.18
Low-pressure preheater and gland seal condenser	I	1.56	0	0	5.93	3.79	5.93	3.79	0	0
	II	1.56	6.25	4.00	6.17	3.95	5.93	3.79	0.24	0.15
Gland seal condenser	I	0.20	0	0	0.5	2.52	0.5	2.52	0	0
	II	0.20	0.13	0.66	0.5	2.52	0.5	2.52	0	0
Deaerator		0.51	0	0	2.97	5.86	2.97	5.86	0	0
Condensate pumps		0.59	0	0	0.01	0.017	0.01	0.017	0	0
Boiler feed pumps		0.93	0	0	4.08	4.39	4.08	4.39	0	0
Overall station		192.9	865.02	4.49	959.51	4.98	894.02	4.63	65.49	0.34

Note: Costs have been modified to 2002 Canadian dollars (as explained in the text).

* In this table as well as Tables 17.9–17.12, the exergy loss rate $\dot{I}_{ex}$ is broken down into internal exergy loss rate $\dot{I}_{ex\text{-}i}$ and external exergy loss rate $\dot{I}_{ex\text{-}e}$. Corresponding values of the ratio R are listed when it is based on internal exergy loss $R_{ex\text{-}i}$ and external exergy loss $R_{ex\text{-}e}$. It is noted that $\dot{I}_{ex} = \dot{I}_{ex\text{-}i} + \dot{I}_{ex\text{-}e}$.

Exergy: Energy, Environment and Sustainable Development

Table 17.9. Device parameter values for Lakeview Coal-Fired Generating Station (using 2002 Canadian dollars).

Device	Case number	K (10^6\$)	$\dot{I}_{en}$ (MW)	R_{en} (W/\$)	$\dot{I}_{ex}$ (MW)	R_{ex} (W/\$)	$\dot{I}_{ex\text{-}i}$ (MW)	$R_{ex\text{-}i}$ (W/\$)	$\dot{I}_{ex\text{-}e}$ (MW)	$R_{ex\text{-}e}$ (W/\$)
Steam generators		29.30	192.60	6.57	555.19	18.95	510.12	17.41	45.07	1.54
Turbines	I	26.19	0	0	36.44	1.39	36.44	1.39	0	0
	II	26.19	371.43	15.38	50.46	1.93	36.44	1.39	14.02	0.53
	III	26.19	591.58	22.58	117.60	4.49	36.44	1.39	81.16	3.08
Preheating section	I	4.16	0	0	10.05	2.41	10.05	2.41	0	0
	II	4.16	2.64	0.63	10.13	2.43	10.05	2.41	0.08	0.019
Condenser		1.35	364.88	269.4	13.94	10.29	8.25	6.09	5.69	4.17
HP preheater	I	1.05	0	0	3.25	3.09	3.25	3.09	0	0
	II	1.05	21.73	20.62	7.68	7.29	3.25	3.09	4.43	4.18
LP preheater (without	I	0.55	0	0	4.03	7.30	4.03	7.30	0	0
gland seal condenser)	II	0.55	2.59	4.69	4.10	7.43	4.03	7.30	0.07	0.13
Deaerator		0.28	0	0	0.56	2.03	0.56	3.03	0	0
Boiler feed pumps		2.31	0	0	2.14	0.93	2.14	3.03	0	0
Overall station		86.09	557.48	6.48	618.62	7.19	567.86	6.56	50.76	0.59

Note: Costs have been modified to 2002 Canadian dollars (as explained in the text).

Table 17.10. Device parameter values for Bruce B Nuclear Generating Station (using 2002 Canadian dollars).

Device	Case number	K (10^6 \$)	$\dot{I}_{en}$ (MW)	R_{en} (W/\$)	$\dot{I}_{ex}$ (MW)	R_{ex} (W/\$)	$\dot{I}_{ex\text{-}i}$ (MW)	$R_{ex\text{-}i}$ (W/\$)	$\dot{I}_{ex\text{-}e}$ (MW)	$R_{ex\text{-}e}$ (W/\$)
Steam generators		6.24	147	2.36	1575.48	25.25	1573.45	25.22	2.03	0.03
Turbines	I	261.5	0	0	41.17	0.16	41.17	0.16	0	0
	II	261.5	1650.05	6.31	138.42	0.53	41.17	0.16	97.25	0.37
	III	261.5	2460.16	9.44	338.31	1.29	41.17	0.16	297.14	1.14
Preheating section	I	13.63	0	0	25.15	1.84	25.15	1.84	0	0
	II	13.63	17.29	1.28	25.71	1.88	25.15	1.84	0.56	0.04
Condenser		5.76	1621.87	281.6	97.16	16.89	74.76	12.98	22.40	3.89
HP preheater	I	5.02	0	0	2.86	0.57	2.86	0.57	0	0
	II	5.02	129.47	25.77	27.43	5.44	2.86	0.57	24.48	4.87
LP preheater	I	3.36	0	0	17.76	5.28	17.76	5.28	0	0
	II	3.36	17.29	5.15	18.32	5.45	17.76	5.28	0.56	0.17
HP and LP preheater	I	8.38	0	0	20.62	2.46	20.62	2.46	0	0
	II	8.38	146.76	17.51	45.66	5.45	20.62	2.46	25.04	2.99
Deaerator		1.82	0	0	1.36	0.75	1.36	0.75	0	0
Condensate pumps		1.54	0	0	0.40	0.26	0.40	0.26	0	0
Boiler feed pumps		1.89	0	0	2.24	1.19	2.24	1.19	0	0
Moisture separator and heat exchanger	I	5.09	0	0	11.25	2.21	11.25	2.21	0	0
	II	5.09	191.35	37.61	61.07	12.00	11.25	2.21	49.82	9.79
Overall station		391.3	1768.87	4.52	1875.68	4.79	1851.25	4.73	24.43	0.06

Note: Costs have been modified to 2002 Canadian dollars (as explained in the text).

Table 17.11. Device parameter values for Nanticoke Coal-Fired Generating Station (using 2002 Canadian $).

Device	Case number	K (10⁶$)	$\dot{I}_{en}$ (MW)	R_{en} (W/$)	$\dot{I}_{ex}$ (MW)	R_{ex} (W/$)	$\dot{I}_{ex\text{-}i}$ (MW)	$R_{ex\text{-}i}$ (W/$)	$\dot{I}_{ex\text{-}e}$ (MW)	$R_{ex\text{-}e}$ (W/$)
Steam generators		188.05	74	0.39	720	3.83	658	3.50	62	0.33
Turbines	I	78.09	0	0	108	1.38	108	1.38	0	0
	II	78.09	775	9.95	162	2.07	108	1.38	54	0.69
	III	78.09	1224	15.67	304	3.90	108	1.38	196	2.51
Preheating section	I	19.56	0	0	23.0	1.18	23.0	1.18	0	0
	II	19.56	11.4	0.58	24.02	1.24	23.0	1.18	1.2	0.06
Condenser		10.61	746	70.42	53	5.02	43	4.05	10	0.94
HP preheater	I	3.91	0	0	6.36	1.63	6.36	1.63	0	0
	II	3.91	50.3	12.89	17.5	1.63	6.36	1.63	11.1	2.84
LP preheater	I	2.68	0	0	10.8	4.03	10.8	4.03	0	0
	II	2.68	11.4	4.25	11.5	4.03	10.8	4.03	0.73	0.27
Deaerator		0.64	0	0	0.56	0.88	0.56	0.88	0	0
Condensate pumps		1.59	0	0	0.07	0.04	0.07	0.04	0	0
Overall station		335.35	820	2.45	916	2.73	843	2.51	73	0.22

Note: Costs have been modified to 2002 Canadian dollars (as explained in the text).

Table 17.12. Statistical data for R values for several electrical generating stations (using 2002 Canadian dollars).

	Based on energy loss	Based on total exergy loss	Based on internal exergy loss	Based on external exergy loss
Lennox (oil)				
Minimum, R_{min}(W/\$)	0	0.017	0.017	0
Maximum, R_{max} (W/\$)	210.25	13.47	12.66	5.11
Mean, R_m (W/\$)	18.54	5.01	4.48	0.53
Standard deviation, SD(R) (W/\$)	54.48	4.22	3.35	1.49
Coefficient of variation, CV(R) (%)	333.00	84	75	281
Lakeview (coal)				
Minimum, R_{min}(W/\$)	0	0.93	0.93	0
Maximum, R_{max} (W/\$)	269.4	18.95	17.44	4.17
Mean, R_m (W/\$)	31.38	5.95	5.25	0.70
Standard deviation, SD(R) (W/\$)	89.29	5.84	5.14	1.41
Coefficient of variation, CV(R) (%)	285	98	98	2001
Bruce B (nuclear)				
Minimum, R_{min}(W/\$)	0	0.16	0.16	0
Maximum, R_{max} (W/\$)	281.6	25.25	25.22	3.89
Mean, R_m (W/\$)	24.52	5.24	4.90	0.38
Standard deviation, SD(R) (W/\$)	82.74	7.98	7.50	1.14
Coefficient of variation, CV(R) (%)	337.00	152.00	153.00	338.00
Nanticoke (coal)				
Minimum, R_{min}(W/\$)	0	0.04	0.04	0
Maximum, R_{max} (W/\$)	70.42	5.02	4.05	0.94
Mean, R_m (W/\$)	8.14	2.30	2.13	0.17
Standard deviation, SD(R) (W/\$)	23.37	1.68	1.46	0.32
Coefficient of variation, CV(R) (%)	287.00	73.00	68.00	191
All stations combined				
Minimum, R_{min}(W/\$)	0	0.017	0.017	0
Maximum, R_{max} (W/\$)	269.4	18.95	17.41	5.11
Mean, R_m (W/\$)	21.00	4.70	4.30	0.45
Standard deviation, SD(R) (W/\$)	63.00	5.10	4.50	1.10
Coefficient of variation, CV(R) (%)	300.00	110.00	100.00	240
Mean of overall-station R values (W/\$)	4.49	4.92	4.61	0.30

based on energy loss is not evident, and based on external exergy loss is likely not evident (although the small magnitudes of the values involved make it difficult to confirm the lack of similarity for external exergy loss).

The observation that the mean R value for the devices in a given station is approximately equal to the overall-station R value (based on total and internal exergy loss) may indicate that devices in a successful station are arranged so as to achieve an 'optimal' overall-station configuration. However, such an indication is evident from the relations for the devices between capital cost and exergy loss (total and internal), but not between energy loss or external exergy loss and capital cost. In other words, the relations between capital cost and total and internal exergy loss suggest that the collective characteristics of the station match and benefit the overall station.

The idea, suggested by the above observations, of optimizing devices (using combined economic and second-law-based thermodynamic methodologies) in balance with the overall system as opposed to in 'isolation,' is not new. Others have emphasized the importance of thermoeconomically optimizing a thermal system based on the interaction among devices and the entire system (Hua et al., 1989).

The above discussions suggest that the relations $R_{m,ex} \approx (R_{ex})_s$ and $R_{m,ex-i} \approx (R_{ex-i})_s$ may be important characteristics of successful electrical generating stations, and may be generalizable to other successful technologies.

Values of the hypothetical terms R_{ex}^* and R_{ex-i}^* for devices in a station are likely similar to values for $R_{m,ex}$, $R_{m,ex-i}$, $(R_{ex})_s$ and $(R_{ex-i})_s$. Note that appropriate values for these terms may vary spatially and temporally. Here, variations are observed for the $(R_{ex})_s$ and $(R_{ex-i})_s$ values in Tables 17.8 through 17.11 and the $R_{m,ex}$ and $R_{m,ex-i}$ values in Table 17.12, which consider generating stations using different fuels and built in different locations and times.

In Table 17.12, where the data for stations are grouped together, the following relations are observed: $R_{m,ex} \approx (R_{m,ex})_s$ and $R_{m,ex-i} \approx (R_{m,ex-i})_s$. Here, $(R_{m,ex})_s$ and $(R_{m,ex-i})_s$ are the means for the four overall-station cases, and are listed at the bottom of Table 17.12. These relations parallel the similar relations identified earlier for individual stations, and suggest that generating stations located within a larger system, here the electrical utility sector, may have characteristics similar to the typical station in that sector. Furthermore, these characteristics are evident from relations between capital costs and total or internal exergy losses, but not energy losses.

It is also noted that the values in Tables 17.6 and 17.12 do not vary significantly when the R values are re-analyzed for two alternate device-groupings: (i) all devices except the overall station and (ii) the devices which represent the most broken down components of the station.

17.6.3. Generalization of results

Before general conclusions can be drawn from the present analysis, data for additional and different technologies must be collected and analyzed. Nevertheless, possible generalities are now discussed.

Based on the results of the present analysis and values of capital costs and thermodynamic losses for other devices (e.g., El-Sayed and Tribus., 1983; Tsatsaronis and Park, 2002), it is suggested that the observed relations between exergy loss rates and capital costs for electrical generating stations may be general. In particular, values of R_{ex}^* may exist for other technologies.

The value of R_{ex}^* may vary for different situations (e.g., time, location, resource costs, knowledge). The values of R_{ex}^* may be different for different technologies. Also, during periods when energy-resource costs increase (as was the case in many locations in the 1970s and in the 2000s), the value of R_{ex}^* likely decreases (i.e., greater capital is invested to reduce losses).

For any technology, it appears that the design of a device may be made more successful if it is modified so that its value of R_{ex} approaches R_{ex}^*. This idea is illustrated in Fig. 17.6, which shows a line intersecting the origin representing R_{ex}^*, and a second curve representing the possible combinations of exergy loss and capital cost for a device. The shape and position of the latter curve illustrate the trade-off between cost and efficiency by showing that losses generally can be reduced through increased capital investment. Specifically, this curve indicates that the total exergy input is wasted if no investment is made, i.e.,

$$\text{Total exergy loss rate} \to \text{Total exergy input rate as Capital cost} \to 0$$

and that performance approaches the ideal if a very large investment is made, i.e.,

$$\text{Total exergy loss rate} \to 0 \text{ as Capital cost} \to \infty$$

A balance is obtained between exergy loss and capital cost in real systems. The expected combination of exergy loss and capital cost is the one for which $R_{ex} = R_{ex}^*$, represented by the intersection of the two curves in Fig. 17.6. If the cost is less, $R_{ex} > R_{ex}^*$, and cost will be increased to reduce loss, and vice versa.

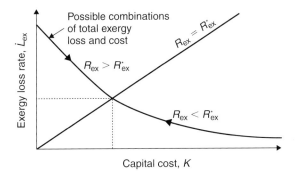

Fig. 17.6. Illustration of the tendency of R_{ex} for a device to approach R_{ex}^*, the value of R_{ex} for which the 'appropriate' trade-off between losses and capital costs is attained.

If successful technologies conform to an appropriate R_{ex}^*, then it follows that technologies which fail in the marketplace may do so because they deviate too far from the appropriate R_{ex}^*. Thus research and development should perhaps strive to identify devices for which the difference between the values of R_{ex} and R_{ex}^* is large, and develop ways to narrow the difference.

The work discussed in this case study can likely be extended to marginal costs. Here, the marginal cost would be the cost increase resulting from saving one unit of energy or exergy (i.e., from reducing the energy or exergy loss by one unit). The results would be expected to indicate for many devices that marginal costs based on exergy have similar values, while marginal costs based on energy vary widely.

The exergoeconomic concepts discussed in this section may prove particularly useful for the introduction of new technologies.

17.7. Exergoeconomics extended: EXCEM analysis

Traditionally, the merit of a system or process has been based on conventional parameters including technical performance and efficiency, economic viability and health and safety implications. In recent years, new concerns like environmental damage and scarcity of resources have increased the considerations involved. The evaluation of the merit of a system or process requires methodologies that take into account all the above factors as well as others. A systems viewpoint is required for completeness.

Here, a methodology for evaluating systems and processes that extends exergoeconomics is described which incorporates four key parameters: exergy, cost, energy and mass. This methodology is referred to as the EXCEM (exergy, cost, energy and mass) analysis and can be useful for the evaluation of systems. The method is intended to form basis of a unified methodology for exergy, energy, economic and environmental decisions. Previous work has been reported on EXCEM analysis (e.g., Rosen, 1986, 1990).

The basic rationale underlying an EXCEM analysis is that an understanding of the performance of a system requires an examination of the flows of each of the quantities represented by EXCEM into, out of and at all points within a system.

17.7.1. The EXCEM analysis concept

The EXCEM analysis concept is illustrated in Fig. 17.7. Of the quantities represented by EXCEM, only mass and energy are subject to conservation laws. Cost increases or remains constant, while exergy decreases or remains constant. Balances can be written for each of the EXCEM quantities.

17.7.2. Development of a code for EXCEM analysis

To make the EXCEM analysis methodology more useful and convenient to apply, it can be applied in a computer code. For example, one EXCEM analysis code, developed by enhancing a state-of-the-art process simulator, Aspen Plus, for EXCEM analysis, has been described elsewhere (e.g., Rosen and Scott, 1985).

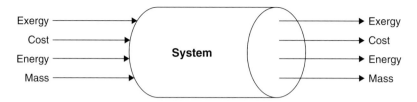

Fig. 17.7. The EXCEM analysis methodology.

That EXCEM code provides valid and accurate results, and is convenient to use and widely applicable. The methodology used to enhance Aspen Plus for EXCEM analysis is general and can be used – with modifications where necessary – to enhance other process-simulation and related codes. Analyses of several mechanical and chemical engineering processes with the code demonstrate that:

- the EXCEM analysis methodology often provides valuable insights into performance and efficiency, economics and the potentials for environmental damage for processes;
- some findings of EXCEM analyses are not obtainable with conventional analyses (e.g., energy analyses and energy economics), while most findings of conventional analyses are obtained more directly and conveniently with EXCEM analyses;
- the exergy-related aspects of EXCEM are often the most revealing.

17.7.3. Illustrative examples of EXCEM analysis

The illustrations considered here are intended to clarify the general concepts associated with EXCEM analysis, particularly cost allocation. Mass, energy and exergy balances are not extensively discussed. The three devices in Fig. 17.8 are considered. The devices are taken to be operating at steady state.

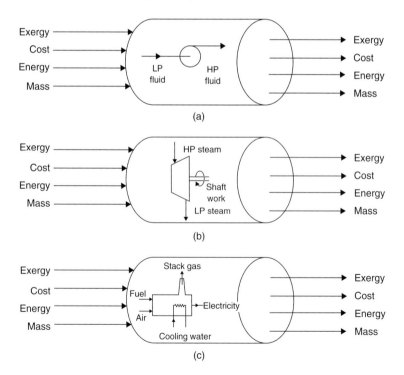

Fig. 17.8. Application of EXCEM analysis to several devices. (a) Pump; (b) steam turbine and (c) coal-fired electrical generating station.

Pump

The application of EXCEM analysis to a pump (Fig. 17.8a) is relatively straightforward. A fluid and electricity are input and the fluid at higher pressure and heat (unless the pump is adiabatic) are output. If the pump is adiabatic all cost associated with inputs and generation are allocated to one output. If the pump is not adiabatic, all input and generation costs are still logically allocated to the output fluid stream, because heat loss is a waste.

Steam turbine

The application of EXCEM analysis to a steam turbine (Fig. 17.8b) requires more thought because a subjective decision must be made regarding the allocation of costs. The shaft work, low-enthalpy steam and heat outputs may each be allocated part of the input and generation costs. The allocations depend on the uses for the outputs. The low-enthalpy steam, for example, may or may not be a waste.

Coal-fired electrical generating station

The application of EXCEM analysis to a coal-fired electrical generating station (Fig. 17.8c) requires more effort than the previous examples because intermediate streams, as well as inputs and outputs, must be examined. The analysis of such a system can be viewed as a set of individual EXCEM analyses of the devices comprising the overall system. As with the steam-turbine example, cost allocations require subjective decisions.

Figure 17.9 illustrates a summary of the results of an EXCEM analysis of a typical coal-fired electrical generating station (the coal-fired Nanticoke Generating Station described in Section 17.5). In Fig. 17.9, the input and generation costs are allocated to the product electricity. The exhaust cooling water is a waste. Alternatively, if the exhaust cooling water is a by-product part of the total costs would be allocated to it.

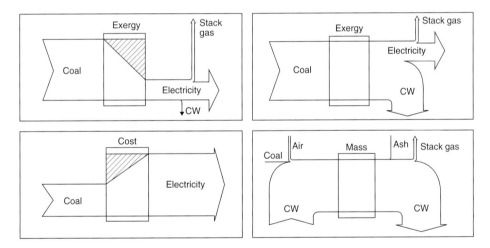

Fig. 17.9. Summary of the results of an EXCEM analysis of a coal-fired electrical generating station. Overall balances of flow rates of exergy, cost, energy and mass are shown. The rectangle in the center of each diagram represents the station. Widths of arrows are proportional to the relative magnitudes of the represented quantities. Rates of exergy consumption and cost creation are denoted by the shaded regions in the appropriate diagrams. CW denotes cooling water.

17.7.4. Exergy loss and cost generation

The relation between exergy and cost is demonstrated using plots of exergy loss as a function of cost generation. Either internal exergy losses (i.e., consumptions) or total exergy losses (i.e., consumptions plus waste emissions) can be considered. The intensive properties of the reference environment need to be completely specified when total exergy losses are considered. Only the temperature of the reference environment need be specified when internal exergy losses are considered. Costs associated with inputs need not be specified.

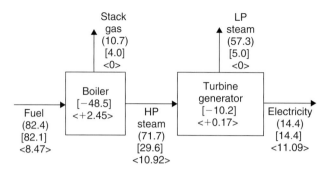

Fig. 17.10. A cogeneration plant for the production of electrical energy and low pressure (LP) steam. Flow rates of energy (values in parentheses), exergy (values in square brackets) and cost (values in angle brackets) are indicated for streams. Rates of exergy consumption (negative values in square brackets) and cost creation (positive values in angle brackets) are indicated for devices. Energy and exergy values are in MW, and cost values are in M$/year. Costs are in 2001 U.S. dollars.

An illustrative example is used in the discussions. The data are drawn from a simplified steady-state analysis of a cogeneration plant for electricity and heat (Reistad and Gaggioli, 1980). The plant is shown in Fig. 17.10, with exergy consumption and cost creation rates; and exergy, energy and cost flow rates. Inputs to the boiler of air and feedwater, for which the associated flow rates of energy, exergy and cost are approximately zero, are omitted from the diagram. The operating condition considered yields 17.9 kg/s (150,000 lb/h) of steam at 44.2 atm (650 psia) and 399°C (750°F) exiting the boiler, and 0.065 atm (0.95 psia) and 37.8°C (100°F) exiting the turbine generator. Energy and exergy values are evaluated relative to a reference environment having a temperature of 10°C, pressure of 1 atm and composition as described in the reference-environment model of Gaggioli and Petit (1977).

To obtain the cost generation rate in this example, the capital cost is multiplied by the ammortization factor and divided by the load factor. The ammortization factor spreads the total costs associated with a device over the life of the device, taking into account the time value of money. Ammortization and load factors of 0.08 and 0.7, respectively, are used.

Exergy loss rates are plotted as a function of cost creation rates for the boiler and turbine generator in Fig. 17.11a. For each device, total and internal exergy loss rates are considered. Figure 17.11b plots, at different points in the plant, cumulative exergy loss rate as a function of cumulative cost creation rate. The slopes and magnitudes of the individual lines indicate characteristics of the corresponding devices. Plots of the type in Figs. 17.11a and 17.11b demonstrate that exergy and cost are the only EXCEM quantities subject to non-conservation laws. Since for any device the associated values of cost creation and exergy loss are positive, the lines in these plots always rise to the right.

The type of plot in Fig. 17.11a showing total exergy loss rate vs. cost creation rate illustrates the trade-off between cost and efficiency. The total exergy input is wasted if no investment (cost generation) is made. Performance approaches the ideal if a very large investment is made.

A balance is obtained between exergy loss and cost creation in real systems. A plot of cumulative energy loss rate vs. cumulative cost creation rate is presented in Fig. 17.11c, and compared with the plot in Fig. 17.11b. The analogous curve in Fig. 17.11c to the broken curve in Fig. 17.11b is a straight line along the x-axis (because energy is conserved), and is not particularly informative. The solid curve in Fig. 17.11c exhibits some similarity to the solid curve in Fig. 17.11b. However, the solid curve in Fig. 17.11c is not as illuminating and can be misleading because it weighs all energy losses equally. Different forms of energy are not necessarily equal in that some forms of energy cannot be completely converted into other forms, even in an ideal process. Different forms of exergy are relatively more equal since one form of exergy can be converted into any other form of exergy in an ideal process.

The idea that costing should be based on exergy rather than energy, because exergy often is a consistent measure of value (i.e., a large quantity of exergy is often associated with a valuable commodity) while energy is only sometimes a consistent measure of value, is supported by the observations made when comparing Fig. 17.11b, c. More general versions of Fig. 17.11a, b, in which flow rates of exergy and cost at different points in the plant are plotted, are shown in Fig. 17.11d. The intensive properties of the reference environment must be completely specified and costs associated with all inputs must be known to construct Fig. 17.11d. A monotonically varying composite line is again traced. However, the line does not necessarily begin at the origin of the plot. The properties of the reference environment and costs associated with inputs determine the origin of the composite line.

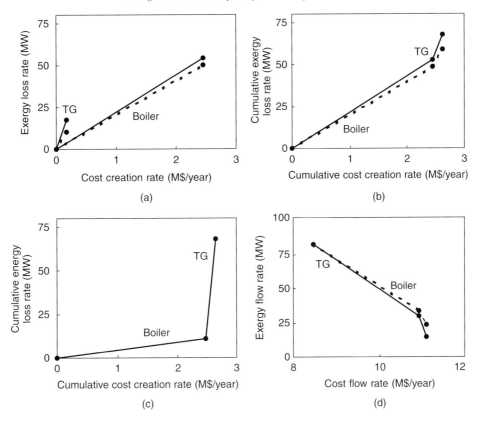

Fig. 17.11. Comparison for a cogeneration plant of rates of cost flow and creation with rates of exergy loss and consumption as well as rates of energy loss. Solid lines denote total losses and broken lines internal losses. TG denotes turbine generator. Costs are in 2001 U.S. dollars.

17.8. Closing remarks

The present chapter provides insights into the relation between energy and exergy losses and capital costs for energy systems. The methods are useful in analysis and design because they integrate thermodynamics and economics.

A systematic correlation often exists between exergy loss rate (total or internal) and capital cost for coal-fired, oil-fired and nuclear generating stations. Furthermore, a correlation appears to exist between the mean thermodynamic-loss-rate-to-capital-cost ratios for all of the devices in a station and the ratios for the overall station, when the ratio is based on total or internal exergy losses, but not when it is based on energy losses. This correlation may imply that devices in successful electrical generating stations are configured so as to achieve an overall optimal design, by appropriately balancing the thermodynamic (exergy-based) and economic characteristics of the overall station and its devices. This idea may extend to the electrical utility sector and the generating stations that comprise it, as well as other technologies (e.g., cogeneration).

Problems

17.1 What is exergoeconomics? What is the difference between exergoeconomics and thermoeconomics?

17.2 What are the advantages of costing based on exergy compared to costing based on energy?

17.3 How are the exergy losses in a system related to economic losses?

17.4 Is it possible to perform an exergoeconomic analysis of a system which has no fuel cost, such as a solar energy system? If so, how useful is such an analysis compared to an exergoeconomic analysis of a fuel-driven system such as a coal power plant?

17.5   Is it possible for a designer to use the recommendations of exergy analysis to improve the performance of a system without conducting an exergy analysis of the system? Explain with examples.

17.6   Explain how the exergy-based price of various fuels can be determined. Provide examples.

17.7   If your local power company decides to price electricity based on exergy, how would your electricity bill change?

17.8   Write general energy, exergy and cost balances and explain their differences. Are there any similarities between cost and entropy balances?

17.9   An energy engineer claims that an exergoeconomic analysis of a fossil-fuel power plant is not advantageous compared to an energy-based economic analysis since the exergy of a fossil fuel is approximately equal to its heating value. Do you agree? Explain.

17.10   In the exergoeconomic case studies in this chapter, what correlations exist between capital cost, exergy loss and energy loss?

17.11   Describe the EXCEM analysis method. Can an energy analysis be performed with this method?

17.12   Obtain a published article on exergoeconomic analysis of a power plant. Using the operating and cost data provided in the article, perform a detailed exergoeconomic analysis of the plant and compare your results to those in the original article.

17.13   Obtain actual operating and cost data from an energy system and perform an exergoeconomic analysis. Discuss the results.

Chapter 18

EXERGY ANALYSIS OF COUNTRIES, REGIONS AND ECONOMIC SECTORS

18.1. Introduction

Designing efficient and cost-effective systems, which also meet environmental and other constraints, is one of the foremost challenges that engineers face. In a world with finite natural resources and large energy demands, it is important to understand mechanisms which degrade energy and resources and to develop systematic approaches for improving systems in terms of such factors as efficiency, cost, environmental impact, etc.

The undesirable effects of poor utilization of energy resources, especially regarding economics and ecology, demonstrate that the designing of appropriate energy systems requires careful analysis and planning. In this regard, exergy analysis is beneficial, providing a useful tool for:

- Improving the efficiency of energy resource utilization.
- Assessing the locations, types and true magnitudes of wastes and losses.
- Distinguishing between high- and low-quality energy resources and services, and better matching the quality of energy required for a service with the quality of the energy supplied.
- Determining whether or not and by how much it is possible to design more efficient energy systems by reducing inefficiencies.
- Reducing the impact of energy resource utilization on the environment.
- Enabling the achievement of some of the criteria for sustainable development, such as a sustainable supply of energy resources that is usable with minimal environmental degradation.

Many of these points are illustrated in Table 18.1, which lists energy and exergy efficiencies for several processes. The energy efficiencies in Table 18.1 represent the ratio of the energy of the useful streams leaving the process to the energy of all input streams, while the exergy efficiencies represent the ratio of the exergy of the products of a process

Table 18.1. Energy and exergy efficiencies for selected devices.

Device	Energy efficiency (%)	Exergy efficiency (%)
Residential space heater (fuel)	60	9
Domestic water heater (fuel)	40	2–3
High-pressure steam boiler	90	50
Tobacco dryer (fuel)	40	4
Coal gasification system	55	46
Petroleum refining unit	~90	10
Steam-heated reboiler	~100	40
Blast furnace	76	46

Source: Gaggioli (1980), Kenney (1984) and Rosen and Dincer (1997a,b).

to the exergy of all inputs. The exergy efficiencies in Table 18.1 are lower than the energy efficiencies, mainly because process irreversibilities destroy some of the input exergy.

Exergy is also useful since it provides a link between engineering and the environment. Exergy analysis determines the true efficiency of systems and processes, making it particularly useful for finding appropriate improvements. For these and other reasons, exergy analysis is strongly recommended by many for use in the design of engineering systems and analyses of regional, national and global energy systems as well as sectors of the economy.

Recently, the use of energy and other resources in the industrial world has reached levels never before observed, leading to reduced supplies of natural resources and increased damage to and pollution of the natural environment. At the same time, energy conversion networks have become more complicated. Sometimes improvement efforts are focused on inappropriate resource conversions, in that the potential to improve the resource use is not significant. By describing the use of energy resources in society in terms of exergy, important knowledge and understanding is gained, and areas are better identified where large improvements can be attained by applying measures to increase efficiency.

18.2. Background and benefits

Analyses of regional and national energy systems provide insights into how effectively a society uses natural resources and balances such factors as economics and efficiency. Such insights can help identify areas in which technical and other improvements should be undertaken, and indicate the priorities which should be assigned to measures. Assessments and comparisons of various societies throughout the world can also be of fundamental interest in efforts to achieve a more equitable distribution of resources.

During the past few decades, exergy has been increasingly applied to the industrial sector and other sectors of the economy, particularly to attain energy savings, and hence financial savings. The energy utilization of a region like a country can be assessed using exergy analysis to gain insights into its efficiency. This approach was first introduced in a landmark paper by Reistad (1975), who applied it to the U.S. Since then, several other countries, e.g., Canada (Rosen, 1992), Japan, Finland and Sweden (Wall, 1990, 1991), Italy (Wall et al., 1994), Turkey (Ozdogan and Arikol, 1995; Rosen and Dincer, 1997b) and others have been examined using modified versions of this approach.

In this chapter, a comprehensive exergy analysis of countries, regions and economic sectors and services is introduced and discussed. Later as an illustration energy and exergy utilization for Saudi Arabia is examined and compared to assessments for Turkey and Canada.

18.3. Applying exergy to macrosystems

Exergy is the 'fuel' of dissipative systems, i.e., systems that are sustained by converting energy and materials. Examples include a living cell, an organism, an ecosystem, and the earth's surface with its material cycles. Societies are also dissipative systems, and can therefore be assessed with exergy analysis. Exergy analysis has mostly been applied industrial systems and processes, but its application can straightforwardly be extended to macrosystems, allowing the examination of regional, national and global energy and material conversions. Such applications describe use of resources and related environmental impacts.

Natural resources are traditionally divided into energy and other resources, especially in regional assessments. The separation often is vague. For example, oil is usually considered an energy resource and wood a material resource. Yet oil can also be converted to useful materials and wood can be used as a fuel. It is more appropriate to assess these resources with one unifying measure, and exergy provides such a resource measure.

18.3.1. Energy and exergy values for commodities in macrosystems

The exergy of an energy resource can for simplicity often be expressed as the product of its energy content and a quality factor (the exergy-to-energy ratio) for the energy resource. Quality factors for some energy forms are listed in Table 18.2.

Energy resources are usually measured in energy units, as are exergy resources. Other resources are usually measured in other quantitative units such as weight or volume. A material can be quantified in exergy units by multiplying its quantity by an exergy-based unit factor for the material. Using such measures could allow for an expanded resource budgeting and provide a first step toward an integration of exergy with traditional energy budgeting. The exergy per unit

Table 18.2. Quality factors for some common energy forms.

Energy form	Quality factor
Mechanical energy	1
Electrical energy	1
Chemical fuel energy	$\sim 1.0^a$
Nuclear energy	0.95
Sunlight	0.9
Hot steam (600°C)	0.6
District heating (90°C)	$0.2–0.3^b$
Moderate heating at room temperature (20°C)	$0–0.2^b$
Thermal radiation from the earth	0

[a] This value may exceed 1, depending on the system definition and state.
[b] This value depends significantly on the environmental temperature.
Source: (Wall, 1986b).

quantity is a measure of the value or usefulness of a resource relative to the environment. This value relates to the price of the material or resource, which is also partly defined by the environment through, for instance, demand.

In assessments of regions and nations, the most common material flows often are hydrocarbon fuels at near ambient conditions. The physical exergy for such material flows is approximately zero, and the specific exergy reduces to the fuel specific chemical exergy ex_f, which can be written as

$$ex_f = \gamma_f H_f \qquad (18.1)$$

where γ_f denotes the exergy grade function for the fuel, defined as the ratio of fuel chemical exergy to fuel higher heating value H_f. Table 18.3 lists typical values of H_f, ex_f and γ_f for fuels typically encountered in regional and national assessments. The specific chemical exergy of a fuel at T_0 and P_0 is usually approximately equal to its higher heating value H_f.

Table 18.3. Properties of selected fuels.*

Fuel	H_f (kJ/kg)	Chemical exergy (kJ/kg)	γ_f
Gasoline	47,849	47,394	0.99
Natural gas	55,448	51,702	0.93
Fuel oil	47,405	47,101	0.99
Kerosene	46,117	45,897	0.99

* For a reference-environment temperature of 25°C, pressure of 1 atm and chemical composition as defined in the text.
Source: Reistad (1975).

18.3.2. The reference environment for macrosystems

The reference environment used in many assessments of macrosystems is based on the model of Gaggioli and Petit (1977), which has a temperature $T_0 = 25°C$, pressure $P_0 = 1$ atm and a chemical composition consisting of air saturated with water vapor, and the following condensed phases at 25°C and 1 atm: water (H_2O), gypsum ($CaSO_4 \cdot 2H_2O$) and limestone ($CaCO_3$). This reference-environment model is used in this chapter, but with a temperature of 10°C.

18.3.3. Efficiencies for devices in macrosystems

Energy η and exergy ψ efficiencies for the principal processes in macrosystems are usually based on standard definitions:

$$\eta = \text{(Energy in products)/(Total energy input)} \qquad (18.2)$$

$$\psi = \text{(Exergy in products)/(Total exergy input)} \qquad (18.3)$$

Exergy efficiencies can often be written as a function of the corresponding energy efficiencies by assuming the energy grade function γ_f to be unity, which is commonly valid for typically encountered fuels (kerosene, gasoline, diesel and natural gas).

Heating

Electric and fossil fuel heating processes are taken to generate product heat Q_p at a constant temperature T_p, either from electrical energy W_e or fuel mass m_f. The efficiencies for electrical heating are

$$\eta_{h,e} = Q_p/W_e \qquad (18.4)$$

and

$$\psi_{h,e} = Ex^{Q_p}/Ex^{W_e} = (1 - T_0/T_p)Q_p/W_e$$

Combining these expressions yields

$$\psi_{h,e} = (1 - T_0/T_p)\eta_{h,e} \qquad (18.5)$$

For fuel heating, these efficiencies are

$$\eta_{h,f} = Q_p/m_f H_f \qquad (18.6)$$

and

$$\psi_{h,f} = Ex^{Q_p}/m_f ex_f$$

or

$$\psi_{h,f} = (1 - T_0/T_p)Q_p/(m_f \gamma_f H_f) \cong (1 - T_0/T_p)\eta_{h,f} \qquad (18.7)$$

where double subscripts indicate processes in which the quantity represented by the first subscript is produced by the quantity represented by the second, e.g., the double subscript h,e means heating with electricity.

Cooling

The efficiencies for electric cooling are

$$\eta_{c,e} = Q_p/W_e \qquad (18.8)$$

$$\psi_{c,e} = Ex^{Q_p}/Ex^{W_e} = (1 - T_0/T_p)Q_p/W_e \qquad (18.9)$$

or

$$\psi_{c,e} = (1 - T_0/T_p)\eta_{c,e} \qquad (18.10)$$

Work production

Electric and fossil fuel work production processes produce shaft work W. The efficiencies for shaft work production from electricity are

$$\eta_{m,e} = W/W_e \qquad (18.11)$$

$$\psi_{m,e} = Ex^W/Ex^{W_e} = W/W_e = \eta_{m,e} \qquad (18.12)$$

For fuel-based work production, these efficiencies are

$$\eta_{m,f} = W/m_f H_f \qquad (18.13)$$

$$\psi_{m,f} = Ex^W/m_f ex_f = W/m_f \gamma_f H_f \cong \eta_{m,f} \qquad (18.14)$$

Electricity generation

The efficiencies for electricity generation from fuel are

$$\eta_{e,f} = W_e/m_f H_f \qquad (18.15)$$

$$\psi_{e,f} = Ex^{W_e}/m_f ex_f = W_e/m_f \gamma_f H_f \cong \eta_{e,f} \qquad (18.16)$$

Kinetic energy production

The efficiencies for the fossil fuel-driven kinetic energy production processes, which occur in some devices in the transportation sector (e.g., turbojet engines and rockets) and which produce a change in kinetic energy Δke in a stream of matter m_s, are as follows:

$$\eta_{ke,f} = m_s \Delta ke_s/m_f H_f \qquad (18.17)$$

$$\psi_{ke,f} = m_s \Delta ke_s/m_f ex_f = m_s \Delta ke_s/m_f \gamma_f H_f \cong \eta_{ke,f} \qquad (18.18)$$

18.4. Case study: energy and exergy utilization in Saudi Arabia

The methodology discussed in previous sections is used to analyze overall and sectoral energy and exergy utilization in Saudi Arabia, which has six economic sectors: residential, public and private, industrial, transportation, agricultural and electrical utility. The country is modeled as a macrosystem as shown in Fig. 18.1. The analysis is carried out for the period 1990–2001 and uses energy data from various local and international sources. Efficiencies are determined to understand how efficiently energy and exergy are used in Saudi Arabia and its sectors. Also, the results for Saudi Arabia for the year 1993 are compared to those for Turkey and Canada.

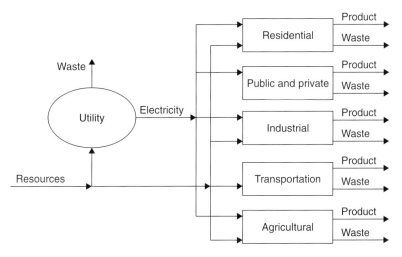

Fig. 18.1. Model for the energy flows in a macrosystem like a country or region. Fossil fuel in the public and private sector is not considered due to lack of data, and electricity use in the Transportation sector is negligibly small.

18.4.1. Analysis of the residential sector

Energy and exergy utilization in the residential sector is evaluated and analyzed.

Energy utilization data for the residential sector

To determine energy and exergy efficiencies for the residential sector, the consumption of total electrical and fossil fuel energy within the sector is determined. In Saudi Arabia, about 70% of the total residential energy consumption is for air conditioning of buildings (Hasnain et al., 2000). The values of total electrical and fossil fuel consumption in the residential sector for a 12-year period (1990–2001) are presented in Table 18.4.

Table 18.4. Energy consumption in the residential sector of Saudi Arabia for 1990–2001.

Year	Device	Breakdown of energy use in sector, by type (PJ)		
		Electrical	LPG	Kerosene
1990	Air conditioning	62.5	–	–
	Lighting	9.7	–	–
	Cooking appliances	7.0	25.0	13.5
	Others*	17.5	–	–
1991	Air conditioning	68.8	–	–
	Lighting	11.5	–	–
	Cooking appliances	8.3	26.3	20.5
	Others*	15.6	–	–
1992	Air conditioning	71.0	–	–
	Lighting	11.8	–	–
	Cooking appliances	8.6	26.0	10.9
	Others*	16.1	–	–
1993	Air conditioning	82.0	–	–
	Lighting	13.5	–	–
	Cooking appliances	9.8	31.4	11.2
	Others*	17.1	–	–
1994	Air conditioning	92.8	–	–
	Lighting	15.2	–	–
	Cooking appliances	11.1	34.3	11.2
	Others*	19.4	–	–
1995	Air conditioning	98.5	–	–
	Lighting	16.1	–	–
	Cooking appliances	11.7	36.1	10.0
	Others*	19.7	–	–

(Continued)

Table 18.4. (Continued)

Year	Device	Breakdown of energy use in sector, by type (PJ)		
		Electrical	LPG	Kerosene
1996	Air conditioning	103.2	–	–
	Lighting	16.7	–	–
	Cooking appliances	12.1	38.8	10.2
	Others*	19.7	–	–
1997	Air conditioning	105.9	–	–
	Lighting	17.1	–	–
	Cooking appliances	12.5	40.6	10.2
	Others*	20.2	–	–
1998	Air conditioning	111.8	–	–
	Lighting	18.0	–	–
	Cooking appliances	13.1	41.4	10.0
	Others*	20.4	–	–
1999	Air conditioning	126.6	–	–
	Lighting	20.2	–	–
	Cooking appliances	14.7	43.0	9.3
	Others*	22.0	–	–
2000	Air conditioning	135.2	–	–
	Lighting	20.3	–	–
	Cooking appliances	15.4	43.3	7.8
	Others*	22.2	–	–
2001	Air conditioning	140.4	–	–
	Lighting	21.1	–	–
	Cooking appliances	16.1	47.5	7.1
	Others*	23.1	–	–

* Others include water heaters, refrigerators, televisions, computers, washing machines, fans, etc.

Efficiencies of principal devices in the residential sector

Energy and exergy efficiencies of principal devices in the residential sector are determined. The energy efficiencies, and process and reference-environment temperatures, are assumed to be the same as those used by Reistad (1975) and Rosen (1992a). The process and operating data of the principal devices in the residential sector for Saudi Arabia are listed in Table 18.5. The device exergy efficiencies are evaluated using these data and following the methodology in Section 18.3.

For air conditioning, we follow the approach of Reistad (1975), who noted that it is reasonable to assume for an air-conditioning system that the 'energy efficiency' is $\eta = 100\%$, the environment temperature is $T_0 = 283$ K and the 'product' heat is delivered at $T_p = 293$ K. Although this treatment of air conditioning does not follow the conventional use of coefficients of performance to evaluate the merit of the device, it facilitates the sectoral assessment considered

Table 18.5. Process and operating data for the residential sector of Saudi Arabia.

| Device | Product-heat temperature T_p (K) | | | Energy and exergy efficiencies (%) | | | |
| | | | | Electrical | | Fuel | |
	Electrical	LPG	Kerosene	η_e	ψ_e	η_f	ψ_f
Air conditioning	293	–	–	100.0	3.4	–	–
Lighting	–	–	–	25.0	24.3	–	–
Cooking appliances	394	374	374	80.0	22.5	65.0	15.8
Others*	–	–	–	–	–	–	–

* Others include water heaters, refrigerators, televisions, computers, washing machines, fans, etc.

here. Using these values and Eqs. (18.8) and (18.10), we find for air conditioning

$$\eta = Q_p/W_e = 1 \text{ (or } 100\%)$$

$$\psi = (1 - T_0/T_p)Q_p/W_e = (1 - 283/293) \times 1 = 0.034 \text{ (or } 3.4\%)$$

For other devices in the residential sector, energy and exergy efficiencies can be obtained following the methodology used for air conditioning or other devices in Section 18.3 and the data in Table 18.5.

Mean efficiencies for the overall residential sector

Weighted mean energy and exergy efficiencies are calculated for the residential sector using a three-step process. First, weighted means are obtained for the electrical energy and exergy efficiencies for the device categories listed in Table 18.5, where the weighting factor is the ratio of electrical energy input to the device category to the total electrical energy input to all device categories in the sector. Second, weighted mean efficiencies for the fossil fuel-driven devices are similarly determined. Third, overall weighted means are obtained for the energy and exergy efficiencies for the electrical and fossil fuel processes, where the weighting factor is the ratio of total fossil fuel or electrical energy input to the residential sector to the total energy input to the sector.

To illustrate, the calculation of the overall weighted mean energy and exergy efficiencies for the principal devices in the residential sector for the year 2000 is shown.

Step 1:

$$\eta_e = [(100 \times 135.2) + (25 \times 20.3) + (80 \times 15.4)]/(135.2 + 20.3 + 15.4) = 89.29\%$$

$$\psi_e = [(3.4 \times 135.2) + (24.3 \times 20.3) + (22.5 \times 15.4)]/(135.2 + 20.3 + 15.4) = 7.61\%$$

Step 2:

$$\eta_f = (65 \times 38.5)/(38.5) = 65\%$$

$$\psi_f = (15.8 \times 38.5)/(38.5) = 15.8\%$$

Step 3:

$$\eta_o = (89.29 \times 0.769) + (65 \times 0.23) = 84.1\%$$

$$\psi_o = (7.61 \times 0.769) + (15.8 \times 0.23) = 9.4\%$$

The weighted mean electrical, fuel and overall energy and exergy efficiencies for the residential sector for the 12 years between 1990 and 2001 are given in Table 18.6. The overall weighted mean energy and exergy efficiencies for the residential sector are illustrated for that period in Fig. 18.2.

Table 18.6. Mean efficiencies for the residential sector in Saudi Arabia for 1990–2001.

Year	Weighted mean electrical efficiencies (%)		Weighted mean fuel efficiencies (%)		Overall efficiencies (%)	
	Energy	Exergy	Energy	Exergy	Energy	Exergy
1990	89.047	7.648	65	15.816	82.199	9.974
1991	88.412	7.902	65	15.816	81.153	10.356
1992	88.412	7.902	65	15.816	82.429	9.925
1993	88.547	7.850	65	15.816	82.466	9.907
1994	88.547	7.850	65	15.816	82.731	9.817
1995	88.613	7.824	65	15.816	82.945	9.742
1996	88.678	7.799	65	15.816	82.901	9.755
1997	88.678	7.799	65	15.816	82.861	9.768
1998	88.743	7.774	65	15.816	83.058	9.699
1999	88.807	7.749	65	15.816	83.525	9.539
2000	89.294	7.606	65	15.816	84.214	9.323
2001	89.294	7.606	65	15.816	84.100	9.361

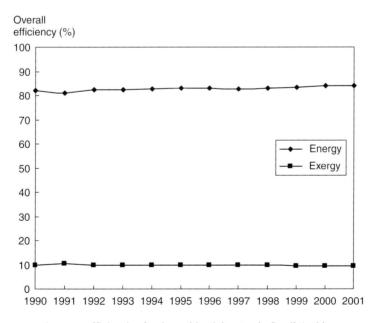

Fig. 18.2. Overall energy and exergy efficiencies for the residential sector in Saudi Arabia.

18.4.2. Analysis of the public and private sector

Energy and exergy utilization in the public and private sector is evaluated and analyzed. This sector is subdivided into commercial, governmental, streets, mosques, hospitals and charity associations.

Energy utilization data for the public and private sector

To determine exergy efficiencies for the public and private sector, the consumptions of electrical and fuel energy within the sector are required. Data on fuel energy use in this sector are not available, so this analysis covers only electrical energy consumption. In Saudi Arabia, about 38% of electrical energy use in the public and private sector is for air conditioning, 42% is for appliances and 20% is for lighting (Ahmad et al., 1994). Annual energy consumption data for the public and private sector for 1990–2001 are presented in Table 18.7.

Table 18.7. Electrical energy consumption (in PJ) in the public and private sector of Saudi Arabia.*

Year	Subsector	Device		
		Air conditioning	Appliances	Lighting
1990	Commercial	6.99	7.72	3.68
	Governmental	11.82	6.57	7.88
	Streets	–	–	3.57
	Mosques	0.76	–	0.50
	Hospitals	1.74	0.95	0.47
	Charity associations	0.13	–	0.09
1991	Commercial	6.54	7.23	3.44
	Governmental	15.28	8.49	10.19
	Streets	–	–	1.46
	Mosques	0.88	–	0.58
	Hospitals	1.88	1.03	0.51
	Charity associations	0.16	–	0.11
1992	Commercial	6.60	7.30	3.47
	Governmental	17.18	9.54	11.45
	Streets	–	–	4.09
	Mosques	0.91	–	0.61
	Hospitals	2.07	1.13	0.56
	Charity associations	0.17	–	0.11
1993	Commercial	7.90	8.73	4.16
	Governmental	17.92	9.96	11.95
	Streets	–	–	3.99
	Mosques	1.02	–	0.68
	Hospitals	2.39	1.30	0.65
	Charity associations	0.18	–	0.12

(*Continued*)

Table 18.7. (Continued)

Year	Subsector	Device		
		Air conditioning	Appliances	Lighting
1994	Commercial	8.73	9.65	4.60
	Governmental	18.79	10.44	12.52
	Streets	–	–	4.39
	Mosques	1.56	–	1.04
	Hospitals	2.49	1.36	0.68
	Charity associations	0.18	–	0.12
1995	Commercial	9.59	10.60	5.05
	Governmental	19.22	10.68	12.81
	Streets	–	–	4.78
	Mosques	1.60	–	1.07
	Hospitals	2.55	1.39	0.69
	Charity associations	0.19	–	0.13
1996	Commercial	10.65	11.77	5.60
	Governmental	19.66	10.92	13.11
	Streets	–	–	4.51
	Mosques	1.83	–	1.22
	Hospitals	2.49	1.36	0.68
	Charity associations	0.21	–	0.14
1997	Commercial	10.53	11.63	5.54
	Governmental	20.71	11.51	13.81
	Streets	–	–	4.66
	Mosques	1.68	–	1.12
	Hospitals	2.33	1.27	0.64
	Charity associations	0.23	–	0.15
1998	Commercial	11.09	12.26	5.84
	Governmental	21.10	11.72	14.07
	Streets	–	–	4.77
	Mosques	1.78	–	1.19
	Hospitals	2.34	1.27	0.64
	Charity associations	0.28	–	0.19

(Continued)

Table 18.7. (Continued)

Year	Subsector	Device		
		Air conditioning	Appliances	Lighting
1999	Commercial	12.64	13.97	6.65
	Governmental	22.76	12.64	15.17
	Streets	–	–	5.04
	Mosques	2.09	–	1.40
	Hospitals	2.87	1.57	0.78
	Charity associations	0.32	–	0.21
2000	Commercial	13.11	14.49	6.90
	Governmental	23.62	13.12	15.75
	Streets	–	–	5.01
	Mosques	2.16	–	1.44
	Hospitals	2.87	1.57	0.78
	Charity associations	0.18	–	0.12
2001	Commercial	13.62	15.05	7.17
	Governmental	24.76	13.75	16.51
	Streets	–	–	5.25
	Mosques	2.36	–	1.57
	Hospitals	2.87	1.56	0.78
	Charity associations	0.28	–	0.18

* All energy consumption is electrical.

Efficiencies of principal devices in the public and private sector

Operating and process data for devices in the public and private sector for Saudi Arabia are listed in Table 18.8. Energy and exergy efficiencies for the devices, evaluated with the methodology in Section 18.3 and these data are listed in Table 18.8 for 1990–2001.

Mean efficiencies for the overall public and private sector

Mean energy and exergy efficiencies for each subsector in the public and private sector are calculated using information from Tables 18.7 and 18.8. These mean efficiencies are then used to determine the overall sector energy and exergy efficiencies (see Table 18.9 and Fig. 18.3).

For illustration, the overall weighted mean energy and exergy efficiencies for the principal devices in the public and private sector for the year 2000 are evaluated as follows:

Step 1: Mean energy and exergy efficiencies of each subsector are evaluated:

Commercial:

$$\eta_c = [(100 \times 13.11) + (80 \times 14.49) + (20 \times 6.9)]/(13.11 + 14.49 + 6.9) = 75.6\%$$

$$\psi_c = [(3.41 \times 13.11) + (1.39 \times 14.49) + (19.5 \times 6.9)]/(13.11 + 14.49 + 6.9) = 5.78\%$$

Table 18.8. Process and operating data for the public and private sector of Saudi Arabia.

Subsector	Devices	Product temperature T_p (K)	Energy and exergy efficiencies (%)	
			η_e	ψ_e
Commercial	Air conditioning	293.00	100.00	3.41
	Appliances	288.00	80.00	1.39
	Lighting	–	20.00	19.50
Governmental	Air conditioning	293.00	100.00	3.41
	Appliances	–	–	–
	Lighting	–	20.00	19.50
Streets	Lighting	–	5.00	4.80
Mosques	Air conditioning	293.00	100.00	3.41
	Lighting	–	20.00	19.50
Hospitals	Air conditioning	293.00	100.00	3.41
	Appliances	–	–	–
	Lighting	–	20.00	19.50
Charity associations	Air conditioning	293.00	100.00	3.41
	Lighting	–	20.00	19.50

Table 18.9. Overall mean efficiencies for the public and private sector in Saudi Arabia.

Year	Overall mean efficiency (%)	
	Energy	Exergy
1990	57.35	6.57
1991	55.89	6.68
1992	55.55	6.71
1993	56.53	6.68
1994	56.81	6.70
1995	57.02	6.67
1996	57.79	6.67
1997	57.42	6.69
1998	57.63	6.69
1999	58.05	6.68
2000	58.13	6.67
2001	58.10	6.69

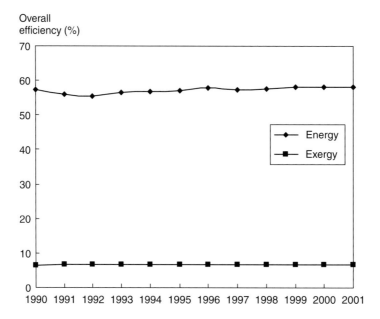

Fig. 18.3. Overall energy and exergy efficiencies for the public and private sector in Saudi Arabia.

Governmental:

$$\eta_g = [(100 \times 23.62) + (20 \times 15.75)]/(23.62 + 15.75) = 68\%$$

$$\psi_g = [(3.41 \times 23.62) + (19.5 \times 15.75)]/(23.62 + 15.75) = 9.85\%$$

Streets:

$$\eta_s = (5 \times 5.01)/(5.01) = 5\%$$

$$\psi_s = (4.8 \times 5.01)/(5.01) = 4.8\%$$

Mosques:

$$\eta_m = [(100 \times 2.16) + (20 \times 1.44)]/(2.16 + 1.44) = 68\%$$

$$\psi_m = [(3.41 \times 2.16) + (19.5 \times 1.44)]/(2.16 + 1.44) = 9.85\%$$

Hospitals:

$$\eta_h = [(100 \times 2.87) + (20 \times 0.78)]/(2.87 + 0.78) = 82.86\%$$

$$\psi_h = [(3.41 \times 2.87) + (19.5 \times 0.78)]/(2.87 + 0.78) = 6.86\%$$

Charity associations:

$$\eta_{ca} = [(100 \times 0.18) + (20 \times 0.12)]/(0.18 + 0.12) = 68\%$$

$$\psi_{ca} = [(3.41 \times 0.18) + (19.5 \times 0.12)]/(0.18 + 0.12) = 9.85\%$$

Step 2: Overall mean energy and exergy efficiencies are calculated for the sector as

$$\eta_o = (75.6 \times 0.341) + (68 \times 0.519) + (5 \times 0.049) + (68 \times 0.035)$$
$$+ (82.86 \times 0.051) + (68 \times 0.0029) = 68.24\%$$
$$\psi_o = (5.78 \times 0.341) + (9.85 \times 0.519) + (4.8 \times 0.049) + (9.85 \times 0.035)$$
$$+ (6.86 \times 0.051) + (9.85 \times 0.0029) = 8.06\%$$

18.4.3. Analysis of the industrial sector

Energy and exergy utilization in the industrial sector is evaluated and analyzed. The industrial sector of Saudi Arabia is composed of many industries. The four most significant are oil and gas, chemical and petrochemical, iron and steel, and cement.

Methodology and energy data for the industrial sector

To simplify the analysis of energy and exergy efficiencies for this complex sector, the four most significant industries, which account for more than 80% of the total sector energy use, are chosen to represent the overall sector.

In the industrial sector of Saudi Arabia, the energy used to generate heat for production processes accounts for 68% of the total energy consumption, and mechanical drives account for 17%. The remaining 15% is divided among lighting, air conditioning, etc. For simplicity, we analyze here heating and mechanical end uses only. This simplification is considered valid since these processes account for 85% of the energy consumption in the industrial sector.

Assumptions and simplifications made for the heating and mechanical processes are as follows:

- Heating processes for each industry are grouped into low-, medium- and high-temperature categories as shown in Table 18.10. The temperature ranges given in Table 18.10 are based on Rosen (1992a) and the heating data are from Brown et al. (1996).
- The efficiencies for the low-temperature category are assumed to be the same as those for heating in the residential sector (Rosen, 1992). The efficiencies for the medium- and high-temperature categories are from Reistad (1975).
- All mechanical drives are assumed to be 90% energy efficient (Wall, 1988).

Table 18.10. Process heating temperatures and efficiencies for the industrial sector.*

T_p category	T_p range (°C)	Heating energy efficiencies (%)	
		Electrical, $\eta_{h,e}$	Fuel, $\eta_{h,f}$
Low	<121	100	65.5
Medium	121–399	90	60
High	>399	70	50

* *Source*: Rosen (1992a).

Three steps are used to derive the overall efficiency of the sector. First, energy and exergy efficiencies are obtained for process heating for each of the product-heat temperature T_p categories. Second, mean heating energy and exergy efficiencies for the four industries are calculated using a two-part procedure: (i) weighted mean efficiencies for electrical heating and fuel heating are evaluated for each industry and (ii) weighted mean efficiencies for all heating processes in each industry are determined with these values, using as weighting factors the ratio of the industry energy consumption (electrical or fuel) to the total consumption of both electrical and fuel energy. Third, weighted mean overall (i.e., heating and mechanical drive) efficiencies for each industry are evaluated using as the weighting factor the fractions of the total sectoral energy input for both heating and mechanical drives.

In the determination of sector efficiencies, weighted means for the weighted mean overall energy and exergy efficiencies for the major industries in the industrial sector are obtained, using as the weighting factor the fraction of the total industrial energy demand supplied to each industry.

For illustration, the efficiencies for the oil and gas industry are calculated in the following subsection.

Process heating efficiencies for the product-heat temperature categories in each industry

Product-heat temperature data for each industry are separated into the categories defined in Table 18.10. The resulting breakdown is shown in Table 18.11, with the percentage of heat in each category supplied by electricity and fossil fuels. The evaluation of efficiencies for electrical and fossil fuel process heating for the oil and gas industry are shown in the next two subsections. The same process is applied to each industry in the industrial sector.

Table 18.11. Process heating data for the industrial sector.*

| Industry | T_p range | Mean T_p in range (°C) | Breakdown of energy used in each T_p range (%) | |
			Electricity	Fuel
Oil and gas	Low	57	10.0	13.8
	Medium	227	9.4	22.6
	High	494	80.4	63.6
Chemical and petrochemical	Low	42	62.5	0.0
	Medium	141	37.5	100.0
	High	494	0.0	0.0
Iron and steel	Low	45	4.2	0.0
	Medium	–	0.0	0.0
	High	983	95.8	100.0
Cement	Low	42	91.7	0.9
	Medium	141	0.0	9.0
	High	–	8.3	90.1

* *Source*: Brown et al. (1996).

Electrical process heating in the oil and gas industry: In the oil and gas industry, electric heating is used to supply all categories of heat as shown in Table 18.11. With Table 18.10 and Eq. (18.15), the energy efficiency for low-temperature electric heating is shown to be

$$\eta = Q_p/W_e = 1 \text{ (or 100\%)}$$

For the medium- and high-temperature categories, the energy efficiencies are similarly found to be 90% and 70%, respectively.

Using Eq. (18.10) with $T_0 = 283$ K, the exergy efficiencies for the three categories are:

- Low temperature: $T_p = 330$ K (mean value in category)

$$\psi = (1 - T_0/T_p)Q_p/W_e = (1 - 283/330) \times 1 = 0.142 \text{ (or 14.24\%)}$$

- Medium temperature: $T_p = 500$ K (mean value in category)

$$\psi = (1 - T_0/T_p)Q_p/W_e = (1 - 283/500) \times 0.9 = 0.3906 \text{ (or 39.06\%)}$$

- High temperature: $T_p = 767$ K (mean value in category)

$$\psi = (1 - T_0/T_p)Q_p/W_e = (1 - 283/767) \times 0.7 = 0.441 \text{ (or 44.1\%)}$$

Fossil fuel process heating in the oil and gas industry: The oil and gas industry requires fossil fuel heating at all ranges of temperatures in Table 18.11. The energy efficiency for low-temperature heating is found using Eq. (18.6) and data from Table 18.10:

$$\eta = Q_p/m_f H_f = 0.65 \text{ (or 65\%)} \tag{18.19}$$

Similarly, the energy efficiency for medium- and high-temperature heating are found to be 60% and 50%, respectively.

The corresponding exergy efficiency for low-temperature process heating is found using Eq. (18.7), a reference-environment temperature T_0 of 283 K and a process heating temperature T_p from Table 18.11 of 330 K, as follows:

$$\psi = (1 - 283/330)Q_p/(m_f \gamma_f H_f) \tag{18.20}$$

Assuming $\gamma_f = 1$, we can combine Eqs. (18.19) and (18.20) to obtain the exergy efficiency for low-temperature process heating as

$$\psi = (1 - 283/330) \times 0.65 = 0.092 \text{ (or 9.25\%)}$$

Similarly, the exergy efficiencies for the medium- and high-temperature process heating are found to be 26.04% and 31.55%, respectively.

Mean process heating efficiencies for each industry of the industrial sector

Prior to obtaining the overall energy and exergy efficiencies for the industrial sector, the overall heating efficiencies for each industry are evaluated. Again, the methodology is illustrated for the oil and gas industry.

A combined mean efficiency for the three temperature categories for electric and fossil fuel processes is evaluated to obtain an overall heating efficiency in a given industry. Using data from Table 18.14, the fraction of total energy utilized by the oil and gas industry for electrical (E_e) and fossil fuel (E_f) heating is found for the year 2000 as follows:

- For electrical energy:

$$E_e = \text{(Electrical energy in)/(Electrical energy in} + \text{Fuel energy in)} = 38.5/(38.5 + 999.47) = 0.037 \text{ (or 3.7\%)}$$

- For fossil fuel energy:

$$E_f = 1 - E_e = 1.00 - 0.037 = 0.963 \text{ (or 96.3\%)}$$

Using energy fractions from Table 18.11 and energy and exergy efficiencies in Table 18.12, an average heating efficiency for the oil and gas industry can be calculated.

The energy efficiency for electrical heating $\eta_{h,e}$ can be evaluated in the oil and gas industry as follows:

$$\eta_{h,e} = \sum \text{(fraction in category)} \times \text{(energy efficiency)} = (0.1 \times 100) + (0.094 \times 90) + (0.805 \times 70) = 74.89\%$$

Similarly, the corresponding exergy efficiency $\psi_{h,e}$ is calculated as

$$\psi_{h,e} = (0.1 \times 14.24) + (0.094 \times 39.06) + (0.805 \times 44.17) = 40.69\%$$

Using data in Tables 18.11 and 18.12, energy and exergy efficiencies for fossil fuel heating in the oil and gas industry for the year 2000 are found as follows:

$$\eta_{h,f} = (0.138 \times 65) + (0.225 \times 60) + (0.635 \times 50) = 54.33\%$$

$$\psi_{h,f} = (0.138 \times 9.26) + (0.225 \times 26.04) + (0.635 \times 31.55) = 27.23\%$$

Table 18.12. Energy and exergy data and efficiencies for all categories of product-heat temperature T_p in the industrial sector of Saudi Arabia.

Industry	T_p range	Breakdown of energy and exergy efficiencies for each T_p category, by type			
		Electrical heating		Fuel heating	
		η_h	ψ_h	η_h	ψ_h
Oil and gas	Low	100.00	14.24	65.00	9.26
	Medium	90.00	39.06	60.00	26.04
	High	70.00	44.17	50.00	31.55
Chemical and petrochemical	Low	100.00	10.16	–	–
	Medium	90.00	28.48	60.00	18.99
	High	–	–	–	–
Iron and steel	Low	100.00	11.01	–	–
	Medium	–	–	–	–
	High	70.00	54.23	50.00	38.73
Cement	Low	100.00	6.91	65.00	4.49
	Medium	–	–	60.00	21.06
	High	70.00	58.71	50.00	41.94

With the energy efficiencies $\eta_{h,e}$ and $\eta_{h,f}$, and the fractions of electrical energy E_e and fossil fuel energy E_f used by the oil and gas industry, overall mean energy and exergy efficiencies for heating can be determined:

$$\eta_h = (0.037 \times 74.89) + (0.963 \times 54.33) = 55.09\%$$

$$\psi_h = (0.037 \times 40.69) + (0.963 \times 27.23) = 28.07\%$$

Following the same methodology, mean heating energy and exergy efficiencies for the other three industries considered are determined (see Table 18.13). The mean heating energy and exergy efficiencies for the year 2000 are illustrated in Fig. 18.4.

Overall efficiencies for the industrial sector

Overall energy and exergy efficiencies for the industrial sector are obtained using process heating efficiencies (see Table 18.13), the mechanical drive efficiency (assumed to be 90%) and the total energy consumption for each industry (see Table 18.14). For 1990–2001, consequently, overall mean heating energy ($\eta_{h,o}$) and exergy efficiencies ($\psi_{h,o}$) are presented in Fig. 18.5, and overall energy and exergy efficiencies for the industrial sector are presented in Fig. 18.6.

18.4.4. Analysis of the transportation sector

Energy and exergy utilization in the transportation sector is evaluated and analyzed. The transportation sector in Saudi Arabia is composed of three main modes: road, air and marine. Mean energy and exergy efficiencies are calculated by multiplying the energy used in each mode by the corresponding efficiency. Then, these values are added to obtain the overall efficiency of the transportation sector.

Table 18.13. Process heating energy and exergy efficiencies for the main industries in the industrial sector of Saudi Arabia.

Year	Industry							
	Oil and gas		Chemical and petrochemical		Iron and steel		Cement	
	η_h	ψ_h	η_h	ψ_h	η_h	ψ_h	η_h	ψ_h
1990	55.05	28.55	61.37	18.78	50.72	40.66	52.85	33.82
1991	55.03	28.47	61.23	18.81	50.74	40.69	52.90	33.73
1992	55.05	28.41	61.15	18.83	50.78	40.67	54.82	34.04
1993	55.04	28.35	61.19	18.84	50.78	40.57	53.02	34.06
1994	55.26	28.30	61.28	18.84	50.88	40.54	53.26	34.16
1995	55.15	28.31	61.29	18.85	50.82	40.39	53.12	34.60
1996	55.11	28.23	61.30	18.85	50.78	40.27	53.02	34.93
1997	55.09	28.15	61.33	18.85	50.77	40.15	52.99	35.28
1998	55.11	28.12	61.31	18.86	50.79	40.11	53.03	35.41
1999	55.11	28.13	61.23	18.87	50.78	40.12	53.00	35.36
2000	55.09	28.07	61.06	18.89	50.75	40.01	52.93	35.69
2001	55.07	28.02	61.17	18.88	50.81	40.08	53.08	35.51

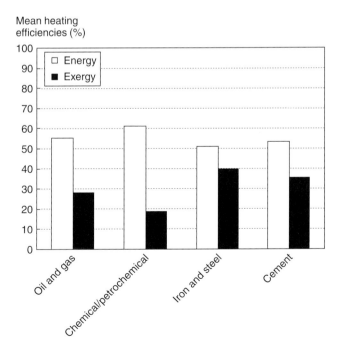

Fig. 18.4. Heating energy and exergy efficiencies for the industrial sector for the year 2000 in Saudi Arabia.

Table 18.14. Energy consumption data (in PJ) for the industrial sector in Saudi Arabia.

Year	Industry							
	Oil and gas		Chemical and petrochemical		Iron and steel		Cement	
	Electrical	Fuel	Electrical	Fuel	Electrical	Fuel	Electrical	Fuel
1990	24.00	660.22	18.00	459.82	12.00	343.57	6.00	147.25
1991	24.47	697.35	18.35	521.18	12.23	340.49	6.12	145.92
1992	26.38	731.60	19.78	602.58	13.19	346.59	13.19	148.54
1993	27.48	763.21	20.61	608.51	13.74	360.16	6.87	154.35
1994	30.11	636.44	22.58	615.97	15.05	349.14	7.53	149.63
1995	30.80	742.40	23.10	623.85	15.40	382.36	7.70	163.87
1996	32.41	821.03	24.31	656.17	16.21	422.75	8.10	181.18
1997	33.73	878.01	25.30	663.32	16.86	448.63	8.43	192.27
1998	36.59	925.79	27.45	733.64	18.30	474.86	9.15	203.51
1999	36.82	929.13	27.61	785.10	18.41	486.55	9.20	208.52
2000	38.50	999.47	28.87	959.91	19.25	526.91	9.62	225.82
2001	40.15	1070.73	30.12	902.62	20.08	508.59	10.04	217.97

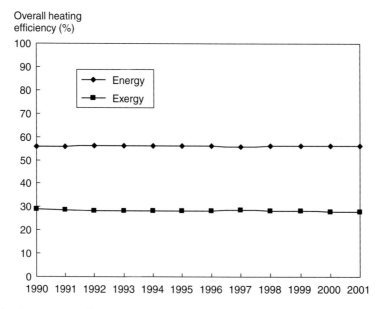

Fig. 18.5. Overall heating energy and exergy efficiencies for the industrial sector in Saudi Arabia.

Energy utilization data for the transportation sector

A breakdown is presented in Table 18.15, by mode of transport, of the energy consumed in the Saudi Arabian transportation sector.

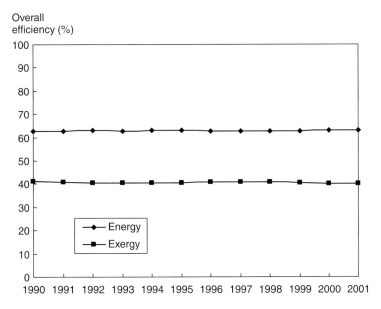

Fig. 18.6. Overall energy and exergy efficiencies for the industrial sector in Saudi Arabia.

Energy efficiencies for the transportation sector

Table 18.15 provides energy efficiencies for the three modes of transport. These values are based on U.S. devices (Reistad, 1975) and are assumed representative of Saudi Arabian devices. Since vehicles generally are not operated at full load, a distinction is made between rated load (full load) and operating (part load) efficiencies (Reistad, 1975).

Table 18.15. Energy consumption and process data for the transportation sector in Saudi Arabia.

Year	Mode of transport	Main fuel types	Energy consumption		Energy efficiencies (%)	
			PJ	%	Rated load	Estimated operating
1990	Road	Gasoline	321.61	43.73	28	22
		Diesel	195.45	26.58	28	22
	Air	Jet kerosene	97.61	13.27	35	28
		Air fuel	8.20	1.12	35	28
	Marine	Ship fuel	112.50	15.30	–	15
		Diesel	0.01	0.00	–	15
1991	Road	Gasoline	337.69	45.65	28	22
		Diesel	206.78	27.95	28	22
	Air	Jet kerosene	99.17	13.41	35	28
		Air fuel	8.34	1.13	35	28
	Marine	Ship fuel	87.75	11.86	–	15
		Diesel	0.01	0.00	–	15

(Continued)

Table 18.15. (Continued)

Year	Mode of transport	Main fuel types	Energy consumption		Energy efficiencies (%)	
			PJ	%	Rated load	Estimated operating
1992	Road	Gasoline	354.57	46.29	28	22
		Diesel	218.78	28.56	28	22
	Air	Jet kerosene	100.76	13.16	35	28
		Air fuel	8.47	1.11	35	28
	Marine	Ship fuel	83.36	10.88	–	15
		Diesel	0.01	0.00	–	15
1993	Road	Gasoline	372.30	46.89	28	22
		Diesel	231.46	29.15	28	22
	Air	Jet kerosene	102.37	12.89	35	28
		Air fuel	8.60	1.08	35	28
	Marine	Ship fuel	79.19	9.97	–	15
		Diesel	0.01	0.00	–	15
1994	Road	Gasoline	390.92	47.45	28	22
		Diesel	244.89	29.73	28	22
	Air	Jet kerosene	104.01	12.63	35	28
		Air fuel	8.74	1.06	35	28
	Marine	Ship fuel	75.23	9.13	–	15
		Diesel	0.01	0.00	–	15
1995	Road	Gasoline	410.46	47.97	28	22
		Diesel	259.09	30.28	28	22
	Air	Jet kerosene	105.68	12.35	35	28
		Air fuel	8.88	1.04	35	28
	Marine	Ship fuel	71.47	8.35	–	15
		Diesel	0.01	0.00	–	15
1996	Road	Gasoline	430.99	48.46	28	22
		Diesel	274.12	30.82	28	22
	Air	Jet kerosene	107.37	12.07	35	28
		Air fuel	9.02	1.01	35	28
	Marine	Ship fuel	67.90	7.63	–	15
		Diesel	0.01	0.00	–	15

(Continued)

Table 18.15. (Continued)

Year	Mode of transport	Main fuel types	Energy consumption		Energy efficiencies (%)	
			PJ	%	Rated load	Estimated operating
1997	Road	Gasoline	441.76	48.64	28	22
		Diesel	278.78	30.70	28	22
	Air	Jet kerosene	109.19	12.02	35	28
		Air fuel	9.18	1.01	35	28
	Marine	Ship fuel	69.25	7.63	–	15
		Diesel	0.01	0.00	–	15
1998	Road	Gasoline	452.81	48.83	28	22
		Diesel	283.52	30.57	28	22
	Air	Jet kerosene	111.05	11.97	35	28
		Air fuel	9.33	1.01	35	28
	Marine	Ship fuel	70.64	7.62	–	15
		Diesel	0.01	0.00	–	15
1999	Road	Gasoline	464.13	49.01	28	22
		Diesel	288.34	30.45	28	22
	Air	Jet kerosene	112.94	11.93	35	28
		Air fuel	9.49	1.00	35	28
	Marine	Ship fuel	72.05	7.61	–	15
		Diesel	0.01	0.00	–	15
2000	Road	Gasoline	475.73	49.20	28	22
		Diesel	293.24	30.33	28	22
	Air	Jet kerosene	114.86	11.88	35	28
		Air fuel	9.65	1.00	35	28
	Marine	Ship fuel	73.49	7.60	–	15
		Diesel	0.01	0.00	–	15
2001	Road	Gasoline	487.62	49.38	28	22
		Diesel	298.23	30.20	28	22
	Air	Jet kerosene	116.81	11.83	35	28
		Air fuel	9.82	0.99	35	28
	Marine	Ships fuel	74.96	7.59	–	15
		Diesel	0.01	0.00	–	15

A weighted mean is obtained for the transportation mode energy efficiencies in Table 18.15, where the weighting factor is the fraction of the total energy input to the sector which is supplied to each transportation mode. The weighted mean overall energy efficiency for the transportation sector for the year 2000, e.g., is calculated as

$$\eta_o = (0.491 \times 22) + (0.303 \times 22) + (0.118 \times 28) + (0.009 \times 28) + (0.076 \times 15) = 22.24\%$$

Exergy efficiencies for the transportation sector

Before evaluating the overall mean exergy efficiencies for the transportation sector, it is noted that the outputs of transportation devices are in the form of kinetic energy (shaft work). The exergy associated with shaft work (W) is by definition equal to the energy, i.e.,

$$Ex^W = W$$

Thus, for electric shaft work production, the energy and exergy efficiencies of transportation devices can be shown to be identical:

$$\eta_{m,e} = W/W_e \tag{18.21}$$

$$\psi_{m,e} = Ex^W / Ex^{W_e} = W/W_e = \eta_{m,e} \tag{18.22}$$

For fossil fueled shaft work production in transportation devices, the exergy efficiency can be shown to be similar to the energy efficiency:

$$\eta_{m,f} = W/m_f H_f \tag{18.23}$$

$$\psi_{m,f} = Ex^W / m_f \gamma_f H_f \tag{18.24}$$

When γ_f is unity, as is often assumed for most fuels Rosen (1992a),

$$\psi_{m,f} = \eta_{m,f} \tag{18.25}$$

Thus, the overall mean exergy efficiencies for the transportation sector are equal to the overall mean energy efficiencies. For the year 2000, for instance,

$$\psi_o = \eta_o = 22.24\%$$

The overall mean energy and exergy efficiencies for the transportation sector for 1990–2001 are illustrated in Fig. 18.7.

18.4.5. Analysis of the agricultural sector

Energy and exergy utilization in the agricultural sector is evaluated and analyzed. The main devices used in the agricultural sector, which we assume to be representative of the sector, are tractors and pumps. Mean energy and exergy efficiencies are calculated by multiplying the energy used in each device type by the corresponding device efficiency. Then, these values are added to obtain the overall efficiency of the agricultural sector.

Energy utilization data for the agricultural sector

The agricultural sector consumes less energy than the other sectors of Saudi Arabia. Diesel fuel and electricity are used in this sector and the breakdown for 1990–2001 is shown in Table 18.16.

Energy and exergy efficiencies for the agricultural sector

To calculate overall energy and exergy efficiencies of the agricultural sector, we assume the energy efficiency for tractors, which operate on diesel fuel, to be 22% under part load conditions and for pumps, which are driven by electric motors, to be 90%.

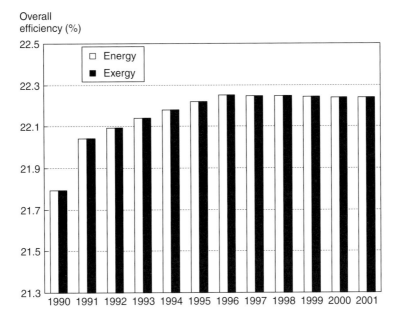

Fig. 18.7. Overall energy and exergy efficiencies of the transportation sector in Saudi Arabia.

Table 18.16. Energy consumption data for the agricultural sector in Saudi Arabia.

Year	Energy consumption (PJ)	
	Diesel	Electrical
1990	238.50	2.70
1991	252.57	3.60
1992	261.74	4.06
1993	266.02	4.81
1994	293.54	5.44
1995	266.02	6.15
1996	179.18	6.53
1997	152.30	6.95
1998	129.46	7.08
1999	110.04	7.29
2000	93.53	7.83
2001	92.98	8.12

Tractors can be treated like other devices in the transportation sector, for which exergy and energy efficiencies are equal.

Overall mean energy and exergy efficiencies for the agricultural sector are determined using a weighting factor, which is the fraction of the total energy input to each device. Overall mean energy and exergy efficiencies for the agricultural sector for 1990–2001 are illustrated in Fig. 18.8. As an example, overall energy and exergy efficiencies for the year 2000 are evaluated as follows:

$$\eta_o = [(22 \times 93.53) + (90 \times 7.83)]/(93.53 + 7.83) = 27.25\%$$

$$\psi_o = [(22 \times 93.53) + (4.53 \times 7.83)]/(93.53 + 7.83) = 20.65\%$$

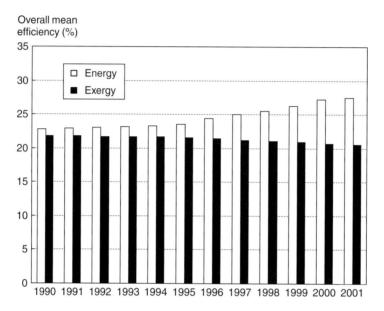

Fig. 18.8. Overall energy and exergy efficiencies for the agricultural sector in Saudi Arabia.

18.4.6. Analysis of the utility sector

Energy and exergy utilization in the utility sector is evaluated and analyzed. The main electricity generation sources in Saudi Arabia are fossil fuels (diesel, crude oil, natural gas, fuel oil). The utility sector also includes electricity generated by desalination plants.

Energy utilization data for the utility sector

For power and desalination plants for 1990–2001, the energy input is listed in Table 18.17 and the electricity generated in Table 18.18. The overall energy efficiency can be determined by dividing total electrical energy produced by the total input energy.

Energy efficiencies for the utility sector

Using data in Tables 18.17 and 18.18, we can determine energy efficiencies for the power and desalination plants. Then, we can calculate the overall mean energy efficiencies of the utility sector.

Sample calculations are shown below for the year 2000. For power plants,

$$\eta_{e,P} = \text{(Electrical energy output)/(Fuel energy input to power plants)} = 350.25\,\text{PJ}/1231.38\,\text{PJ} = 28.44\%$$

and for desalination plants,

$$\eta_{e,d} = \text{(Electrical energy output)/(Fuel energy input to desalination plants)} = 78.2\,\text{PJ}/167.09\,\text{PJ} = 46.8\%$$

Table 18.17. Energy consumption data for the utility sector in Saudi Arabia.

Year	Energy Consumed (PJ)	
	Power plants	Desalination plants
1990	591.24	141.28
1991	648.12	139.82
1992	694.36	148.43
1993	779.78	142.99
1994	893.05	156.12
1995	912.21	159.50
1996	958.95	153.36
1997	1011.98	179.88
1998	1095.59	172.77
1999	1195.56	171.12
2000	1231.38	167.10
2001	1298.81	179.22

Table 18.18. Energy generation data for the utility sector in Saudi Arabia.

Year	Electricity generated (PJ)	
	Power plants	Desalination plants
1990	162.96	70.68
1991	179.51	69.64
1992	194.76	71.68
1993	222.26	73.63
1994	251.86	75.88
1995	259.01	79.11
1996	273.40	78.84
1997	291.89	72.12
1998	311.74	75.42
1999	337.31	75.33
2000	350.25	78.20
2001	371.14	78.50

The overall mean energy efficiency is then

$$\eta_o = (0.817 \times 28.44) + (0.182 \times 46.8) = 31.75\%$$

Exergy efficiencies for the utility sector

Since for fossil fuel energy we assume $\gamma_f = 1$, the exergy efficiencies for electricity generation from power and desalination plants are the same as the energy efficiencies. This equivalence is shown earlier for the industrial sector.

Thus, the mean overall exergy efficiency is equal to the mean overall energy efficiency. That is, for the year 2000,

$$\psi_o = \eta_o = 31.75\%$$

Overall mean energy and exergy efficiencies for the utility sector for 1990–2001 are shown in Figs. 18.9 and 18.10.

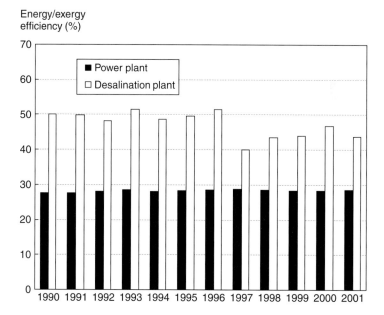

Fig. 18.9. Energy and exergy efficiencies of the two main plant types in the utility sector in Saudi Arabia.

18.4.7. Energy and exergy efficiencies and flows for the sectors and country

Overall energy and exergy efficiencies for all sectors of Saudi Arabia are evaluated. Using the efficiencies from previous sections and energy consumption data in each sector, energy and exergy flow diagrams are constructed for the year 2000 (see Figs. 18.11 and 18.12). Overall energy and exergy efficiencies for the Saudi Arabian economy for 1990–2001 are shown in Fig. 18.13. Energy and exergy efficiencies for the six sectors and the overall Saudi Arabian economy for the year 2000 are illustrated in Fig. 18.14.

For illustration, overall energy and exergy flows and efficiencies for the sectors in Saudi Arabia for the year 2000 are evaluated as follows:

$$\text{Total energy input} = 7922.1 \text{ PJ}$$

$$\text{Residential sector product energy} = \eta_o \times \text{Total energy input to residential sector} = 0.84 \times 244.2 = 205.12 \text{ PJ}$$

$$\text{Public and private sector product energy} = \eta_o \times \text{Total energy input to public and private sector}$$
$$= 0.58 \times 101.13 = 58.65 \text{ PJ}$$

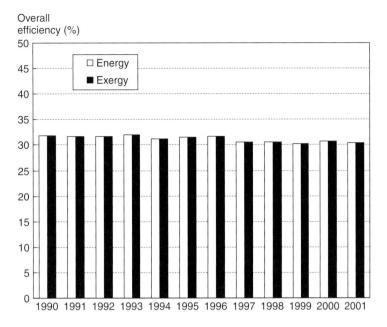

Fig. 18.10. Overall energy and exergy efficiencies for the utility sector in Saudi Arabia.

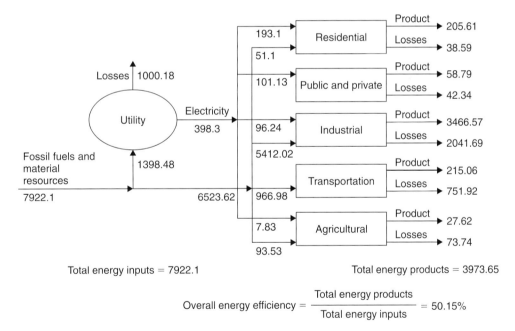

Fig. 18.11. Energy flow diagram for Saudi Arabia for the year 2000. Numerical values are in PJ/yr. Losses represent waste energy emissions. Electricity use in the public and private sector is not considered as data were not available.

Industrial sector product energy $= \eta_o \times$ Total energy input to industrial sector $= 0.62 \times 5412.02 = 3466.57$ PJ

Transportation sector product energy $= \eta_o \times$ Total energy input to transportation sector $= 0.22 \times 966.98 = 215.06$ PJ

Agricultural sector product energy $= \eta_o \times$ Total energy input to agricultural sector $= 0.27 \times 93.53 = 27.62$ PJ

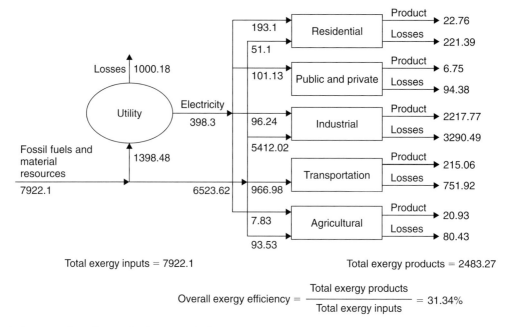

Total exergy inputs = 7922.1   Total exergy products = 2483.27

$$\text{Overall exergy efficiency} = \frac{\text{Total exergy products}}{\text{Total exergy inputs}} = 31.34\%$$

Fig. 18.12. Exergy flow diagram for Saudi Arabia for the year 2000. Numerical values are in PJ/yr. Losses represent waste exergy emissions and internal exergy consumptions. Electricity use in the public and private sector is not considered as data were not available.

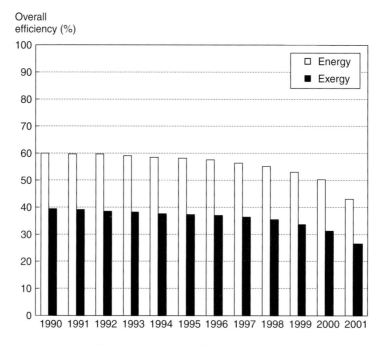

Fig. 18.13. Overall energy and exergy efficiencies for the Saudi Arabian economy.

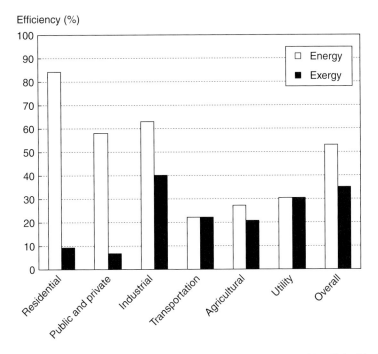

Fig. 18.14. Energy and exergy efficiencies for the sectors and the overall economy of Saudi Arabia for 2000.

$$\text{Total product energy} = 205.12 + 58.65 + 3466.57 + 215.06 + 27.62 = 3973.65 \text{ PJ}$$

$$\text{Overall energy efficiency} = 3973.65/7922.1 = 50.15\%$$

Similarly, exergy flows and efficiencies are evaluated for the year 2000 as follows, with the assumption that the fuel energy grade function is unity, so exergy inputs are equal to energy inputs:

$$\text{Residential sector product exergy} = \psi_o \times \text{Total exergy input to residential sector} = 0.093 \times 244.2 = 21.97 \text{ PJ}$$

$$\text{Public and private sector product exergy} = \psi_o \times \text{Total exergy input to public and private sector}$$
$$= 0.06 \times 101.13 = 6.06 \text{ PJ}$$

$$\text{Industrial sector product exergy} = \psi_o \times \text{Total exergy input to industrial sector} = 0.4 \times 5412.02 = 2164.8 \text{ PJ}$$

$$\text{Transportation sector product exergy} = \psi_o \times \text{Total exergy input to transportation sector} = 0.22 \times 966.98 = 212.73 \text{ PJ}$$

$$\text{Agricultural sector product exergy} = \psi_o \times \text{Total exergy input to agricultural sector} = 0.20 \times 93.53 = 18.7 \text{ PJ}$$

$$\text{Total product exergy} = 21.97 + 6.06 + 2164.8 + 212.73 + 18.7 = 2483.27 \text{ PJ}$$

$$\text{Overall exergy efficiency} = 2483.27/7922.1 = 31.34\%$$

18.4.8. Discussion

Exergy analysis indicates a less efficient picture of energy use in Saudi Arabia than does energy analysis.

The residential sector has the lowest exergy efficiency of all sectors, followed closely by the public and private sector. The reason for the low exergy efficiencies in these sectors is inefficient utilization of the work potential or quality of the input energy. In these sectors, the primary use of energy is to produce cold or heat at near environmental temperatures. With the production of such products from a fossil fuel or electrical energy source, there is a loss in energy quality that can only be reflected with exergy analysis. The nearer to the temperature of the environment is the temperature of the heat produced, the lower is the exergy efficiency. The residential, public and private and industrial sectors exhibit wide variations between energy and exergy efficiencies. This is attributable to the extent to which heating processes occur in these sectors.

An energy analysis of Saudi Arabian energy utilization does not provide a true picture of how well the economy utilizes its energy resources. An assessment based on energy can be misleading because it often indicates the main inefficiencies to be in the wrong sectors, and a state of technological efficiency higher than actually exists. In order to accurately assess the true efficiency of energy utilization, exergy analysis must be used. Exergy flow diagrams are a powerful tool for indicating to industry and government where emphasis should be placed in programs to improve the use of the exergy associated with the main energy resources (e.g., oil). Furthermore, the results provide important insights for future research and development allocations and projects.

Energy utilization also causes environmental concerns such as global warming, air pollution, acid rain and stratospheric ozone depletion. These issues must be addressed if humanity is to achieve a sustainable energy future. Since all energy use leads to some environmental impact, some environmental concerns can be overcome through increased efficiency and some through use of sustainable energy resources. The former method is used in this chapter to evaluate and understand the efficiency of a country and to assist in increasing efficiency and reducing environmental impact.

18.4.9. Summary of key findings

The overall and sectoral energy and exergy assessments of Saudi Arabia and its main economic sectors have yielded several interesting findings:

- The overall energy and exergy efficiencies for the year 2000, at 53% and 35%, respectively, differ significantly.
- Sectoral energy and exergy efficiencies, respectively, for the year 2000 are 84% and 9% for the residential sector, 58% and 7% for the public and private sector, 63% and 40% for the industrial sector, 22% and 22% for the transportation sector, 27% and 21% for the agricultural sector, and 31% and 31% for the utility sector. Thus, the most energy efficient sector is the residential sector (84%) and the most exergy efficient sector is the industrial sector (40%).
- In analyzing the relationship between energy and exergy losses (which can be viewed as representing perceived and actual inefficiencies, respectively), it is seen that actual inefficiencies in the residential, public and private, industrial and agricultural sectors are much higher than the perceived inefficiencies. For the transportation and utility sectors, the actual inefficiencies are the same as the perceived inefficiencies.

18.5. Comparison of different countries

Sector and overall energy and exergy efficiencies for Saudi Arabia, Turkey and Canada are compared (see Figs. 18.15 and 18.16). The sectors which are common to these countries are residential–commercial, industrial, utility and transportation. The comparison is based on previous studies, and the data used is for the year 1993 for Saudi Arabia and Turkey and 1986 for Canada. The efficiencies differ slightly, but the main trends described earlier in this chapter regarding the differences between energy and exergy efficiencies are exhibited by each country.

18.6. Closing remarks

Exergy analyses provide useful information about sectoral and overall energy and exergy utilization in a macrosystem like a country, and can consequently help achieve energy savings through efficiency and/or conservation measures. Exergy analyses can also help in establishing standards to facilitate energy planning in the entire macrosystem and its sectors. The sample assessment carried out here demonstrates the simplicity and value of using exergy when analyzing industrial processes and economic sectors of a region or country.

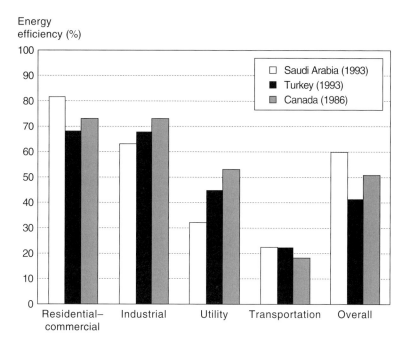

Fig. 18.15. Comparison of several sector and overall energy efficiencies for Saudi Arabia, Turkey and Canada.

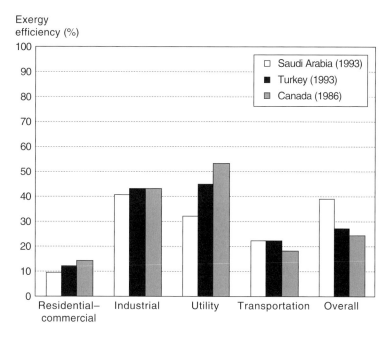

Fig. 18.16. Comparison of several sector and overall exergy efficiencies for Saudi Arabia, Turkey and Canada.

Problems

18.1 How can exergy be used in sectoral, regional and national energy assessments? What are the advantages of sectoral, regional and national exergy analyses?

18.2 Provide typical values of energy and exergy efficiencies for the transportation, industrial and residential sectors.

18.3 Propose methods of increasing the exergy efficiency of the transportation, industrial and residential sectors.

18.4 Provide expressions for the exergy of energy and material resources commonly encountered in sectoral, regional and national energy assessments?

18.5 When analyzing the exergy flow in a country, one often finds that the exergy efficiencies of fossil fuel power plants are greater than those of power plants using renewable energy sources. Does this mean that fossil fuel power plants contribute more to resource sustainability for the country? Explain.

18.6 Obtain a published article on sectoral exergy analysis of a region or country. Using the data provided in the article, try to duplicate the results. Compare your results to those in the original article.

18.7 Perform a sectoral exergy analysis of your town, city, region or country.

Chapter 19

EXERGETIC LIFE CYCLE ASSESSMENT

19.1. Introduction

As environmental awareness grows, society has become increasingly concerned about the issues of natural resource depletion and environmental degradation. Industries and businesses have responded to this awareness by assessing how their activities affect the environment and, in many cases, providing 'greener' products and using 'greener' processes. The environmental performance of products and processes has become an important concern, leading many companies to investigate ways to reduce or minimize their impacts on the environment. Many companies have found it advantageous to explore ways of moving *beyond* compliance using pollution prevention strategies and environmental management systems to improve their environmental performance. One valuable tool for this work is life cycle assessment (LCA), a technique in which the entire life cycle of a product is considered (Curran, 1996).

The environmental impact and efficiency of technologies depend on the characteristics of the many steps and chains involved over their lifetimes, from natural resource extraction and plant construction to distribution and final product utilization. Adequate evaluation of environmental impact and energy use throughout the overall production and utilization life cycle ('from cradle to grave') is critical for the proper evaluation of technologies.

LCA is a methodology for this type of assessment, and represents a systematic set of procedures for compiling and examining the inputs and outputs of materials and energy, and the associated environmental impacts, directly attributable to a product or service throughout its life cycle. A life cycle is the interlinked stages of a product or service system, from the extraction of natural resources to final disposal (ISO, 1997).

In this chapter, LCA is modified and extended by considering exergy. Exergetic LCA (ExLCA) is described and, as a case study, applied to several hydrogen production technologies.

19.2. Life cycle assessment

An LCA consists of three main steps, once the goal and scope of the analysis is defined: (i) determination of the mass and energy flows into, out of and through all stages of the life cycle, including production through utilization to final disposal, of the chosen product or service (inventory step), (ii) evaluation of the environmental impacts associated with the mass and energy flows determined in the previous stage (impact assessment step) and (iii) determination of reasonable ways to decrease the environmental, economic and other burdens (improvement step). Mass and energy balances are often used extensively in the first stage.

The ability to track and document shifts in environmental impacts can help engineers as well as decision and policy makers fully characterize the environmental trade-offs associated with product or process alternatives. Performing an LCA allows one to:

- Quantify environmental releases to air, water and land in relation to each life cycle stage and/or major contributing process.
- Evaluate systematically the environmental consequences associated with a given product or process.
- Assist in identifying significant shifts in environmental impacts between life cycle stages and environmental media.
- Assess the human and ecological effects of material and energy consumption and environmental releases to the local community, region and world.
- Compare the health and ecological impacts of alternative products and processes.

- Identify impacts related to specific environmental areas of concern.
- Analyze the environmental trade-offs associated with one or more specific products/processes to help gain stakeholder (region, community, etc.) acceptance for a planned action.

The importance of LCA becomes apparent if one considers industrial processes (metallurgical, chemical, etc.) for products (metals, plastics, glass, etc.) and services and notes that almost all currently rely on fossil fuels, the consumption of which leads to a range of environmental impacts.

Within an LCA, mass and energy flows and environmental impacts related to plant construction, utilization and dismantling stages are accounted for. The determination of all input and output flows is often a very complicated task, so simplifications and assumptions are often made to facilitate LCA. The challenge is to ensure the assumptions and simplifications (e.g., simplified models of processes) retain the main characteristics of the actual system or process being analyzed.

19.3. Exergetic LCA

In this section, we extend LCA, which aims to reduce material and energy use and environmental and ecological impacts while increasing product quality and sustainability, to ExLCA. In addition to the aims of LCA, ExLCA examines exergy flows and seeks to reduce exergy destructions and improve the efficiency and effectiveness of processes and systems.

In an LCA of a system involving several technological steps, the ith technological step is evaluated by its material and energy flows (e.g., fossil fuel consumption) and environmental impacts.

The exergy consumption rate corresponding to fossil fuel use can be evaluated with the following expression:

$$\dot{Ex}^i_{LFC} = \dot{Ex}^i_{dir} + \dot{Ex}^i_{dir} + \Delta\dot{Ex}^i_{ind} \tag{19.1}$$

where $\dot{Ex}^i_{LFC}$ is the life cycle fossil fuel exergy consumption rate, $\dot{Ex}^i_{dir}$ is the rate fuel exergy is directly transformed into final products, $\Delta\dot{Ex}^i_{dir}$ is the rate fuel exergy is consumed to perform the transformation, $\Delta\dot{Ex}^i_{ind}$ is the rate fuel exergy is consumed through being embodied in construction materials and equipment, and during installation, operation, maintenance, decommissioning, etc.

The difference between $\dot{Ex}^i_{dir}$ and $\Delta\dot{Ex}^i_{dir}$ can be explained by considering the example of natural gas reforming, which is often the first stage in large-scale manufacturing of ammonia, methanol and other synthetic fuels. The sum of the reactions to produce hydrogen through natural gas reforming is the following:

$$CH_4 + 2H_2O \xrightarrow{T\approx950°C} 4H_2 + CO_2 - 165\,kJ \tag{19.2}$$

As seen in the reaction in Eq. (19.2), which is endothermic, a flow of methane is directly converted to hydrogen. The reaction is driven by high-temperature heat, which is typically supplied by another flow of methane being combusted according to

$$CH_4 + O_2 \rightarrow CO_2 + 2H_2O + 802.6\,kJ \tag{19.3}$$

In this example, $\dot{Ex}_{dir}$ includes the exergy of the methane utilized in the reaction in Eq. (19.2) and $\Delta\dot{Ex}_{dir}$ includes the exergy of the methane employed in Eq. (19.3).

The standard exergies of most fuels are similar to their lower heating values (LHVs). The LHV is equal to the heat released by the complete burning of all fuel components to CO_2 and H_2O in the form of a vapor. The standard exergy of fuels (e.g., hydrogen, methane, gasoline) Ex^0_f is equal to the maximum work obtainable (or the work obtainable in an ideal fuel cell), and can be evaluated as the negative of the standard Gibbs free energy change ΔG^0 (at $p_o = 1$ atm, $T_o = 298$ K) for the fuel combustion reaction:

$$Ex^0_f = -\Delta G^0 = -(\Delta H^0 - T_0\Delta S^0) \tag{19.4}$$

Here, ΔH^0 and ΔS^0 are respectively the change of standard enthalpy and entropy in this reaction. For the standard exergy calculation H_2O can be considered a liquid or steam. The LHVs and standard chemical exergies, along with the ratios of standard chemical exergy to LHV, are presented for several fuels in Table 19.1. In this table, the resulting water is considered a vapor.

Table 19.1. Values of standard exergy and LHV and their ratio for different fuels.

Fuel	LHV (MJ/kg)	Standard exergy Ex_f^0 (MJ/kg)	$\dfrac{Ex_f^0}{\text{LHV}}$
Hydrogen	121.0	118.2	0.977
Natural gas	50.1	52.1	1.04
Conventional gasoline	43.7*	46.8	1.07
Conventional diesel	41.8*	44.7	1.07
Crude oil	42.8*	45.8	1.07

* From Wang (1999).

When electricity, hydrogen or other manufactured secondary energy carriers are the input exergy, the generally accepted efficiencies are usually applied to evaluate the direct input fossil fuel exergy rates $\dot{Ex}_{dir}^i$ and $\Delta\dot{Ex}_{dir}^i$. For use of electricity, for example, which is often generated from fossil fuels, $\Delta\dot{Ex}_{dir}^i$ is expressible as

$$\Delta\dot{Ex}_{dir}^i = \frac{\dot{W}_i}{\psi} \tag{19.5}$$

where ψ is the exergy efficiency for electricity generation from a fossil fuel.

If the fossil fuel exergy and LHV values are similar, the exergy and energy efficiencies for the processes of electricity, mechanical work and hydrogen generation do not differ significantly, where the energy efficiency is

$$\eta = \frac{\dot{W}_i}{\text{LHV}_i}$$

The indirect exergy $\Delta\dot{Ex}_{ind}$ cannot be treated as equal to the embodied exergy (i.e., the exergy required to produce a given material or device) or energy. The embodied exergy (energy) adequately reflects the environmental impact of the material extraction and material and device production stages, but it is inconsistent with the economic cost of these products. Note that construction materials are also produced from mineral sources (ores, limestone, etc.) which, like fossil fuels, have value; their exergy (energy) contents are much lower than their real economic values. To account for this, the exergy (energy) equivalent of construction materials and devices (Granovskii et al., 2006a, b) is calculated by dividing the cost of materials or devices utilized in a given technological stage by the cost of a unit of fossil fuel exergy (energy). Then, the indirect exergy consumption rate can be evaluated as

$$\Delta\dot{Ex}_{ind} = \frac{\sum\text{EEQ} + \text{EOP}}{\text{LFT}} \tag{19.6}$$

where $\sum$EEQ is the sum of the exergy equivalents of construction materials and devices related to a given technological operation, EOP is the operation exergy, i.e., the fossil fuel exergy required for installation, construction, operation, maintenance, decommissioning, etc. of equipment, and LFT is the lifetime of the unit performing a technological operation.

The variability of data, efficiencies, costs, etc. introduces some uncertainties into LCA and ExLCA, but they nonetheless are powerful tools for evaluating and comparing the exergy (energy) efficiencies of technological chains, including their construction and operating stages and environmental impacts.

19.4. Case study: exergetic life cycle analysis

An ExLCA is presented of four technologies (two using fossil fuels and two renewable energy forms) for producing gasoline and hydrogen and their use in internal combustion (gasoline) or fuel cell (hydrogen) vehicles. Life cycle exergy efficiencies, capital investment efficiency factors and environmental impacts are examined.

Although numerous LCAs of gasoline and hydrogen vehicles have been reported, the need to consider exergy and energy losses throughout the life cycle of fuels, starting from production and leading to utilization in a vehicle, have not been carefully considered. Such comprehensive assessments can help explain why renewable technologies for hydrogen production are economically less attractive than traditional ones.

The principal technological steps to produce gasoline from crude oil, and hydrogen from natural gas and renewable energy (solar and wind), are presented in Fig. 19.1. Gasoline is utilized in an internal combustion engine (ICE) and hydrogen in a fuel cell vehicle.

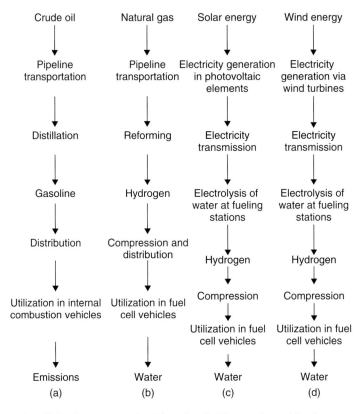

Fig. 19.1. Principal steps in utilizing in transportation (a) crude oil, (b) natural gas, (c) solar energy and (d) wind energy.

19.4.1. Natural gas and crude oil transport

To evaluate and compare the exergy consumption and environmental impact of transporting natural gas and crude oil by pipeline, equal lengths of pipelines (1000 km) are considered. Typical characteristics for transporting crude oil and natural gas via pipeline from several sources (Kirk and Othmer, 1998; Meier, 2002; Cleveland, 2004) are listed in Table 19.2. The energy values embodied in the materials and devices are evaluated and used to obtain the exergy values assuming that the only fossil fuel employed in their production is natural gas. The exergies embodied in the pipeline materials, compressors and pumps, and the exergy equivalents, are presented in Table 19.3. It is assumed that the operation exergy (EOP) to install, maintain and operate the equipment is equal to the embodied exergy to produce it.

The mechanical work or electricity required for pipeline transport is assumed produced by a gas turbine unit with an average exergy efficiency $\psi^{gt} = 0.33$ (Cleveland, 2004). This assumption permits evaluation of the direct exergy consumption rate. Table 19.4 lists the direct and indirect exergy consumption rates to transport an amount of natural gas and crude oil with an exergy flow rate $\dot{Ex}_{dir}$ of 1 MJ/s.

19.4.2. Natural gas reforming and crude oil distillation

The exergy losses in natural gas reforming, where natural gas is the only source of exergy input, comprise approximately 22% of the total exergy input of natural gas (Rosen, 1996). The exergy efficiency and environmental impact to produce

Table 19.2. Typical characteristics for crude oil and natural gas pipeline transportation.

Characteristic	Natural gas	Crude oil
Velocity in pipeline w (m/s)	7.0	2.0
Diameter of pipeline d (m)	0.8	0.4
Length of pipeline L (m)	1.0×10^6	1.0×10^6
Viscosity of crude oil μ (mPa s)	0.011	60.0
Efficiency of isothermal compressors (natural gas) and pumps (crude oil)	0.65	0.65
Maximum pressure in natural gas pipeline p_{max} (atm)	70.0	–
Minimum pressure in natural gas pipeline p_{min} (atm)	50.0	–
Exergy rate of input flow (MJ/s)	6914	90,849
Mass of pipeline (tons)	126,102	64,739
Embodied exergy in pipeline (GJ)	4,551,428	2,316,103
Lifetime of pipeline (years)	80	80
Embodied exergy in compressors and pumps (GJ)	1,574,277	4,174,729
Lifetime of pumps and compressors (years)	20	20

Source: Cleveland (2004); Kirk and Othmer (1998); Meier (2002).

Table 19.3. Embodied exergy, exergy equivalent (EEQ) and operation exergy (EOP) for natural gas and crude oil pipeline transportation.

Materials and equipment	Embodied exergy per second of lifetime (MJ/s)	Exergy equivalent per second of lifetime (MJ/s)	Operation exergy per second of lifetime (MJ/s)
Natural gas pipeline	1.79	11.3	n/a
Natural gas compressors	2.50	77.4	n/a
Total	4.29	88.7	4.29
Crude oil pipeline	0.92	5.82	n/a
Crude oil pumps	6.62	205.4	n/a
Total	7.54	211.2	7.54

Table 19.4. Indirect and direct exergy consumption rate to transport a quantity of natural gas and crude oil with an exergy content of 1 MJ/s.

Transportation	$\Delta \dot{E}x_{dir}$ (kJ/s)	$\Delta \dot{E}x_{ind}$ (kJ/s)
Natural gas pipeline transportation	95.3	13.0
Crude oil pipeline transportation	16.2	2.2

Table 19.5. Hydrogen plant material requirements (base case).*

Material	Quantity required (tons)	Embodied exergy (GJ/ton)	Embodied exergy consumption per second of lifetime (MJ/s)	Exergy equivalent per second of lifetime (MJ/s)	Operation exergy per second of lifetime (MJ/s)	Indirect exergy rate $\Delta \dot{E}x_{ind}$ (MJ/s)
Concrete	10,242	1.5	0.0236	0.361	n/a	n/a
Steel	3272	35.8	0.185	1.180	n/a	n/a
Aluminum	27	209.5	0.00896	0.0581	n/a	n/a
Iron	40	24.4	0.00155	0.00986	n/a	n/a
Total	13,581	271.1	0.219	1.600	0.219	1.83

* Assumes a 20-year lifetime, a 1.5 million Nm^3/day hydrogen production capacity and a hydrogen exergy production rate of 183.8 MJ/s.

Table 19.6. Total rate of direct exergy consumption (in MJ/s of fuel exergy produced) for natural gas and crude oil transportation and reforming (distillation) processes.

Fuel	$\Delta \dot{E}x_{dir}$	$\Delta \dot{E}x_{ind}$
Hydrogen	0.391	0.025
Gasoline	0.168	n/a

1 MJ of exergy of gasoline have been estimated according to the energy consumption of all petroleum refineries in the US in 1996 (Energetics, 1998). The overall direct exergy rates in the reforming and transportation stages are presented in Table 19.6 (column 2).

The indirect exergy for natural gas reforming is based on data of Spath and Mann (2001). In Table 19.5, the material requirements of a natural gas reforming plant are presented. The values of the energy embodied in materials, taken from Spath and Mann (2001), have been used to obtain the values of embodied exergies assuming that embodied energy relates to the LHV of natural gas. In Table 19.6 (column 3), the resulting indirect exergy values are presented. It has been assumed that the operation exergy to install, maintain and operate equipment is equal to the embodied exergy consumed to produce it. Comparing the values of direct and indirect exergies reveals that the indirect exergy rate ($\Delta \dot{E}x_{ind}$) is more than 10 times less than the direct exergy rate ($\Delta \dot{E}x_{dir}$). In the following calculations, therefore, the indirect exergy consumption rate is neglected.

Data to calculate the indirect exergy consumption for crude oil refining are not available. However, as shown by Lange and Tijm (1996), the capital cost of crude oil distillation is lower than that for natural gas reforming. As in the case of natural gas reforming, the indirect exergy consumption for crude oil refining is negligible compared to the direct exergy consumption.

19.4.3. Hydrogen production from renewable energy

Hydrogen production via wind energy

The system considered here for producing hydrogen from wind energy involves two main devices: a wind turbine that produces electricity which in turn drives a water electrolysis unit that produces hydrogen. Wind energy is converted to mechanical work by wind turbines and then transformed by an alternator to alternating current (AC) electricity which is transmitted to the power grid (Fig. 19.1). The efficiency of wind turbines depends on location, with wind energy applications normally making sense only in areas with high wind activity. Data for a 6 MW wind power generation plant (White and Kulcinski, 2000) are used. Table 19.7 presents the material requirements and indirect exergy consumption for this plant.

Table 19.7. Material requirements and corresponding rate of indirect energy consumption $\Delta \dot{E}x_{ind}$ for a 6 MW wind power generation plant coupled with an electrolyzer to produce hydrogen.

Materials and processes	Quantity required (tons)	Embodied exergy (GJ/ton)	Embodied exergy consumption per second of lifetime (MJ/s)	Exergy equivalent per second of lifetime (MJ/s)	Operation exergy per second of lifetime (MJ/s)	Indirect exergy rate $\Delta \dot{E}x_{ind}$ (MJ/s)
Concrete	7647.3	1.46	0.0141	0.216	n/a	n/a
Copper	5.275	136	0.000911	0.0119	n/a	n/a
Fiberglass	496.6	13.5	0.00851	0.122	n/a	n/a
Steel-carbon/low alloy	1888.0	35.8	0.0857	0.545	n/a	n/a
Steel-stainless	226.2	55.1	0.0158	0.101	n/a	n/a
Total	10,263.4	n/a	n/a	0.994	0.136	1.130
Electrolysis	n/a	n/a	n/a	n/a	n/a	0.075
Total	n/a	n/a	n/a	n/a	n/a	1.21
Total for 1 MJ/s of hydrogen exergy	n/a	n/a	n/a	n/a	n/a	0.301

Based on data of Spath and Mann (2004), for electrolysis to produce hydrogen with a 72% efficiency (on an exergy basis), the indirect exergy (energy) is 6.61% of that for a wind power generation plant. Accounting for 7% electricity loss during transmission, the efficiency of hydrogen production is 66.9%. Thus, a 6 MW wind power plant combined with water electrolysis can produce 3.93 MJ/s of exergy in the form of hydrogen. With these data, the indirect energy consumption rate in a wind power plant coupled with water electrolysis for 1 MJ of exergy in the form of hydrogen is evaluated (see Table 19.7, column 7).

Hydrogen production via solar energy

The production of hydrogen using solar energy considered here involves two main systems: a solar photovoltaic system that produces electricity which in turn drives a water electrolysis unit that produces hydrogen. The photovoltaic elements convert solar energy into direct current (DC) electricity, which is transformed by inverters to AC electricity and transmitted to the power grid. At fueling stations, AC electricity is used to electrolyze water to produce hydrogen (Fig. 19.1). Data are considered here for a 1.231 kW building-integrated photovoltaic system in Silverthorne, Colorado (Meier, 2002), which utilizes thin-film amorphous silicon technology and for which indirect exergy consumption has been evaluated. Tables 19.8 and 19.9 present the material requirements and indirect exergy consumption for hydrogen production by photovoltaic power generation and water electrolysis. A procedure similar to that used for the wind power plant in the previous section is applied to evaluate the indirect exergy consumption associated with electrolysis. Taking into account the efficiency of electrolysis and transmission losses, the 1.231 kW photovoltaic system combined with water electrolysis can produce 807.3 J/s of hydrogen exergy.

19.4.4. Hydrogen compression

The density of hydrogen at standard conditions is low. To assist in storage and utilization as a fuel, its density is often increased via compression. Neglecting the indirect exergy consumption ΔEx_{ind}, the total and direct fossil fuel (natural gas) exergy consumption ΔEx_{dir}^{cmp} to compress isothermally 1 mol of hydrogen can be expressed, assuming ideal gas behavior, as

$$\Delta Ex_{dir}^{cmp} = \frac{RT_0}{\eta^{cmp}\psi^{gt}} \ln\left(\frac{p_{max}}{p_{min}}\right) \tag{19.7}$$

Table 19.8. Exergy equivalents for thin film photovoltaic solar cell block with 157.2 m² of surface area in a 1.231 kW photovoltaic system.

Material	Embodied exergy (MJ/m²)	Exergy equivalent (MJ/m²)	Embodied exergy in manufacturing (MJ/m²)	Exergy equivalent (GJ/unit)
Encapsulation	0.220×10^3	3.472×10^3	0.143×10^3	568.12
Substrate	0.0266×10^3	0.170×10^3	0.0587×10^3	35.84
Deposition materials	0.0196×10^3	0.308×10^3	0.0962×10^3	63.53
Busbar	0.00530×10^3	0.0835×10^3	0	13.14
Back reflector	0.000728×10^3	0.0114×10^3	0.0770×10^3	13.90
Grid	n/a	n/a	0.0356×10^3	5.60
Conductive oxide	n/a	n/a	0.101×10^3	15.84
Total for solar cell block	0.272×10^3	4.044×10^3	0.511×10^3	716.0

Table 19.9. Indirect exergy consumption rate for the units of a 1.231 kW thin film photovoltaic system with a lifetime of 30 years.

Unit	Embodied exergy (GJ/unit)	Exergy equivalent (GJ/unit)	Operation exergy per second of lifetime (J/s)	Indirect exergy rate $\Delta \dot{Ex}_{ind}$ (J/s)
Inverters	41.6	115.9	n/a	n/a
Wiring	3.02	48.6	n/a	n/a
Solar cell block	123.1	716.0	n/a	n/a
Total	167.8	880.4	82.12	1012.7
Electrolysis	n/a	n/a	n/a	67.0
Total per unit	n/a	n/a	n/a	1079.6
Total for 1 MJ/s of hydrogen exergy	n/a	n/a	n/a	1337.8

where $T_0 = 298$ K is the standard environmental temperature and $R = 8.314$ kJ/mol K is the universal gas constant. Also, η^{cmp} denotes the isothermal compression efficiency and ψ^{gt} the gas-turbine power plant efficiency.

The direct exergy consumed in compressing hydrogen is shown in Table 19.10 and is evaluated assuming an isothermal compression efficiency η_{cmp} of 0.65 and a typical gas-turbine power plant exergy efficiency ψ^{gt} of 0.33. A maximum pressure $p_{max} = 350$ atm in the tank of the fuel cell vehicle is considered (Wilson, 2002). Minimum pressures before compression of $p_{min} = 1$ atm and $p_{min} = 20$ atm are taken for hydrogen production via electrolysis and natural gas reforming (Spath and Mann, 2001), respectively.

19.4.5. Hydrogen and gasoline distribution

Hydrogen distribution is replaced by electricity distribution in cases using wind and solar energy (Fig. 19.1) and such distribution has been accounted for in hydrogen production. The distribution of compressed hydrogen after its production via natural gas reforming is similar to that for liquid gasoline, but compressed hydrogen is characterized by a lower volumetric energy capacity and higher material requirements for a hydrogen tank.

Table 19.10. Direct exergy consumption rates for 1 MJ of chemical exergy of hydrogen and gasoline for compression $\Delta \dot{Ex}_{dir}^{cmp}$ and distribution to refueling stations $\Delta \dot{Ex}_{dir}^{distr}$.

Energy carriers	$\Delta \dot{Ex}_{dir}^{cmp}$ (MJ/s)	$\Delta \dot{Ex}_{dir}^{distr}$ (MJ/s)
Hydrogen from natural gas, $p_{min} = 20$ atm, $p_{max} = 350$ atm	0.144	0.025
Hydrogen from wind energy, $p_{min} = 1$ atm, $p_{max} = 350$ atm	0.289	n/a
Hydrogen from solar energy, $p_{min} = 1$ atm, $p_{max} = 350$ atm	0.289	n/a
Gasoline	n/a	0.0025

According to the 1997 Vehicle Inventory and Use Survey, the average 'heavy–heavy' truck in the US traveled 6.1 miles per gallon of diesel fuel (Charles River Associates, 2000). Neglecting the indirect exergy consumption rate $\Delta \dot{Ex}_{ind}^{distr}$, the total and direct fuel (diesel) exergy consumption rate $\Delta \dot{Ex}_{dir}^{distr}$ is evaluated assuming a distance of 300 km is traveled before refueling for a truck with a 50 m^3 tank (see Table 19.10).

19.4.6. Life cycle exergy efficiencies

The overall results of the LCAs are summarized in Table 19.11. The life cycle exergy efficiency of fossil fuel and mineral resource utilization is defined as

$$\psi_{H_2}^{LFC} = \frac{\dot{Ex}_{H_2}}{\dot{Ex}_{LFC}^{H_2}} \tag{19.8}$$

for hydrogen production technologies and

$$\psi_{g}^{LFC} = \frac{\dot{Ex}_{g}}{\dot{Ex}_{LFC}^{g}} \tag{19.9}$$

for gasoline production from crude oil. Here, $\dot{Ex}_{H_2}$ and $\dot{Ex}_g$ are the exergies of hydrogen and gasoline, and $\dot{Ex}_{LFC}^{H_2}$ and $\dot{Ex}_{LFC}^{g}$ are the overall life cycle fossil fuel and mineral exergy consumption rates to produce hydrogen and gasoline, respectively.

Table 19.11. LCA of the exergy efficiency of fossil fuel and mineral resource utilization to produce 1 MJ/s of chemical exergy of hydrogen and gasoline.

Energy carriers	$\dot{Ex}_{dir}$ (MJ/s)	$\sum \dot{Ex}_{dir}$ (MJ/s)	$\sum \Delta \dot{Ex}_{ind}$ (MJ/s)	Total $\dot{Ex}_{LFC}$ (MJ/s)	ψ^{LFC}
Hydrogen from natural gas, $p = 350$ atm	1	0.560	n/a*	1.560	0.64
Hydrogen from wind energy, $p = 350$ atm	n/a	0.289	0.301	0.590	1.69
Hydrogen from solar energy, $p = 350$ atm	n/a	0.289	1.338	1.627	0.62
Gasoline	1	0.171	n/a*	1.171	0.85

*For fossil fuel technologies, the indirect exergy consumption rate is considered negligible relative to the direct exergy consumption rate.

The LCA of the exergy efficiency of fossil fuel and mineral resource utilization to produce compressed hydrogen from wind energy ψ^{LFC} reaches 1.69, meaning that the consumed fossil fuel exergy (embodied in materials, equipment, etc.) is 1.69 times less than the exergy of the hydrogen produced. A value of ψ^{LFC} greater than 1 occurs because the exergy of wind is considered 'free' and is not included in the expression for ψ^{LFC}. This value should not be confused with the exergy efficiencies of wind power generation plants, which are about 12–25% and usually calculated as the ratio of electricity produced to the sum of all sources of input exergy (mainly kinetic exergy of wind).

The life cycle exergy efficiency to produce hydrogen from solar energy also accounts for solar energy being 'free,' but in this case ψ^{LFC} is less than 1 because valuable materials are employed in photovoltaic solar cells, rendering the indirect fossil fuel and mineral exergy consumption high.

The chemical exergies of gasoline and hydrogen are converted to work with different efficiencies in an ICE vehicle and a proton exchange membrane (PEM) fuel cell vehicle. The efficiency ranges from 0.2 to 0.3 for an ICE (Cleveland, 2004) and from 0.4 to 0.6 for a fuel cell engine (Larminie and Dicks, 2003). The efficiency of fossil fuel energy consumption in a vehicle ψ^{VCL} can be expressed as the product of the life cycle ψ^{LFC} and engine ψ^{eng} efficiencies:

$$\psi^{\text{VCL}} = \psi^{\text{LFC}} \psi^{\text{eng}} \tag{19.10}$$

Figure 19.2 shows the mechanical work produced per unit of life cycle fossil fuel exergy consumption as a function of engine efficiency. Note that the curves for hydrogen from natural gas and solar energy coincide in this scale. This figure indicates that the efficiency of a fuel cell vehicle operating on hydrogen from natural gas must be at least 25–30% greater than that for an internal combustion gasoline engine to be competitive. The application of hydrogen from wind energy in a fuel cell vehicle is extremely efficient with respect to fossil and mineral resource utilization.

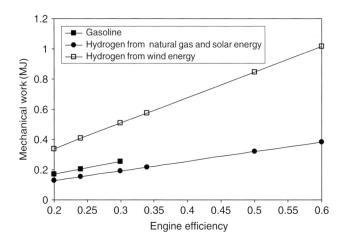

Fig. 19.2. Mechanical work per exergy of fossil fuels consumed to produce 1 MJ of exergy of gasoline and hydrogen as a function of engine efficiency for internal combustion (gasoline) and fuel cell (hydrogen) engines.

19.5. Economic implications of ExLCA

Fossil fuel and renewable energy technologies for hydrogen production are generally distinguished by (1) source of energy consumed, (2) efficiency of hydrogen production per unit of energy consumed and (3) capital investments made per unit of hydrogen produced. To account for these factors, we can utilize a quantity called the capital investment efficiency factor γ as a measure of economic efficiency (Granovskii et al., 2006a, b). This indicator is proportional to the relationship between gain and investment and is expressible as

$$\gamma = \frac{\dot{E}x_{\text{H}_2}(\alpha - 1/\psi^{\text{LFC}})}{\Delta \dot{E}x_{\text{ind}}} \tag{19.11}$$

Here, the numerator is proportional to the gain from the exploitation of a technology and the denominator to the investments made in it. Also, α denotes the ratio in costs of hydrogen (C_{H_2}) and natural gas (C_{ng}):

$$\alpha = \frac{C_{\text{H}_2}}{C_{\text{ng}}} \tag{19.12}$$

Furthermore, $\dot{Ex}_{H_2}$ is the capacity of hydrogen production, expressed in units of exergy of hydrogen per unit time, $\Delta \dot{Ex}_{ind}$ is the indirect exergy rate which is proportional to the capital investments in a technology and ψ^{LFC} is the life cycle exergy efficiency of fossil fuel and mineral resource utilization (Eq. 19.8). The input solar and wind energies do not have any direct cost, so they are not included in the denominator of Eq. (19.8) for renewable technologies. As a result, the value of ψ^{LFC} for renewable technologies can exceed 1.

Technologies for hydrogen production via wind and solar energy, although increasing, are not yet widespread due to economic challenges. Figure 19.3 presents the capital investment efficiency factor γ, as a function of the cost ratio α for hydrogen and natural gas, for life cycle exergy efficiencies of $\psi^{LFC} = 0.72$ for hydrogen via natural gas, $\psi^{LFC} = 3.32$ for hydrogen via wind energy and $\psi^{LFC} = 0.75$ for hydrogen via solar energy. Here, the compression stages are excluded for all technologies, and the distribution stage is excluded for hydrogen via natural gas. Since the cost of 1 MJ of hydrogen exergy is presently about two times more than that of natural gas (Padro and Putsche, 1999), it follows from Fig. 19.3 at $\alpha = 2$ that the capital investment efficiency factor for hydrogen production via natural gas is about five times higher than that to produce hydrogen via wind energy. This situation can be altered by reducing the construction material requirements of wind per unit of electricity generated. A fair assessment when comparing different renewable technologies requires consideration of both energy efficiency (ability to convert renewable energy to mechanical work or electricity) and the efficiency of construction materials and equipment exploitation.

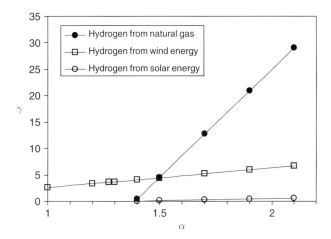

Fig. 19.3. Capital investment efficiency factor γ for several hydrogen production technologies as a function of the cost ratio α for hydrogen and natural gas.

19.6. LCA and environmental impact

Some applications of LCA to assess and improve environmental impact are described.

19.6.1. Power generation and transportation

Expansions of modern power generation and transportation systems should account simultaneously for economic growth and environmental impact. The latter corresponds mostly to the combustion of hydrocarbon fuels and the accompanying emissions of large quantities of greenhouse gases (GHGs) and air pollutants (see Fig. 19.4).

The main cause of global climate change is generally accepted to be increasing emissions of GHGs as a result of increased use of fossil fuels (Wuebbles and Atul, 2001). The effects of other emissions to air are significant as well. Nitrogen oxides in combination with volatile organic compounds (VOCs) cause the formation of ground-level ozone and smog, exposure to which can lead to eye irritation and a decrease in lung function. Elevated levels of ozone can also cause lung and respiratory disorders and noticeable leaf damage in many crops, plants and trees. NO_x and VOCs react in the presence of sunlight to produce ozone. CO emissions impact the ability of red blood cells to transport oxygen to body tissues (e.g., EC, 2005). Numerous other environmental impacts are associated with emissions of NO_x, VOCs and CO (Dincer, 2002; Rosen, 2002, 2004).

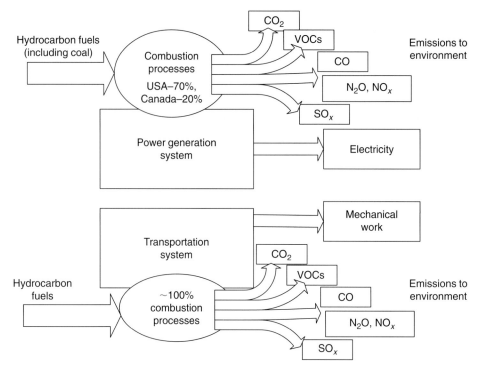

Fig. 19.4. Environmental impact of modern power generation and transportation systems.

Rising concerns about the effects of global warming, air pollution (AP) and declining fossil fuel stocks have led to increased interest in renewable energy sources such as wind and solar energies. An environmentally improved scheme for power generation and transportation systems based on renewable technologies and hydrogen is presented in Fig. 19.5.

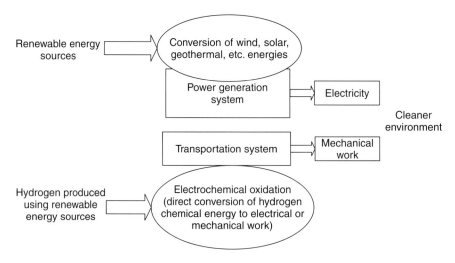

Fig. 19.5. An environmentally improved scheme for power generation and transportation systems.

The prospects for generating electricity, hydrogen or synthetic fuels by employing only renewable energy sources are good. In some ways, electricity generation technologies including wind turbines and photovoltaic cells are as developed as hydrogen production via water electrolysis. Pure hydrogen can be used as a fuel for fuel cell vehicles, which are rapidly improving, or converted into synthetic liquid fuels by means of such processes as Fischer–Tropsch reactions (Dry, 1999).

Maack and Skulason (2006) report that in an Icelandic community the use of renewable energy and tests with a clean domestic fuel (often referred to as the fuel of the future) have become points of focus. Hydrogen is used currently as an energy carrier in the public transportation system and is electrolyzed from water using hydroelectric power. The exhaust is water. Icelandic New Energy Ltd. has been working on projects related to hydrogen as an energy carrier since 1999, while a number of projects and feasibility studies are currently being carried out in Reykjavik on producing hydrogen domestically from water and hydroelectric and geothermal power, abundant local resources.

The use of hydrogen as a fuel for fuel cell vehicles can lead to significant improvements in AP and GHG emissions. In a fuel cell stack, electricity (which is converted to mechanical work in electrical motors with efficiencies higher than 90%) is generated via the following electrochemical reactions:

$$\text{Anode: } 2H_2 \rightarrow 4H^+ + 4e^-$$

$$\text{Cathode: } O_2 + 4H^+ + 4e^- \rightarrow 2H_2O \tag{19.13}$$

These reactions occur in a PEM fuel cell stack at low temperature ($<100°C$) and involve separation of oxygen from air at the cathode. At these conditions the formation of harmful nitrogen oxides is inhibited and only water is produced during power generation. Thus, the utilization of hydrogen in fuel cell vehicles can be considered as ecologically benign, regarding direct vehicle emissions. Any associated emissions of pollutants and GHGs are associated with hydrogen production.

AP and GHG emissions are associated with gasoline production and its utilization in ICE vehicles. In such vehicles, gasoline (a mixture of hydrocarbons) is combusted in air. The combustion reaction can be expressed for a general hydrocarbon C_nH_m as

$$C_nH_m + (n + m/4)O_2 \rightarrow nCO_2 + m/2H_2O + Q \tag{19.14}$$

The heat Q released during this exothermic reaction is in part converted to mechanical work. According to the Carnot principle, the higher is the temperature of fuel combustion the greater is the mechanical work that can be extracted theoretically. The average temperature of the combusting mixture of gasoline and air is about 1300°C. At such high temperatures the formation of nitrogen oxides is promoted. Evaporation of gasoline and incomplete combustion lead to emissions of VOCs and carbon monoxide.

In previous sections, ExLCA of wind and solar technologies for electricity and hydrogen generation, as well as of hydrogen production from natural gas and gasoline from crude oil, are described. By introducing a capital investment efficiency factor it is shown that 'renewable' hydrogen is economically less attractive (i.e., it has a higher cost) than hydrogen produced via reforming of natural gas.

19.6.2. Environmental-impact reduction by substitution of renewables for fossil fuels

We consider the reduction of environmental impact related to the introduction of wind- and solar-based technologies. The direct and indirect fossil fuel exergy consumptions $\dot{Ex}_{dir}$, $\Delta\dot{Ex}_{dir}$ and $\Delta\dot{Ex}_{ind}$ lead to different kinds of harmful emissions, which are divided in this section into GHG and AP emissions. A GHG indicator can be used to assess GHGs according to the values of their global warming potentials. Airborne pollutants are analogously combined into a generalized indicator of AP in line with their impact weighting coefficients (relative to NO_x) as follows:

$$AP = \sum_1^3 m_i w_i \tag{19.15}$$

where m_i is the mass of air pollutant i and w_i is the corresponding weighting coefficient. For simplicity, we consider here only three pollutants (CO, NO_x, VOCs). Weighting coefficients are used from the Australian Environment Protection Authority (Beer et al., 2006), obtained using cost–benefit analyses of health effects. The weighting coefficients for GHGs, based on global warming potentials relative to carbon dioxide which is assigned a value of unity, and air pollutants are listed in Table 19.12.

Although wind and solar energy can be considered 'free,' the quantity of construction materials consumed per unit of electricity or hydrogen produced for a 'renewable' plant is often much higher than that for more traditional technologies for electricity and hydrogen production from natural gas. Taking into account AP emissions from the construction and operation stages of power or hydrogen generation plants, and their lifetimes and capacities, the indirect GHG and AP

Table 19.12. Weighting coefficients for GHGs and airborne pollutants.

Emission	Compound	Weighting coefficient
GHGs	CO_2	1
	CH_4	21
	N_2O	310
Airborne pollutants	CO	0.017
	NO_x	1
	VOCs	0.64

emissions per unit of produced energy can be calculated. For fossil fuel technologies, these indirect life cycle emissions are small with respect to the direct emissions related to fuel combustion or removing carbon from natural gas to produce hydrogen.

Assuming that embodied energy is related to the natural gas combustion energy, GHG and AP emissions per MJ of produced electricity, hydrogen and gasoline from previous LCA studies (Granovskii et al., 2006a, b, c, d) are presented (Table 19.13). The GHG and AP emissions from producing a unit of electricity from natural gas are calculated assuming that electricity is generated from natural gas with an average efficiency of 40% (which is reasonable since the efficiency of electricity production from natural gas varies from 33% for gas turbine units to 55% for combined-cycle power plants, with about 7% of the electricity dissipated during transmission).

To transmit hydrogen or use it in a fuel cell vehicle, it needs to be compressed to an appropriate volumetric energy density. For instance, the pressure of gaseous hydrogen in the tank of Honda's fuel cell car is about 350 atm (Wilson, 2002). Data regarding hydrogen compression in Table 19.2 have been obtained assuming that electricity for 'renewable' hydrogen compression is derived from the same renewable energy sources and electricity for compression of hydrogen from natural gas is generated in a natural gas power plant.

The electrical energy E_{el} required to compress 1 mol of hydrogen is calculated according to the formula for isothermal compression with a compressor efficiency $\eta_{cmp} = 0.65$:

$$E_{el} = \frac{RT_0}{\eta_{cmp}} \ln\left(\frac{p_{max}}{p_{atm}}\right) \tag{19.16}$$

where the reference-environment temperature is $T_0 = 298$ K, R is the universal gas constant, p_{max} is the required pressure of hydrogen and the atmospheric pressure is $p_{atm} = 1$ atm. In Table 19.13, the environmental impact of hydrogen compression using renewable-based electricity is seen to be very small compared to that for the stages of electricity production and electrolysis.

The improvement in environmental impact (i.e., reduced GHG and AP emissions in the present case) as a result of introducing a renewable technology depends on the replaced technology. The efficiency of such an introduction can be determined as the cost of GHG and AP emissions reduction per unit mass (C_{GHG} and C_{AP}), with the following expressions:

$$C_{GHG} = \frac{1000}{GHG_{ng} - GHG_R}(C_R - C_{ng}) \tag{19.17}$$

$$C_{AP} = \frac{1000}{AP_{ng} - AP_R}(C_R - C_{ng}) \tag{19.18}$$

where GHG_{ng}, GHG_R, AP_{ng} and AP_R are GHG and AP emissions (in grams per MJ of electricity or energy of hydrogen) produced using natural gas and renewable technologies, respectively, and C_{ng} and C_R are the costs per MJ of electricity or hydrogen produced using natural gas and renewable technologies, respectively.

Figure 19.6 shows the costs of the major energy carriers (per MJ of electricity or LHV) for 1999–2004 based on the data taken from the EIA (2005). The contemporary cost of fossil fuel-based electricity assumes that the electricity cost in Fig. 19.6 is consistent with its generation from natural gas with an average efficiency of 40% (as assumed in the environmental-impact evaluation). Data are not widely available for the cost of hydrogen, but according to one analysis

Table 19.13. GHG and AP emissions (in g/MJ of electricity or LHV of hydrogen and gasoline) for various production technologies.

Technology	m_{GHG}	m_{CO}	m_{NOx}	m_{VOC}	AP
Electricity from natural gas					
Electricity from natural gas with a thermal efficiency $\eta = 40\%$	149.9	0.094	0.11	0.72	0.57
Hydrogen from natural gas					
Natural gas pipeline transportation and reforming to produce hydrogen at pressure $p = 20$ atm[a]	75.7	0.022	0.026	0.054	0.061
Hydrogen compression from 20 to 350 atm	6.8	0.0042	0.0050	0.032	0.026
Hydrogen delivery to fueling stations ($p = 350$ atm)	3.1	0.0072	0.045	0.00135	
Total for $p = 350$ atm		0.026	0.031	0.086	0.087
Electricity and hydrogen from wind energy					
Electricity generation	4.34	0.0030	0.0035	0.00027	0.0038
Hydrogen production via electrolysis	2.51	0.0017	0.0020	0.000159	0.0022
Hydrogen compression to $p = 20$ atm	0.20	0.00014	0.00017	1.3×10^{-5}	0.00018
Hydrogen compression to $p = 350$ atm	0.40	0.00027	0.00033	2.54×10^{-5}	0.00035
Total for $p = 20$ atm	7.05	0.0048	0.0057	0.00044	0.0062
Total for $p = 350$ atm	7.25	0.0050	0.0058	0.00045	0.0063
Electricity and hydrogen from solar energy					
Electricity generation	10.7	0.0073	0.0087	0.00068	0.0092
Hydrogen production via electrolysis	6.18	0.0042	0.0050	0.00039	0.0053
Hydrogen compression to $p = 20$ atm	0.50	0.00034	0.00041	3.19×10^{-5}	0.00044
Hydrogen compression to $p = 350$ atm	1.0	0.00067	0.00080	6.23×10^{-5}	0.00085
Total for $p = 20$ atm	17.4	0.012	0.014	0.0011	0.015
Total for $p = 350$ atm	17.9	0.012	0.015	0.0011	0.015
Gasoline from crude oil					
Crude oil pipeline transportation and distillation to produce gasoline	12.1	0.012	0.061	0.023	0.015
Gasoline delivery to fuelling stations	0.19	0.00044	0.0028	8.26×10^{-5}	0.11
Gasoline utilization in ICE vehicles[b]	71.7	0.86	0.05	0.15	0.11
Total	84.0	0.87	0.11	0.17	0.24

[a] Hydrogen is produced by natural gas reforming at a typical pressure of 20 atm.
[b] From Walwijk et al. (1999).

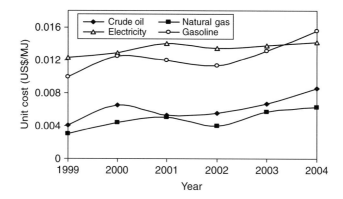

Fig. 19.6. Unit costs of selected energy carriers from 1999 to 2004. Based on the data from EIA (2005).

(Padro and Putsche, 1999), the ratio of the cost of hydrogen to its LHV (or exergy value) is about two times that of natural gas. In Fig. 19.6, the cost of gasoline is observed to be about two times that of crude oil. The efficiency of producing gasoline from crude oil is slightly higher than that for hydrogen from natural gas (Granovskii et al., 2006a). As the relative cost of natural gas is slightly lower than that of crude oil (see Fig. 19.6), we assume here that the ratio of the cost to exergy of hydrogen produced by natural gas reforming at a typical pressure (e.g., 20 atm) is equal to that of gasoline. The average unit costs of natural gas, crude oil, gasoline, hydrogen and electricity for 1999–2004 that are employed here are listed in Table 19.14.

Table 19.14. Average unit costs (in US$ per MJ of energy) of several energy carriers for 1999–2004.*

Energy carrier	Average unit cost (US$/MJ)
Natural gas	0.00473
Crude oil	0.00611
Gasoline	0.0124
Electricity	0.0134
Hydrogen (at 20 atm)	0.0124

* Based in part on data in Fig. 19.6.

Figures 19.7 and 19.8 present the cost (per kg) of reducing GHG and AP emissions, as a result of the substitution of wind and solar energies for natural gas to produce electricity and compressed hydrogen, as a function of the ratio of the costs of electricity:

$$\beta_w = \frac{EL_w}{EL_{ng}}; \quad \beta_s = \frac{EL_s}{EL_{ng}} \tag{19.19}$$

Here, β_w and β_s are the ratios in costs of electricity produced from wind and solar energy sources to the costs of electricity produced from natural gas respectively, EL_w and EL_s are the costs of electricity generated from wind and solar energy sources respectively, and EL_{ng} is the cost of electricity produced from natural gas. The cost of natural gas-derived electricity is assumed to be equal to the cost of electricity in Fig. 19.6. For the exergy efficiency calculation for 'renewable' hydrogen production, the cost of hydrogen is determined in line with the efficiency of low-temperature water electrolysis (e.g., 72%).

A comparison of Figs. 19.7 and 19.8 shows that wind-derived electricity allows less expensive abatement of GHG and AP emissions. Replacement of natural gas-derived electricity by renewable-derived electricity is more favorable than the same replacement for hydrogen. Elevating pressure favors renewable technologies because the cost of hydrogen includes the cost of electricity required for its compression. The range of contemporary ratios between production costs

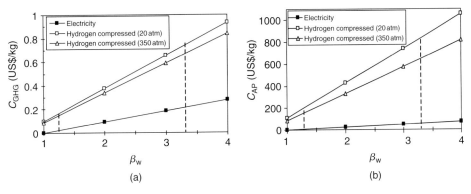

Fig. 19.7. Unit cost of (a) GHG and (b) AP emissions reduction as a result of wind energy substitution for natural gas to produce electricity and compressed hydrogen, as a function of the ratio in electricity costs β_w. The range of present ratios between production costs of 'wind' and 'natural gas' electricity is shown by dashed lines. Based on data from Newton and Hopewell (2002).

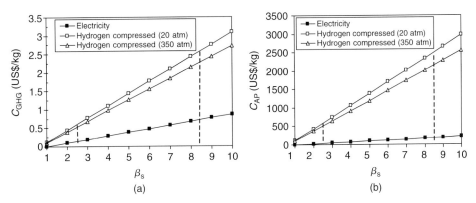

Fig. 19.8. Unit cost of (a) GHG and (b) AP emissions reduction as a result of solar energy substitution for natural gas to produce electricity and compressed hydrogen, as a function of the ratio in electricity costs β_s. The range of present ratios between production costs of 'solar' and 'natural gas' electricity is shown by dashed lines. Based on data from Newton and Hopewell (2002).

of renewable- and natural gas-based electricity are shown in Figs. 19.7 and 19.8 by dashed lines, based on data of Newton and Hopewell (2002).

The cost of reducing AP emissions by introducing hydrogen as a fuel for a fuel cell vehicle instead of gasoline is evaluated using average cost values for wind and solar electricity of $\beta_w = 2.25$ and $\beta_s = 5.25$. For this evaluation, Eqs. (19.17) and (19.18) have been modified as follows:

$$C_{GHG} = \frac{1000}{GHG_g - \dfrac{GHG_{H_2}}{\varepsilon}} \left(C_g - \frac{C_{H_2}}{\varepsilon} \right) \tag{19.20}$$

$$C_{AP} = \frac{1000}{AP_g - \dfrac{AP_{H_2}}{\varepsilon}} \left(C_g - \frac{C_{H_2}}{\varepsilon} \right) \tag{19.21}$$

where ε is the ratio in efficiencies of fuel cell and internal combustion vehicles; C_g, GHG_g and AP_g are the cost per MJ of gasoline and the corresponding GHG and AP emissions; and C_{H_2}, GHG_{H_2} and AP_{H_2} are the cost per MJ of compressed (350 atm) hydrogen and the corresponding GHG and AP emissions.

The cost of reducing GHG and AP emissions (per kg) as a result of gasoline substitution with hydrogen is presented in Fig. 19.9 as a function of the ratio in efficiencies ε of fuel cell (hydrogen powered) and internal combustion (gasoline

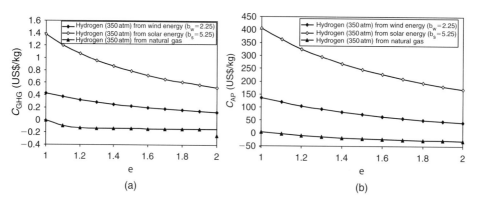

Fig. 19.9. Unit cost of (a) GHG and (b) AP emissions reduction as a result of hydrogen substitution for gasoline, as a function of the ratio in efficiencies ε of internal combustion (gasoline powered) and fuel cell (hydrogen powered) vehicles.

powered) vehicles. It can be seen that when 'renewable' hydrogen is used instead of gasoline, the cost of AP emissions abatement approaches that for 'renewable' electricity only if the efficiency of the fuel cell vehicle exceeds significantly (about two times) that of an ICE. On the contrary, the low positive and negative values for hydrogen from natural gas point out that its application in fuel cell vehicles allows a reduction in AP emissions almost without any financial expenditure connected to the fuel production technology. Present efficiencies (mechanical work per exergy of fuels) of an ICE and a fuel cell engine, respectively, average about 0.25 (Cleveland, 2004) and 0.35 (Larminie and Dicks, 2003). The latter value is evaluated as the product of the efficiencies of a fuel cell stack (about 0.4) and a device for electrical energy conversion into mechanical work (about 0.9).

The respective reductions of GHG and AP emissions from gasoline substitution with hydrogen (GHG_g/GHG_{H_2} and AP_g/AP_{H_2}) are presented as a function of ε in Fig. 19.10. 'Renewable' hydrogen substitution for gasoline is observed to lead to:

- A reduction in GHG emissions of more than 5 times (from 12 to 23 times for hydrogen derived from wind and from 5 to 8 times for hydrogen derived from solar energy).
- A reduction in AP of more than 10 times (from 38 to 76 times for hydrogen derived from wind and from 16 to 32 times for hydrogen derived from solar energy).

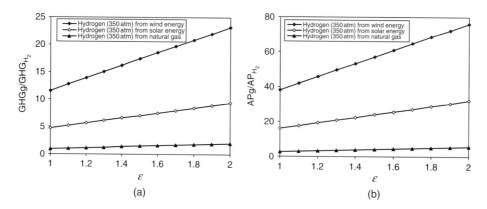

Fig. 19.10. Reductions of (a) GHG and (b) AP emissions as a result of hydrogen substitution for gasoline, as a function of the ratio in efficiencies ε of internal combustion (gasoline powered) and fuel cell (hydrogen powered) vehicles.

It can be seen that gasoline substitution with hydrogen from natural gas allows a relatively smaller reduction in GHG and AP emissions (about 2.5–5 times). The data in Fig. 19.10 suggest that 'renewable' hydrogen represents a potential long-term solution to environmentally related transportation problems.

19.6.3. Main findings and extensions

Several important findings can be drawn from the ExLCA performed here to evaluate exergy and economic efficiencies and environmental impacts from substituting renewable wind and solar energy for fossil fuels to produce electricity and hydrogen. In the analysis, fossil fuel technologies for hydrogen and electricity production from natural gas and gasoline from crude oil are contrasted with renewable ones, hydrogen is considered as a fuel for fuel cell vehicles and a substitute for gasoline, and exergy efficiencies and GHG and AP emissions are evaluated. Emissions are determined during all process steps, including crude oil and natural gas pipeline transportation, crude oil distillation and natural gas reforming, wind and solar electricity generation, hydrogen production through water electrolysis, and gasoline and hydrogen distribution and utilization. The key findings follow:

- The use of wind and solar power to produce electricity and hydrogen via electrolysis, and its application in a fuel cell vehicle, exhibits the lowest GHG and AP emissions.
- The economic attractiveness (capital investment efficiency factor) of renewable technologies depends significantly on the ratio in costs for hydrogen and natural gas. At the present cost ratio of about 2 (per unit of LHV or exergy), for example, capital investments are about five times lower to produce hydrogen via natural gas than to produce hydrogen via wind energy, rendering the cost of wind- and solar-based electricity and hydrogen substantially higher than that of natural gas.
- Implementing wind- and solar-based electricity for GHG and AP emissions, mitigation is less costly than introducing wind- and solar-based hydrogen. With present costs of wind and solar electricity, when electricity from renewable sources replaces electricity from natural gas, the cost of GHG and AP emissions abatement is more than 4 and 10 times less, respectively, than the cost if hydrogen from renewable sources replaces hydrogen from natural gas. Introducing 'renewable' hydrogen as a fuel for fuel cell vehicles instead of gasoline can lead to economically effective reductions of GHG and AP emissions only if the efficiency of a fuel cell vehicle is about two times higher than that of an internal combustion one.
- Substituting gasoline with 'renewable' hydrogen leads to GHG emissions reductions of up to 23 times for hydrogen from wind and 8 times for hydrogen from solar energy, and AP emissions reductions of up to 76 times for hydrogen from wind and 32 times for hydrogen from solar energy. By comparison, gasoline substitution with hydrogen from natural gas allows reductions in GHG and AP emissions of up to only 5 times.

The data presented in this section can be applied and extrapolated to make useful predictions. For instance, Canada needs to reduce its GHG emissions by approximately 270 megatons annually during the period 2008–2012 to meet its Kyoto commitments. According to the data presented here, when 6000 wind turbines (Kenetech KVS-33), with a capacity of 350 kW and a capacity factor 24%, replace a 500 MW gas-fired power generation plant with an electricity generation efficiency of 40%, annual GHG emissions are reduced by 2.3 megatons and an additional annual cost is incurred (at an average $\beta_w = 2.25$) of about US$280 million. According to Canada's per capita electricity consumption this amount of electricity corresponds to the needs of 280,000 Canadians. These data suggest that 'renewable' hydrogen represents a potential long-term solution to environmentally related problems. The approach outlined here can assist in developing strategies for reducing GHG and AP emissions.

19.7. Closing remarks

LCA can be extended to consider exergy. ExLCA can be usefully applied to various energy processes. The approach can be used to evaluate exergy and economic efficiencies and environmental impacts for various processes and measures, including fuel substitution, power generation, cogeneration and fuel processing.

Problems

19.1 What is exergetic life cycle assessment? How does it differ from energetic life cycle assessment?

19.2 What is the difference between an exergoeconomic analysis and an exergetic life cycle assessment?

19.3 An environmental engineer claims that exergetic life cycle assessment of a fossil fuel energy system is no different than an energetic life cycle assessment since the exergy of a fossil fuel is approximately equal to its heating value. Do you agree? Explain.

19.4 Explain what steps are involved in a life cycle assessment.

19.5 How does a life cycle assessment help evaluate the impact of energy systems on the environment?

19.6 Describe the usefulness of exergetic life cycle assessment in evaluating hydrogen production processes using fossil fuels and renewable energy sources.

19.7 Compare the results of exergetic life cycle assessments of hydrogen production using fossil fuel and renewable energy sources.

19.8 How can an exergetic life cycle assessment assist in developing strategies for reducing greenhouse gas and air pollution emissions? Give examples.

19.9 Obtain a published article on exergetic life cycle assessment. Using the data provided in the article, try to duplicate the results. Compare your results to those in the original article.

Chapter 20

EXERGY AND INDUSTRIAL ECOLOGY

20.1. Introduction

Industrial ecology is an approach to designing industrial systems that promotes systems that are less damaging to the environment. The approach seeks a reasonable balance between industrial profit and environmental stewardship and thereby can contribute to sustainable development. Industrial ecology methods can beneficially incorporate exergy to provide more powerful tools. Exergy analysis pinpoints significant process and device exergy losses, or non-recoverable losses of fuel exergy. It is generally accepted that an increase in efficiency of fossil fuel utilization makes industrial technologies more ecologically benign and safe. Therefore, exergy methods can help in rational modification of contemporary technologies.

Szargut (2005) cites the following example. In a combined power plant equipped with a coal boiler and gas turbine, the heat-transfer exergy losses in the heat recovery boiler of the gas turbine can be reduced by shifting the steam superheater from the coal boiler to the heat recovery boiler of the gas turbine. In another example, from the chemical industry, energy and exergy analyses of a traditional one-stage crude oil distillation unit and a newly proposed two-stage unit are conducted to investigate the efficiencies and exergy losses (Al-Muslim et al., 2003). The results are compared for both one- and two-stage distillation units. The proposed two-stage distillation unit exhibits a 43.8% decrease in overall exergy losses and 125% increase in the overall exergy efficiency, leading to the recommendation to perform distillation in two stages rather than one to reduce the heat duty of the heating furnace and thus reduce irreversible losses.

In this chapter, the relation between industrial ecology and exergy is described and illustrated, and the enhancements possible of industrial ecology methods through inclusion of exergy concepts are highlighted.

20.2. Industrial ecology

Industrial ecology is concerned with shifting industrial processes from linear (open loop) systems, in which resource and capital investments move through the system to become wastes, to closed loop systems where wastes become inputs for other processes (Graedel and Allenby, 1995). Industrial ecology was popularized by Frosch and Gallopoulos (1989) who asked why industrial systems do not behave like an ecosystem, where wastes of one species are a resource to another species. Why should not the outputs of one industry be the inputs of another, thereby reducing the use of raw materials and pollution, and saving on waste treatment? Lowe and Evans (1995) note that industrial ecology suggests using the design of ecosystems to guide the redesign of industrial systems to achieve a better balance between industrial performance and ecological constraints and consequently to determine a path to sustainable development.

According to this concept, modern industrial technologies should be designed like ecosystems where (i) input mass and energy flows are minimized and (ii) energy supply is provided by renewable energy sources. Minimization of the fossil fuel energy consumption in industrial processes implies eliminating output waste energy flows or the emission of wastes that are in equilibrium with the conditions (pressure, temperature, composition) of the environment.

Applying these principles to industrial processes, like power generation and transportation, leads to several interesting observations. The technical ability to transform renewable energy to electricity for industrial and other needs is developed, but the relevant technologies involve significant consumptions of resources such as construction materials per unit of output generated and are often less attractive economically and sometimes less attractive environmentally than traditional fossil fuel plants.

20.3. Linkage between exergy and industrial ecology

Graedel (1996) writes, 'The term industrial ecology was conceived to suggest that industrial activity can be thought of and approached in much the same way as a biological ecosystem and that in its ideal form it would strive toward integration of activities and cyclization of resources, as do natural ecosystems.' He goes on to note that little has been done to explore the usefulness of the analogy.

The use of exergy in conjunction with industrial ecology can provide a useful tool that permits practical applications (Connelly and Koshland, 2001a,b; Dewulf and Van Langenhove, 2002; Kay, 2002). Waste exergy emissions and exergy destructions, unlike energy losses, can account for the environmental impacts of energy utilization (Dincer and Rosen, 2005). Szargut et al. (2002) suggest that the cumulative consumption of non-renewable exergy provides a measure of the depletion of non-renewable natural resources.

Reducing entropy generation leads to a decline in exergy destruction (losses) $\dot{E}x_\mathrm{D}$ due to reducing the irreversibility of the processes constituting an industrial system. According to the Gouy–Stodola formula,

$$\dot{E}x_\mathrm{D} = T_0 \dot{S}_\mathrm{gen} \tag{20.1}$$

where T_0 is the reference-environment temperature (often fixed at 298 K or the local temperature), and $\dot{S}_\mathrm{gen}$ is the entropy generation rate in a process or device.

20.3.1. Depletion number

Connelly and Koshland (2001a,b) suggest that the efficiency of fossil fuel consumption be characterized by a depletion number D_p:

$$D_\mathrm{p} = \frac{\dot{E}x_\mathrm{D}}{\dot{E}x_\mathrm{in}} \tag{20.2}$$

which represents the relation between the exergy destruction rate $\dot{E}x_\mathrm{D}$ and total exergy consumption rate $\dot{E}x_\mathrm{in}$ (in this chapter only direct exergies are considered).

In line with the definition of exergy efficiency, if there are no waste exergy emissions the exergy efficiency ψ is expressible as follows:

$$\psi = 1 - D_\mathrm{p} \tag{20.3}$$

The exergy efficiency is always a measure of how nearly a process approaches the ideal.

20.3.2. Integrated systems

The efficiency of integrated or combined technologies (e.g., cogeneration) can be evaluated and compared by examining the depletion numbers D_p for the separate and combined technologies (see Fig. 20.1).

The consumption of non-renewable energy resources corresponds to lower depletion numbers (see Eq. (20.2)). Consequently, the depletion number for an advanced combined technology $D_\mathrm{p}^\mathrm{comb}$ should be lower than the weighted sum of the depletion numbers $D_\mathrm{p}^\mathrm{sep}$ for the separate technologies. For the system in Fig. 20.1, $D_\mathrm{p}^\mathrm{sep}$ is expressible as follows:

$$D_\mathrm{p}^\mathrm{sep} = \frac{\dot{E}x_\mathrm{p1}^\mathrm{comb}}{\dot{E}x_\mathrm{p1}^\mathrm{comb} + \dot{E}x_\mathrm{p2}^\mathrm{comb}} D_\mathrm{p}^{(1)} + \frac{\dot{E}x_\mathrm{p2}^\mathrm{comb}}{\dot{E}x_\mathrm{p1}^\mathrm{comb} + \dot{E}x_\mathrm{p2}^\mathrm{comb}} D_\mathrm{p}^{(2)} \tag{20.4}$$

where $D_\mathrm{p}^{(1)}$ and $D_\mathrm{p}^{(2)}$ are depletion numbers for two separate technologies and $\dot{E}x_\mathrm{p1}^\mathrm{comb}$ and $\dot{E}x_\mathrm{p2}^\mathrm{comb}$ are the rates of output exergy flows for products 1 and 2, respectively.

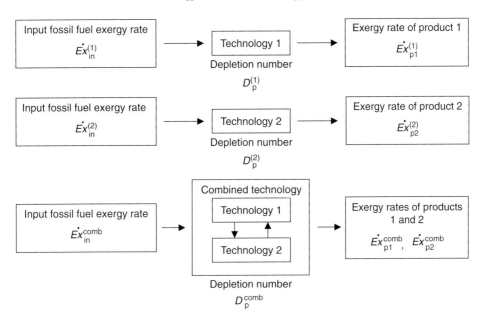

Fig. 20.1. Input and output exergy rates for separate and combined technologies to produce two products.

20.4. Illustrative example

The principles discussed in this chapter are demonstrated for a combined gas-turbine cycle with a hydrogen generation unit (Granovskii et al., 2008). This design includes two important technologies: a solid oxide fuel cell (SOFC) with internal natural gas reforming and a membrane reactor (MR), and their combination with a hydrogen generation unit.

A common feature of SOFCs and MRs is their utilization of high-temperature oxygen ion-conductive membranes. Such membranes are conductive to negatively charged ions of oxygen and permit the separation of oxygen from air. This property accounts for their application as an electrolyte in SOFCs, where the chemical exergy of methane, through an intermediate stage involving its conversion to hydrogen and carbon monoxide and electrochemical oxidation with oxygen, is transformed into electrical work. In an MR, the membrane conducts both oxygen ions and electrons in opposite directions; such membranes are consequently often called mixed conducting membranes. In the present case, electrical work is not generated, but oxygen is separated from air and fuel combustion proceeds in an atmosphere of oxygen.

Oxygen ion-conductive membranes are made of ceramic materials (usually zirconia oxides) and have good performance characteristics at temperatures higher than 700°C. An SOFC stack is often introduced into traditional power generation cycles, where it operates at temperatures of 800–1100°C (e.g., Chan et al., 2002; Kuchonthara et al., 2003a,b). An MR is being developed for operation up to 1250°C, as a substitute for combustion chambers in advanced zero-emission power plants (e.g., Sundkvist et al., 2001). New materials for the anodes of SOFCs contain a catalyst for the methane reforming process, allowing methane conversion to a mixture of hydrogen and carbon monoxide directly on the surface of the anode (Eguchi et al., 2002; Weber et al., 2002). SOFCs thereby become more flexible, compact and effective, and avoid the need for preliminary reforming of methane.

20.4.1. The considered gas-turbine combined cycle with hydrogen generation

A combined gas-turbine cycle with a hydrogen generation unit is presented in Fig. 20.2. The initial stream of natural gas, after heating in device 14 (in order to achieve after compression the temperature of combustion products) and compression in device 15, is divided into two flows. The first is mixed with combustion products (carbon dioxide and steam) and directed to the anodes of the SOFC stack (device 4), where two processes occur simultaneously: conversion of methane to a mixture of carbon monoxide and hydrogen on the surface of the anodes and electrochemical oxidation of the resultant mixture with oxygen. The oxygen reduction is accompanied by electricity generation in the SOFCs. The gaseous mixture from the anodes (conversion and combustion products) is cooled in a heat exchanger (device 10), compressed in device

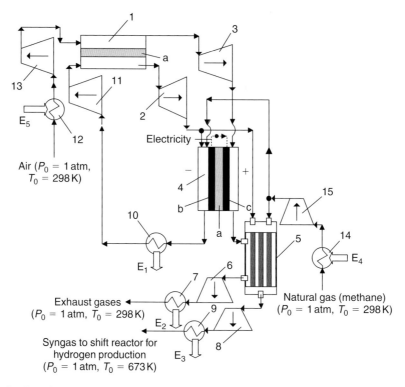

Fig. 20.2. An application of an SOFC and MR in a combined gas turbine cycle with a hydrogen generation unit. Numbers indicate devices according to the following legend: 1: MR; 2, 3, 6, 8: turbines; 11, 13, 15: compressors; 4: SOFC stack; 5: methane converter; 7, 9, 10, 12, 14: heat exchangers; a: oxygen ion-conductive membranes; b, c: anode and cathode of SOFC stack, respectively.

11, and directed to the MR (device 1), where the remainder of the conversion products combust in oxygen, and then expand in a turbine (device 2).

The combustion products are then divided into two flows. The first is mixed with the initial flow of methane and directed to the SOFC stack, while the other is mixed with the second flow of methane and enters the catalytic methane converter (device 5). After methane conversion to hydrogen and carbon monoxide in device 5, the gaseous mixture is expanded in a turbine (device 8), cooled in a heat exchanger (device 9) and directed to the shift reactor, where the remainder of the carbon monoxide and steam is converted to hydrogen.

Air is heated in device 12, compressed in device 13, directed to the MR (device 1), where some quantity of oxygen is transferred through the oxygen ion-conductive membrane and combusted with fuel. The air heating in device 12 is required in order to achieve after compression the temperature of the fuel flow which is directed, like air, to the MR.

The temperature of air reaches its maximum at the MR (device 1) outlet, at which point it is expanded in the turbine (device 3) and directed to the cathodes of the SOFCs (device 4). In the SOFCs, the oxygen concentration in the air decreases, and the air is heated and enters the space between pipes in the catalytic converter (device 5). In device 5, heat is transferred from the air to the reaction mixture in the pipes. The mixture is then expanded in the turbine (device 6) and cooled in the heat exchanger (device 7).

The power generation design combines a traditional gas-turbine cycle – which consists of compressors (devices 11 and 13), a combustion chamber (which is represented by the MR, device 1) and turbines (devices 2 and 3) – with the SOFC stack (device 4) and the methane converter (device 5). Heat exchangers are conditionally divided into the heat releasing (devices 7, 9 and 10) and heat receiving (devices 12 and 14) types. Mechanical work is produced in the turbines and consumed in the compressors. The work is transformed into electrical energy, which is also directly generated in the SOFC stack. The endothermic process of methane conversion to hydrogen (via a synthesis gas) in device 5 is implemented into the power generation cycle.

20.4.2. Exergy analysis of the gas-turbine combined cycle with hydrogen generation

The general assumptions applied in the exergy analysis of the proposed design follow: (i) gases are modeled as ideal; (ii) energy losses due to mechanical friction are negligible; (iii) thermodynamic and chemical equilibria are achieved at the outlet of the SOFC stack and methane converter and (iv) all combustible components are combusted completely in the MR. The general parameters used in the combined power generation cycle are listed in Table 20.1. Values for the parameters η_t, η_{cmp}, P_{max}, P_{min} and T_{max} are often cited (e.g., Kirillin et al., 1979).

Table 20.1. General parameter values for the combined power generation cycle in Fig. 20.2.

Parameter	Value
Isentropic efficiency of turbines η_t	0.93
Isentropic efficiency of compressors η_{cmp}	0.85
Operational circuit voltage of the SOFC stack (V)	0.85
Maximum pressure in the gas turbine cycle p_{max} (atm)	10
Minimum pressure in the gas turbine cycle p_{min} (atm)	1
Maximum temperature in the cycle (at the MR outlet) T_{max} (K)	1573
Temperature of fuel at the inlet of the SOFC stack T_s (K)	1273
Temperature of fuel and air at the outlet of the SOFC stack T_s (K)	1273
Ratio of methane combusted in the power generation cycle to the methane converted	1.0:0.7
Molar ratio of combustion products after the MR to methane combusted in the power generation cycle	6
Ratio of amounts of combustion products directed to SOFC and methane converter	1:1
Standard temperature T_0 (K)	298
Standard pressure p_0 (atm)	1
Air composition (vol %)	21% O_2, 79% N_2

An exergy balance of a system permits evaluation of the efficiency with which input energy flows are utilized. For the power generation scheme presented in Fig. 20.2 the exergy balance can be expressed as

$$\Delta \dot{E}x = \dot{E}x_{in} - \dot{E}x_{out} = \sum \dot{W}_i + \Delta \dot{E}x_T + \sum \dot{E}x_{D_i} \qquad (20.5)$$

where $\Delta \dot{E}x$ is the rate of exergy change in the system, $\dot{E}x_{in}$ is the sum of the exergy rates of the input flows of methane and air, $\dot{E}x_{out}$ is the sum of the exergy rates of the output flows of conversion products (synthesis gas) directed to a shift converter and exhaust gases, $\sum \dot{W}_i$ is the sum of powers generated in the turbines and in SOFCs, and consumed in the compressors (with a negative sign), $\Delta \dot{E}x_T$ is the sum of thermal exergy rates released in heat exchangers 7, 9 and 10 and consumed in 4 and 12 (with a negative sign), and $\sum \dot{E}x_{D_i}$ is the sum of the exergy loss rates in the devices of the system.

20.4.3. Results

The analysis results are presented in Tables 20.2–20.4.

Table 20.3 presents the mechanical and electrical work generated in the turbines and SOFC stack, the mechanical work consumed in the compressors (with a negative sign), and the exergy losses accompanying these processes. Table 20.4 presents the exergy losses in the MR and methane converter. Table 20.4 also lists the exergy losses $Ex_{D_{tr}}$ accompanying the heat transfer from hot to cool flows, and the excess of thermal exergy ΔEx_T which can be converted to mechanical

Table 20.2. Generated work and exergy losses for the processes in the combined gas-turbine cycle in Fig. 20.2.*

Device number in Fig. 20.2	W (kJ/mol)	Ex_D (kJ/mol)
2	89.7	1.6
3	207.1	4.1
4	497.4	29.4
6	85.0	2.3
8	35.6	0.2
11	−89.8	4.2
13	−324.4	22.3
15	−18.8	0.7
Total	481.8	64.8

* Data are given per mole of methane combusted in the power generation cycle.

Table 20.3. Exergy losses in the MR and methane converter.*

Device number in Fig. 20.2	Ex_D (kJ/mol)
1	27.6
5	15.9
Methane mixing	10.0
Total	53.5

* Data are given per mole of methane combusted in the power generation cycle, which corresponds to 0.7 mol of methane converted in methane converter 5.

Table 20.4. Released thermal exergy ΔEx_T and its utilization in the Rankine bottoming cycle.*

ΔEx_T (kJ/mol)	$Ex_{D_{tr}}$ (kJ/mol)	W_R (kJ/mol)	Ex_{D_R} (kJ/mol)
58.4	26.3	35.0	23.4

* Data are given per mole of methane combusted in the power generation cycle.

work in a bottoming steam–water (Rankine) cycle (not shown in Fig. 20.2) with an exergy efficiency ψ_R of about 60% (Cengel and Turner, 2005), so that

$$W_R = \eta_R \Delta Ex_T \quad \text{and} \quad Ex_{D_R} = \Delta Ex_T - W_R \tag{20.6}$$

After substituting W_R and Ex_{D_R} into Eq. (20.5) instead of ΔE_T, the exergy change $\Delta Ex = 684.8$ kJ/mol in the system is distributed only between work $W = 516.8$ kJ/mol and the exergy losses (destruction) $Ex_D = 168.0$ kJ/mol.

Since data are calculated per mole of methane combusted to generate electricity and 0.7 mol of methane converted to hydrogen, and the value of standard exergy of methane $Ex_{CH_4}^0 = 831.7$ kJ/mol (Szargut et al., 1988), the depletion number of the combined system D_p^{comb} becomes

$$D_p^{comb} = \frac{\sum Ex_{D_i}}{1.7 Ex_{CH_4}^0} = \frac{168.0}{1414.0} = 0.12 \tag{20.7}$$

The combined system yields two products: electricity and synthesis gas (a mixture of carbon monoxide and hydrogen). The exergy of electrical work is equal to its energy and standard exergies of carbon monoxide and hydrogen are $Ex_{H_2}^0 = 236.1$ kJ/mol and $Ex_{CO}^0 = 275.1$ kJ/mol (Szargut et al., 1988). Then the exergy of the synthesis gas directed to the shift reactor to produce hydrogen (see Fig. 20.1) is $Ex_{SG}^0 = 656.1$ kJ/mol (for 1 mol of methane combusted and 0.7 mol of methane converted). The exergy efficiency of a combined gas-turbine steam power cycle where only electrical work is generated is taken to be $\psi^{(1)} = 0.54$ (e.g., Cleveland, 2004) and the exergy efficiency of methane conversion to synthesis gas is $\psi^{(2)} = 0.84$ (e.g., Rosen and Scott, 1998). With Eq. (20.3), the depletion numbers are calculated as $D_p^{(1)} = 0.46$ and $D_p^{(2)} = 0.16$. Substitution of these values into the expression for D_p^{sep} (Eq. (20.4)) yields the following:

$$D_p^{sep} = \frac{W}{W + Ex_{SG}^0} D_p^{(1)} + \frac{Ex_{SG}^0}{W + Ex_{SG}^0} D_p^{(2)} = 0.29 \tag{20.8}$$

The depletion number for the separate technologies D_p^{sep} is seen to be more than two times greater than that for the combined system D_p^{comb}. The implication is that the combined technology is more environmentally benign (and behaves more like an ecosystem) than the separate devices, and requires combustion of less natural gas.

The limiting value of $D_p^{(1)}$ for the separate electricity generation process can be obtained by equalizing $D_p^{comb} = D_p^{sep}$ with the given value of $D_p^{(2)}$. In this case, the limiting value is found to be $D_p^{(1)} = 0.068$, which corresponds to an exergy efficiency of electricity generation $\psi^{(1)} = 0.93$ (Eq. (20.3)). This value is unrealistic, as it exceeds even the highest SOFC efficiency obtained in laboratory experiments (e.g., Larminie and Dicks, 2003). Thus, this magnitude of efficiency can be attained only through an integrated process like cogeneration.

The conducted analysis confirms that integrated energy systems, developed via an appropriate combination of technologies, represent an important opportunity for increasing the utilization efficiency of natural resources and thereby achieving the aims of industrial ecology.

20.5. Closing remarks

Industrial ecology is an approach that suggests designing industrial systems like ecosystems, where the wastes of one species are often the resource of another. Important ways of implementing industrial ecology include the appropriate combination of separate technologies in order to match the waste outputs of one with the inputs of the other, and the introduction of processes that reduce non-renewable energy consumption. Exergy analysis can help in designing industrial systems that follow the principles of industrial ecology, and in the evaluation of the efficiencies and losses for such activities. One such evaluation measure is the depletion number, which relates the exergy destruction and exergy input for a system.

An example has been used to illustrate how to compare depletion numbers for separate and combined technologies, so as to assess the effectiveness of their integration. The analysis suggests that an exergy-based approach to industrial ecology can be advantageous in the creation and modification of industrial systems, through integrating separate technologies and other measures.

Problems

20.1 What is industrial ecology? How is it related to exergy?
20.2 What is the difference between exergetic life cycle assessment and industrial ecology?
20.3 What is the relationship between industrial ecology and the environmental impact of energy systems?
20.4 What is the depletion number? How is it related to industrial ecology?
20.5 What is the relationship between depletion number and exergy efficiency?
20.6 What is the effect of depletion number on resource sustainability?
20.7 Are the depletion numbers for renewable energy systems zero? Explain.
20.8 Obtain a published article on industrial ecology. Using the data provided in the article, try to duplicate the results. Compare your results to those in the original article.

Chapter 21

CLOSING REMARKS AND FUTURE EXPECTATIONS

Exergy is a fascinating and curious aspect of the Second Law of Thermodynamics, in that on a practical basis it provides a very useful tool while on a more abstract level it provides the basis for endless contemplation. In this book, we have focussed on the former, although we have delved into the latter at many times.

The understanding provided by exergy and its potential benefits are what has attracted to the field in the past numerous researchers, who have contributed immeasurably to the development and application of exergy methods. These attributes of exergy will, we are confident, continue to attract researchers and practitioners well into the future, and exergy methods will continue to evolve.

In short, this book has sought to provide the reader with a broad working familiarity and understanding of exergy and analysis techniques based upon it, by:

- identifying and explaining exergy and its use in improving efficiency and reducing losses, and in designing 'better' processes and systems.
- discussing the broader aspects of exergy, in areas such as economics and environment as well as many others, in addition to corresponding assessment and improvement methods.
- illustrating applications of exergy to many different systems in hopes of explaining how to apply exergy methods beneficially. The applications presented have by no means been comprehensive, as many more applications exist, but have been selected so as to provide a representative cross-section and to cover various important fields.
- showing the many ways in which exergy can assist in efforts to achieve more sustainable development, in terms of many factors including use of sustainable resources, increased efficiency, reduced environmental impact, improved economics, etc.

We feel that the understanding and benefits provided by exergy of industrial and other processes and systems are clearly profound in many ways. It is, therefore, worth developing an appreciation and working knowledge of exergy methods.

Much has been done to help increase the use of exergy methods over the last couple decades, including the establishment of the *International Journal of Exergy* and the organization of many regional and international conferences, short courses and workshops on exergy. These activities complement well the impressive developments occurring in almost every area of science and engineering, from mechanical engineering to biology and from chemical engineering to economics, and should ensure that exergy is utilized as a tool for enhancing advances in science, engineering and other disciplines.

After writing a book such as this, we can not help but contemplate about what the future may hold in store from an exergy perspective. Some expectations follow on how exergy may be utilized, or how we hope it will be utilized, in the future so as to reap the greatest benefits:

- Exergy will become increasingly utilized, across a diverse array of fields and throughout the developing and developed world, to increase efficiency, reduce wastes and losses and improve processes and systems.
- A standard nomenclature and terminology for exergy will coalesce, making applications more convenient and straightforward.
- Exergy will become more integrated with economics and applied more broadly, and disciplines like thermoeconomics and exergoeconomics will become more mature.
- Exergy will become more integrated with environmental and ecological assessments, through tools like industrial ecology, life cycle assessment and others, and applied more widely. Disciplines like environomics will become more mature.
- Exergy will be used to foster sustainability and contribute to making development more sustainable.

- Exergy will become a 'conventional' design tool for engineers of energy and other systems.
- Exergy use will grow and expand outside its normal sphere of applications, in areas like management methods, information and communications, and biology and ecology.
- Exergy will become more broadly covered in educational programs, and used as a basis for explaining and giving practical meaning to the Second Law of Thermodynamics.
- Exergy will achieve a broader understanding and appreciation by the public and the media, allowing its broader application.
- Exergy will be increasingly utilized in decision and policy making, by leaders in industry and government.

If these expectations come to pass, even in part, the potential benefits to humanity through improved processes, technologies and systems, will be both exciting and wonderful to observe.

NOMENCLATURE

a	speed of sound
A	surface area
AF	air–fuel ratio
AP	air pollution
Ar	Archimedes number
c	specific heat; velocity relative to a fixed reference environment
c_p	specific heat at constant pressure
c_v	specific heat at constant volume
C	heat capacity rate; concentration; cost
COP	coefficient of performance
CV	coefficient of variation
D	moisture diffusivity; depletion number
e	specific energy
E	energy; voltage
EEQ	exergy equivalent of materials and devices
El	electricity cost
EOP	operation exergy
ex	specific exergy (flow or non-flow)
$\dot{E}_{design}$	heat requirement rate for colder/winter months
$\dot{E}_{smr}$	heat requirement rate for hot water during warmer/summer months
Ex	exergy
$\dot{Ex}$	exergy rate
ex_{dest}	specific irreversibility (exergy destruction)
Ex_{dest}	irreversibility (exergy destruction)
$\dot{Ex}_{dest}$	irreversibility rate (exergy destruction)
Ex^Q	exergy transfer associated with heat transfer
Ex^W	exergy transfer associated with work
f	fraction; mean height fraction; exergetic factor; figure of merit; fuel-to-air ratio
F	fraction of storage-fluid mass in liquid phase; Faraday constant
$\dot{F}$	fuel exergy rate
FF	fill factor
g	gravitational acceleration
G	Gibbs free energy; natural gas consumption
GHG	greenhouse gas
h	specific enthalpy; specific base enthalpy; height (relative to thermal energy storage (TES) bottom); solar radiation ratio
$h°$	specific enthalpy of formation
H	enthalpy; TES fluid height
HHV	higher heating value
i	specific irreversibility (exergy destruction); time step increment
I	irreversibility (exergy destruction); current
$\dot{I}$	irreversibility rate (exergy destruction rate)
IP	improvement potential
k	thermal conductivity; number of zones

K	capital cost
ke	specific kinetic energy
KE	kinetic energy
L	ratio of lost exergy to incoming exergy; thermodynamic loss; width
LHV	lower heating value
m	mass
M	molar mass; moisture content
$\dot{m}$	mass flow rate
n	number of values in a set
N	number of moles
N_{dw}	number of dwellings
N_{per}	number of persons per dwelling
N_{warm}	number of warmer of summer days
p	pressure
P	power; pressure
$\dot{P}_{\text{F}}$	product exergy
P_{T}	thrust power
pe	specific potential energy
PE	potential energy
q	specific heat transfer; electron charge
Q	heat transfer
$\dot{Q}$	heat transfer rate
R	gas constant; thermal resistance; energy grade function; ratio of thermodynamic loss rate to capital cost
Re	Reynolds number
RH	relative humidity
RI	relative irreversibility
R_{energy}	net station condenser energy (heat) rejection rate
R_{exergy}	net station condenser exergy rejection rate
s	specific entropy
S	entropy; thickness; stoichiometry; average daily usage of sanitary hot water per person
s°	specific entropy of formation
S_{T}	hourly measured total solar irradiation
SD	standard deviation
SExI	specific exergy index
t	time
T	temperature
T_{indoor}	indoor design temperature
T_{outdoor}	outdoor design temperature
T_{R}	temperature ratio
u	specific internal energy; superficial gas velocity
U	total internal energy; overall heat-transfer coefficient; flight speed
v	specific volume
V	volume; velocity; voltage
$\dot{V}$	volumetric flow rate
w	specific work
W	work; weight
$\dot{W}$	work rate or power
x	mass fraction; mole fraction; vapor fraction
X	exergy transfer associated with heat transfer; exergy loss; humidity ratio; thickness
y	molal fraction; fraction of gas liquefied
Z	elevation

Greek letters

α constant parameter; thermal diffusivity; ratio of costs for hydrogen and natural gas

β ratio of electricity production prices; fuel consumption reduction; mass constituent

δ thickness; fuel depletion rate

ε heat exchanger effectiveness; ratio of efficiencies

λ excess air fraction

η energy efficiency; overpotential

γ capital investments efficiency factor; oxygen fraction; exergy grade function

ψ exergy efficiency

ϕ zone temperature distribution; relative humidity

ρ density

τ exergetic temperature factor

μ chemical potential

θ parameter; incidence angle

ω specific humidity or humidity ratio

ξ productivity lack

ζ specific exergy function

χ relative irreversibility

Π entropy creation

Superscripts

·   rate with respect to time

–   mean

$\prime$ modified case

REFERENCES

Ahmad A, Abdelrahman MA, Billote WC. 1994. Analysis of electric energy conservation in a commercial building in Saudi Arabia. *Proceedings of the 2nd Saudi Symposium on Energy Utilization and Conservation.* November 27–30, Dhahran, Saudi Arabia.

Ahrendts J. 1980. Reference states. *Energy-The International Journal* 5:667–678.

Al-Bagawi JJ. 1994. *Energy and Exergy Analysis of Ghazlan Power Plant.* M.Sc. Thesis, Mechanical Engineering Department, King Fahd University of Petroleum and Minerals, Dhahran, Saudi Arabia.

Al-Muslim H. 2002. *Energy and Exergy Analyses of Crude Oil Distillation Plants.* M.Sc. Thesis, Mechanical Engineering Department, King Fahd University of Petroleum and Minerals, Dhahran, Saudi Arabia.

Al-Muslim H. and Dincer I. 2005. Thermodynamic analysis of crude oil distillation systems. *International Journal of Energy Research* 29(7):637–655.

Al-Muslim H, Dincer I, Zubair SM. 2003. Exergy analysis of single- and two-stage crude oil distillation unit. *ASME Journal of Energy Resources Technology* 125(3):199–207.

Al-Muslim H, Dincer I, Zubair SM. 2005. Effect of reference state on exergy efficiencies of one- and two-stage crude oil distillation plants, *International Journal of Thermal Sciences* 44(1):65–73.

Anaya A, Caraveo LM, Carcia VH, Mendoza A. 1990. Energetic optimization analysis of a petroleum refinery applying second law of thermodynamics, *Proceedings of the Computer-Aided Energy Systems Analysis*, Vol. 21 American Society of Mechanical Engineers, New York, 49–54.

ASHRAE. 2006. *Handbook of Refrigeration.* American Society of Heating, Refrigerating and Air Conditioning Engineers: Atlanta, GA.

Baehr HD, Schmidt EF. 1963. Definition und berechnung von brennstoffexergien (Definition and calculation of fuel exergy). *Brennst-Waerme-Kraft* 15:375–381.

Bailey FG. 1981. Steam turbines, *Encyclopedia of Energy*, 2nd ed. McGraw-Hill: Toronto.

Ballard. 2004. *HY-80 Fuel Cell Engine Brochure.* Ballard Power Systems. Extracted from http://www.ballard.com/be_a_customer/transportation/fuel_cell_engines/light-duty_fuel_cell _engines.

Barbier E. 2002. Geothermal energy technology and current status: an overview. *Renewable and Sustainable Energy Reviews* 6:3–65.

Barker J, Norton P. 2003. *Building America System Performance Test Practices: Part 1-Photovoltaic Systems.* National Renewable Energy Laboratory (NREL), NREL/TP: 550–30301.

Barnes SL. 1964. A technique for maximizing details in numerical weather map analysis. *Journal of Applied Meteorology* 3:396–409.

Baschuk JJ, Li X. 2003. Mathematical model of a PEM fuel cell incorporating CO poisoning and O_2 (air) bleeding. *International Journal of Global Energy Issues* 20(3):245–276.

Bechtel Power Corporation. 1982. *Harry Allen Station-Preliminary Design Report.* Appendix 1, Technical Scope.

Becken DW. 1960. Thermodrying in fluidized beds. *British Chemical Engineering* 5:484–495.

Beer T, Grant T, Morgan G, Lapszewicz J, Anyon P, Edwards J, Nelson P, Watson H, Williams D. 2006. Weighting methodologies for emissions from transport fuels. *Australian Greenhouse Office.* www.greenhouse.gov.au/transport/comparison/ pubs/3ch1.pdf.

Bejan A. 1982. *Entropy Generation through Heat and Fluid Flow.* Wiley: Toronto.

Bejan A. 1988. A second look at the second law. *Mechanical Engineering* 110(5):58–65.

Bejan A. 1996. *Entropy Generation Minimization.* CRC Press: FL.

Bejan A, Tsatsaronis G, Moran M. 1996. *Thermal Design and Optimization.* John Wiley & Sons: New York.

Bejan A. 2001. New century, new methods. *Exergy, An International Journal* 1:2.

Bejan A. 2006. *Advanced Engineering Thermodynamics*, 3rd ed. John Wiley & Sons: New York.

Berg CA. 1974. A technical basis for energy conservation. *Mechanical Engineering*: 30–42.

Berg CA. 1980. Process integration and the second law of thermodynamics: future possibilities. *Energy-The International Journal* 5:733–742.

Betz A. 1946. Windenergie und ihre Ausnutzung durch Windmühlen. Gttingen: Vandenhoek and Ruprecht, Göttingen.

Bisquert J, Cahen D, Hodes G, Ruhle S, Zaban A. 2004. Physical chemical principles of photovoltaic conversion with nanoparticulate, mesoporous dye-sensitized solar cells. *Journal of Physical Chemistry B* 108(24):8106–8118.

Bodvarsson G, Eggers DE. 1972. The exergy of thermal power. *Geothermics* 1:93–95.

Bosnjakovic F. 1963. Bezugszustand der exergie eines reagiernden systems (Reference states of the exergy in a reacting system). *Forschung Im Ingenieurwesen Engineering Research* 20:151–152.

Brilliant HM. 1995a. Analysis of scramjet engines using exergy methods. *AIAA Paper* 95–2767.

Brilliant HM. 1995b. Second law analysis of present and future turbine engines. *AIAA Paper* 95–3030.

Brown HL, Hamel BB, Hedman BA. 1996. *Analysis of 108 Industrial Processes*. Fairmount Press: Lilburn, GA.

Bryant HC, Colbeck I. 1977. A solar pond for London. *Solar Energy* 19:321.

Cengel YA. 1996. An intuitive and unified approach to teaching thermodynamics. *Proceedings of the ASME Advanced Energy Systems Division*, AES-Vol. 36, American Society of Mechanical Engineers: New York, pp. 251–260.

Cengel YA, Boles MA. 2006. *Thermodynamics: An Engineering Approach*, 5th ed. McGraw-Hill: New York.

Cengel Y, Turner R. 2005. *Fundamentals of Thermal-fluid Science*, 2nd ed. McGraw-Hill: New York.

CETC. 2003. Facts Sheet, No: 31901, CANMET Energy Technology Center, Ottawa, Canada.

Chan SH, Ho H, Tian Y. 2002a. Modelling of simple hybrid solid oxide fuel cell and gas turbine power plant. *Journal of Power Sources* 109:111–120.

Chan SH, Low CF, Ding OL. 2002b. Energy and exergy analysis of simple solid-oxide fuel cell (SOFC) power systems. *Journal of Power Sources* 103:188–200.

Charles River Associates. 2000. *Diesel Technology and the American Economy*. Report No. D032378-00, Washington, DC.

Cherry NJ. 1980. Wind energy resources methodology. *Journal of Industrial Aerodynamics* 5:247–280.

Chu ST, Hustrulid A. 1968. Numerical solution of diffusion equations. *Transactions of the ASAE* 11:705–708.

Clarke JM, Horlock JH. 1975. Availability and propulsion. *Journal of Mechanical Engineering Science* 17(4):223–232.

Cleveland CJ. 2004. *Encyclopedia of Energy*. Elsevier: New York.

Colen HR. 1990. *HVAC Systems Evaluation*. R.S. Means Company: Kingston, MA.

Collins S. 1992. Worlds first nuclear-to-coal conversion a success. *Power* 136(4):73–80.

Connelly L, Koshland CP. 2001a. Exergy and industrial ecology. Part 1: An exergy-based definition of consumption and a thermodynamic interpretation of ecosystem evolution. *Exergy, An International Journal* 1:146–165.

Connelly L, Koshland CP. 2001b. Exergy and industrial ecology. Part 2: A non-dimensional analysis of means to reduce resource depletion. *Exergy, An International Journal* 1:234–255.

Cornelissen RL. 1997. *Thermodynamics and Sustainable Development*. Ph.D. Thesis, University of Twente, The Netherlands.

Cownden R, Nahon M, Rosen MA. 2001. Exergy analysis of a fuel cell power system for transportation applications. *Exergy, An International Journal* 1:112–121.

Crane P, Scott DS, Rosen MA. 1992. Comparison of exergy of emissions from two energy conversion technologies, considering potential for environmental impact. *International Journal of Hydrogen Energy* 17:345–350.

Cressman GP. 1959. An operational objective analysis system. *Monthly Weather Review* 87(10):367–374.

Curran ET. 1973. *The Use of Stream Thrust Concepts for the Approximate Evaluation of Hypersonic Ramjet Engine Performance*. US Air Force Aero Propulsion Laboratory Report, pp. 73–38.

Curran MA. 1996. *Environmental Life-Cycle Assessment*. McGraw-Hill, New York.

Czysz P, Murthy SNB. 1991. Energy analysis of high-speed flight systems, *High Speed Flight Propulsion Systems*, Vol. 137, Progress in Astronautics and Aeronautics, AIAA: Washington, DC, pp. 143–235.

Davis JC. 1986. *Statistics and Data Analysis in Geology*. John Wiley & Sons: New York.

De Nevers W, Seader JD. 1980. Lost work: a measure of thermodynamic efficiency. *Energy The International Journal* 5:757–770.

Dewulf JP, Van Langenhove HR. 2002. Quantitative assessment of solid waste treatment systems in the industrial ecology perspective by exergy analysis. *Environmental Science and Technology* 36(5):1130–1135.

Dicks A. 1998. Advances in catalysts for internal reforming in high temperature fuel cells. *Journal of Power Sources* 71:111–122.

Dickson MH, Fanelli M. 1990. Geothermal energy and its utilization, in: Dickson MH, Fanelli M (eds.), *Small Geothermal Resources*. UNITAR/UNDP Centre for Small Energy Resources: Rome, Italy, pp. 1–29.

Dimattia DG. 1993. *Onset of Slugging in Fluidized Bed Drying of Large Particles.* M.A.Sc. Thesis, Technical University of Nova Scotia, Halifax, Canada.

Dincer I. 1998. Energy and environmental impacts: present and future perspectives. *Energy Sources* 20(4/5):427–453.

Dincer I. 2000. Renewable energy and sustainable development: a crucial review. *Renewable and Sustainable Energy Reviews* 4(2):157–175.

Dincer I. 2002a. Exergy and sustainability. *Proceedings of SET-2002: First International Conference on Sustainable Energy Technologies* 1–10, 12–14 June 2002, Porto, Portugal.

Dincer I. 2002b. On energetic, exergetic and environmental aspects of drying systems. *International Journal of Energy Research* 26(8):717–727.

Dincer I. 2002c. Technical, environmental and exergetic aspects of hydrogen energy systems. *International Journal of Hydrogen Energy* 27:265–285.

Dincer I. 2002d. The role of exergy in energy policy making. *Energy Policy* 30:137–149.

Dincer I. 2003. *Refrigeration Systems and Applications.* Wiley: New York.

Dincer I, Al-Muslim H. 2001. Thermodynamic analysis of reheat cycle steam power plants. *International Journal of Energy Research* 25(8):727–739.

Dincer I, Rosen MA. 1999a. The intimate connection between exergy and the environment. *Proceeding of Thermodynamic Optimization of Complex Energy Systems* 221–230, Kluwer Academic Publishers, the Netherlands.

Dincer I, Rosen MA. 1999b. Energy, environment and sustainable development. *Applied Energy* 64(1–4):427–440.

Dincer I, Rosen MA. 2002. *Thermal Energy Storage: Systems and Applications.* Wiley: London.

Dincer I, Hussain MM, Al-Zaharnah I. 2003. Energy and exergy use in industrial sector of Saudi Arabia. *IMechE-Part A: Journal of Power and Energy* 217(5):481–492.

Dincer I, Rosen MA. 2005. Thermodynamic aspects of renewables and sustainable development. *Renewable and Sustainable Energy Reviews* 9:169–189.

Dipippo R. 1994. Second law analysis of flash-binary and multilevel binary geothermal plants. *Geothermal Resources Council Transactions* 18:505–510.

Dipippo R, Marcille DF. 1984. Exergy analysis of geothermal power plants. *Geothermal Resources Council Transactions* 8:47–52.

Dixon JR. 1975. *Thermodynamics I: An Introduction to Energy.* Prentice-Hall: Englewood Cliffs, NJ.

DOE. 2000. *Combined Heat and Power: A Federal Manager's Resource Guide* (Final Report). US Department of Energy, Federal Energy Management Program.

DOE. 2003. *Cogeneration or Combined Heat and Power.* Office of Energy Efficiency and Renewable Energy: US Department of Energy.

Dry M. 1999. Fischer-Tropsch reactions and the environment. *Applied Catalysis A: General* 189:185–190.

Dunbar WR, Lior N. 1992. Teaching power cycles by both first- and second-law analysis of their evolution, in: Boehm RF (ed.), *Thermodynamics and the Design, Analysis and Improvement of Energy Systems,* AES-Vol. 27/HTD-Vol. 228, American Society of Mechanical Engineers: New York, pp. 221–230.

Ebmpapst. 2006. *Airflow Performance and Speed at a Glance: Fan Curves and Optimum Operating Points.* http://www.ebmpapst.us.

EC. 2005. Air Pollution. *Environment Canada.* www.ec.gc.ca/air/air_pollution_e.html.

Edgerton RH. 1982. *Available Energy and Environmental Economics.* DC, Heath: Toronto.

Edmonton Power. 1991. City of Edmonton District Energy Development Phase: Section 2. Engineering Report.

Eguchi K, Kojo H, Takeguchi T, Kikuchi R, Sasaki K. 2002. Fuel flexibility in power generation by solid oxide fuel cells. *Solid State Ionics* 152–153:411–416.

EIA. 2005. Official energy statistics from the US Government. Energy Information Administration, US Department of Energy. http://tonto.eia.doe.gov/steo_query/app/pricepage.htm.

EI-Sayed YM, Tribus M. 1983. Strategic use of thermoeconomics for system improvement, in: Gaggioli RA, (ed.), *Efficiency and Costing: Second Law Analysis of Processes,* ACS Symposium Series 235, American Chemical Society: Washington, DC, pp. 215–238.

El-Sayed YM, Gaggioli RA, Ringhausen DP, Smith Jr. JL, Gyftopoulos E, Sama D. Fiszdon J. 1997. A second law in engineering education, in: Ramalingam ML, Lage JL, Mei VC, Chapman JN. (eds.), *Proceeding of ASME Advanced Energy Systems Division,* AES-Vol. 37, American Society of Mechanical Engineers: New York, pp. 77–85.

Energetics. 1998. *Energy and Environmental Profile of the US Petroleum Refining Industry.* Report, Energetics Inc.: Columbia, MD.

EOLSS. 1998. *Encyclopedia of Life Support Systems: Conceptual Framework.* EOLSS Publishers: Oxford, UK.

Etele J, Rosen MA. 1999. The impact of reference environment selection on the exergy efficiencies of aerospace engines. *Proceedings of the ASME Advanced Energy Systems Division*, American Society of Mechanical Engineers 39:583–591.

Etele J, Rosen MA. 2000. Exergy losses for aerospace engines: effect of reference-environments on assessment accuracy. *AIAA Paper* 00-0744.

Evans RB, Tribus M. 1962. A contribution to the theory of thermoeconomics. *UCLA Department of Energy*: 62–63.

Finn AJ, Johnson GL, Tomlinson TR. 1999. Developments in natural gas liquefaction. *Hydrocarbon Processing* 78(4), 19–26.

Foerg W. 2002. History of cryogenics: the epoch of the pioneers from the beginning to the year 1911. *International Journal of Refrigeration* 25(3):283–292.

Freris LL. 1981. *Meteorological Aspects of the Utilization of Wind as an Energy Source*. WMO-Report No: 575: Switzerland.

Freris LL. 1990. *Wind Energy Conversion Systems*. Prentice Hall International Ltd: UK.

Fridleifsson IB. 2001. Geothermal energy for the benefit of the people. *Renewable and Sustainable Energy Reviews* 5: 299–312.

Friend DG. 1992. NIST Mixture Property Database. Version 9.08. Fluid Mixtures Data Center. Thermophysics Division. National Institute of Standards and Technology: Boulder, CO 80303.

Frosh D, Gallopoulos N. 1989. Strategies for manufacturing. *Scientific American* 261:94–102.

Froude RE. 1889. On the part played in propulsion by differences of fluid pressure. *Transactions of the Institute of Naval Architects* 30:390.

Funk JE. 1985. *Personal communication with author*, 24 October.

Gaggioli RA. 1980. Second law analysis for process and energy engineering. Paper presented at the AIChE Winter Meeting, Washington, DC.

Gaggioli RA. 1983a. *Efficiency and Costing: Second Law Analysis of Processes*. ACS Symposium Series 235, American Chemical Society: Washington, DC.

Gaggioli RA. 1983b. Second law analysis to improve process and energy engineering, in *Efficiency And Costing: Second Law Analysis of Processes*. ACS Symposium Series 235, pp. 3–50, American Chemical Society, Washington, DC.

Gaggioli RA. 1998. Available energy and exergy. *International Journal of Applied Thermodynamics* 1:1–8.

Gaggioli RA, Petit PJ. 1977. Use the second law first. *Chemtech* 7:496–506.

Gaggioli RA, Sama AD, Qian S, El-Sayed YM. 1991. Integration of a new process into an existing site: a case study in the application of exergy analysis. *Journal of Engineering for Gas Turbines and Power* 113:170–180.

Gandin LS. 1963. *Objective Analysis of Meteorological Fields*. Gidrometeorologicheskoe Izdatel'stvo, Leningrad, translated from Russian in 1965 by Israel Program for Scientific Translations, Jerusalem.

Georgescu-Roegen N. 1971. *The Entropy Law and Economic Process*. Harvard University Press: Cambridge.

Giner SA, Calvelo A. 1987. Modeling of wheat drying in fluidized bed. *Journal of Food Science* 52:1358–1363.

Gipe P. 1995. *Wind Energy Comes Of Age*. Wiley: New York.

Goff LH, Hasert UF, Goff PL. 1999. A new source of renewable energy: the coldness of the wind. *International Thermal Science* 38:916–924.

Goldemberg J, Johansson TB, Reddy AKN, Williams RH. 1988. *Energy for a Sustainable World*. Wiley: New York.

Golding EW. 1955. *The Generation of Electricity by Wind Power*. E.&F. N. Spon Limited: London.

Graedel TE. 1996. On the concept of industrial ecology. *Annual Review of Energy and the Environment* 21:69–98.

Graedel TE, Allenby BR. 1995. *Industrial Ecology*. Prentice Hall, Englewood Cliffs: NJ.

Granovskii M, Dincer I, Rosen MA. 2006a. Life cycle assessment of hydrogen fuel cell and gasoline vehicles. *International Journal of Hydrogen Energy* 31:337–352.

Granovskii M, Dincer I, Rosen MA. 2006b. Environmental and economic aspects of hydrogen production and utilization in fuel cell vehicles. *Journal of Power Sources* 157:411–421.

Granovskii M, Dincer I, Rosen MA. 2006c. Application of oxygen ion-conductive membranes for simultaneous electricity and hydrogen generation. *Chemical Engineering Journal* 120:193–202.

Granovskii M, Dincer I, Rosen MA. 2008. Exergy analysis of gas turbine cycle with steam generation for methane conversion within solid oxide fuel cells. *ASME-Journal of Fuel Cell Science and Technology* (in press).

Granovskii M, Dincer I, Rosen MA. 2007. Greenhouse gas emissions reduction by use of wind and solar energies for hydrogen and electricity production: economic factors. *International Journal of Hydrogen Energy* (in press).

Grundfos. 2006. Circulating Pump, Type: UPS 25-42 180, Product No. 59544561, http://www.grundfos.com/web/hometr.nsf, February 10, 2006 (in Turkish).

Gunnewiek LH, Rosen MA. 1998. Relation between the exergy of waste emissions and measures of environmental impact. *International Journal of Environment and Pollution* 10(2):261–272.

Habib MA, Said SAM, Al-Bagawi JJ. 1995. Thermodynamic performance analysis of the Ghazlan power plant. *Energy* 20: 1121–1130.

Habib MA, Said SAM, Al-Zaharna I. 1999. Thermodynamic optimization of reheat regenerative thermal power plants. *Applied Energy* 63:17–34.

Hafele W. 1981. *Energy in a Finite World: A Global Systems Analysis*. Ballinger: Toronto.

Hahne E, Kubler R, Kallewit J. 1989. The evaluation of thermal stratification by exergy. *Energy Storage Systems*, Kluwer Academic Publishers: Dordecht, 465–485.

Haile S. 2003. Fuel cell materials and components. *Acta Materialia* 51:5981–6000.

Hajidavalloo E. 1998. *Hydrodynamic and Thermal Analysis of a Fluidized Bed Drying System*. Ph.D. Thesis, Dalhousie University-Daltech, Halifax, Canada.

Hammond GP, Stapleton AJ. 2001. Exergy analysis of the United Kingdom energy system. *Proceedings of the Institution of Mechanical Engineers* 215(2):141–62.

Hasnain SM, Smiai MS, Al-Ibrahim AM, Al-Awaji SH. 2000. Analysis of electric energy consumption in an office building in Saudi Arabia. *ASHRAE Transactions* 106:173–184.

Hawlader MNA. 1980. The influence of the extinction coefficient on the effectiveness of solar ponds. *Solar Energy* 25:461–464.

Hengyong T, Stimming U. 2004. Advances, aging mechanism and lifetime in solid-oxide fuel cells. *Journal of Power Sources* 127:284–293.

Hepbasli A, Akdemir O. 2004. Energy and exergy analysis of a ground source (geothermal) heat pump system. *Energy Conversion and Management* 45:737–753.

Hepbasli A, Canakci C. 2003. Geothermal district heating applications in Turkey: A case study of Izmir-Balcova. *Energy Conversion and Management* 44(8):1285–1301.

Hinderink A, Kerkhof F, Lie A, Arons J, Kooi H. 1996. Exergy analysis with a flowsheeting simulator – II. application: synthesis gas production from natural gas. *Chemical Engineering Science* 51:4701–4715.

Holman JP. 2001. *Experimental Methods for Engineers*. McGraw-Hill: New York.

Horlock JH. 1987. *Some Practical CHP Schemes*. Pergamon Press: Oxford.

Hoyer MC, Walton M, Kanivetsky R, Holm TR. 1985. Short-term aquifer thermal energy storage (ATES) test cycles, St. Paul, Minnesota, USA. *Proceeding of 3rd International Conference on Energy Storage for Building Heating and Cooling*, Toronto, pp. 75–79.

Hua B, Yin Q, Wu G. 1989. Energy optimization through exergy-economic evaluation. *ASME Journal of Energy Resources Technology* 111:148–153.

Hui SCM. 1997. From renewable energy to sustainability: the challenge for Hong Kong. *Hong Kong Institution of Engineers*: 351–358.

Hussain MM, Baschuk JJ, Li X, Dincer I. 2005. Thermodynamic modeling of a PEM fuel cell power system, *International Journal of Thermal Sciences* 44(9):903–911.

ISO. 1997. ISO 14040: Environmental management: Life cycle assessment-Principles and framework, International Standards Organization.

Jaefarzadeh MR. 2004. Thermal behavior of a small salinity-gradient solar pond with wall shading effect. *Solar Energy* 77: 281–290.

Jenne EA. 1992. *Aquifer Thermal Energy (Heat and Chill) Storage*. Pacific Northwest Lab: Richland, WA.

Jia GZ, Wang XY, Wu GM. 2004. Investigation on wind energy-compressed air power system. *Journal of Zhejiang University Science* 5(3):290–295.

Justus CG. 1978. *Wind and Wind System Performance*. Franklin Institute Press: Philadelphia, PA.

Kanoglu M. 2000. Uncertainty analysis of cryogenic turbine efficiency. *Mathematical and Computational Applications* 5(3):169–177.

Kanoglu M. 2001. Cryogenic turbine efficiencies. *Exergy, An International Journal* 1:202–208.

Kanoglu M. 2002a. Exergy analysis of a dual-level binary geothermal power plant. *Geothermics* 31:709–724.

Kanoglu M. 2002b. Exergy analysis of multistage cascade refrigeration cycle used for natural gas liquefaction. *International Journal of Energy Research* 26:763–774.

Kanoglu M, Cengel YA. 1999. Improving the performance of an existing binary geothermal power plant: a case study. *Transactions of the ASME, Journal of Energy Resources Technology* 121(3):196–202.

Kanoglu M, Carpinlioglu MO, Yildirim M. 2004. Energy and exergy analyses of an experimental open-cycle desiccant cooling system. *Applied Thermal Engineering* 24:919–932.

Kanoglu M, Cengel YA, Turner RH. 1998. Incorporating a district heating/cooling system into an existing geothermal power plant. *Transactions of the ASME, Journal of Energy Resources Technology* 120(2):179–184.

Karakilcik M. 1998. *Determination of the Performance of an Insulated Prototype Solar Pond*. Ph.D. Thesis, University of Cukurova: Adana, Turkey (in Turkish).

Karakilcik M, Dincer I, Rosen MA. 2006a. Performance investigation of a solar pond. *Applied Thermal Engineering* 26:727–735.

Karakilcik M, Kiymac K, Dincer I. 2006b. Experimental and theoretical temperature distributions in a solar pond. *International Journal of Heat and Mass Transfer* 49:825–835.

Karnøe P, Jørgensen U, 1995. Samfundsmæssig værdi af vindkraft: Dansk vindmølleindustris internationale position og udviklingsbetingelser [Socio-economic assessment of wind power: the Danish wind turbine manufacturers' position and condition for development]. AKF Forlaget: Copenhagen, Denmark.

Kay J. 2002. On complexity theory, exergy, and industrial ecology: some implications for construction ecology. in: C. Kibert, J. Sendzimir, B. Guy. *Construction Ecology: Nature as a Basis for Green Buildings*, (eds), London: Spon Press, pp. 72–107.

Kazarian EA, Hall CW. 1965. Thermal properties of grain. *Transaction of the ASAE* 8:33–37.

Keenan JH, Gyftopoulos EP, Hatsopoulos GN. 1973. The fuel shortage and thermodynamics: the entropy crisis. *Energy: Demand, Conservation, and Institutional Problems*, MIT Press: Cambridge, MA, 455–466.

Kerkhof PJAM. 1994. The role of theoretical and mathematical modeling. *Drying Technology* 12(1&2):235–257.

Kestin J. 1980. Availability: the concept and associated terminology. *Energy-The International Journal* 5:679–692.

Kirillin V, Sychev V, Sheindlin A. 1979. *Engineering Thermodynamics*. Nauka: Moscow.

Kirk RE, Othmer DF. 1998. *Kirk-Othmer Encyclopedia of Chemical Technology*. John Wiley & Sons: New York.

Klein SA. 2006. *EES-Engineering Equation Solver*. F-Chart Software.

Kodama A, Jin W, Goto M, Hirose T, Pons M. 2000. Entropic analysis of adsorption open cycles for air conditioning. Part 2: interpretation of experimental data. *International Journal of Energy Research* 24:263–278.

Koroneos C, Spachos N, Moussiopoulos N. 2003. Exergy analysis of renewable energy sources. *Renewable Energy* 28:295–310.

Kotas TJ. 1995. *The Exergy Method of Thermal Plant Analysis*. Krieger: Malabar, Florida.

Kotas TJ, Raichura RC, Mayhew YR. 1987. Nomenclature for exergy analysis. *Second Law Analysis of Thermal Systems*. Mechanical Engineering, 171–176, ASME, New York.

Krane RJ, Krane MJM. 1991. The optimum design of stratified thermal energy storage systems. *Proceedings of International Conference on the Analysis of Thermal and Energy Systems*: 197–218, Athens, Greece.

Kresta GR. 1992. *A Comparison using Availability Analysis of a Common Core Turbojet and Turbofan Engine*, Thesis, Department of Mechanical Engineering, Ryerson Polytechnic University, Toronto.

Krige DG. 1951. A statistical approach to some basic mine evaluation problems on the Witwateround. *Journal of the Chemical Mettulogical and Mining Society South-Africa* 52:119–139.

Krishna SM, Murthy SS. 1989. Experiments on silica gel rotary dehumidifier. *Heat Recovery Systems and CHP* 9:467–473.

Krokida MK, Kiranoudis CT. 2000. Product quality multi-objective optimization of fluidized bed dryers. *Drying Technology* 18:143–163.

Kuchontchara P, Bhattacharya S, Tsutsumi A. 2003a. Energy recuperation in solid oxide fuel cell (SOFC) and gas turbine (GT) combined system. *Journal of Power Sources*. 117:7–13.

Kuchontchara P, Bhattacharya S, Tsutsumi A. 2003b. Combinations of solid oxide fuel cell and several gas turbine cycles. *Journal of Power Sources* 124: 65–75.

Kunii D, Levenspiel O. 1991. *Fluidization Engineering*. Butterworth-Heinemann: Boston.

Landsberg PT, Markvart T. 1998. The Carnot factor in solar-cell theory. *Solid-State Electronics* 42(4):657–659.

Lange JP, Tijm PJA. 1996. Processes for converting methane to liquid fuels: economic screening through energy management. *Chemical Engineering Science* 51(9):2379–2387.

Langrish TAG, Harvey AC. 2000. A flowsheet model of a well-mixed fluidized bed dryer: applications in controllability assessment and optimization. *Drying Technology* 18:185–198.

Larminie J, Dicks A. 2003. *Fuel Cell Systems Explained*, 2nd ed. John Wiley & Sons: Chichester, England.

Lavan Z, Monnier JB, Worek WM. 1982. Second law analysis of desiccant cooling systems. *ASME Journal of Solar Energy Engineering* 104:229–236.

Lee KC. 2001. Classification of geothermal resources by exergy. *Geothermics* 30:431–442.

Lewis III JH. 1976. Propulsive efficiency from an energy utilization standpoint. *Journal of Aircraft* 13(4):299–302.

Lowe EA, Evans LK. 1995. Industrial ecology and industrial ecosystems. *Journal of Cleaner Production* 3(1):47–53.

Lucca G. 1990. The exergy analysis: role and didactic importance of a standard use of basic concepts, terms and symbols. *A Future for Energy: Proceedings of the Florence World Energy Research Symposium*: 295–308.

Lundin M, Bengtsson AM, Molander S. 2000. Life cycle assessment of wastewater systems: influence of system boundaries and scale on calculated environmental loads. *Environmental Science Technology* 34:180–186.

Maack MH, Skulason JB. 2006. Implementing the hydrogen economy. *Journal of Cleaner Production* 14:52–64.

Maclaine-cross IL. 1985. High performance adiabatic desiccant open cooling cycles. *ASME Journal of Solar Energy Engineering* 107:102–104.

MacRae KM. 1992. Realizing the Benefits of Community Integrated Energy Systems. Canadian Energy Research Institute: Calgary, Alberta.

Malinovskii KA. 1984. Exergy analysis of TJE cycle. *Izvestiya VUZ. Aviatsionnaya Tekhnika* 27(1):34–40.

Malmo Energi AB. 1988. *District Heating in Malmo*. Information brochure.

Markovich SJ. 1978. Autonomous living in the Ouroboros house. *Solar Energy Handbook, Popular Science*: 46–48.

Matheron G. 1963. Principles of geostatistics. *Economic Geology* 58:1246–1266.

Meier PJ. 2002. *Life-cycle assessment of electricity generation systems and applications for climate change policy analysis*. Ph.D. Thesis, University of Wisconsin: Madison.

Moran MJ. 1989. *Availability Analysis: A Guide to Efficient Energy Use*. American Society of Mechanical Engineers: New York.

Moran MJ, Shapiro HN. 2007. *Fundamentals of Engineering Thermodynamics*, 6th ed., Wiley: New York.

Mujumdar AS. 1995. *Handbook of Industrial Drying*. Second and revised edition. Vol. 2. Marcel Dekker: New York.

Murthy SNB. 1994. Effectiveness of a scram engine. *AIAA Paper* 94–3087.

Nagano K, Mochida T, Shimakura K, Murashita K, Takeda S. 2003. Development of thermal-photovoltaic hybrid exterior wallboards incorporating PV cells in and their winter performances. *Solar Energy Materials and Solar Cells* 77: 265–282.

Neij L. 1999. Cost dynamics of wind power. *Energy* 24:375–389.

Newton M, Hopewell P. 2002. Costs of sustainable electricity generation. *Engineering Science and Education Journal* April: 49–55.

OECD. 1996. *Pollution Prevention and Control: Environmental Criteria for Sustainable Transport*. Organisation for Economic Co-operation and Development Report No: 96–136.

Oktay Z, Coskun C, Dincer I. 2007. Energetic and exergetic performance investigation of the Bigadic geothermal district heating system. *Energy and Buildings* (in press).

Ontario Hydro. 1969, Internal and External Manhour and Cost Comparisons for Fossil and Nuclear Projects.

Ontario Hydro. 1973. Nanticoke, data sheet no. PRD2387-73-10m.

Ontario Hydro. 1979. Reports Lennox Generating Station. Lakeview Generating Station Units 7 & 8 Completion Costs.

Ontario Hydro. 1983. Nanticoke Generating Station, data sheet.

Ontario Hydro. 1985. Pickering Generation Station Data: Units 1–4, data sheet no. 966 A119:0.

Ontario Hydro. 1991. Final Costs. Information document.

Ontario Hydro. 1996. Nanticoke Generating Station Technical Data. Information document.

Ontario Weather Data. 2004. http://www.theweathernetwork.com/weather/stats/northamerica.htm.

Osczevski RJ. 2000. Windward cooling: an overlooked factor in the calculation of wind chill. *Bulletin of the American Meteorological Society* 81(12):2975–2978.

Ozdogan O, Arikol M. 1995. Energy and exergy analyses of selected Turkish industries, *Energy-The International Journal* 20:73–80.

Ozgener L, Hepbasli A, Dincer I. 2004. Thermo-mechanical exergy analysis of Balcova geothermal district heating system in Izmir, Turkey. *Transactions of the ASME, Journal of Energy Resources Technology*, 126(4): 293–301.

Ozgener L, Hepbasli A, Dincer I. 2006. Performance investigation of two geothermal district heating systems for building applications: energy analysis. *Energy and Buildings* 38(4): 286–292.

Padro CEG, Putsche V. 1999. *Survey of the Economics of Hydrogen Technologies*. National Renewable Energy Laboratory: US Department of Energy, pp. 570–27079.

Pedersen TF, Petersen SM, Paulsen US, Fabian O, Pedersen BM, Velk P, Brink M, Gjerding J, Frandsen S, Olesen J, Budtz L, Nielsen MA, Stiesdal H, Petersen KØ, Danwin PL, Danwin LJ, Friis P. 1992. Recommendation for wind turbine power curve measurements to be used for type approval of wind turbines in relation to technical requirements for type approval and certification of wind turbines in Denmark. Danish Energy Agency: Denmark.

Perry RH, Green DW, Maloney JO. 1997. *Perry's Chemical Engineers Handbook*. McGraw Hill: New York.

Petala R. 2003. Exergy of undiluted thermal radiations. *Solar Energy* 74:469–488.

Petersen EL, Troen I, Frandsen S, Hedegaard K. 1981. *Wind Atlas for Denmark: a Rational Method for Wind Energy Siting*. Report No. Riso-R-428, Riso National Laboratory: Roskilde, Denmark.

Petersen EL, Mortensen NG, Landberg L, Hojstrup J, Frank HP. 1998. Wind power meteorology – part I: climate and turbulence. *Wind Energy* 1:25–45.

Pierce MA. 2002. *District Energy in Denmark*. http://www.energy.rochester.edu/dk/

Pons M, Kodama A. 2000. Entropic analysis of adsorption open cycles for air conditioning. Part 1: first and second law analyses. *International Journal of Energy Research* 24:251–262.

Quijano J. 2000. Exergy analysis for the Ahuachapan and Berlin geothermal fields, El Salvador. *Proceedings World Geothermal Congress*, May 28–June 10, Kyushu-Tohoku, Japan.

Reistad GM. 1970. *Availability: Concepts and Applications*. Ph.D. Thesis, University of Wisconsin: Madison.

Reistad GM. 1975. Available energy conversion and utilization in the United States. *J.Eng. Power* 97:429–434.

Reistad GM, Gaggioli RA. 1980. Available energy costing. *Thermodynamics: Second Law Analysis*. American Chemical Society Symposium Series 122:143–159.

Rodriguez LSJ. 1980. Calculation of available-energy quantities. *Thermodynamics: Second Law Analysis*. ACS Symposium Series 122:39–60.

Rosen MA. 1986. *The Development and Application of a Process Analysis Methodology and Code Based on Exergy, Cost, Energy and Mass*, Ph.D. Thesis, Department of Mechanical Engineering, University of Toronto, Toronto.

Rosen MA. 1990. The relation between thermodynamic losses and capital costs for a modern coal-fired electrical generation station. *Computer-Aided Energy Systems Analysis*, AES-Vol. 21, American Society of Mechanical Engineers: New York, pp. 69–78.

Rosen MA. 1991. An investigation of the relation between capital costs and selected thermodynamic losses. *Second Law Analysis: Industrial and Environmental Applications*, AES-Vol. 25/HTD-Vol. 191, American Society of Mechanical Engineers: New York, pp. 55–62.

Rosen MA. 1992a. Evaluation of energy utilization efficiency in Canada using energy and exergy analyses. *Energy-The International Journal* 17:339–350.

Rosen MA. 1992b. Appropriate thermodynamic performance measures for closed systems for thermal energy storage. *ASME Journal of Solar Energy Engineering* 114:100–105.

Rosen MA. 1996. Comparative assessment of thermodynamic efficiencies and losses for natural gas-based production processes for hydrogen, ammonia and methanol. *Energy Conversion and Management* 37:359–367.

Rosen MA. 2001. Energy- and exergy-based comparison of coal-fired and nuclear steam power plants. *Exergy, An International Journal* 1:180–192.

Rosen MA. 2002. Assessing energy technologies and environmental impacts with the principles of thermodynamics. *Applied Energy* 72(1):427–441.

Rosen MA. 2003a. Can exergy help us understand and address environmental concerns? *Exergy, An International Journal* 2(4): 214–217.

Rosen MA. 2003b. Exergy and economics: is exergy profitable? *Exergy, An International Journal* 2(4): 218–220.

Rosen MA. 2004. Energy considerations in design for environment: appropriate energy selection and energy efficiency. *International Journal of Green Energy* 1:21–45.

Rosen MA. 2007. Allocating carbon dioxide emissions from cogeneration systems: descriptions of selected output-based methods. *Journal of Cleaner Production* (in press).

Rosen MA, Dimitriu J. 1993. Potential benefits of utility-based cogeneration in Ontario. *Proceedings of the International Symposium on CO₂ Fixation and Efficient Utilization of Energy*, Tokyo, pp. 147–156.

Rosen MA, Dincer I. 1997a. On exergy and environmental impact. *International Journal of Energy Research* 21:643–654.

Rosen MA, Dincer I. 1997b. Sectoral energy and exergy modeling of Turkey. *ASME Journal of Energy Resources Technology* 119:200–204.

Rosen MA, Dincer I. 1999. Exergy analysis of waste emissions. *International Journal of Energy Research* 23(13):1153–1163.

Rosen MA, Dincer I. 2001. Exergy as the confluence of energy, environment and sustainable development. *Exergy, An International Journal* 1(1):3–13.

Rosen MA, Dincer I. 2002. Energy and exergy analyses of thermal energy storage systems. Chapter 10 of *Thermal Energy Storage: Systems and Applications*, Wiley: London, pp. 411–510.

Rosen MA, Dincer I. 2003a. Exergoeconomic analysis of power plants operating on various fuels. *Applied Thermal Engineering* 23(6):643–658.

Rosen MA, Dincer I. 2003b. Thermoeconomic analysis of power plants: an application to a coal-fired electrical generating station. *Energy Conversion and Management* 44(17):2743–2761.

Rosen MA, Dincer I. 2003c. A survey of thermodynamic methods to improve the efficiency of coal-fired electricity generation. *IMechE-Part A: Journal of Power and Energy* 217(1):63–73.

Rosen MA, Dincer I. 2003d. Exergy-cost-energy-mass analysis of thermal systems and processes. *Energy Conversion and Management* 44(10):1633–1651.

Rosen MA, Dincer I. 2004a. Effect of varying dead-state properties on energy and exergy analyses of thermal systems. *International Journal of Thermal Sciences* 43(2):121–133.

Rosen MA, Dincer I. 2004b. A study of industrial steam process heating through exergy analysis. *International Journal of Energy Research* 28(10):917–930.

Rosen MA, Etele J. 2004. Aerospace systems and exergy analysis: applications and methodology development needs. *International Journal of Exergy* 1(4):411–425.

Rosen MA, Horazak DA. 1995. Energy and exergy analyses of PFBC power plants. in: Alvarez Cuenca M, Anthony EJ (eds.), *Pressurized Fluidized Bed Combustion*, Chapman and Hall: London, Chapter 11, pp. 419–448.

Rosen MA, Le M. 1994. Assessment of the potential cumulative benefits of applying utility-based cogeneration in Ontario. *Energy Studies Review* 6:154–163.

Rosen MA, Le MN. 1995. Efficiency measures for processes integrating combined heat and power and district cooling. *Thermodynamics and the Design, Analysis and Improvement of Energy Systems*, AES-Vol. 35, American Society of Mechanical Engineers: New York, pp. 423–434.

Rosen MA, Le MN. 1998. Thermodynamic assessment of the components comprising an integrated system for cogeneration and district heating and cooling. *Proceedings of the ASME Advanced Energy Systems Division*, American Society of Mechanical Engineers, 38:3–11.

Rosen MA, Scott DS. 1985. The enhancement of a process simulator for complete energy-exergy analysis. *Analysis of Energy Systems – Design and Operation*, AES-Vol. 1, American Society of Mechanical Engineers: New York, pp. 71–80.

Rosen MA, Scott DS. 1998. Comparative efficiency assessments for a range of hydrogen production processes. *International Journal of Hydrogen Energy* 23:653–659.

Rosen MA, Dimitriu J, Le M. 1993. *Opportunities for utility-based district cooling in Ontario*. Report by Ryerson Polytechnical Institute: Toronto.

Rosen MA, Le MN, Dincer I. 2005. Efficiency analysis of a cogeneration and district energy system. *Applied Thermal Engineering* 25(1):147–159.

Rosen MA, Leong WH, Le MN. 2001. Modelling and analysis of building systems that integrate cogeneration and district heating and cooling. *Proceedings of the Canadian Conference on Building Energy Simulation*, June 13–14, pp. 187–194.

Rosen MA, Tang R, Dincer I. 2004. Effect of stratification on energy and exergy capacities in thermal storage systems. *International Journal of Energy Research* 28:177–193.

Sahin AD. 2002. *Spatio-Temporal Modelling Winds of Turkey*. Ph.D. Dissertation, ITU, Institute of Science and Technology, Istanbul (in Turkish).

Salle SA. Reardon D. Leithead WE, Grimble MJ. 1990. Review of wind turbine control. *International Journal of Control* 52(6): 1295–1310.

Santarelli M, Macagno SA. 2004. A thermoeconomic analysis of a PV-hydrogen system feeding the energy requests of a residential building in an isolated valley of the Alps. *Energy Conversion and Management* 45:427–451.

Schlatter TW. 1988. Past and present trends in the objective analysis of meteorological data for nowcasting and numerical forecasting. *Eighth Conference on Numerical Weather Prediction*. American Meteorological Society, 9–25.

Sciubba E. 1999. Exergy as a direct measure of environmental impact. *Proceedings of the ASME Advanced Energy Systems Division*, American Society of Mechanical Engineers, 39:573–581.

Sciubba E. 2001. Beyond thermoeconomics? The concept of extended exergy accounting and its application to the analysis and design of thermal systems. *Exergy-An International Journal* 1(2):68–84.

Scott DS. 2000. Conservation, confusion and language. *International Journal of Hydrogen Energy* 25:697–703.

Sen Z, Sahin AD. 1998. Regional wind energy evaluation in some parts of Turkey. *J. Wind Engineering Industrial Aerodynamics* 37(7):740–741.

Senadeera W, Bhandari BR, Young G, Wijesinghe B. 2000. Methods for effective fluidization of particulate food materials. *Drying Technology* 18:1537–1557.

Sener AC. 2003. *Optimization of Balcova-Narlidere Geothermal District Deating System*. M.Sc. Thesis, Graduate School of Natural and Applied Sciences, Izmir Institute of Technology, Izmir, Turkey.

Serpen U. 2004. Hydrogeological investigations on Balçova geothermal system in Turkey. *Geothermics* 33(3): 309–335.

Shen CM, Worek WM. 1996. The second law analysis of a recirculation cycle desiccant cooling system: cosorption of water vapor and carbon dioxide. *Atmospheric Environment* 20:1429–1435.

Silvestri GJ, Bannister RL, Fujikawa T, Hizume A. 1992. Optimization of advanced steam condition power plants. *Journal of Engineering for Gas Turbines and Power* 114(3):612–620.

SimSci/PRO. 2000. *II*. Simulation Sciences Inc., Lake Forest: California.

Sonntag RE, Borgnakke C, van Wylen G. 2003. *Fundamentals of Thermodynamics*, 6th ed. John Wiley: New York.

Spath PL, Mann MK. 2001. *Life cycle assessment of hydrogen production via natural gas steam reforming*. Report 570-27637, National Renewable Energy Laboratory, US Department of Energy.

Spath PL, Mann MK. 2004. *Life cycle assessment of renewable hydrogen production via wind/electrolysis*. Report 560-35404, National Renewable Energy Laboratory, US Department of Energy.

Speight JG. 1996. *Environmental Technology Handbook*. Taylor & Francis: Washington, DC.

Spurr M. 2003. *District Energy/Cogeneration in US Climate Change Strategy*. International District Energy Association: Westborough, MA. Via http://www.energy.rochester.edu/us/climate/.

Stull R. 2000. *Meteorology for Scientists and Engineers*, 2nd ed. Brooks/Cole Thomson Learning: Belmont, CA.

Sundkvist S, Griffin T, Thourshaug N. 2001. AZEP – Development of an integrated air separation membrane – gas turbine. *Second Nordic Minisymposium on Carbon Dioxide Capture and Storage*, Goteborg. Via http://www.entek.chalmers.se/~any/symp/ symp2001.html.

Sussman MV. 1980. Steady-flow availability and the standard chemical availability. *Energy-The International Journal* 5:793–804.

Sussman MV. 1981. Second law efficiencies and reference states for exergy analysis. *Proceeding of the 2nd World Congress Chemical Engineering*, Canadian Society of Chemical Engineers, Montreal: 420–421.

Syahrul S. 2000. *Exergy Analysis of Fluidized Bed Drying Process*. M.A.Sc. Thesis, Dalhousie University-Daltech, Halifax, Canada.

Syahrul S, Hamdullahpur F, Dincer I. 2002a. Energy analysis in fluidized bed drying of wet particles. *International Journal of Energy Research* 26(6):507–525.

Syahrul S, Hamdullahpur F, Dincer I. 2002b. Exergy analysis of fluidized bed drying of moist particles. *Exergy, An International Journal* 2(2):87–98.

Syahrul S, Hamdullahpur F, Dincer I. 2002c. Thermal analysis in fluidized bed drying of moist particles. *Applied Thermal Engineering* 22(15):1763–1775.

Szargut J. 1967. Grezen fuer die anwendungsmoeglichkeiten des exergiebegriffs (Limits of the applicability of the exergy concept). *Brennst.-Waerme-Kraft* 19:309–313.

Szargut J. 1980. International progress in second law analysis. *Energy* 5:709–718.

Szargut J. 2005a. Exergy analysis. *The magazine of the Polish Academy of Sciences* 7:31–33.

Szargut J. 2005b. *Exergy Method: Technical and Ecological Applications*. WIT Press: Southampton, UK.

Szargut J, Morris DR, Steward FR. 1988. *Exergy Analysis of Thermal, Chemical, and Metallurgical Processes*. Hemisphere: New York.

Szargut J, Ziebik A, Stanek W. 2002. Depletion of the non-renewable natural resources as a measure of the ecological cost. *Energy Conversion and Management* 43:1149–1163.

Tangsubkul N, Waite TD, Schäfer AI, Lundie S. 2002. *Applications of Life Cycle Assessment in Evaluating the Sustainability of Wastewater Recycling Systems*, Research Report. The University of New South Wales, Sydney, Australia.

The Weather Network. 2006. http://historical.farmzone.com/climate/historical.asp, February 4.

Topic R. 1995. Mathematical model for exergy analysis of drying plants. *Drying Technology* 13:437–445.

Tribus M, McIrivne EC. 1971. Energy and information. *Sci. American* 225(3):179–188.

Troen I, Petersen EL. 1989. *European Wind Atlas*. Commission of the European Communities, Riso National Laboratory, Roskilde, Denmark.

Tsatsaronis G. 1987. A review of exergoeconomic methodologies. *Second Law Analysis of Thermal Systems*, American Society of Mechanical Engineers: New York, 81–87.

Tsatsaronis G, Winhold M. 1985. Exergoeconomic analysis and evaluation of energy conversion plants, parts 1 and 2. *Energy-The International Journal* 10:69–94.

Tsatsaronis G, Valero A. 1989. Thermodynamics meets economics. *Mechanical Engineering*: August, 84–86.

Tsatsaronis G, Park MH. 2002. On avoidable and unavoidable exergy destructions and investment costs in thermal systems. *Energy Conversion and Management* 43:1259–1270.

Van Den Bulck E, Klein SA, Mitchell JW. 1988. Second law analysis of solid desiccant rotary dehumidifiers. *ASME Journal of Heat Transfer* 110:2–9.

van Gool W. 1997. Energy policy: fairly tales and factualities. *Innovation and Technology-Strategies and Policies* 93–105.

van Schijndel PPAJ, Den Boer J, Janssen FJJG, Mrema GD, Mwaba MG. 1998. Exergy analysis as a tool for energy efficiency improvements in the Tanzanian and Zambian industries. *International Conference on Engineering for Sustainable Development*, July 27–29. University of Dar Es Salaam, Tanzania.

Verkoelen J. 1996. Initial Experience with LNG/MCR Expanders in MLNG-Dua. *17th International LNG/LPG Conference: Gastech'96*, Vienna, Vol. 2, Session 7, Part 1.

Wall G. 1988. Exergy flows through industrial processes. *Energy-The International Journal* 13:197–208.

Wall G. 1990. Exergy conversion in the Japanese society. *Energy-The International Journal* 15:435–444.

Wall G. 1991. Exergy conversions in the Finnish, Japanese and Swedish societies. *OPUSCULA Exergy Papers*: 1–11.

Wall G. 1993. Exergy, ecology and democracy-concepts of a vital society. *ENSEC'93: International Conference on Energy Systems and Ecology*, July 5–9, Cracow, Poland, pp. 111–121.

Wall G. 1997. Exergy use in the Swedish society 1994. *Proceedings of the International Conference on Thermodynamic Analysis and Improvement of Energy Systems (TAIES'97)*, June 10–13, Beijing, China, pp. 403–413.

Wall G, Sciubba E, Naso V. 1994. Exergy use in the Italian society. *Energy-The International Journal* 19:1267–1274.

Wang M. 1999. *GREET 1.5-Transportation Fuel-Cycle Model, Vol. 1: Methodology, Development, Use and Results.* Argonne National Laboratory, Report No. PB2000–101157INW.

Wark KJ. 1995. *Advanced Thermodynamics for Engineers.* McGraw-Hill: New York.

Weber A, Sauer B, Muller A, Herbstritt D, Ivers-Tiffee E. 2002. Oxydation of H_2, CO and methane in SOFCs with Ni/YSZ-cermet anodes. *Solid State Ionics* 152–153:543–550.

Wepfer WJ, Gaggioli RA. 1980. Reference datums for available energy. *Thermodynamics: Second Law Analysis.* ACS Symposium Series 122, American Chemical Society: Washington, DC. 77–92.

Wepfer WJ, Gaggioli RA, Obert EF. 1979. Proper evaluation of available energy for HVAC. *ASHRAE Trans.* 85:214–230.

White SW, Kulcinski GL. 2000. Birth to death analysis of the energy payback ratio and CO_2 gas emission rates from coal, fission, wind, and DT-fusion electrical power plants. *Fusion Engineering and Design* 48: 473–481.

Wilson G. 2002. Preview: Honda FCX fuel cell car. *Canada's Online Auto Magazine*, July 30. www.canadiandriver.com/previews/ fcx-v4.htm.

Wuebbles DJ, Atul KJ. 2001. Concerns about climate change and the role of fossil fuel use. *Fuel Processing Technology* 71:99–119.

Xiang JY, Cali M, Santarelli M. 2004. Calculation for physical and chemical exergy of flows in systems elaborating mixed-phase flows and a case study in an IRSOFC plant. *International Journal of Energy Research* 28:101–115.

Yantovskii EI. 1994. *Energy and Exergy Currents.* NOVA: New York.

Zecher JBM. 1999. A new approach to an accurate wind chill factor. *Bulletin of the American Meteorological Society* 80(9): 1893–1899.

Appendix A

GLOSSARY OF SELECTED TERMINOLOGY

This glossary identifies exergy-related terminology from the literature and categorizes the terms by area. Most exergy terminology has only recently been adopted and is still evolving. Often more than one name is assigned to the same quantity, and more than one quantity to the same name. Only exergy-related definitions are given for terms having multiple meanings. The glossary is based in part on previously developed broader glossaries (Kestin, 1980; Kotas et al., 1987; Kotas, 1995; Dincer and Rosen, 2002).

General thermodynamic terms

Control mass. A closed system containing a fixed quantity of matter, in which no matter enters or exits.
Control volume. An open system in which matter is allowed to enter and/or exit.
Entropy. A measure of disorder, which always increases for the universe.
Heat. A form of energy transfer between systems due to a temperature difference. Heat is a flow quantity (i.e., energy in transit). By convention in analysis, heat input to a system is considered positive while heat exiting is negative.
Heat capacity. Ratio of the heat absorbed in a substance to the resulting increase in temperature. The change in temperature depends on the heating process, with the most common being constant volume or constant pressure.
Internal energy. Sum of all forms of microscopic energy for matter.
Irreversible process. A process in which both the system and its surroundings cannot be returned to their initial state(s) through a subsequent reversible process.
Kinetic energy. Energy of a system as a result of a change in its motion relative to a reference frame.
Latent energy. Internal energy associated with a phase change of a system.
Potential energy (gravitational). Energy of a system as a result of a change of its elevation relative to a reference frame in a gravitational field.
Process. An action that results in a change in the state of a system.
Property. Any characteristic of a system.
Reversible process. A process in which both the system and its surroundings can be returned to their initial state(s) with no observable effects.
Sensible energy. Internal energy of a system associated with a change in the kinetic energies of its molecules, without phase change.
State. The condition of a system specified by the values of its properties.
System. A quantity of matter or any region of space (also thermodynamic system).
Work. A form of energy transfer. Thermodynamic work can be in various forms (e.g., mechanical, electrical, magnetic). By convention in analysis, work done on a system is considered negative while work done by the system is positive.

Exergy quantities

Available energy. See exergy.
Available work. See exergy.
Availability. See exergy.
Base enthalpy. The enthalpy of a compound (at T_0 and P_0) evaluated relative to the stable components of the reference environment (i.e., relative to the dead state).

Chemical exergy. The maximum work obtainable from a substance when it is brought from the environmental state to the dead state by means of processes involving interaction only with the environment.

Essergy. See exergy. Derived from essence of energy.

Exergy. (1) A general term for the maximum work potential of a system, stream of matter or a heat interaction in relation to the reference environment as the datum state. Also known as available energy; availability; essergy; technical work capacity; usable energy; utilizable energy; work capability; work potential; exergy. (2) The unqualified term exergy or exergy flow is the maximum amount of shaft work obtainable when a steady stream of matter is brought from its initial state to the dead state by means of processes involving interactions only with the reference environment.

Negentropy. A quantity defined such that the negentropy consumption during a process is equal to the negative of the entropy creation. Its value is not defined, but is a measure of order.

Non-flow exergy. The exergy of a closed system, i.e., the maximum net usable work obtainable when the system under consideration is brought from its initial state to the dead state by means of processes involving interactions only with the environment.

Physical exergy. The maximum amount of shaft work obtainable from a substance when it is brought from its initial state to the environmental state by means of physical processes involving interaction only with the environment. Also known as thermomechanical exergy.

Technical work capacity. See exergy.

Thermal exergy. The exergy associated with a heat interaction, i.e., the maximum amount of shaft work obtainable from a given heat interaction using the environment as a thermal energy reservoir.

Thermomechanical exergy. See physical exergy.

Usable energy. See exergy.

Useful energy. See exergy.

Utilizable energy. See exergy.

Work capability. See exergy.

Work potential. See exergy.

Xergy. See exergy.

Exergy consumption, energy degradation and irreversibility

Degradation of energy. The loss of work potential of a system which occurs during an irreversible process.

Dissipation. See exergy consumption.

Entropy creation. See entropy production.

Entropy generation. See entropy production.

Entropy production. A quantity equal to the entropy increase of an isolated system (associated with a process) consisting of all systems involved in the process. Also known as entropy creation; entropy generation.

Exergy consumption. The exergy consumed or destroyed during a process due to irreversibilities within the system boundaries. Also known as dissipation; irreversibility; lost work.

External irreversibility. The portion of the total irreversibility for a system and its surroundings occurring outside the system boundary.

Internal irreversibility. The portion of the total irreversibility for a system and its surroundings occurring within the system boundary.

Irreversibility. (1) An effect which makes a process non-ideal or irreversible. (2) See exergy consumption.

Environment and reference environment

Dead state. The state of a system when it is in thermal, mechanical and chemical equilibrium with a conceptual reference environment (having intensive properties pressure P_o, temperature T_o, and chemical potential μ_{ioo} for each of the reference substances in their respective dead states).

Environment. See reference environment.

Environmental state. The state of a system when it is in thermal and mechanical equilibrium with the reference environment, i.e., at pressure P_o and temperature T_o of the reference environment.

Ground state. See reference state.

Reference environment. An idealization of the natural environment which is characterized by a perfect state of equilibrium, i.e., absence of any gradients or differences involving pressure, temperature, chemical potential, kinetic energy and potential energy. The environment constitutes a natural reference medium with respect to which the exergy of different systems is evaluated.

Reference state. A state with respect to which values of exergy are evaluated. Several reference states are used, including environmental state, dead state, standard environmental state and standard dead state. Also known as ground state.

Reference substance. A substance with reference to which the chemical exergy of a chemical element is calculated. Reference substances are often selected to be common, valueless environmental substances of low chemical potential.

Resource. A material found in nature or created artificially in a state of disequilibrium with the environment.

Restricted equilibrium. See thermomechanical equilibrium.

Thermomechanical equilibrium. Thermal and mechanical equilibrium.

Unrestricted equilibrium. Complete (thermal, mechanical and chemical) equilibrium.

Efficiencies and other measures

Effectiveness. See second-law efficiency.

Energy efficiency. An efficiency determined using ratios of energy. Also known as thermal efficiency; first-law efficiency.

Energy grade function. The ratio of exergy to energy for a stream or system.

Exergetic temperature factor. A dimensionless function of the temperature T and environmental temperature T_o given by $(1 - T_o/T)$.

Exergy efficiency. A second-law efficiency determined using ratios of exergy.

First-law efficiency. See energy efficiency.

Rational efficiency. A measure of performance for a device given by the ratio of the exergy associated with all outputs to the exergy associated with all inputs.

Second-law efficiency. A general name for any efficiency based on a second-law analysis (e.g., exergy efficiency, effectiveness, utilization factor, rational efficiency, task efficiency). Often loosely applied to specific second-law efficiency definitions.

Task efficiency. See second-law efficiency.

Thermal efficiency. See energy efficiency.

Utilization factor. See second-law efficiency.

Energy and exergy methods

Energy analysis. A general name for any technique for analyzing processes based solely on the first law of thermodynamics. Also known as first-law analysis.

Exergy analysis. An analysis technique in which process performance is assessed by examining exergy balances. A type of second-law analysis.

First-law analysis. See energy analysis.

Second-law analysis. A general name for any technique for analyzing process performance based solely or partly on the second law of thermodynamics.

Economics and exergy

Exergoeconomics. See thermoeconomics.

Thermoeconomics. A techno-economic method for assessing and designing systems and processes that combines economics with exergy parameters.

Appendix B

CONVERSION FACTORS

Table B.1. Conversion factors for commonly used quantities.

Quantity	SI to English	English to SI
Area	$1\,m^2 = 10.764\,ft^2$ $= 1550.0\,in.^2$	$1\,ft^2 = 0.00929\,m^2$ $1\,in^2 = 6.452 \times 10^{-4}\,m^2$
Density	$1\,kg/m^3 = 0.06243\,lb_m/ft^3$	$1\,lb_m/ft^3 = 16.018\,kg/m^3$ $1\,slug/ft^3 = 515.379\,kg/m^3$
Energy	$1\,J = 9.4787 \times 10^{-4}\,Btu$	$1\,Btu = 1055.056\,J$ $1\,cal = 4.1868\,J$ $1\,lb_f\,ft = 1.3558\,J$ $1\,hp\,h = 2.685 \times 10^6\,J$
Energy per unit mass	$1\,J/kg = 4.2995 \times 10^{-4}\,Btu/lb_m$	$1\,Btu/lb_m = 2326\,J/kg$
Force	$1\,N = 0.22481\,lb_f$	$1\,lb_f = 4.448\,N$ $1\,pdl = 0.1382\,N$
Gravitation	$g = 9.80665\,m/s^2$	$g = 32.17405\,ft/s^2$
Heat flux	$1\,W/m^2 = 0.3171\,Btu/h\,ft^2$	$1\,Btu/h\,ft^2 = 3.1525\,W/m^2$ $1\,kcal/h\,m^2 = 1.163\,W/m^2$ $1\,cal/s\,cm^2 = 41{,}870.0\,W/m^2$
Heat generation (volume)	$1\,W/m^3 = 0.09665\,Btu/h\,ft^3$	$1\,Btu/h\,ft^3 = 10.343\,W/m^3$
Heat-transfer coefficient	$1\,W/m^2\,K = 0.1761\,Btu/h\,ft^2\,{}^\circ F$	$1\,Btu/h\,ft^2\,{}^\circ F = 5.678\,W/m^2\,K$ $1\,kcal/h\,m^2\,{}^\circ C = 1.163\,W/m^2\,K$ $1\,cal/s\,m^2\,{}^\circ C = 41{,}870.0\,W/m^2\,K$
Heat-transfer rate	$1\,W = 3.4123\,Btu/h$	$1\,Btu/h = 0.2931\,W$
Length	$1\,m = 3.2808\,ft$ $= 39.370\,in.$ $1\,km = 0.621\,371\,mi$	$1\,ft = 0.3048\,m$ $1\,in. = 2.54\,cm = 0.0254\,m$ $1\,mi = 1.609344\,km$ $1\,yd = 0.9144\,m$
Mass	$1\,kg = 2.2046\,lb_m$ $1\,ton\,(metric) = 1000\,kg$ $1\,grain = 6.47989 \times 10^{-5}\,kg$	$1\,lb_m = 0.4536\,kg$ $1\,slug = 14.594\,kg$
Mass flow rate	$1\,kg/s = 7936.6\,lb_m/h$ $= 2.2046\,lb_m/s$	$1\,lb_m/h = 0.000126\,kg/s$ $1\,lb_m/s = 0.4536\,kg/s$

(Continued)

Table B.1. (Continued)

Quantity	SI to English	English to SI
Power	$1\,\text{W} = 1\,\text{J/s} = 3.4123\,\text{Btu/h}$ $= 0.737\,562\,\text{lb}_\text{f}\,\text{ft/s}$ $1\,\text{hp (metric)} = 0.735\,499\,\text{kW}$ $1\,\text{ton of refrigeration} = 3.516\,85\,\text{kW}$	$1\,\text{Btu/h} = 0.2931\,\text{W}$ $1\,\text{Btu/s} = 1055.1\,\text{W}$ $1\,\text{lb}_\text{f}\,\text{ft/s} = 1.3558\,\text{W}$ $1\,\text{hp}^\text{UK} = 745.7\,\text{W}$
Pressure and stress $(\text{Pa} = \text{N/m}^2)$	$1\,\text{Pa} = 0.020886\,\text{lb}_\text{f}/\text{ft}^2$ $= 1.4504 \times 10^{-4}\,\text{lb}_\text{f}/\text{in}^2$ $= 4.015 \times 10^{-3}\,\text{in water}$ $= 2.953 \times 10^{-4}\,\text{in Hg}$	$1\,\text{lb}_\text{f}/\text{ft}^2 = 47.88\,\text{Pa}$ $1\,\text{lb}_\text{f}/\text{in}^2 = 1\,\text{psi} = 6894.8\,\text{Pa}$ $1\,\text{stand. atm.} = 1.0133 \times 10^5\,\text{Pa}$ $1\,\text{bar} = 1 \times 10^5\,\text{Pa}$
Specific heat	$1\,\text{J/kg\,K} = 2.3886 \times 10^{-4}\,\text{Btu/lb}_\text{m}\,{}^\circ\text{F}$	$1\,\text{Btu/lb}_\text{m}\,{}^\circ\text{F} = 4187\,\text{J/kg\,K}$
Surface tension	$1\,\text{N/m} = 0.06852\,\text{lb}_\text{f}/\text{ft}$	$1\,\text{lb}_\text{f}/\text{ft} = 14.594\,\text{N/m}$ $1\,\text{dyn/cm} = 1 \times 10^{-3}\,\text{N/m}$
Temperature	$T(\text{K}) = T(^\circ\text{C}) + 273.15$ $= T(^\circ\text{R})/1.8$ $= [T(^\circ\text{F}) + 459.67]/1.8$ $T(^\circ\text{C}) = [T(^\circ\text{F}) - 32]/1.8$	$T(^\circ\text{R}) = 1.8T(\text{K})$ $= T(^\circ\text{F}) + 459.67$ $= 1.8T(^\circ\text{C}) + 32$ $= 1.8[T(\text{K}) - 273.15] + 32$
Temperature difference	$1\,\text{K} = 1^\circ\text{C} = 1.8^\circ\text{R} = 1.8^\circ\text{F}$	$1^\circ\text{R} = 1^\circ\text{F} = 1\,\text{K}/1.8 = 1^\circ\text{C}/1.8$
Thermal conductivity	$1\,\text{W/m\,K} = 0.57782\,\text{Btu/h\,ft}\,{}^\circ\text{F}$	$1\,\text{Btu/h\,ft}\,{}^\circ\text{F} = 1.731\,\text{W/m\,K}$ $1\,\text{kcal/h\,m}\,{}^\circ\text{C} = 1.163\,\text{W/m\,K}$ $1\,\text{cal/s\,cm}\,{}^\circ\text{C} = 418.7\,\text{W/m\,K}$
Thermal diffusivity	$1\,\text{m}^2/\text{s} = 10.7639\,\text{ft}^2/\text{s}$	$1\,\text{ft}^2/\text{s} = 0.0929\,\text{m}^2/\text{s}$ $1\,\text{ft}^2/\text{h} = 2.581 \times 10^{-5}\,\text{m}^2/\text{s}$
Thermal resistance	$1\,\text{K/W} = 0.52750^\circ\text{F\,h/Btu}$	$1^\circ\text{F\,h/Btu} = 1.8958\,\text{K/W}$
Velocity	$1\,\text{m/s} = 3.2808\,\text{ft/s}$ $1\,\text{km/h} = 0.62137\,\text{mi/h}$	$1\,\text{ft/s} = 0.3048\,\text{m/s}$ $1\,\text{ft/min} = 5.08 \times 10^{-3}\,\text{m/s}$
Viscosity (dynamic) $(\text{kg/m\,s} = \text{N\,s/m}^2)$	$1\,\text{kg/m\,s} = 0.672\,\text{lb}_\text{m}/\text{ft\,s}$ $= 2419.1\,\text{lb}_\text{m}/\text{fh\,h}$	$1\,\text{lb}_\text{m}/\text{ft\,s} = 1.4881\,\text{kg/m\,s}$ $1\,\text{lb}_\text{m}/\text{ft\,h} = 4.133 \times 10^{-4}\,\text{kg/m\,s}$ $1\,\text{centipoise (cP)} = 10^{-2}\,\text{poise}$ $= 1 \times 10^{-3}\,\text{kg/m\,s}$
Viscosity (kinematic)	$1\,\text{m}^2/\text{s} = 10.7639\,\text{ft}^2/\text{s}$ $= 1 \times 10^4\,\text{stokes}$	$1\,\text{ft}^2/\text{s} = 0.0929\,\text{m}^2/\text{s}$ $1\,\text{ft}^2/\text{h} = 2.581 \times 10^{-5}\,\text{m}^2/\text{s}$ $1\,\text{stoke} = 1\,\text{cm}^2/\text{s}$
Volume	$1\,\text{m}^3 = 35.3134\,\text{ft}^3$ $1\,\text{L} = 1\,\text{dm}^3 = 0.001\,\text{m}^3$	$1\,\text{ft}^3 = 0.02832\,\text{m}^3$ $1\,\text{in}^3 = 1.6387 \times 10^{-5}\,\text{m}^3$ $1\,\text{gal}^\text{US} = 0.003785\,\text{m}^3$ $1\,\text{gal}^\text{UK} = 0.004546\,\text{m}^3$
Volumetric flow rate	$1\,\text{m}^3/\text{s} = 35.3134\,\text{ft}^3/\text{s}$ $= 1.2713 \times 10^5\,\text{ft}^3/\text{h}$	$1\,\text{ft}^3/\text{s} = 2.8317 \times 10^{-2}\,\text{m}^3/\text{s}$ $1\,\text{ft}^3/\text{min} = 4.72 \times 10^{-4}\,\text{m}^3/\text{s}$ $1\,\text{ft}^3/\text{h} = 7.8658 \times 10^{-6}\,\text{m}^3/\text{s}$ $1\,\text{gal}^\text{US}/\text{min} = 6.309 \times 10^{-5}\,\text{m}^3/\text{s}$

Appendix C

THERMOPHYSICAL PROPERTIES

Table C.1. Thermophysical properties of pure water at atmospheric pressure.

T (°C)	ρ (kg/m^3)	$\mu \times 10^3$ (kg/m s)	$\nu \times 10^6$ (m^2/s)	k (W/m K)	$\beta \times 10^5$ (1/K)	c_p (J/kg K)	Pr
0	999.84	1.7531	1.7533	0.5687	−6.8140	4209.3	12.976
5	999.96	1.5012	1.5013	0.5780	1.5980	4201.0	10.911
10	999.70	1.2995	1.2999	0.5869	8.7900	4194.1	9.2860
15	999.10	1.1360	1.1370	0.5953	15.073	4188.5	7.9910
20	998.20	1.0017	1.0035	0.6034	20.661	4184.1	6.9460
25	997.07	0.8904	0.8930	0.6110	20.570	4180.9	6.0930
30	995.65	0.7972	0.8007	0.6182	30.314	4178.8	5.3880
35	994.30	0.7185	0.7228	0.6251	34.571	4177.7	4.8020
40	992.21	0.6517	0.6565	0.6351	38.530	4177.6	4.3090
45	990.22	0.5939	0.5997	0.6376	42.260	4178.3	3.8920
50	988.04	0.5442	0.5507	0.6432	45.780	4179.7	3.5350
60	983.19	0.4631	0.4710	0.6535	52.330	4184.8	2.9650
70	977.76	0.4004	0.4095	0.6623	58.400	4192.0	2.5340
80	971.79	0.3509	0.3611	0.6698	64.130	4200.1	2.2010
90	965.31	0.3113	0.3225	0.6759	69.620	4210.7	1.9390
100	958.35	0.2789	0.2911	0.6807	75.000	4221.0	1.7290

Source: Kukulka DJ. 1981. *Thermodynamic and Transport Properties of Pure and Saline Water*. MSc Thesis, State University of New York at Buffalo.

Table C.2. Thermophysical properties of air at atmospheric pressure.

T (K)	ρ (kg/m^3)	c_p (J/kg K)	$\mu \times 10^7$ (kg/m s)	$\nu \times 10^6$ (m^2/s)	$k \times 10^3$ (W/m K)	$a \times 10^6$ (m^2/s)	Pr
200	1.7458	1.007	132.5	7.59	18.10	10.30	0.737
250	1.3947	1.006	159.6	11.44	22.30	15.90	0.720
300	1.1614	1.007	184.6	15.89	26.30	22.50	0.707
350	0.9950	1.009	208.2	20.92	30.00	29.90	0.700
400	0.8711	1.014	230.1	26.41	33.80	38.30	0.690
450	0.7740	1.021	250.7	32.39	37.30	47.20	0.686
500	0.6964	1.030	270.1	38.79	40.70	56.70	0.684
550	0.6329	1.040	288.4	45.57	43.90	66.70	0.683
600	0.5804	1.051	305.8	52.69	46.90	76.90	0.685
650	0.5356	1.063	322.5	60.21	49.70	87.30	0.690
700	0.4975	1.075	338.8	68.10	52.40	98.00	0.695
750	0.4643	1.087	354.6	76.37	54.90	109.00	0.702
800	0.4354	1.099	369.8	84.93	57.30	120.00	0.709
850	0.4097	1.110	384.3	93.80	59.60	131.00	0.716
900	0.3868	1.121	398.1	102.90	62.00	143.00	0.720
950	0.3666	1.131	411.3	112.20	64.30	155.00	0.723

Source: Dincer I. 1997. *Heat Transfer in Food Cooling Applications*. Taylor & Francis: Washington, DC; Borgnakke C, Sonntag RE. 1997. *Thermodynamic and Transport Properties*. Wiley: New York.

Table C.3. Thermophysical properties of ammonia (NH$_3$) gas at atmospheric pressure.

T (K)	ρ (kg/m^3)	c_p (J/kg K)	$\mu \times 10^7$ (kg/m s)	$\nu \times 10^6$ (m^2/s)	$k \times 10^3$ (W/m K)	$a \times 10^6$ (m^2/s)	Pr
300	0.6994	2.158	101.5	14.70	24.70	16.66	0.887
320	0.6468	2.170	109.0	16.90	27.20	19.40	0.870
340	0.6059	2.192	116.5	19.20	29.30	22.10	0.872
360	0.5716	2.221	124.0	21.70	31.60	24.90	0.870
380	0.5410	2.254	131.0	24.20	34.00	27.90	0.869
400	0.5136	2.287	138.0	26.90	37.00	31.50	0.853
420	0.4888	2.322	145.0	29.70	40.40	35.60	0.833
440	0.4664	2.357	152.5	32.70	43.50	39.60	0.826
460	0.4460	2.393	159.0	35.70	46.30	43.40	0.822
480	0.4273	2.430	166.5	39.00	49.20	47.40	0.822
500	0.4101	2.467	173.0	42.20	52.50	51.90	0.813
520	0.3942	2.504	180.0	45.70	54.50	55.20	0.827
540	0.3795	2.540	186.5	49.10	57.50	59.70	0.824
560	0.3708	2.577	193.5	52.00	60.60	63.40	0.827
580	0.3533	2.613	199.5	56.50	63.68	69.10	0.817

Source: Dincer I. 1997. *Heat Transfer in Food Cooling Applications*. Taylor & Francis: Washington, DC; Borgnakke C, Sonntag RE. 1997. *Thermodynamic and Transport Properties*. Wiley: New York.

Table C.4. Thermophysical properties of carbon dioxide (CO_2) gas at atmospheric pressure.

T (K)	ρ (kg/m^3)	c_p (J/kg K)	$\mu \times 10^7$ (kg/m s)	$\nu \times 10^6$ (m^2/s)	$k \times 10^3$ (W/m K)	$a \times 10^6$ (m^2/s)	Pr
280	1.9022	0.830	140.0	7.36	15.20	9.63	0.765
300	1.7730	0.851	149.0	8.40	16.55	11.00	0.766
320	1.6609	0.872	156.0	9.39	18.05	12.50	0.754
340	1.5618	0.891	165.0	10.60	19.70	14.20	0.746
360	1.4743	0.908	173.0	11.70	21.20	15.80	0.741
380	1.3961	0.926	181.0	13.00	22.75	17.60	0.737
400	1.3257	0.942	190.0	14.30	24.30	19.50	0.737
450	1.1782	0.981	210.0	17.80	28.20	24.50	0.728
500	1.0594	1.020	231.0	21.80	32.50	30.10	0.725
550	0.9625	1.050	251.0	26.10	36.60	36.20	0.721
600	0.8826	1.080	270.0	30.60	40.70	42.70	0.717
650	0.8143	1.100	288.0	35.40	44.50	49.70	0.712
700	0.7564	1.130	305.0	40.30	48.10	56.30	0.717
750	0.7057	1.150	321.0	45.50	51.70	63.70	0.714
800	0.6614	1.170	337.0	51.00	55.10	71.20	0.716

Source: Dincer I. 1997. *Heat Transfer in Food Cooling Applications*. Taylor & Francis: Washington, DC; Borgnakke C, Sonntag RE. 1997. *Thermodynamic and Transport Properties*. Wiley: New York.

Table C.5. Thermophysical properties of hydrogen (H_2) gas at atmospheric pressure.

T (K)	ρ (kg/m^3)	c_p (J/kg K)	$\mu \times 10^7$ (kg/m s)	$\nu \times 10^6$ (m^2/s)	$k \times 10^3$ (W/m K)	$a \times 10^6$ (m^2/s)	Pr
100	0.2425	11.23	42.1	17.40	67.00	24.60	0.707
150	0.1615	12.60	56.0	34.70	101.00	49.60	0.699
200	0.1211	13.54	68.1	56.20	131.00	79.90	0.704
250	0.0969	14.06	78.9	81.40	157.00	115.00	0.707
300	0.0808	14.31	89.6	111.00	183.00	158.00	0.701
350	0.0692	14.43	98.8	143.00	204.00	204.00	0.700
400	0.0606	14.48	108.2	179.00	226.00	258.00	0.695
450	0.0538	14.50	117.2	218.00	247.00	316.00	0.689
500	0.0485	14.52	126.4	261.00	266.00	378.00	0.691
550	0.0440	14.53	134.3	305.00	285.00	445.00	0.685
600	0.0404	14.55	142.4	352.00	305.00	519.00	0.678
700	0.0346	14.61	157.8	456.00	342.00	676.00	0.675
800	0.0303	14.70	172.4	569.00	378.00	849.00	0.670
900	0.0269	14.83	186.5	692.00	412.00	1030.00	0.671

Source: Dincer I. 1997. *Heat Transfer in Food Cooling Applications*. Taylor & Francis: Washington, DC; Borgnakke C, Sonntag RE. 1997. *Thermodynamic and Transport Properties*. Wiley: New York.

Table C.6. Thermophysical properties of oxygen (O$_2$) gas at atmospheric pressure.

T (K)	ρ (kg/m^3)	c_p (J/kg K)	$\mu \times 10^7$ (kg/m s)	$\nu \times 10^6$ (m^2/s)	$k \times 10^3$ (W/m K)	$a \times 10^6$ (m^2/s)	Pr
100	3.9450	0.962	76.4	1.94	9.25	2.44	0.796
150	2.5850	0.921	114.8	4.44	13.80	5.80	0.766
200	1.9300	0.915	147.5	7.64	18.30	10.40	0.737
250	1.5420	0.915	178.6	11.58	22.60	16.00	0.723
300	1.2840	0.920	207.2	16.14	26.80	22.70	0.711
350	1.1000	0.929	233.5	21.23	29.60	29.00	0.733
400	0.9620	0.942	258.2	26.84	33.00	36.40	0.737
450	0.8554	0.956	281.4	32.90	36.30	44.40	0.741
500	0.7698	0.972	303.3	39.40	41.20	55.10	0.716
550	0.6998	0.988	324.0	46.30	44.10	63.80	0.726
600	0.6414	1.003	343.7	53.59	47.30	73.50	0.729
700	0.5498	1.031	380.8	69.26	52.80	93.10	0.744
800	0.4810	1.054	415.2	86.32	58.90	116.00	0.743
900	0.4275	1.074	447.2	104.60	64.90	141.00	0.740

Source: Dincer I. 1997. *Heat Transfer in Food Cooling Applications*. Taylor & Francis: Washington, DC; Borgnakke C, Sonntag RE. 1997. *Thermodynamic and Transport Properties*. Wiley: New York.

Table C.7. Thermophysical properties of water vapor (steam) gas at atmospheric pressure.

T (K)	ρ (kg/m^3)	c_p (J/kg K)	$\mu \times 10^7$ (kg/m s)	$\nu \times 10^6$ (m^2/s)	$k \times 10^3$ (W/m K)	$a \times 10^6$ (m^2/s)	Pr
380	0.5863	2.060	127.1	21.68	24.60	20.40	1.060
400	0.5542	2.014	134.4	24.25	26.10	23.40	1.040
450	0.4902	1.980	152.5	31.11	29.90	30.80	1.010
500	0.4405	1.985	170.4	38.68	33.90	38.80	0.998
550	0.4005	1.997	188.4	47.04	37.90	47.40	0.993
600	0.3652	2.026	206.7	56.60	42.20	57.00	0.993
650	0.3380	2.056	224.7	66.48	46.40	66.80	0.996
700	0.3140	2.085	242.6	77.26	50.50	77.10	1.000
750	0.2931	2.119	260.4	88.84	54.90	88.40	1.000
800	0.2739	2.152	278.6	101.70	59.20	100.00	1.010
850	0.2579	2.186	296.9	115.10	63.70	113.00	1.020

Source: Dincer I. 1997. *Heat Transfer in Food Cooling Applications*. Taylor & Francis: Washington, DC; Borgnakke C, Sonntag RE. 1997. *Thermodynamic and Transport Properties*. Wiley: New York.

Table C.8. Thermophysical properties of some solid materials.

Composition	T (K)	ρ (kg/m^3)	k (W/m K)	c_p (J/kg K)
Aluminum	273–673	2720	204–250	895
Asphalt	300	2115	0.0662	920
Bakelite	300	1300	1.4	1465
Brass (70% Cu + 30% Zn)	373–573	8520	104–147	380
Brick, refractory				
Carborundum	872	–	18.5	–
Chrome brick	473	3010	2.3	835
	823	–	2.5	–
Bronze (75% Cu + 25% Sn)	273–373	8670	26.0	340
Clay	300	1460	1.3	880
Coal (anthracite)	300	1350	0.26	1260
Concrete (stone mix)	300	2300	1.4	880
Constantan (60% Cu + 40% Ni)	273–373	8920	22–26	420
Copper	273–873	8950	385–350	380
Cotton	300	80	0.06	1300
Diatomaceous silica, fired	478	–	0.25	–
Fire clay brick	478	2645	1.0	960
	922	–	1.5	–
Glass				
Plate (soda lime)	300	2500	1.4	750
Pyrex	300	2225	1.4	835
Ice	253	–	2.03	1945
	273	920	1.88	2040
Iron (C ≈ 4% cast)	273–1273	7260	52–35	420
Iron (C ≈ 0.5% wrought)	273–1273	7850	59–35	460
Lead	273–573	–	–	–
Leather (sole)	300	998	0.159	–
Magnesium	273–573	1750	171–157	1010
Mercury	273–573	13,400	8–10	125
Molybdenum	273–1273	10,220	125–99	251
Nickel	273–673	8900	93–59	450
Paper	300	930	0.18	1340
Paraffin	300	900	0.24	2890

(Continued)

Table C.8. (Continued)

Composition	T (K)	ρ (kg/m^3)	k (W/m K)	c_p (J/kg K)
Platinum	273–1273	21,400	70–75	240
Rock				
Granite, Barre	300	2630	2.79	775
Limestone, Salem	300	2320	2.15	810
Marble, Halston	300	2680	2.80	830
Soft Sandstone, Berea	300	2150	2.90	745
Rubber, vulcanized				
Soft	300	1100	0.13	2010
Hard	300	1190	0.16	–
Sand	300	1515	0.27	800
Silver	273–673	10,520	410–360	230
Soil	300	2050	0.52	1840
Snow	273	110	0.049	–
Steel (C $\approx$ 1%)	273–1273	7800	43–28	470
Steel (Cr $\approx$ 1%)	273–1273	7860	62–33	460
Steel (18% Cr + 8% Ni)	273–1273	7810	16–26	460
Teflon	300	2200	0.35	–
Tin	273–473	7300	65–57	230
Tissue, human				
Fat layer (adipose)	300	–	0.2	–
Muscle	300	–	0.41	–
Skin	300	–	0.37	–
Tungsten	273–1273	19,350	166–76	130
Wood, cross grain				
Fir	300	415	0.11	2720
Oak	300	545	0.17	2385
Yellow pine	300	640	0.15	2805
White pine	300	435	0.11	–
Wood, radial				
Fir	300	420	0.14	2720
Oak	300	545	0.19	2385
Zinc	273–673	7140	112–93	380

Source: Dincer I. 1997. *Heat Transfer in Food Cooling Applications*. Taylor & Francis: Washington, DC; Incropera FP, DeWitt DP. 1998. *Fundamentals of Heat and Mass Transfer*. Wiley: New York.

INDEX